An Introduction to
Modern Vehicle Design

An Introduction to
Modern Vehicle Design

Edited by
Julian Happian-Smith
PhD, MSc, BTech, Cert Ed HE, MSAE

LEARNING
RESOURCES
CENTRE

Oxford Auckland Boston Johannesburg Melbourne New Delhi

Butterworth-Heinemann
Linacre House, Jordan Hill, Oxford OX2 8DP
225 Wildwood Avenue, Woburn, MA 01801-2041
A division of Reed Educational and Professional Publishing Ltd

 A member of the Reed Elsevier plc group

First published 2001

British Library Cataloguing in Publication Data
A catalogue record for this book is available from the British Library

Library of Congress Cataloging in Publication Data
A catalogue record for this book is available from the Library of Congress

ISBN 07506 5044 3

Typeset at Replika Press Pvt Ltd, 100% EOU, Delhi 110 040, India

Transferred to digital printing 2006

Contents

Preface

There have not been many books published that concern themselves with the analytical design of the complete motor vehicle. My source of inspiration for this work was Janusz Pawlowski's most interesting and informative Vehicle Body Engineering. However, this classic book is now only of historical interest and it is the editor's hope that this book may well take the place of that book on the bookshelves of current motor vehicle designers. A change from this classic book is that it is now impossible for one person to write knowledgeably about all aspects of vehicle design. This reason has dictated that specialists in each field covered by this book have written an appropriate chapter. This is a sign of how times have changed since the days of Pawlowski, and is a trend that can only continue.

The text is intended to provide the reader with an introduction to most of the topics that are of concern when a vehicle is being designed from the 'clean sheet of paper' stage. There are a wide range of references alluded to within the text that the reader can draw upon for more detailed information at the end of each chapter. Some of these references are drawn from the list and briefly summarized indicating particular texts that the contributor has found interesting. It is hoped that this will help the reader that any especial interest further.

It is hoped that this text will help to inspire engineers new to Automotive Engineering to take up career paths in this field of engineering as I believe that all branches of engineering are now involved with vehicle design.

Readers' comments on the contents of this text will be welcomed so that their observations will be of great assistance when the text is revised.

Julian Happian-Smith

Acknowledgements

It is inevitable that when compiling a manuscript of this type which involves many contributors, that the editor of such a work in indebted to a wide variety of people and organizations.

First, I must thank Sunderland University and especially Mr. Stathis Lertas without whom this task would have been most onerous. Secondly, I must thank all the contributors and their respective organizations without whom this task would have been impossible. I would also like to make particular mention of Prof. David Crolla at Leeds University who has always been very helpful and encouraging.

However, many of the contributors are also indebted to organizations for their generous supply of support information and figures. Especial mention should be made of the following organizations:

Bosch
FIAT
Ford Motor Company Ltd
Hyundai
London Transport Museum
Mercedes Benz
Motorola Ltd
Munro and Associates Inc.
National Motor Museum, Beaulieu
PSA Peugeot Citroen
Renault Cars Ltd
Rover Group Ltd
SAE
Simpson International (UK) Ltd
TRL Ltd

My sincere thanks
Julian Happian-Smith

1. Automotive engineering development

R.H. Barnard, PhD, CEng, FRAeS

The aim of this chapter is to:

- Introduce the wide range of skills required for vehicle design and manufacture;
- Briefly set the historical scene and development of vehicles and their design;
- Introduce the vast range of possibilities for vehicle design;
- Demonstrate the interactivity of processes within the design and manufacture of vehicles.

1.1 Introduction

In the development of the motor vehicle, there are three readily identifiable groups of activities.

- technical innovation and refinement
- construction, configuration and styling
- methods of production, and manufacturing systems.

To the layman, the most obvious aspects of progress are technical innovations and styling changes, but from a professional engineering viewpoint, the major achievements lie as much in the areas of refinement and systems of manufacture. Innovations can be important in giving manufacturers a competitive advantage, but new ideas often make their debut many decades before they are widely adopted. It is the processes of refinement and production development that make new technical features reliable and cheap enough for use in mass-produced vehicles.

1.2 Innovations and inventions

Engineering history is bedevilled by rival and sometimes false claims to particular inventions. In reality, innovative developments have often been the work of several different engineers working in parallel but quite independently, and the recognized inventor is simply one whose name is well known, or who has been championed for nationalistic reasons. Many apparently new inventions are, in any case, simply adaptations from different technologies. The differential mechanism, for example, was used by watchmakers before being adapted for automotive purposes. It is frequently difficult to trace the earliest examples of the use of a particular device or mechanism. J. Ickx, 1992, describes how the Bollées (father and two sons) invented or adapted an amazing array of devices in the late 19th century, including all-round independent suspension, and power steering (originally applied to steam-powered vehicles). In 1894, the younger Amédée produced a gas turbine, and later went on to invent fuel injection, supercharging, and hydraulic valve lifters. All these devices are usually ascribed to other, later inventors.

1.2.1 The first major technical breakthrough

It is a little surprising that road vehicle transport lagged so far behind the development of the railways. Steam locomotives appeared early in the 19th century, and by the time the first really practical road vehicles emerged over half a century later, rail transport had become a mature technology with large networks covering many countries. The problem of road transport development lay in the combination of the heavy cumbersome steam engine and poorly surfaced roads. By the end of the 19th century, significant developments of the steam engine had taken place such as the use of oil or paraffin instead of coal as the fuel, and the development of the lighter more compact 'flash' boiler system in which steam was generated by passing water through heated tubes rather than boiling it up in a pressure vessel. Practical steam-powered road vehicles started to appear in small numbers, and indeed for commercial vehicles, the line of development was not finally terminated until the 1950s. Some impression of the level of refinement of steam cars may be drawn from the elegant 1905 Stanley shown in Figure 1.1. Two major drawbacks to automotive steam propulsion were the long start-up time required, and the high rate of water consumption.

Figure 1.1 A Stanley steam car of 1905. This elegant vehicle is far removed from the lumbering smoky traction engines that nowadays chug their way to nostalgic steam rallies. Steam cars were much quieter and smoother-running than their petrol engined contemporaries, but took some time to fire up. They also needed frequent intakes of water.

A major change of direction and a spur to progress, occurred in the 1870s with the appearance of gas-fuelled reciprocating internal combustion engines, notably those patented and produced by Dr A.N. Otto in Germany. Gas engines were originally used as static units for driving machinery, and usually ran on the common domestic or 'town' gas, but several engineers started experimenting with the use of vaporized petroleum spirit instead, as this offered the possibility of a mobile engine. Petroleum spirit was at that time a somewhat useless by-product of the process of manufacturing paraffin which was widely used in lamps. In 1885 Gottlieb Daimler modified an Otto four-stroke gas engine to run on petroleum vapour, and fitted it to a crude bicycle with a stabilizing outrigger wheel. One year later, he modified a horse carriage to produce what is now generally recognized as the forerunner of the modern motor car. The invention of the petrol-engined motor car is, however, one of the classic examples of parallel development, and there are many rival claimants, chief amongst these being Karl Benz, who produced a powered tricycle in 1885. A replica of the 1886 version is shown in Figure 1.2. Following the introduction of the petrol engine, road vehicle technology progressed rapidly, but it was the development of mass production techniques rather than any technical innovation that provided the next major step.

Figure 1.2 An 1896 Benz tricycle replica where the influence of bicycle technology is clearly evident. From the collection of the National Motor Museum, Beaulieu.

1.3 Mass production

Most early cars were produced by the same techniques of hand craftsmanship that had been used for centuries for the construction of horse-drawn carriages. Cars required the manufacture of a large number of components, and each item was individually made and fitted by skilled craftsmen. Unlike the modern processes of assembly that simply rely on joining items by bolting or welding, fitting usually involved using hand tools to cut or file components to make them fit together. The great leap in automotive production engineering came when Henry Ford started to develop the techniques of mass production. Ford did not invent the idea; indeed it had been used many years earlier during the American Civil War for the production of rifles. The vehicle that really launched his advanced approach was the Model T (Figure 1.3) which first

Figure 1.3 The Ford Model T. This example is from 1913. Note the single transverse front spring and the starting handle, which was the only means of starting. In addition to factory-built vehicles, independent coachbuilders used the Model T chassis as the basis for a wide range of bodywork styles, from trucks and charabanc buses to elegant coachbuilt family cars. The 2898cc petrol engine gave adequate power for use in quite large commercial vehicles. The spindly chassis was deceptively strong, being made of a vanadium steel alloy. (Photo courtesy of Ford Motor Company Ltd.)

appeared in 1909. Ford had produced many previous models, working his way through the alphabet from the Model A, and had been gradually honing his production methods. The Model T was one of the first cars whose design was primarily dictated by the requirements of manufacture, and thus it represents an early major example of the application of the concept of 'design for production'.

The principle of mass production is that each worker only has to perform either one, or a very limited number of tasks, usually involving very little skill: bolting on the steering wheel for example. To keep the workers continuously busy, the volume of production has to be large.

There must always be another vehicle just ready for its steering wheel. Interestingly, although hand-crafting is always associated in the public's mind with high quality, mass production actually requires higher standards of accuracy and consistency of dimension, because in mass production, all similar parts must be completely interchangeable. Hand-built cars may look superficially identical, but there are often large differences in the dimensions of individual components. It was the achievement of dimensional accuracy and interchangeability that made mass production possible.

Ford initially assembled the vehicles on fixed stands, but in 1913 he opened his large new Highland Park plant in Detroit (Figure 1.4), and this featured another major innovation, the moving production line. Workers no longer had to move from one task to another; the vehicles simply came to them along a track at an unending steady stream, thereby taking control of the rate of assembly away from the shop-floor workers.

Figure 1.4 Early mass production at Ford's Highland Park plant in Detroit in 1914: the fuel tank assembly station. The chassis are moved on a track, and the cylindrical fuel tanks are supplied to the assemblers from an overhead store. The production techniques may look somewhat rudimentary by modern standards, but were innovative in their time. (Photo courtesy of Ford Motor Company Ltd.)

Apart from developing the idea of design for production, Henry Ford was also conscious of the need to design for maintainability, and the importance of ergonomic considerations. The Model T was almost the ultimate in simplicity. Initially it had no instruments, and to make driving easier, it had no clutch pedal or gear lever, gear changing being effected by pedals. The owner was supplied with a comprehensive handbook that set out in simple terms how to perform a wide range of maintenance and repair tasks. The construction and layout of the mechanical parts were designed to make most jobs easy, thereby dispensing with the need for a skilled mechanic. The bodywork was minimal and rudimentary. Only one basic chassis was produced, and body colour schemes were initially limited, and finally restricted to one, thereby conforming to the famous slogan 'any colour you like, as long as it is black'. The black paint was chosen not for aesthetic reasons, but simply because it dried quickly. Ford was also aware of the advantages of using advanced materials, and employed vanadium steel for the chassis, thereby producing a relatively light vehicle.

Like their horse-drawn predecessors, most early cars were expensive, both to purchase and to run, and their ownership was almost entirely restricted to the very wealthy. The major attraction of Ford's Model T was that its method of production made it much cheaper than competing hand-crafted vehicles. The simplicity of its controls and the fact that it was designed to be readily maintained by an unskilled owner were also good selling points. As a consequence, the Ford T opened up automotive ownership to a new mass market, and by 1923, production had reached a peak of over two million cars per year. Apart from production in the United States, Ford plants were opened in Europe, including one at Trafford Park in England in 1911.

Ford's enthusiasm for mass production led to his attempting to apply the same principles to a wide range of products, including aeroplanes. He also decided to bring all the stages of car production under his control, not just the final assembly (Ford originally bought in his engines and other components). At Ford's massive new Rouge plant in Detroit, opened in 1927, raw materials went in one end, and finished cars emerged at the other. Other manufacturers started to copy and even develop these ideas, both in Europe and America, but European cars retained a much higher level of craftsmanship until the outbreak of the Second World War. The requirements of armament production then led to the almost universal acceptance of the principles of mass production.

Mass production made cars available to a large section of the public, but it was soon found to have disadvantages. The hard tedious repetitive work was resented by the assembly workers, who were forced to accept it for want of a comparably paid alternative. The huge plants became organizationally complex and bureaucratic. Worker dissatisfaction made itself apparent in a rash of strikes, as the labour force tried to compensate for the working conditions by seeking ever higher wages and shorter hours. Resentment generated an us-and-them war between shop-floor and management that resulted in some workers taking pleasure in poor workmanship and occasionally, in deliberate sabotage. The resulting products though relatively cheap, were of poor quality, and by the early 1970s, most cars were badly finished, unreliable and prone to rusting. To make matters worse, manufacturers adopted the principle of built-in obsolescence, believing that the faster a vehicle deteriorated, the quicker its owner would need to buy a replacement, thereby increasing sales. There were exceptions to this trend towards poor quality, one of the most notable being the little Volkswagen 'Beetle'. This vehicle was designed by Ferdinand Porsche in the late 1930s at the behest of Hitler, and although innovative in many respects, it had little in the way of refinement. By the 1970s, its styling was quite antiquated, and its air-cooled engine noisy, yet it sold in extremely large numbers throughout the world. Its success in the USA was particularly surprising, as the American public generally considered European cars to be too small to be either practical or safe. Despite its lack of refinement, the Volkswagen had two great virtues, it was mechanically reliable, and it did not rust quickly. Other manufacturers were slow to learn the lessons, but eventually it became apparent that systematic quality control was of major importance in automobile manufacture.

Although the example of Volkswagen was important, it did not question the underlying principles of mass production, and the real challenge to this concept came from Japan. The growing Japanese penetration of the traditional American and European markets, starting roughly in the 1960s, was initially ascribed to low wage rates, automation and a disciplined society. All of these aspects were important factors, but a major component in the Japanese success story was the adoption of a new system of production, where workers instead of being assigned to a single task, worked collaboratively in teams. Production was also flexible, and machinery could be rapidly switched from one task to another. Quality became paramount, and the system used made it financially beneficial to the workers to get the job right first time, rather than pass off poor work that would later have to be rectified. The philosophy and techniques of this system, which is often now referred to as 'lean' production, were introduced and developed by the Toyota company to cover not just the basic manufacture, but all aspects of automotive production, including the relationships between assembler and component suppliers, which were more co-operative. A major feature of this flexible approach to manufacture is that it is possible to have relatively short production runs, and a wide range of models and variants can be accommodated. Details of this production system and its history are given by Womack *et al.* (1990).

1.4 The development of the world motor industry

The motor industry originated in small workshops producing hand-built vehicles tailor-made to the customers' specification, but Henry Ford's mass production techniques were soon copied by others. Throughout the 1920s and 1930s, small low volume manufacturers of coachbuilt vehicles were able to co-exist with the large mass production companies such as Ford, Chrysler, Morris and Fiat. The smaller firms were, however, gradually forced to merge or to be swallowed up by the large companies, or to simply disappear. After the Second World War the trend accelerated, until by the 1970s, only a few specialist companies such as Rolls-Royce remained.

The process of absorption brought its own problems. Large organizations that bought up a failing company often found that they had bought its weaknesses as well. All too often, there was a failure to rationalize. A good example was the British Motor Corporation (BMC) which was formed from the merging of the two major British Motor manufacturers Austin and Morris, and a number of smaller companies such as MG. At one stage in the 1950s this resulted in its trying to cope with having to stock over 100 000 different components. A further series of mergers resulted in the formation in 1968 of British Leyland (BL), which comprised nearly the whole British motor industry, and was the fifth largest motor manufacturer in the world. Lack of rationalization resulted in its having 46 different models at one time. Similar mergers took place elsewhere, and by the 1980s, most European countries had only one or two major native motor manufacturers, and these were often kept alive by being nationalized and subsidized. In Britain, the Conservative Government removed the protection of nationalization, and a few years later, the Rover Group, a late manifestation of the BL empire, was sold to BMW. After a few years of disastrous performance, it was returned to British ownership, becoming the MG Rover Group. Apart from this company, the British motor industry now comprises several large plants belonging to multi-national manufacturers, and a number of major component suppliers. The demise of the native British car assemblers is well described by Wood (1988).

American manufacturers also had to rationalize, but were in a rather different position, as a significant part of their operations was carried out in subsidiary plants abroad. This was the result of their attempting to overcome import barriers that had been erected in the early days, when the success of Ford and Chrysler had threatened to overwhelm the European industry. The American conglomerates discovered the advantages of moving parts of their operations around the globe to take advantage of local conditions. Japanese manufacturers faced with similar restrictions on their exports, developed the same strategy. Nowadays most of the major manufacturers operate as multi-national organizations, producing vehicles for a world market, and making use of facilities and suppliers throughout the world. Ford now operates several design offices in different countries, each one concentrating on a particular class of vehicle.

1.4.1 Construction development

Most early car manufacturers adopted the construction methods of horse-drawn carriages for the upper bodywork, but bicycle technology was also used to some extent, and the wire-spoked cycle type wheels eventually replaced the wooden-spoked carriage wheels. The construction of horse-drawn vehicles was of necessity light. Above a wooden chassis, sat a light wooden framework that was covered with a skin of sheet metal, wood or fabric. The largely wooden construction was less suitable for motor vehicles that travelled at much higher speeds, thereby giving rise to higher shock loads. The motor vehicles also had to sustain the loads and vibrations of the engine and transmission, and therefore, a much more substantial metal chassis frame was usually employed. For many years, the upper bodywork retained the wooden framework, usually in ash, but the wooden or fabric skinning soon gave way to sheet metal. A few fabric and wooden bodied vehicles were still produced as late as the 1930s by specialist coachbuilders, but this was mainly because the antiquated style conveyed an air of past elegance. The combination of steel chassis, wooden framework and sheet metal skinning was used for most vehicles, whether mass produced or coachbuilt, until the late 1930s, with aluminium often being used for the more expensive and high-performance vehicles (Figure 1.5). Aluminium has a lower density than steel and produced a lighter body with better resistance to corrosion. It was however, more expensive, and was more difficult to weld, particularly in the higher strength alloys. It also tends to stretch when dented, making minor repairs more difficult. After the Second World War, the wood frame and metal skin form of construction became restricted to specialist coachbuilt vehicles, and indeed it is still used for the Morgan sports cars.

Figure 1.5 This 1935 Railton Carbodies saloon combines classic features of both American and British design. The engine is a Hudson unit with a three-speed crash gearbox. Sixteen-inch American size wheels are used, but the hand-built coachwork with ash frame, aluminium panelling and steel wings is typically British. The long bonnet is actually justified in this case, because of the straight-eight 4.2 litre engine. Note the small boot, which had only recently evolved from a separate trunk. Windscreen wipers had been standard for several years, but the flat front screen could still be hinged open.

In the 1930s, increasing use was made of pressed-steel skin panels in place of flat sheets or hand beaten or wheeled panels. Sheets of steel were pressed in moulds to produce complex shapes with multiple curvature. This process enabled the economic production of the bulbous styling forms that became popular, particularly in the USA. The multiple curvature also made the panels much stiffer, and the skin could then take a significant part of the loads. Some manufacturers began to dispense with the wooden frame, and to use either a metal frame or even no framework at all, relying on the panels and formed sheet steel stiffening elements to provide all the rigidity necessary for the upper body. A substantial lower chassis frame was initially retained, but the separate chassis began to disappear, being replaced by a stiff floor 'pan' that was fabricated from welded (usually spot welded) shaped sheet elements. The floor pan was welded to the upper shell, and much of the stress could then be carried by the upper body shell. By the 1950s, this 'unitary' type of construction had been almost universally adopted for mass-produced cars. In recent years, the shell construction has been refined to produce a smooth aerodynamically optimized shape with a minimum of protrusions or gaps. More recently, attention has been paid to the contouring of the underside.

A great disadvantage of early unitary construction was the problem of severe corrosion that rapidly developed around the welds and in inaccessible areas. It took some time for really effective anti-corrosion treatments to be developed, and even longer for some manufacturers to shake off their belief in the advantages of built-in obsolescence.

Composite construction, originally in fibreglass and resin was developed soon after the war. It has a number of advantages including the lack of corrosion, and the ability to produce complex shapes cheaply. The tooling costs of composite construction are very much lower than for pressed steel, making composites attractive for small-scale manufacture or short production runs. The techniques of composite vehicle body construction have been developed notably by Lotus, and applied to their sports cars. Disadvantages of the material include the difficulty of attaching metal components, and high material costs. Increasing use is being made of composite and plastic materials for body components, but their use for the main shell is generally restricted to specialist high-performance vehicles (Figure 1.6).

1.4.2 Styling development

Many early motor cars were essentially powered versions of horse-drawn vehicles (Figure 1.7), and the bodywork retained the forms and names of carriage styles such as Phaeton and Landaulette. In his first cars, Daimler placed the engine under the bodywork at the rear, an arrangement that was often used in the early days, although this was not a convenient location, as the engines needed frequent attention. In 1890–91, the Panhard-Levassor company produced a vehicle that had a front-mounted engine driving through a clutch and gearbox. This so-called 'système Panhard', which was in fact originated by the partner Emile Levassor, quickly became the main conventional layout for a car. None of these features was an original idea; front-engined steam cars had been built, and clutches and gearboxes were used on machine tools; it was the combination that was original and innovative.

With the transfer of the engine to the front, the characteristic engine cover, the bonnet (or hood in the U.S.A.) emerged. Since the size of the engine cover indicated the size and hence potency of the motor, a large bonnet became an important styling feature. By the 1930s, excessively elongated engine covers had become 'de rigeur' for powerful cars with sporting aspirations.

Figure 1.6 Composite material construction of the main bodywork shell has been highly developed by Lotus for its specialist low production volume sports cars.

Figure 1.7 Early motor vehicles were often simply adaptations of horsedrawn vehicles, as may be seen in this 1897 Bersley electrically propelled cab. Only the absence of shafts for the horses betrays the fact that it is a motor vehicle. Note how the driver is totally exposed to the elements. (From the collection of the National Motor Museum, Beaulieu).

At first, luggage was usually strapped to the rear of the vehicle on an external rack, a feature that was still found on popular European cars in the early 1930s. The racks were eventually replaced by an integral trunk or boot, and thus the basic three-box saloon or sedan form became established as the standard arrangement. Initially, the rear box was quite small (Figure 1.5), but in the 'fifties, 'sixties and 'seventies, rear trunks of extraordinary size complemented the equally exaggerated engine covers on popular American vehicles (Figure 1.8).

Figure 1.8 Post-war American exaggerated styling. The engine, though large, is in fact a compact vee-configuration unit. This Lincoln incorporated a wide range of refinements including electric window lifts.

With the rapidly increasing speed of motor cars it soon became apparent that a greater degree of protection from the elements was required. The first innovation was the provision of a front windscreen, something that was impractical on horse-drawn vehicles where the driver had to hold the reins and control the source of motive power by a combination of using the reins and verbal commands. Until the introduction of the windscreen wiper in the mid 1920s, the only means of dealing with rain was to hinge the screen, either folding it flat from the bottom, or hinging it further up, so that the bottom edge could be tilted forward.

The latter method provided more protection from the elements. Even after the general introduction of the windscreen wiper, a windscreen tilting mechanism was provided on many cars until the end of the 'thirties (Figure 1.5). Refinements such as hot-air demisting did not become standard on most vehicles until the 1950s.

In horse-drawn vehicles, the driver of necessity sat in the open, and few other than large carriages hauled by several horses had a permanently enclosed passenger compartment. On most carriages, protection from the weather was provided by a folding hood. Motor vehicles continued the essentially open top condition for many years, and it was not until the 1930s that the closed saloon or sedan became the dominant type of body style. Even then, many sporting and luxury vehicles were produced in cabriolet or drophead configuration which, being lighter, gave improved performance.

1.5 Streamlining

The important influence of aerodynamic drag on speed and performance was appreciated by more enlightened constructors at a very early stage, and in the late 1890s, Amédée Bollée the younger produced torpedo-shaped semi-streamlined vehicles that even featured a raked windscreen. Truly scientific streamlining was developed after the First World War by several engineers, including Rumpler and Kamm. The most notable proponent, however, was Paul Jaray, an Austrian engineer who worked initially for Count von Zeppelin on airship design. Jaray's designs, patents and ideas were employed by several major manufacturers in the 'twenties and 'thirties. The attractive Czech Tatra of 1937 (Figure 1.9) designed by the Austrian Hans Ledwinka is a classic example of a truly streamlined vehicle of this period. Its styling and layout foreshadowed the Volkswagen Beetle.

Figure 1.9 True aerodynamic design. The Czech Tatra of 1937 with air-cooled rear-engined V-8 was a very advanced vehicle for its time.

The introduction of a large network of Autobahns in Germany in the 1930s meant that high speed road travel became a practical possibility in that country long before most others, and since streamlining produced significant advantages, it was generally more highly developed in Germany than elsewhere. The Volkswagen 'Beetle' designed in the late 30s may not look very streamlined by modern standards, but it was a considerable improvement on the box-like vehicles that were popular in the UK and much of the rest of Europe.

Many pseudo-streamlined vehicles appeared in the USA in the 30s, but these were largely exercises in styling, with no scientific basis. Any potential aerodynamic advantages in these styles, which simply reflected contemporary aeronautical forms, were usually destroyed by highly obtrusive front-end decorative elements. One exception was the Chrysler Airflow (Figure 1.10), where some attempt to use Jaray's principles was made. The American public did not like such a radical styling development, however, and few consider it a truly attractive vehicle even now. The commercial failure of this car made the American motor industry wary of experimenting with real streamlining for several decades.

Figure 1.10 The Chrysler Airflow incorporated a number of advanced features for the 1930s, including aerodynamic styling and semi-unitary construction. Though popular now at shows because of its rarity, it was a commercial failure in its time. Note the split windscreen and almost blended headlamps.

For many years, aerodynamic design was considered an impediment to commercially attractive styling. In the 1960s and early 70s, the preferred style was decidedly poor in terms of aerodynamic drag, being highly angular. Apart from styling considerations, these box-like forms were popular with the manufacturers, as they lent themselves well to cheap production and assembly. The stylists' unfavourable attitude to aerodynamic forms was only reluctantly abandoned in the mid 70s when the oil-exporting countries arranged a cartel which drastically raised the price of crude oil. Low fuel consumption suddenly became a major selling point, and manufacturers started to refine their shapes to reduce the drag. The most obvious change was the rounding of the front end, with the consequential abandonment of the vertical radiator grille, which sadly meant the end of the primary means of distinguishing one manufacturer's products from those of another. A major milestone was the bold introduction by Ford in Europe of its Sierra model (Figure 1.11) which was designed on aerodynamic principles. The Sierra's radically different appearance produced some initial consumer resistance, and it had to be heavily discounted at first. The Sierra eventually became very popular, and since that time, aerodynamic considerations have tended to dominate in motor vehicle styling. The improvements in fuel consumption produced by aerodynamic design are readily apparent, particularly in motorway cruising. Further details on road vehicle aerodynamic design may be found in Chapter 5 and in Barnard 1996.

1.6 Commercial vehicles

Although steam-powered carriages were little more than an experimental rarity in the early nineteenth century, steam-engined road vehicles based on railway technology were commonly used for pulling and powering agricultural and fairground equipment; fairgrounds represented a significant part of the entertainment industry at that time.

Figure 1.11 The Ford Sierra that introduced truly aerodynamic styling in the 1980s. The public took some time to adjust to the styling, which nevertheless set the trend for the next two decades.

By the end of the 19th century, steam-powered lorries and buses had begun to appear, but these were immediately challenged by petrol-engined vehicles that were developed concurrently with the motor cars. Steam engine technology was particularly highly developed in Britain, and steam lorry manufacturers such as Foden and Sentinel were loath to abandon their expertise and experience. By the 1930s, Sentinel was producing highly sophisticated steam lorries, but nevertheless fighting a losing battle. In Britain, coal-fired steam propulsion did have one significant advantage over petrol or diesel power in that coal was an indigenous fuel, and relatively cheap. The final blow to the steam lorry in the UK was the introduction in the early 1930s of legislation designed to protect the railways from competition from road transport. Vehicles over 4 tons were subjected to heavy taxation, and this effectively ruled out the heavy steam vehicles. Sentinel did, however, produce a few vehicles for export to South America as late as the 1950s.

As 1914 approached, the threat of a war in Europe loomed, and lorry production (by then mostly petrol-engined) was increased in anticipation. The British Government offered a subsidy to lorry purchasers who bought vehicles designed to a military specification and agreed to subsequently purchase them at a good price when they were needed for the war effort.

This subsidy may have helped to encourage the development of the commercial vehicle industry in Britain, but the war itself resulted in a massive production of commercial vehicles throughout Europe. After the war, the huge fleet of surplus military vehicles helped to fuel an expansion in the use of commercial road transport. A similar spur to the development of the road haulage industry occurred after the Second World War, particularly in continental Europe which had suffered massive damage to the rail system. The size and engine power of large trucks rapidly increased due to the building of autobahn-style motorways throughout Europe, and to the relaxation of restrictions on maximum weight and speed. The historical development of commercial vehicles including buses is well described by Nicholas Faith, 1995.

Apart from its contribution to the evolution of road haulage, the internal combustion engine facilitated the development of motor buses (Figure 1.12) which rapidly ousted their horsedrawn rivals. By the 1950s, buses had also almost completely displaced the electric tram and the later

Figure 1.12 A 1911 London B-type bus. Petrol-engined buses soon displaced the earlier horsedrawn vehicles. Note how the driver and upper deck passengers are totally exposed to the weather. The tyres are solid rubber. (From the collection of the London Transport Museum.)

electric trolley bus. In addition to having a major impact on urban public transport, buses were able to fill the gaps in the rail network, particularly in rural districts.

The building of the Autobahns in Germany in the 1930s encouraged a new form of passenger transport: the long-distance high-speed coach service. Taking advantage of the wide dual carriageway roads, special fast streamlined buses were built. The combination of streamlining and high-powered engines resulted in vehicles with top speed of well over 100 km/h (62 mph). Refinements such as toilets were also incorporated. The buses were operated by the German state railway company, with integrated bus and rail services. At the same time in America, long-distance bus services were also expanding rapidly and challenged the railroads which were suffering from the expense of track and rolling stock maintenance. The Greyhound Bus Company developed a nation-wide network by a process of absorption and collaboration with competitors. After the Second World War, the combination of road and air transport in America threatened the very survival of the railways which did not traditionally have the protection of nationalization. By the 1960s, long-distance American buses incorporated tinted windows, toilets, air-conditioning and pneumatic suspension, and were able to cruise comfortably at the legal speed limit.

1.7 Engine developments

Following the early vehicles of Daimler and Benz in 1895, engine developments rapidly ensued, and by 1888 Daimler had produced a vee-twin engine. Improvements in the ignition and carburation system followed, together with more sophisticated valve and cooling arrangements. In 1910, Ettore Bugatti was using an overhead camshaft on his Type 13 (Figure 1.13). This remarkable car had an engine of only 1327cc and yet managed 100 km/h. By the outbreak of

Figure 1.13 The Bugatti type 13 of 1910. This elegant little car had an engine of only 1327cc, but its advanced overhead camshaft design gave the vehicle a remarkable top speed of 60mph. (From the collection of the National Motor Museum, Beaulieu.)

the First World War, the petrol engine had evolved to a form that was little different from the modern unit. From that point on, there has been a process of continuous refinement. The most obvious innovations have been in the areas of fuel injection, electronic combustion management, catalytic converters and a limited amount of variable geometry. Compact vee-configuration engines have also become common.

Despite the advanced features of Bugatti's engines, side valves were mostly used for popular cars until after the Second World War. These were then replaced by overhead valves driven by pushrods via a rocker shaft, and later by overhead camshafts. Improvements in materials technology have permitted higher speeds, temperatures and compression ratios to be used, and this has resulted in much greater efficiency and power-to-weight ratio. The most significant achievement, however, is that with automation and advanced production methods, it is now possible to produce an engine of great complexity at an amazingly low cost, and with a level of reliability that would have seemed impossible only a few decades ago.

1.7.1 The diesel engine

Apart from the steam engine, the main rival to the petrol engine has been the diesel. Outwardly the engines are similar, and retain many common mechanical features. The diesel engine, however, works by the spontaneous combustion of fuel in the presence of compressed air, rather than ignition by electric spark. The diesel engine eliminates the need for an electrical spark ignition system and a carburettor, two of the weak points in petrol engines. As a result, diesel engines tend to be more reliable under adverse conditions. The diesel engine is more economical than the petrol engine, but generally has an inferior power-to-weight ratio, although turbocharging narrows the gap. These factors led to its initial adoption being in heavy commercial and military vehicles. In 1923 and 1924, diesel-engined trucks were introduced by the German manufacturers Benz, Daimler and MAN. Diesel power gradually took over for the propulsion of large commercial vehicles and buses, although, according to Faith (1995), in Britain in the mid-1960s, still only a third of large commercial vehicles was diesel powered.

As the cost of crude oil rose, particularly in the 1970s, the higher efficiency of the diesel engine began to make it an attractive alternative for domestic cars. When combined with turbocharging, the performance of diesel-engined cars becomes comparable with petrol engined vehicles. Continuous development has increased the power-to-weight ratio and smoothness of running. The diesel engine has the added attraction of lower emissions of some of the noxious gases, although this is offset by higher particulate emissions that have recently been recognized as representing a major health hazard.

1.7.2 Supercharging and turbocharging

A considerable improvement in the power-to-weight ratio of an internal combustion engine can be obtained if the air is compressed before entry to the cylinders. In the 1930s it became commonplace for racing and sports cars to be fitted with a supercharger which consisted of a compressor driven mechanically by the engine. The expense of the supercharger coupled with a significant increase in fuel consumption soon led to its demise on production cars, however.

After the Second World War, turbochargers were introduced. In the turbocharger, the compressor is driven by a turbine which is powered by the exhaust from the engine. The turbocharger therefore makes use of energy that would otherwise be wasted, and is much more efficient than a mechanically driven supercharger. Despite this improvement, turbochargers are still expensive, and for petrol-engined vehicles, are generally only used for racing and high performance. On diesels, however, turbochargers are much more commonly used, as they produce a worthwhile improvement in the power-to-weight ratio. They also improve the torque characteristics, and produce a smoother quieter running engine.

1.7.3 Two-strokes and unconventional petrol engines

Like the original Otto gas engine, most car engines work on a four-stroke cycle. The alternative two-stroke, with one firing stroke per revolution, has the theoretical advantage of potentially producing twice as much power for a given speed and capacity. In simple unsupercharged petrol engines, however, it is difficult to scavenge or drive the exhaust gases out without losing some of the incoming fuel-air mixture. On most small two-stroke petrol engines the fuel/air mixture was initially taken into the crank-case, where the pressure rise produced by the descending piston was used to force the mixture into the cylinder. This necessitated mixing lubricating oil with the fuel, which resulted in a smoky exhaust that was disapproved of, even before the public became conscious of the problems of pollution. There were some notably successful small two-stroke cars such as the three-cylinder Swedish SAAB, which became the Scandinavian equivalent of the Volkswagen 'Beetle', but the petrol two-stroke arrangement was always restricted to small car and motor cycle engines.

In the post Second World War era there has been some considerable development of the two-stroke, petrol engine, particularly by Japanese motor cycle manufacturers, but although emissions have been greatly reduced, it is difficult to provide the required level of combustion control. Two-stroke diesel engines, however, have enjoyed a greater popularity, and the two-stroke cycle is still used on extremely large engines for marine and railway applications. The use of turbo or supercharging allows the exhaust gases to be driven out fully, and as the fuel is injected as a spray, rather than being pre-mixed with air, no fuel is lost when excess air is used for purging. Two-stroke diesel truck engines were once quite common; an interesting example being the post-war British Commer engine which used an opposed piston arrangement with two pistons in each cylinder being driven towards each other with a crank arrangement at each end. The Commer engine is described by Newton *et al.* (1983). Vehicles with these engines gave a characteristic rasp that sounded more like a sports car than a heavy truck. The famous 'Tiger' tank of the Second World War also employed a two-stroke that produced a readily recognizable sound.

The shelves of the world's patent offices are littered with numerous unconventional engine arrangements, but only the Wankel rotary engine has made any impact. Considerable resources went into developing this promising engine, which has fewer moving parts and a semi-rotary rather than reciprocating motion. The problem of wear on the tip seals of the rotor proved to be a major stumbling block, however, and just as this appeared to be nearing solution, another inherent weakness appeared, namely the problems of emissions associated with its two-stroke cycle.

The gas turbine, which has become the universal power plant of all but the smallest aircraft, has not yet made any real impact on road vehicles, despite some enthusiastic developments, notably by Rover in the early post-war years. The gas turbine is ideal for high-speed flight, where it combines good system efficiency and a high thrust-to-weight ratio with excellent reliability, but compared to the petrol engine, it is less efficient at low speeds. It also has a poor response rate. The gas turbine does, however, possess some potential advantages in terms of emissions, as the maximum temperatures reached are lower than in reciprocating engines, and it becomes more efficient when run at constant speed and power. For this reason, it is being considered as a serious candidate for the prime mover in hybrid propulsion systems where, as described below, it would be used in conjunction with an energy storage device.

1.7.4 Electric and hybrid propulsion

Electric power has been used for automotive propulsion from the earliest times (Figure 1.7); indeed in April 1899, Jenatzy's electric-powered 'La Jamais Contente', a crudely streamlined torpedo-shaped car, was the first road vehicle to exceed 100 km/h. The short range and excessive weight of electric vehicles have hitherto limited their use primarily to local goods delivery, most notably for the daily fresh milk deliveries in the UK. The rising problem of urban pollution has, however, forced a re-evaluation of the electric vehicle, particularly in southern California where a vast urban sprawl and particular climatic conditions cause a major problem of smog generation. Improvements in lead-acid battery construction and developments of more exotic types of battery have led to the limited production of practical electric cars for urban use, and fleets of electric buses are currently in use in several large cities around the world.

One important negative aspect of electric vehicles is that although the effect on pollution may be reduced locally, the problem has simply been shifted to the 'backyard' of the power station. The overall system efficiency, including initial electrical power generation and distribution, is low, and there may consequently be no improvement in the total amount of pollutants released, unless the electricity is generated by non-combusting energy sources such as wind or nuclear power.

The limited range that can be provided by battery storage has more recently led to the development of hybrid vehicles where the batteries or other energy storage devices such as a flywheel or compact 'ultracapacitors' can be recharged by a small hard-working petrol, diesel or even gas turbine engine. All of these engines tend to be more efficient when working hard, and the energy storage system can allow energy from braking to be recovered, resulting in vehicles that are potentially much more efficient than current conventional types. Such hybrid arrangements should not be confused with the early petrol-electric drive systems which were purely used to provide a simple stepless and clutchless transmission, as described in the next section.

1.8 Transmission system development

The steam engine does possess a number of advantages compared to the petrol engine. Amongst these are the fact that once adequate steam pressure has been achieved, the engine can be stopped and instantly re-started as required, and it can produce full torque from rest. Steam-engined cars therefore required no clutch or gearbox, and were almost silent in operation, which was one reason why they persisted for so long. On early petrol engined cars, various methods of decoupling the engine from the drive were initially used, including belts that could be slipped on and off pulleys, and various types of clutch. The single plate dry clutch eventually predominated, and has been the standard mechanism on manual gearboxes for cars for many years. Where very large amounts of torque have to be transmitted, as on racing cars and some heavy commercial and military vehicles, multiple plate clutches are used.

Many different forms of gearbox have been employed, but the modern arrangement of input, output and layshaft quickly emerged as the dominant type. In early versions, the gears themselves were slid in and out of mesh by moving them along splined shafts, but this arrangement was replaced by designs where most of the gears were in constant mesh, but were locked to or unlocked from their shaft by a system of sliding toothed dog-clutches. The dog-clutches bore the abrading effects of remeshing, thereby avoiding damage to the carefully machined gear teeth. This type is known as a 'crash' gearbox, because of the characteristic crash of gears that occurs with inexpert use. Changing gear, particularly changing down, was not easy. The clutch pedal had to be depressed whilst one ratio was disengaged, then let up again so that a touch on the throttle could be used to bring the shaft up to the correct speed for meshing the new ratio. The clutch was then depressed once more as the new gear was engaged, and finally let up again. This system of double declutching required some skill, as the correct meshing speed could only be judged by ear and experience. By the 'thirties, an ingenious arrangement of sychromesh cone clutches began to be added to the gearbox so that the shaft was automatically pulled up to the correct meshing speed before the dog-clutch engaged. Some American cars retained the older 'crash' gearbox even after the Second World War for a while, because it was cheaper and lighter, but gear changing was made relatively easy by the fact that the large low-revving engines only required three ratios. The lowest ratio was normally used for pulling away or climbing exceptionally steep hills, so most driving was done with just two gears.

An alternative type of gearbox favoured particularly by the great inventor F.W. Lanchester and used on the Lanchester and British Daimler company's products until the 1960s, was the epicyclic type. As described in Chapter 13, and Newton *et al.*, 1983, this comprises a number of gear assemblies each consisting of an inner sun gear meshing continuously with a set of planet wheels, which in turn engaged on an outer toothed ring. The ratios are changed by locking and unlocking various of the outer rings with brake bands, or sometimes by linking other elements together by clutches. Because all the gears are constantly in mesh, no dog-clutches are needed, and thus gear changing is simple and quiet. The epicyclic gearbox was employed by Lanchester as early as 1895, but its most significant early use was on the ubiquitous Model T Ford.

Gear selection on the Model T was effected by use of foot pedals, one to hold low gear, and another for reverse. Lanchester, however, developed a more sophisticated system of 'preselection' which was introduced in 1901 and subsequently used on Lanchester and British Daimler cars until after the second world war. In the preselector arrangement employed on these cars, a small hand lever was used to preselect the ratio which was subsequently engaged by depressing a foot-operated button. A fluid coupling was used instead of a mechanical clutch. Vehicles fitted with this type of gearbox were pleasant to drive, particularly in heavy traffic, and had many of the characteristics of a modern automatic. The disadvantages of the system were that the gearbox was heavy, and the friction losses were high, particularly in the fluid clutch. Preselector gearboxes were not used for popular mass-produced cars, but they did find widespread application on buses, as they eliminated the hard work associated with the frequent gear shifting and clutch operation of conventional transmission systems.

In the 1940s, the epicyclic type of gearbox was developed in the USA to produce a fully automatic arrangement similar in principle to most modern designs. It first appeared in general use as an Oldsmobile option in 1940. The simple fluid coupling was soon replaced by a fluid torque converter, which allowed a limited range of continuously variable speed and torque ratios in addition to the fixed gear steps. This is the configuration found on most modern automatics.

Automatic gearboxes rapidly grew in popularity in America, until they became by far more common than the 'standard' mechanical shift. European manufacturers were slow to follow this lead. Engines in European cars were smaller, and higher revving, so transmission losses were much more noticeable, and four ratios were really required. Early automatics had a poorer performance and higher fuel consumption than corresponding manual gearbox models. The high cost of fuel in Europe made the low efficiency important, and the poor performance gave automatics an 'auntie' image. For many years they were only offered as an option on expensive vehicles. More recently, four or even five-speed units suitable for the small European and oriental cars have been introduced. These now feature a lock-up facility in top gear, whereby the transmission drives directly, and there are no losses associated with slip in the torque converter. Torque-converter lock-up was first introduced in 1949 on the Packard Ultramatic drive, but its use did not become widespread for some time. The unsporty 'thirsty' image of the automatic still persists in parts of Europe, and by 1996 still only 2.4% of French and Italian cars were automatics. In Germany, the figure had reached 18%, the discrepancy being partly attributable to different social attitudes to driving.

Apart from automatic gearboxes there has always been an interest in gearless or stepless continuously variable (CVT) transmissions. Large amounts of time and money have been developed attempting to produce an efficient practical device. One of the earliest examples was Ferdinand Porsche's 'mixte' system for which the original patents were filed in 1897. In this method, a petrol engine drove an electrical generator which in turn drove electric motors on the wheels. Some early buses, notably those produced by Tilling-Stevens in the U.K., used essentially the same arrangement which had the great advantage then that it could be operated with minimal training by former drivers of horse-drawn vehicles. Although this system provided a smooth stepless transmission that made driving very easy, it was killed off by its poor efficiency.

Semi-stepless transmissions reappeared briefly in America in the early post-war period in the form of fluid torque-converter boxes, and again somewhat later (1955) in the ingenious Van Doorne variable diameter pulley and belt system used initially on the Dutch DAF cars. A steel belt development of the same basic system has more recently been used for small European vehicles. A stepless transmission system is inherently more efficient than a stepped box, as it should allow the engine to run at its optimum speed regardless of road speed, but in practice, the efficiency of the unit itself has tended to be relatively low. Another disadvantage is that whereas problems in a conventional gearbox are usually apparent and progressive, failure of the DAF boxes could be inconveniently unexpected and sudden.

1.9 Steering

On horsedrawn vehicles, both the front and rear pairs of wheels are usually mounted on simple beam axles. The front axle is pivoted about a vertical axis at its centre, and attached to the shafts to which the horse is harnessed. Steering is thus effected by the horse being encouraged to turn in the required direction by a pull on the reins. On very early motor vehicles, the same pivoting axle was initially used, often with some form of tiller for manual steering. This was found to produce a dangerously unstable arrangement, however, because the horse in its shafts had provided a stabilizing moment. Numerous technical solutions were tried, but the Ackerman linkage used on modern vehicles was quickly adopted. This allows the two front wheels to pivot about their own separate axes, and for the inner wheel to be turned more than the outer, so that the two wheels have a common turn centre. The geometric arrangement of the mechanism incorporates a degree of caster which makes the wheels tend to naturally return to a central or straight ahead position, thereby making the system directionally stable.

The tiller soon gave way to the steering wheel, and various mechanisms were used to connect the steering linkage to the steering wheel. There was little change to the overall mechanism for several decades, and a major advance did not occur until the introduction of power steering. Crude forms of power steering had been used with steam-driven vehicles in the 19th century, but it did not become common on domestic cars until after the Second World War, appearing as a standard feature on the 1951 Chrysler Crown Imperial. Four-wheel steering is now available on some models, but this is still an unusual arrangement.

1.10 Suspension

For many years, the semi-elliptic leaf springs used on carts and carriages were the most common method of providing suspension springing. The springs were also used to provide the means of locating the axle. This was a neat and simple arrangement, but it unfortunately produced a number of problems such as a tendency of the axle to wind up around the springs on braking or acceleration. Gradually, suspension mechanisms were developed, in which the wheel was located by a number of links, and constrained to move in a predominantly vertical direction. The springing could then be provided by other types of device such as torsion bars, which were developed by Ferdinand Porsche before the war, and also used on a number of the post-war vehicles, notably the Morris Minor. Pneumatic variable height suspension was developed for family cars by Citroën, and has also been used on commercial vehicles. The simple coil spring gradually became the norm, however, initially for the independently sprung front axles, but later for rear axles as well. A more recent development is active suspension, where the wheel vertical movement is controlled by power jacks.

The use of beam axles and cart springs provided a simple and robust arrangement, but it was soon found that allowing the wheels to move independently of each other improved the roadholding, steering and ride comfort. The improvements are partly a function of the geometry of movement, and partly due to a reduction in the ratio of unsprung to sprung mass. Although independently suspended wheels were used on an early Bolleé steam vehicle, independent suspension did not come into widespread use until the late 1930s, and then mostly for just the front wheels. At that time almost all popular cars except the Citroën 'Traction Avante' had rear wheel drive. Providing independent suspension for the rear wheels made the final drive arrangements much more complicated, as universal joints and other items were required. Volkswagen and Tatra partially solved the problem by mounting the engine in the rear. By the 1950s and 1960s, some sporting prestige vehicles such as the Jaguar (Figure 1.14) had all-round independent suspension, and by the 1990s most cars had adopted the system: a move hastened by the popularity of front-wheel drive, which meant that the rear wheels could be independently sprung quite simply. Further

Figure 1.14 The Jaguar 3.4 of the late 'fifties featured all round independent suspension, an unusual feature on a production saloon car at that time

refinement of the suspension system has come by the use of relatively complex mechanisms, and by reduction of the unsprung weight, partly achieved by the adoption of lighter wheels and tyres.

1.10.1 Wheels and tyres

Most horsedrawn carriages used wooden-spoked wheels with a tyre consisting of a simple metal hoop. This arrangement was satisfactory for slow vehicles, but poor adhesion and lack of shock absorption made them unsuitable for the faster motor cars. Various methods were employed to soften the ride, such as the use of tyres made of a hemp rope or solid rubber. These approaches were not very effective, and suffered from poor durability. The important breakthrough was the development of a practical inflatable rubber or pneumatic tyre. A crude form of pneumatic tyre had been constructed by R.W. Thompson of Britain for a horse-drawn carriage in 1846. This consisted of a rubberized inner tube, and an outer cover of riveted leather segments. The true precursor of the modern tyre was, however, invented by J.B. Dunlop, whose inflatable tyre contributed greatly to the late 19th century popularity of the bicycle. Dunlop did not initially think that his inflatable tyres would be suitable for the heavy motor vehicles, but in 1895 the Michelin brothers fitted a Peugeot car with inflatable tyres and competed in the Paris-Bordeaux-Paris race. Although they failed to complete the race, the improvement in roadholding was readily apparent, and pneumatic tyres soon began to displace the solid tyres except for large commercial vehicles which still commonly used the solid type until the late 1920s.

Pneumatic tyres were initially something of a mixed blessing, as punctures were frequent, due largely to the presence in the roadways of old horseshoe nails. Journeys of any distance invariably involved a number of punctures that had to be repaired on the spot by the owner, or more commonly by his mechanic-cum-chauffeur. It was a surprisingly long time before it dawned on motorists that it would be a good idea to carry a spare rim and tyre or later a spare wheel.

On rutted pot-holed roads, there was an advantage in using large diameter wheels, but gradually, as roads improved, wheel diameters decreased. By the 1930s, American cars had more or less standardized on a 16 inch diameter rim (see Figure 1.5). European manufacturers were slower to follow, and the little Austin 10 of 1934 still had 18 inch wheels. After the war, wheel diameters decreased further, as the smaller wheels improved the ride and suspension dynamics. It was also found that using wider tyres improved the roadholding and braking, and there has been a progressive trend to ever wider tyres.

Tyres changed radically during the early post-war period, firstly by the general introduction of tubeless tyres, where the outer casing forms an airtight seal with the wheel rim, eliminating the need for an inner tube. Tubeless tyres are less prone to explosive puncturing than the older tubed variety, as the thick rubber of the tyre tends to form a seal around any sharp penetrating object. The durability of tyres was also increased by the introduction of new synthetic mixes. A further improvement was effected by the adoption of a different arrangement of the fibre or wire reinforcement filaments. The newer radial-ply tyres rapidly displaced the older cross-ply type. The radial-ply tyres gave an improved grip whilst reducing the rolling resistance. A more recent development has been the introduction of 'low profile' tyres where the ratio of outer to inner diameter is decreased. This has led to rim sizes becoming larger again, although the overall wheel diameter remains unchanged.

1.11 Brakes

Although horsedrawn vehicles usually had some form of brake, the retardation was partly provided by the horse, so it became necessary to devise much more effective brakes when motor vehicles were introduced. Various arrangements of rim brakes, brake belts and drum brakes were developed, but gradually the drum brake with a pair of internal brake shoes evolved as the dominant type, and this arrangement is found on many vehicles even now.

A major problem with the drum type of brakes is that the linings tend to overheat with prolonged or rapidly repeated use, causing loss of effectiveness or 'fading'. The solution was provided by the development of disc brakes, where a metal disc is squeezed between a pair of brake pads in a similar manner to the action of calliper brakes on a bicycle. The disc is exposed to the air flow on both sides, and is hence cooled more effectively than a drum. Disc brakes had been used in crude form on some early vehicles, and appeared in a more refined arrangement on the 1949 Chrysler Crown Imperial. Rival designs by Lockheed and Girling, similar to those used today, were presented at the 1952 London Motor Show, and this type of brake was gradually adopted, initially for the front wheels, and finally for all four wheels. The reluctance to fit them to the rear wheels stemmed from difficulties in getting them to hold effectively on the mechanical parking brake. Disc brakes provided a major improvement in braking force and resistance to fade. The brake pads are generally easier to check and replace than the brake shoes. A recent development, employed in racing cars, has been the introduction of carbon based components which will withstand being heated to red or orange heat.

Until the 1920s, most cars had rear wheel brakes only. Experience with flying over the handlebars of bicycles due to over enthusiastic application of the front wheel brake convinced people that front wheel brakes on cars would be dangerous. Although rolling a car over forwards due to the application of brakes was shown to be virtually impossible, it did appear logical that stopping the front wheels would make the rear of the vehicle slew round out of control. In reality, a skid is far more likely to develop from the application of the rear brakes, since the steering can be used to correct any front wheel pull; nevertheless, prejudice prevailed for some time. Front wheel braking is far more effective than rear wheel, since the inertia of deceleration increases the vertical reaction on the front wheels, thus improving the grip on the road. When vehicles with four-wheel brakes were first introduced, they were vulnerable to being hit in the rear by the less effective two-wheel braked competition. It was customary therefore to carry a warning triangle on the rear.

Mechanically operated brakes using rods or cables needed frequent adjustment by skilled mechanics. When incorrectly adjusted, the braking effect could be different on each wheel, and the car would tend to swing on braking. The introduction of hydraulically operated brakes in the 1930s was a great improvement. In particular, hydraulic operation ensured that the brake actuating forces were applied equally on both sides of the car. One inherent danger of early hydraulic brakes was the fact that any large leak or fracture would mean the loss of all but the mechanically operated parking brake, which was not usually very effective. This major defect was not rectified until the late 1960s when dual circuit systems started to be introduced.

On heavy vehicles, mechanical or hydraulic operation of the brakes required a large pedal force, and various forms of mechanical servo system were introduced. One of the most popular was the floating shoe type. In this system, only one shoe, the primary shoe was brought into contact with the drum by the normal mechanical or hydraulic linkage. The primary shoe was dragged round by the drum, and forced the secondary shoe into contact. Although very effective, this design could cause a dangerous lock-up condition if badly adjusted or worn. A better method of reducing the pedal load is the use of power assistance, normally provided by using the engine manifold vacuum to produce an actuating force via a piston. Initially introduced on large expensive vehicles, power braking is now used even on small cars.

Anti-lock (ABS) brakes which contain a mechanism that prevents the wheels locking up and hence generating a skid, were originally developed for aircraft, but have become increasingly common on road vehicles. One of the best known early applications was on the British Jensen FF in the 1960s. This vehicle also featured four-wheel drive.

1.12 Interior refinement

The most noticeable feature of recent car development has been in the area of refinement. Modern family cars are not greatly superior in performance to specialist high-powered vehicles of the 1930s, but they are much quieter and easier to drive. The quietness is largely due to advances in the techniques of dynamic analysis and design: an area where mathematical analysis has had a major impact. Sound insulation and the isolation of reciprocating or vibrating components have become increasingly effective. One of the most important factors, however, was the replacement of a direct belt drive from the engine to the cooling fan, by an intermittent electrical drive.

Arrangements for heating and demisting were fairly rudimentary until the late 1930s, but after the war, the provision of hot air for both purposes became an important aspect of interior design. Air cooling or air conditioning was introduced by Packard in 1940 and gradually gained in popularity in the USA. As with automatic transmission, its adoption was much slower in Europe, and it only started to appear on medium-priced vehicles in the mid 1990s.

The widespread use of electrical components such as window lifts has been made possible by a combination of improved quality control and solid state electronics. Until about the 1980s, electrical devices were often of poor quality, and owners preferred the reliability of mechanical systems.

1.13 Safety design

It is a sad fact that the invention of the motor car has produced more deaths and injuries annually than almost any other human invention. During the Second World War, the number of fatalities and injuries sustained by American forces in any great battle, rarely exceeded the monthly civilian road-accident toll back home. Despite this, very little effort was made in terms of safety design until the 1950s and 1960s. In the 1930s the streamlined Chrysler Airflow (Figure 1.10) incorporated a level of unitary construction that made it relatively resistant to impact. The manufacturers tried to exploit this feature in its advertisements which showed the vehicle escaping with surprisingly little damage after being driven over a cliff. The public, however, did not wish to be reminded of the dangers of motoring, and negative reaction to the advertisement produced another blow to the sales of this vehicle. With such a public attitude, it is not surprising that manufacturers did not see safety engineering as a selling point.

The most dangerous item on pre-war cars was the steering column, which pointed like a spear at the heart of the driver. Despite an increasing awareness in the police and safety services, collapsible steering columns did not start to appear until the 1960s. This feature subsequently became mandatory in most advanced countries. In 1950, Nash offered seat belts on its reclining seat option, but belts even for the front-seat occupants did not become standard or mandatory in most countries for several years. Regulations requiring the fitting of rear seat belts appeared even later. Air bags were first introduced in the United States, but their use elsewhere lagged by at least a decade. Safety design is now a major consideration, and all vehicles have to demonstrate adequate energy absorption in front, rear side and quarter impacts. Despite the major advances in safety design that have taken place, the effect on road accident injuries has been disappointing, as drivers have seemingly adjusted their driving habits to maintain the level of risk.

1.14 Too much innovation

Innovation is not always a key to financial success, indeed, Lord Montagu and Michael Sedgwick in their book *Lost Causes of Motoring* (1960) identified a number of factors that led to motor manufacturers becoming lost causes. These include wasting money defending patents, too many models, and too many technical innovations. Radical technical innovations are usually expensive to develop and can produce a backlash of unfavourable customer reaction if the inevitable teething troubles are not quickly remedied. Alec Issigonis's revolutionary Mini with its package of front wheel drive, transverse engine and rubber-bush suspension was a great success, but the manufacturer's attempts to follow this up with even more technological advances such as hydroelastic suspension and five speed gearboxes, were less successful, at least in financial terms. It could be argued that the company simply did not have the development resources or production methods to enable it to produce sufficiently refined and reliable vehicles.

Although too much innovation can cause financial problems, being too conservative can be equally damaging, and successful companies have been those who have managed to find the right blend of creativity and caution. The key is to pick the technical winner like disc brakes and power steering at the right time, and reject the losers like the Wankel engine. Above all, innovative developments have to be properly costed.

Technical developments are still appearing, but these tend now to be more in the form of refinements rather than major changes. However, a revolution is probably just around the corner. The challenge of emissions and the problems of relying on fossil fuels suggests that radical changes will soon become necessary, and a whole new era of technical development is about to begin, probably centred on electric propulsion.

1.15 References and further reading

Books on motoring are numerous, but unfortunately, they only seem to stay in print for a relatively short time. Wherever possible, we have tried to select books that should be readily available in a good academic library. The remainder should be obtainable on the inter-library loan system.
Barnard, R.H. (1996). *Road Vehicle Aerodynamic Design*. Longman.

A description of the basic principles of the subject aimed particularly at undergraduate engineers, with emphasis on the physical principles, and with a minimum of mathematical content.

Faith, N. (1995). *Classic Trucks: Power on the Move* (accompanied the Television Channel Four series). Boxtree (ISBN 0-7522-1021-1).

A very readable text which gives a good outline of the history of commercial vehicles, with emphasis on developments in Britain.

Ickx, J. (1992). 'The Bollées', in Barker, R. and Harding, A. (eds.), *Automobile Design: Twelve Great Designers and their Work*. SAE (ISBN 1-56091-210-3).

Describes in detail the achievements of some of the great automotive designers from around the world. It also traces the origins of many important innovations.

Lord Montagu of Beaulieu and Sedgwick, M. (1960). *Lost Causes of Motoring*, Cassell, London.

An excellent description of the rise and fall of all the famous marques in British motoring history with an analysis of the reasons for their demise. There are two companion volumes: Lost Causes of Motoring – Europe, volumes 1 and 2.

Newton, K., Steeds, W. and Garrett, T.K. (1983). *The Motor Vehicle* (10th edn). Butterworths (ISBN 0-408-01118-1 (hard cover) and 0-408-01157-2 (soft cover)).

Detailed descriptions of the workings of automotive components both current and historical.

Womack, J.P., Jones, D.T. and Roos, D. (1990). *The Machine that Changed the World*. Maxwell Macmillan International.

The result of a large-scale research exercise, this book traces the development of vehicle manufacturing systems and expounds the advantages of the lean production system which originated in Japan.

Wood, J. (1998). *Wheels of Misfortune: the Rise and Fall of the British Motor Industry*. Sidgwick and Jackson, London (ISBN 0-283-99527-0).

An analysis of the reasons for the failure of the indigenous British motor manufacturing industry.

Further reading

'Automotive Milestones'. *Automotive Engineering*, September 1996.

Key dates for the introduction of technical innovations in the USA, with other historical articles.

Howard, G. (1986) *Automobile Aerodynamics*. Osprey (ISBN 0-850445-665-7).

An illustrated history of the subject of automobile aerodynamics.

Pawlowski, J. (1969). *Vehicle Body Engineering*. Business Books Ltd (ISBN 0-220-68916-4).

A good general description of the subject of automotive engineering design with some historical material.

Whyte, A. (1984) *The Centenary of the Car*,1885–1985. Octopus Books (ISBN 0-7064-2006-3).

A well-illustrated general history.

2. Modern materials and their incorporation into vehicle design

Rob Hutchinson, BSc, MSc, MRIC, CChem, MIM, CEng

The aim of this chapter is to:

- Introduce the broad range of materials that designers can draw upon;
- Introduce the properties of materials that are required for vehicle design;
- Demonstrate particular uses of material properties by case studies;
- Demonstrate the material selection process and its interactivity with design.

2.1 Introduction

The main theme of this chapter will be the study of the various inter-relationships between the structure of engineering materials, the methods of component manufacture and their ultimate designed behaviour in service. The four major groups of engineering materials are metals and alloys; ceramics and glasses; plastics and polymers and modern composites, such as silicon carbide reinforced aluminium alloys. Illustrative case studies will make up a significant section of this chapter.

The full range of these engineering materials is used in the construction of motor vehicles. It is a common myth that the aerospace, defence and nuclear industries lead the way in the use of materials for aggressive environments and loading regimes. The automotive industry has its own agenda with the added criteria of consumer demands of acceptable costs as well as critical environmental issues. Engineers, in general, are familiar with metals since they have the all-round properties, which are required for load bearing and other applications. This situation is helped economically by the fact that of the hundred or so elements within the earth's crust, the majority are metals. This means that whilst some are more difficult to extract than others, a wide range of metals is available to supplement iron, aluminium, copper and their wide-ranging alloys. Metals have adequate strength, stiffness and ductility under both static and dynamic conditions. Other physical properties are also acceptable such as fracture toughness, density, expansion coefficient, electrical conductivity and corrosion and environmental stability. A wide range of forming and manufacturing processes have been developed as well as an extensive database of design properties (Timmings and May, 1990). There is also a well-established scrap and recycling business.

Only when extreme properties such as low density, low thermal and electrical conductivity, high transparency or high temperature and chemical resistance are required, and where ease of manufacture and perhaps low cost are important, do engineers consider fundamentally different materials, such as polymers and ceramics. These two groups of materials have alternative engineering limitations such as low strength or brittleness. Consequently, combinations of these three materials groups have been used to form the fourth group of engineering materials known as composites, of which the major tonnage group is glass reinforced polymers. Ceramic reinforced metals also form a significant technical group of composite materials (Sheldon, 1982). All four groups of these materials have an essential part to play in the design, construction and service use in vehicle engineering.

In addition to the direct engineering issues, the vehicle designer needs to consider the political issues such as pollution and recycling due to the vast quantities of materials used in automotive manufacture. In Western Europe, the EC politicians now expect vehicles to be clean, safe, energy efficient, affordable and also 'intelligent', which means that they should be able to anticipate the actions of the driver and other road users. This has lead to significant research funding which, in the materials area, has involved work in the areas of combustion engine materials, batteries and fuel cells, wear resistant materials and light weight vehicle body materials. Such work is expected to continue. However, whilst engineering, environmental and safety issues will be of general concern, the manufacturer will continue to be motivated by profit whilst the driver will still expect personal freedom. On this issue government is caught between the environmental lobbies and the car industry, which makes a considerable contribution to gross domestic product. Thus, road usage is likely to continue to increase, so that some form of overall traffic management may well become essential as road building programmes are scaled down due to economic and environmental pressures.

2.2 Structure and manufacturing technology of automotive materials

Engineering materials are evolving rapidly, enabling new vehicle component designs, for load bearing structures and bodywork, engines, fuel supply, exhaust systems, electrical and electronic devices and manufacturing systems. Modern materials include fibre composites, technical ceramics, engineering polymers and high temperature metal alloys (Ashby *et al.*, 1985). The vehicle designer must be aware of these developments and be able to select the correct material for a given application, balancing properties with processing, using a basic understanding of the structural inter-relationships.

2.2.1 Metals and alloys

Many metals are not abundant and so can only be used for specialist applications such as in catalytic converters and powerful permanent magnets. In contrast, iron, copper and aluminium are very abundant and more easily obtained and so are widely used in both pure and alloy forms (Cottrell, 1985).

Iron-based or ferrous metals are the cheapest and the most widely used at present. For low load applications, such as bodywork and wheels, mild or low carbon steel is sufficiently strong with yield strengths varying between 220 and 300 MPa. It is also easy to cut, bend, machine and weld. For drive shafts and gear wheels, the higher loads require medium carbon, high carbon or alloy steels, which have yield strengths of about 400 MPa. Higher strength and wear resistance are needed for bearing surfaces. Medium and high carbon steels can be hardened by heat treatment and quenching to increase the yield strengths to about 1000 MPa. Unfortunately, these hardened steels become brittle following this heat treatment, so that a further mild re-heating, called tempering, is required. This reduces the brittleness whilst maintaining most of the strength and hardness. Stainless steels are alloys with a variety of forms, Austenitic, Ferritic, Martensitic and the newer Duplex steels. A common composition contains 18% chromium and 8% nickel, as shown in BS 970, 1991. Their corrosion resistance and creep resistance are superior to plain carbon steels, particularly at high temperatures. However, higher material and manufacturing costs limit their use in vehicle engineering to specialist applications such as longer life exhaust systems. Cast irons have 2 to 4% carbon, in contrast to the 1% or less for other ferrous metals mentioned above. This makes it brittle, with poor impact properties, unless heat-treated to produce ductile iron. It is more readily cast than steel, since the higher carbon content reduces the melting point, making pouring into complex shaped moulds much easier. In addition, the carbon in the form of graphite makes an ideal boundary lubricant, so that cylinders and pistons have good wear characteristics, for use in diesel engines. However, it is now largely replaced by the much lighter aluminium alloys for these applications in petrol engines.

Copper and its alloys form a second group of vehicle engineering metals, including copper itself, brass, bronze and the cupro-nickels. Copper is more expensive than steel, but is ductile and easily shaped. It also has high thermal conductivity, giving good heat-transfer for radiators, although more recently replaced by the lighter aluminium in this application. Its high electrical conductivity is made use of in wiring and cabling systems. Brass is a copper alloy, commonly with 35% zinc, which makes it easier to machine yet stronger than pure copper. Thus, complex shapes can be produced for electrical fittings. However, such alloys suffer from a long term problem, known as 'dezincification', in water. Corrosion can be minimized by using the more expensive copper alloy, bronze, where tin is the alloying element, although this material may be harder to machine. Copper-nickel alloys have good creep resistance at high temperatures where they are also corrosion resistant. The latter property is made use of in brake fluid pipe-work.

Aluminium and its alloys have a major advantage over steels and copper alloys, as vehicle engineering materials. Their much lower densities lead to lower weight components and consequent fuel energy savings. Whilst aluminium ores are abundant, the extraction of pure aluminium is very energy intensive, being electro-chemical in nature rather than the purely chemical process used for steels. Copper occupies an intermediate position on this point. Thus, pure aluminium is more expensive than iron and copper and has lower inherent strength and stiffness. However, it does have corrosion resistance with good thermal and electrical conductivity. A wide range of alloys is now available with various heat treatments and manufacturing opportunities. These materials have now replaced steels and copper alloys in many vehicle component applications, where their higher materials costs can be designed out, see Figure 2.1. Nevertheless, materials developments are such that aluminium alloys are themselves in competition with polymers and composite materials for such applications as vehicle bodywork, see Figure 2.2.

Considerable price fluctuations in materials occur from time to time due to fuel price variations so that the above values should be considered in relative terms.

The selection of a metal for a design application requires experimental data. The first stage will determine which group of metals should be used, steels, copper or aluminium (see Table 2.1). Then a specific selection will require more detailed information. Testing of materials and components will therefore be required. Some properties are largely independent of composition, microstructure and processing. These include density, modulus, thermal expansion and specific heat. However, many properties are very dependent on alloy composition, microstructure, heat-

Figure 2.1 A typical aluminium extrusion framework for automotive use

Figure 2.2 A typical plastics composite body panel

treatment and mechanical history. These properties include yield and tensile strength, ductility, fracture toughness, creep and fatigue strength, so that specific information is required (Smith, 1993).

2.2.2 Plastics and polymers

Animal and vegetable materials are composed of a wide range of natural polymer molecules, such as proteins, fats and carbohydrates. These occur in the structures of timber, leather, rubber, cotton, wool and silk, which are all load bearing, in service. These natural polymers are widely

Table 2.1 Material properties

Material	Density (Mgm^{-3})	Cost/Tonne (£)	Yield Strength (MPa)	Modulus (GPa)
Aluminium and steel alloys	2.7–2.9	1000–1500	40–600	69–79
Mild steel and steel alloys	7.5–8.3	200–1800	220–1300	190–209
Copper and alloys	8.5–8.9	750–1500	60–960	120–150

used in engineering and technically demanding applications, such as building products, sports equipment, vehicle tyres and internal car trims.

About 100 years ago, the first man-made or synthetic polymers were produced, such as cellulose products and the phenolics. These are still used in fabrics and electrical products respectively. There are now about 30 different groups of polymeric materials in common usage, many of which find application in vehicle engineering (Brydson, 1995). These materials are less strong than metals and alloys by a factor of 10, although they can be reinforced by fibrous and particulate materials, such as glass, carbon and aramid (Kevlar) fibres. These again replicate natural materials, such as wood, which is a two-phase composite. Composites are stiff, strong, ductile and light-weight, and although expensive in some cases, are used extensively in vehicle engineering for such applications as bodywork, bumpers, prop-shafts and fuel inlet manifolds, see Figure 2.3.

Figure 2.3 A typical lightweight polymer, truck air filter container

The wide range of commercial polymers has resulted from a greater understanding of polymer structures, from atomic through molecular to solid state levels, such that 'tailor-making' of polymers is now possible, at a price. Thus, polymers are currently available which will process readily and have the required properties and behaviour in service. Polymers are also less stiff than metals, by a factor of 100, so that their use requires new design procedures. Polymers and particularly their composites can also be very anisotropic in behaviour, leading to directional properties. They are also much more temperature, time and frequency sensitive than metals and ceramics (Hall, 1989). Again price values are subject to considerable fluctuations and the property values should be considered as a ranking due to wide range of available grades within each polymer group.

Polymeric materials are made up of very long chain molecules with a backbone of principally carbon atoms, which are held together by primary forces or bonds, comparable in strength to those in metals. Silicone polymers have a silicon oxide backbone. However, these long chain molecules are held together, in thermoplastics, by much weaker, secondary forces producing a more open structure, which leads to the inherently lower density, strength and stiffness values, as shown in Tables 2.1 and 2.2. In thermosetting polymers, such as phenolics, these weaker secondary forces are chemically supplemented with stronger primary forces, during curing, forming a three-dimensional molecular network, thereby increasing their strength and stiffness. Rubbery materials for tyres, hoses, belting and engine mountings are similarly cross-linked by a vulcanization process with sulphur, but since they are already above their softening or melting points at room temperature, they remain typically flexible. Thus, many of the general properties of polymers are those of materials near their softening points. They creep under load in service, a problem which requires a more complex, pseudo-elastic design approach from that used for metals and ceramics (Hall, 1989).

Table 2.2 Properties of plastics

Material	Density (Mgm^{-3})	Cost/Tonne (£)	Yield strength (MPa)	Modulus (GPa)
Polypropylene	0.91	500	30–50	1.0
U–PVC	1.4	500	40–50	1.5
Acrylics	1.2	2500	40–50	3.5
Polycarbonate	1.2	1500	50–60	3.0
Nylons	1–2	2000	50–90	2–4
Natural rubber	0.8–0.9	650	3–30	0.01–0.1
Phenolics	1–2	600	30–100	20
Polymer composites	1–2	800–8000	100–600	20–200

Most polymers are now made from oil and natural gas and form the basis of several major industries, namely plastics, rubbers, fibres, adhesives and coatings, all of which supply the motor vehicle industry with a range of products. The plastics group is commonly divided into thermoplastics, which soften on heating and re-harden again on cooling, and thermosetting polymers or resins, which are not softened on re-heating after the original forming process.

Thermoplastics may have a crystalline phase melting point, T_m, above room temperature, as well as an amorphous phase softening point, T_g. The latter is below room temperature in ductile thermoplastics, such as polyethylene, but well above room temperature in brittle thermoplastics, such as polystyrene. Natural rubber and synthetic elastomers have melting and softening points well below room temperature and so are normally flexible, the polymer chains being held together by a loose arrangement of strong cross-links. Textile fibres including many natural polymers such as cotton, wool and silk are thermoplastic in nature. Synthetic fibres such as nylon, polyesters and polypropylene are extruded to form very fine filaments, as required.

They are additionally, mechanically drawn, during cold or hot processing, to orient the polymer molecules along the axis of the fibre to give additional strength and stiffness, which results in anisotropic properties. Adhesives and coatings can be considered as thin films of either thermoplastic or thermosetting polymers. With the correct formulation they can form corrosion resistant, decorative barriers as well as structural joining materials (Mills, 1986).

Commercially, thermoplastics can be divided into two groups. There is the tonnage or commodity group, involving the polyolefins, such as low and high density polyethylenes, PE, polypropylene, PP, polyvinyl chlorides, PVC, both plasticized and unplasticized and the polystyrenes, PS, including the general purpose and high impact grades. Engineering thermoplastics are used in smaller quantities for more demanding applications. Such materials include nylons, PES, PTFE, PEEK and polyacetals, such as Delrin. Because of the long names of polymeric materials, an internationally recognized system of letter symbols is used, as indicated above and also PES for polyethersulphone, PTFE for polytetrafluoroethylene and PEEK for polyetheretherketone (Brydson, 1995).

Thermoplastics molecules are linear or branched. Chemical engineers produce polymer molecules by the chemical process of polymerization. Polymer molecules are easily melted to viscous fluids and can be processed by a range of techniques into complex shapes. Processes include injection moulding, extrusion, and thermoforming, and there are well established welding and joining techniques (Kalpakjian, 1991). The molecules have a range of lengths within a broad band and they may solidify to amorphous solids (PVC and PS) or partially crystalline solids (PE, PP and nylons). This range of molecular and solid state structures means that melting points, Tm, and other temperature transitions, T_g, are not sharp, in contrast to those for metals.

Common thermosets are phenolics (Bakelite), epoxy resins (Araldite) and unsaturated polyesters used in GRP composites. These materials also find use in coatings and adhesives (Kennedy, 1993). Thermosets are normally made by mixing two components, a resin and a hardener, which react and harden at room temperature or on heating. The hardened or cured resin material consists of polymer molecules, heavily cross-linked to form a three-dimensional molecular network. This complex polymerization and cross-linking process prevents crystallization, leaving the solid material amorphous, like inorganic glasses, and so are inherently brittle, requiring reinforcement with wood-flour, paper, glass or mica, depending on the end use. Thus, in contrast to thermoplastics, re-heating causes minimal softening and the extensive cross-linked network structure prevents melting or viscous flow. Consequently, these materials cannot be hot-worked or recycled. Excessive heating will, of course, lead to decomposition as with thermoplastics.

The term 'rubber' normally refers to natural rubber, whereas 'elastomer' is a term usually reserved for synthetic rubbers, such as chloroprene (Neoprene), nitrile and butadiene rubbers, widely used in tyres, hoses, seals and belting, as well as general mechanicals such as engine mountings. These materials consist of very high molecular length polymer molecules with occasional, chemical and physical cross-links, giving a very loose and open network. At room temperature, the materials are well above their softening points, (T_g) and melting points, (T_m). Thus, they would be viscous liquids but for the cross-links. The latter, however, lead to flexible solids with an active 'memory', which returns them to their original shape, rapidly and completely on unloading.

Textile fibres are made from both natural and synthetic polymeric raw materials. Synthetic fibres consist of simple, thermoplastic polymer molecules, such as nylons, polyesters and acrylics. They are characterized by being very anisotropic, where the physical and mechanical properties are very directional. The strength and stiffness values along the fibre are very much greater than across the fibre. This is a potential problem since axial shrinkage at high temperatures is considerable, again due to the 'memory' effect. This has significance in fabric and clothes washing. Textile and packaging films are commonly biaxially oriented during manufacture, again for improved properties and for greater economies, since thinner films can be used. Significant shrinkage is again a potential problem. In contrast, thermoplastic bulk mouldings must be manufactured to give isotropic components to avoid dimensional distortion and poor impact properties.

The properties and behaviour in service of polymeric materials are more dependent on their molecular structures and methods of manufacture, which may introduce significant anisotropy, in comparison to metals. The property values are also affected by the methods of testing, particularly the test temperature and the rate or frequency of mechanical loading. Thus, whilst it is possible to make general comparisons between one group of polymers and another, say between polypropylene, polyvinyl chloride and nylons, as in Table 2.2, information on a specific polymer grade must be obtained for design purposes (Turner, 1983).

Temperature and time (rate or frequency) effects can be explained structurally as follows. Crystalline materials, commonly metals, have a characteristic melting point, T_m. They may also have phase changes in the solid state, such as the body centred cubic to face centred cubic crystal structure change, at 910 °C, in pure iron (Cottrell, 1985). Amorphous materials such as glasses and many polymers, such as acrylics, also have solid state transitions. In polymers, the main amorphous transition is known as the glass/rubber or brittle/ductile transition at the T_g temperature (Brydson, 1995). Thus, semi-crystalline materials such as ceramics and some polymers, typically polypropylene, nylons and to a minor extent PVC, will have both crystalline and amorphous phase temperature transitions. With metals, glasses and ceramics, the T_m and T_g values are well above the room or service temperatures, although it should be noted that the melting point of tin is low at 232 °C, lead at 327 °C and solders at around 180 °C. Thus, the properties of most load bearing metals and ceramics, in service, are largely unaffected by temperature and temperature change. In contrast, the melting points of semi-crystalline polymers are lower, with PTFE at 327 °C, nylon 66 at 260 °C, PP and PVC at 175 °C, PE at 143 °C and natural rubber at −39 °C. Their T_g values are even lower, a little above room temperature for thermoplastics and well below room temperature for rubbery materials. Hence, thermoplastic polymers are generally ductile and their properties are very much affected by small temperature changes. Consequently, the 'elastic' modulus values for thermoplastic polymers at room temperatures are considerably lower than for metals and ceramics, at between 1 and 3 GPa. Additionally, under constant load at room temperature, significant creep occurs due to slip between polymer molecules. As a result, the modulus value falls with time, making it time dependent. Literature values of modulus are normally quoted at short loading times of 100 seconds. For longer loading times, say 1000 hours, the modulus value could drop to 1/3 of the short-term value. At temperatures increasing from the T_g to the T_m, in the rubbery state, the modulus drops dramatically, from about 1 to 3 GPa down to 1 to 3 MPa.

These temperature and time effects have considerable influence on the design procedures used for polymeric materials. Thermoplastics materials are said to be 'visco-elastic' and a 'pseudo-elastic' approach is used to design significantly loaded components such as pressure pipe-work for gas and water distribution. The design principles and equations are the same as for metals, but the temperature and time effects must be part of the property gathering procedures, so that the appropriate materials data are used in the design calculations (Powell, 1983). Other physical properties of polymers are also influenced by their structure, as well as temperature and time. In contrast to metals, mechanical strains of polymers are higher but recoverable, specific heats are larger, which influences processing and their coefficients of thermal expansion are higher, which may influence the interaction with other materials. Their thermal and electrical conductivity, however, are much smaller than for metals so they may be used as insulating materials. The structure/processing/property relationships are more complex than those for metals and ceramics. However, when these relationships are understood, the potential for new designs, new processes and new products is considerable, beyond being direct substitute materials for existing metallic or ceramic components.

2.2.3 Ceramics and glasses

These materials are brittle, so do they have a place in engineering let alone automotive engineering? The facts are that the Pyramids and the Great Wall of China still stand, as do the Norman cathedrals, making use of granite, sandstone and the less durable limestone. Some early clay pottery still survives as do early weapons and cutting tools made of flint. Glass and ceramics are not tough and ductile like metals and some polymers. Their inter-atomic bond structures do not normally permit the operation of enough slip systems to give a general change of shape or plastic deformation. This leaves only bond deformation under external loading and inter-atomic bonds break at very low strains. Even taking this bonding factor into account, the strengths and stiffness of ceramics are still much lower than might be theoretically expected, although this is now fully understood (Kingery, 1986).

In contrast to metals and polymers, however, ceramics do have in their favour good wear and chemical resistance in corrosive environments. They have high temperature resistance and are good electrical insulators. Vehicle component design has need of such material properties, so that the brittle behaviour is being resolved, firstly by understanding the nature of this engineering problem. In addition to the fundamental bonding problem, brittle failure is exacerbated by the presence of defects, at all structural levels, including those caused by manufacture. Control of the atomic composition, the thermal history and the manufacturing methods can minimize the size, the size distribution, the number and the shape of the defects, so reducing the potential for brittle failure at low stress levels. Such control is now implemented in the production of technical ceramics. However, this still leaves the fundamental cause of brittleness, limited bond deformation, to be overcome or avoided. The latter tactic is used at present by using ceramics in both metal and polymer composites. Thus, some of the advantageous physical and mechanical properties of ceramics are now being utilized. Manufacturing processes are making use of ceramic cutting tools made from silicon carbide, silicon nitride, Sialon (Si/Al/O/N), zirconia and dense alumina (aluminium oxide with minimal porosity), which can run at higher temperatures and speeds with lower wear rates. Their higher melting points mean that they can be used in engine components, which can run at higher temperatures to give higher fuel efficiency. This is particularly the case with diesel engine parts in the cylinder and piston head regions and valve seats. The same advantage is not available at the moment with standard petrol engines since the higher running temperatures would lead to pre-ignition and 'knocking'.

Ceramics can be classified into several groups; glasses, vitreous-ceramics, technical ceramics, and modern composite materials. Glasses can be considered as amorphous ceramics, based on silicon dioxide or silica, with additional metal oxides to reduce the melting or softening points of the formulated mixtures. Glasses are widely used in building and construction as well as other load bearing applications such as vehicle windscreens. The latter are designed in a laminated form and/or manufactured to leave the glass skin in compression, so that missile impact cracks do not spread before repair or replacement can be effected. There are two main material types, soda-lime and boro-silicate glasses. Soda-lime glass is used for windscreens and boro-silicate glass finds application in technical glassware, where the higher silica content results in a higher softening point, lower coefficient of expansion and good thermal shock resistance (Doremus, 1991). Vitreous ceramics have two constituent phases, consisting of a vitreous or glassy phase and a ceramic or crystalline phase. Engineering products include electrical porcelains and pipe-work, as well as structural and refractory bricks. During component manufacture the firing process forms a glassy phase, which melts and spreads around the surfaces of the inert but strong crystalline phase particles, bonding them together with some localized interaction. Diamond has established engineering applications for cutting tools, rock drills, dies for electrical wire drawing and abrasives. However, it is expensive and is being supplemented by engineering or technical ceramics such as dense alumina, silicon carbide and silicon nitride, Sialon and zirconia. These ceramics simulate the diamond crystal structure, with a narrower distribution of smaller micro-defects than traditional ceramics, leading to superior mechanical properties such as higher fracture toughness. They are used as coatings for engine bearings and upper diesel engine parts, as well as for machine cutting tools and modern personal body armour (Lenoe *et al.*, 1983).

The properties of ceramics are dominated by these materials forming hard and brittle components. They fail in a brittle manner or by thermal shock, in contrast to most engineering metals, which generally fail by plastic deformation, fatigue or corrosion. Thus, whilst the tensile modulus and strength are of concern, of greater importance in the design of ceramic vehicle components, are the bend strength or 'modulus of rupture' and 'thermal shock resistance'. As with metals, ceramics show general property bands of behaviour. For specific ceramic materials, test data are needed for design purposes and final quality assurance tests are essential on the finished products. However, property variations within a production batch are much greater than for metal components. The structure-insensitive properties, such as theoretical density, elastic modulus and melting point may vary by about 10%. In contrast, structure-sensitive properties, such as fracture toughness, modulus of rupture and some thermal properties are much more variable within a product batch, requiring detailed statistical analysis. Consequently, whilst there appeared to be exciting engineering possibilities for technical ceramics, in the 1980s, their potential has not yet been realized in vehicle design.

2.2.4 Composite materials

A composite material is a combination of two materials, with its own distinctive properties. Its strength or other desirable quality is better or very different from either of its components working alone. The principal attraction of composite materials is that they are lighter, stiffer and stronger than most other structural materials. They were developed to meet the severe demands of supersonic flight, space exploration and deep water applications, but are now used in general engineering including automotive applications. Composite materials imitate nature. Wood is a composite of cellulose and lignin; cellulose fibres are strong in tension but flexible and lignin acts to cement the fibres together to create a material with stiffness. Man-made composites achieve similar results by combining strong fibres such as carbon or glass, in a softer matrix such as epoxy or polyester resin.

In the broadest sense, most engineering materials are composites; for example, steels are painted to prevent rusting of valuable structural components. The more usual concept is illustrated by the bi-metallic strip used in water thermostats. Firstly, neither the iron nor the brass alone would be useful in this application. The combination of the two has an entirely new property. Secondly, the two components act together to equalize their different strains. This property of combined action is most important in the design of composite materials and components.

The ideal load-bearing component or structure is made of a material that is light in weight, strong in tension and not easily corroded. It must expand very little with changes in temperature, with a high resistance to abrasion and a high softening or melting point. In vehicle engineering, high strength and stiffness per unit weight or density are all important 'design properties'. Materials with these properties are ceramics such as glass, boron carbide, alumina, silicon carbide, as well as carbon. They also have high softening or melting points and low coefficients of expansion. In addition, they can be made from inexpensive raw materials such as sand, coke and coal. Metals are usually poor on a unit weight basis apart from magnesium and titanium. Polymers are satisfactory on a strength-to-weight basis but are poor in terms of stiffness-to-weight ratio.

However, the reason ceramics have had limited use as direct engineering materials is that they are brittle. Their high strength and stiffness are only realized under special conditions where there are no internal or surface cracks, notches or other defects. Normal processing and environmental conditions produce cracks in all materials. However, metals and polymers are less sensitive to the presence of defects, in that they can, in practice, withstand much higher loadings, without defect propagation leading to fracture. Metals and polymers have structural bonding, which can accommodate deformation, leading to crack blunting at the macroscopic level. Ceramics have no such accommodation, so that cracks move easily through these materials at low stresses.

Composites normally combine the potential, reinforcing strength and stiffness of glass or ceramics with the ductility of metals or polymers, although zirconia-reinforced alumina is also of technical interest (See Transformation Toughening, *J Mat Sci*, 1982). The reinforcement is commonly divided into small particles or longer fibres, so that any cracks present cannot find a continuous path through the composite material. The properties of the matrix are therefore of equal and vital importance. Firstly, it must not allow fibre damage by rubbing and scratching. Secondly, it must act as a medium to transmit the external forces as stresses on to the fibres. Thus, there needs to be some adhesion between the matrix and the fibres, usually assisted by the use of chemical coupling agents. Thirdly, the matrix must deflect and control the cracks in the overall composite material (Mayer, 1993).

The matrix properties of polymers, such as epoxy and polyester resins, and more recently thermoplastics such as the nylons, and those of ductile metals such as copper, aluminium and cobalt, are weak in shear and do not scratch fibres or allow them to rub against each other. The other two functional requirements call for a compromise in properties. Internal stress originates from the externally applied force. The matrix transmits the stress on to the fibres. Composites of the highest strength have all the fibres aligned in the direction of the external loading. In this case, the principle of combined action comes into play, with the strains in the fibres and the matrix being virtually equal. However, due to the difference in stiffness values, the major part of the stress in the composite will be carried by the fibres. Thus, any cracks in the fibre will propagate and lead to fibre breakage. The crack reaches the ductile matrix interface, where it becomes blunted and so is less easily propagated.

Two other factors prevent cracks running through the composite. Firstly, reinforcing fibres do not all break in the same plane. Thus, considerable pull-out forces are required to fracture the component. This pull-out work contributes to the work of fracture which does not arise in homogeneous materials. The second, crack controlling effect is the regulation of the degree of adhesion, via the coupling agent, between the fibres and the matrix. If the adhesion is not too high, the composite material will be weak in a direction at right angles to the fibres. This is an advantage, when the fibre crack runs in this direction, it will reach and be deflected along the weak fibre-matrix interface and become blunted.

Reinforcement theory, initially, uses a model for a composite material consisting of long aligned fibres in a ductile matrix. This model, however, is too simple in that many composites have short fibres arranged in a variety of orientations, in three dimensions. The response to external loading is, therefore, complex but allows vehicle engineers considerable potential in designing both simple and complex structures. Most automotive components, in service, suffer a range of external forces, tensile, compression and shear, in a variety of directions. Thus, it is useful with composites to arrange for the reinforcing fibres to be oriented in the most favourable directions. It is also found that when fibres are stressed to fracture, the broken pieces can still carry loads and so remain useful, which has two benefits. Firstly, reinforced thermoplastics can be processed using conventional techniques. Secondly, the strongest materials can only be obtained as short, single crystal filaments, known as whiskers, from materials such as alumina and silicon nitride (Somiya *et al.*, 1989).

The largest tonnage composites, at present, are glass fibre reinforced polyester resin materials, GRP, due to the relatively low cost of these raw materials. Glass can easily be drawn to give high strength filaments, although they need a protective coating within the coupling agent system to prevent surface cracking. Unsaturated polyester resins can be cured at low temperatures and pressures. The combinations of fibre and resin can give limitless shapes, the largest of which are naval minesweeper hulls, so that vehicle body parts present no problems of scale. Curing times are being reduced to from hours to minutes, resulting in an economical manufacturing operation (Harrison, 1997). However, glass fibre composites do have limitations. Whilst glass fibres are strong they are not stiff and the polyester resin degrades above 200°C. Thus, for high modulus components, carbon and boron/tungsten fibres are used with newer polymers such as epoxy resins and polyimides. These composite materials have high strength and stiffness-to-weight ratios, compared with steels. They were initially developed by the aerospace industry for such applications as compressor blades in jet engines, carbon in the UK and boron/tungsten in the USA. Such materials are now more economically available for racing car bodies, as well as for a wide range of sporting goods. In contrast, thermoplastic matrices, such as glass reinforced nylons, make for easier component manufacture.

For high temperature applications, up to 1000 °C and above, tungsten fibre reinforced cobalt and nickel are used. The major problem with these composites is that of the fabrication of the component, even with simple shapes. In some cases this may be overcome by forming the fibre whiskers in situ within certain alloy systems. For example, niobium carbide whiskers can be produced in a niobium matrix, which gives a composite with high strength and heat resisting properties up to 1650 °C. From their early beginnings in the aerospace industry, the potential of composite materials for use in automotive engineering is being realized.

2.3 Mechanical and physical properties of automotive materials

The product designer and the manufacturer both need to have a thorough knowledge of the properties and terminology associated with materials in order to select and use them more effectively. Every material has certain properties, which make it more suitable for some applications than others. Construction materials, in general, must be able to withstand the action of forces without undergoing significant distortion, and should incorporate a high level of operational safety. This is particularly important where structures in vehicles and other forms of transportation are concerned. Vehicle component manufacture, however, requires different and sometimes conflicting materials properties. These must permit the permanent deformation of materials to enable the components to be shaped easily and economically, with the least amount of energy (Bolton, 1989).

Mechanical properties are associated with the behaviour of a material when linked to the application of a force. It is these properties with which the vehicle designer is initially concerned when considering a material for a specific duty, such as the chassis, bodywork and suspension systems. Testing is used to determine such properties (Montgomery, 1991). However, additional characteristics may also be critical for some components, such as electrical and electronic control systems, which may also incorporate magnetic components. Optical, thermal and chemical properties will be important for such components as windscreens, heat exchangers and anti-corrosion systems. These non-mechanical properties are generally classified as physical properties.

2.3.1 Mechanical properties

A sufficiently strong force will produce a definite amount of deformation, either temporary (elastic) or permanent (plastic), in a material. Strength is defined as the ability of a material to withstand a force without breaking or permanently deforming. Different forces will require different types of strength to resist them. Tensile strength is the ability to resist stretching or pulling, as in a towing bar. Compressive strength is the ability to withstand a pushing force, which tries to compress or shorten, as in an engine connecting rod. Torsional strength is the ability to withstand twisting forces, as in the prop shaft, cylinder head bolts or indeed the whole body shell structure. Strength values of materials range from 10 to 1000 MPa, from polymers through metals to ceramics.

Elasticity is the ability to stretch and bend when subjected to the various forces above and then to regain the original shape and size, when these forces have been removed, rather like an elastic rubber band. All vehicle components need to possess some degree of elasticity, which is quantified in terms of the material's modulus, defined as the elastic stress divided by the elastic strain. Modulus values range from kPa for rubbers and plastics through to GPa for metals and ceramics. A broader term, called 'stiffness', takes into account the shape and design of the component as well as the inherent modulus of the component material.

Plasticity is the ability of a material to be changed permanently in shape, by external forces or blows, without cracking or breaking. Some materials are more 'plastic' when heated. Two subsidiary terms are 'malleability' and 'ductility'. Malleability refers to the extent to which a material can undergo permanent deformation in all directions, under compression, by hammering, pressing or rolling, without rupture or cracking, as in forging or sheet manufacture. Plasticity is essential but malleable materials need not be strong. Malleability increases with temperature. Ductility on the other hand is the ability to undergo cold plastic deformation in bending, torsion or more usually in tension. A permanent reduction in cross-section can be achieved by pulling a rod through a die to produce a wire, without breaking, as in the manufacture of electrical cables. Ductility, in contrast to malleability, decreases with temperature.

Hardness is a complex property. It is the ability of a material to resist both abrasive wear and/ or indentation. It is an important quality in bearing materials, as well as for drills and other machine tools. Toughness is a term usually used to denote the ability of a material to withstand sudden shocks or blows without fracture, as required say of a hammer head. It also includes resistance to cracking when subjected to bending or shear loads. In contrast to toughness is the property of brittleness, which is a tendency to show little or no strain or plastic deformation before fracture. If a material is brittle, such as glasses and ceramics, as well as some metals and amorphous plastics, such as cast iron and polystyrene, it will show no ductility and only limited deformation before fracture (Atkinson *et al.*, 1985).

Dimensional stability is the resistance to changes in size and shape. Plastics at room temperature and metals at high temperature also gradually deform with time and may eventually fail, when subjected to a steady or constant force for long periods. This gradual deformation, at constant load, is known as 'creep'. Creep resistant materials must, therefore, be used when high loads are applied for long times at high temperatures, as in engine cylinder head bolts. Fatigue failure is caused by repeated or reversed stress cycles in any of the above stressing modes usually at stress levels, which would not have caused failure under static conditions. Such cycling is frequently found in vehicle body structures and components such as crank shafts, connecting rods and tyres. Fatigue failure may be accelerated by corrosion, higher temperatures and poor surface finish (Smallman, 1985).

Durability is the ability of a material to withstand long-term weathering and corrosion and the deterioration that these may cause. It often involves changes in appearance, but changes in mechanical and physical properties are of more concern. Wet corrosion and oxidation is common in metals, such as steels, with the exception of gold. The combined effects can lead to mechanical failures, which may be disastrous in vehicle components. Sunlight, particularly UV radiation, oxidation and some chemicals can also cause the deterioration of plastics and other polymers, some more than others. Commonly, stabilizers are added to enhance the processing and in-service lifetime. In contrast, glasses and ceramics are inherently more stable than metals and polymers to aggressive environments but are still brittle (West, 1986).

2.3.2 Physical properties

There are several physical or non-mechanical properties of interest in vehicle design and manufacture. The fusibility of engineering materials, metals, ceramics and polymers, is the ability to change into a liquid or molten state when heated above specific temperatures, known as the melting point, T_m, in crystalline materials or the softening point, T_g, in amorphous materials (Smith, 1993). Semi-crystalline materials will exhibit both T_m and T_g transitions, on heating. These temperatures vary considerably between materials but are important properties in the casting of aluminium pistons, injection moulding of polypropylene bumpers, welding of steel sub-assemblies and soldering of electrical components onto PCB's.

Density has become of considerable significance in vehicle design since it predetermines the final mass of the component, its behaviour and efficiency. Thus, aluminium and composite materials are now serious competitors to traditional steels, which have much higher relative densities due the heavier iron atoms and closer atomic packing.

All materials restrict the flow of electricity to some extent but those used in vehicle design show a complete range of electrical properties. Metals, especially gold, silver and copper, are generally good electrical conductors. Copper is commonly used in cable manufacture for electrical wiring harnesses for vehicle control systems. Gold, silver and other precious metals such as platinum, being more expensive, have a more restricted use in electronic control devices. Electrolytes, some gases, liquids and certain solid ceramics also allow current to pass through them easily. Liquid electrolytes, such as sulphuric acid, are used in the lead-acid battery to store chemical energy, traditionally used to start the vehicle, although battery technology has now developed as a primary driving force in the electric powered vehicle, such as the Nissan Altra EV. Fuel cell and solar energy power sources are being actively researched at this time.

Non-metals are generally good electrical insulators but again vary in their ability to resist the flow of electricity. Ceramics are normally good insulators, as are glasses and many plastics. Such materials are also used to store electrical energy in capacitors for electrical control systems. The insulation around the copper conducting wire in an electrical cable is commonly plasticized PVC, polyethylene or the more recent fire resisting compounds. Semi-conductors range in electrical properties between the two conducting extremes, allowing current to flow only under certain conditions. Silicon and germanium in their pure state are poor conductors but their electrical resistance can be altered by the addition of small quantities of 'additive doping' materials. Semi-conductors are widely used in electrical control devices on all types of vehicle (Callister, 1987).

Thermal properties are of concern for materials used in engine construction and exhaust systems, which need to withstand temperatures up to 1000°C. In contrast, the materials used in vehicle air conditioning systems will need to perform at low refrigerant temperatures and be chemically resistant. Ceramic materials are good at high temperatures but are brittle, metals have lower temperature capabilities but can be heavy, whereas polymers are poor at high temperatures but can be flexible at low temperatures. Metals, especially copper, possess high thermal conductivity. Vehicle radiators and other heat exchangers make use of copper, aluminium, steel as well as plastics in their construction. In contrast, thermal insulators are generally non-metallic materials with low values of thermal conductivity. They are used to prevent heat gains or losses, such as for the shrouds around exhaust systems adjacent to car bodywork. Air is actually one of the best thermal insulators. Materials, which can trap air, such as foams and open fibrous composites, are used to prevent heat as well as sound transfer away from or towards thermally sensitive areas such as car and cab interiors. Thermal expansion occurs when materials get hot whilst shrinkage takes place on cooling. Expansion values vary considerably between the differing materials groups, ceramics, metals and polymers, where the ratio of expansion coefficient is of the order of 1 to 10 to 100, respectively. Consequently, there may be design problems when these materials are brought together in a device where large temperature variations are involved, such as around a vehicle engine. Conversely, expansion effects can be used in control mechanisms such as radiator thermostats (Adler, 1993).

Optical properties vary widely such that materials may reflect, radiate or absorb light energy and may be opaque, translucent or transparent. Colour is also a significant property, acting as a means of identification as well as being decorative. Glass is still the favoured material for windshields at present, due to legislation, low raw materials costs and easy fabrication of curved shapes. Significant weight savings would be made by reducing the thickness of the glass windshield or by replacing glass with a plastics alternative. The reflectors and lens components, for vehicle lighting, are already been manufactured from polymeric materials, such as metallized thermosetting compounds, ABS, acrylics and polycarbonates, due to their lower densities.

2.4 Materials selection for automotive components

The science and technology of materials, as outlined above, are essential tools in rational vehicle design, to counteract the empirical view that metals are materials of the past, plastics are materials of the present and ceramics are materials of the future. An understanding of the behaviour of the various types of materials forms a basis by which comparisons can be made. Thus, informed choices can be made regarding materials selection for a particular engineering design and its realization. Designers in general have always experimented with different materials and production methods, to make improved products. Engineering designers are no exception, although current vehicle engineering problems are more complex than in the past. Fortunately, there is now a wider range of materials and production techniques available, particularly with composites.

2.4.1 The design process

The design of vehicles and their components is covered in detail elsewhere in this book but a summary of the design process would be useful in understanding the complexity of the related materials selection operation. Design is an activity, which uses a wide range of experiences to find the best solution to an engineering problem within certain constraints. Ideally, it is creative rather than just problem solving, involving the whole process of producing a solution from conception to evaluation, including elements such as aesthetics, ergonomics, manufacture and cost. Designs change with time due to the changing needs of the customer, such as the trend towards smaller and lighter cars, and the development of new computing technologies. Designing is an integrated, multi-stage operation, which must be flexible enough to allow modifications for specific problems as they arise during the design process.

Design is usually initiated by recognizing or accepting that there is a problem, by preparing a design brief or questionnaire with the client, which should identify the real task involved. The brief must not be too vague so that the designer has no idea where to start. On the other hand, the brief must not be so precise, such that the designer has no room for innovation. Having obtained the design brief on the new product proposals, the designer must fully understand the client's requirements and design limits. This analysis will lead to some investigations or research, which could involve a study of former and existing products and further discussions to produce a supplementary questionnaire. This will generate a good understanding of the problem from which the exact limits and constraints can be set out and formally agreed. This agreement creates the specification, which helps to focus on the key aspects of the problem, such as size, shape, function and appearance. Other factors will include materials, manufacture, finish, maintenance, reliability, cost, safety and ergonomics. Some factors will be conflicting, so that balancing or compromise will be required before moving on to the next stage.

Generating ideas is the creative area of the design activity. Ideally, ideas come from thinking and sketching and storing for future use, since they do not automatically appear to order. Unfortunately, solutions to engineering design problems cannot wait for ideas to just arrive. They must be worked at to determine solutions. A number of techniques are used here but 'brainstorming' between a group of designers is usually synergistic and generally the most profitable, in engineering and vehicle design, when time constraint is part of the design brief. The evaluation of these ideas is a critical stage from which a proposed solution should emerge, which will satisfy the design brief and specification as well as manufacturing and cost constraints.

The proposed solution or solutions must then be converted into reality. This normally involves producing component or product models using traditional model making or rapid prototyping. Several questions arise at this stage regarding construction materials. The requirements of the component behaviour must be known so that a material with the appropriate properties can be selected. Properties of interest in automotive engineering will include weight, strength and a range of physical properties such as corrosion and thermal resistance. Size and shape, ergonomics, aesthetics and the appropriate safety standards must always be applied to the vehicle design.

Having developed and refined the proposed solution, its realization must be planned. Planning involves the creation of presentation and production drawings and the organization of the realization. It identifies in advance the materials and specialist equipment required. The realization of the solution is usually the most interesting but time consuming stage. It can also be the most frustrating if the planning stage has not been done thoroughly. Prototypes and finished products will be produced. The testing stage will discover how well the solution works under a variety of loading regimes and environmental conditions. Aesthetic and ergonomic factors will be included in the testing programme. The testing results may indicate some further redesigning, to correct faults or to improve the solution further. Finally, the new component design and its realization must be evaluated, in a constructively critical manner, in order to answer to the overall question of how would a similar problem be approached in future.

2.4.1 Materials selection

Materials selection depends very much on the skills and experience of the design team although materials databases are now available to help in this process. To make successful choices requires knowledge, understanding and experience of working with a wide range of materials. Steel, concrete, glass and timber will remain the major materials for civil engineering. Mechanical and automotive engineers can afford to look at a wider range of metals, as well as polymers, composites and some ceramics materials. Electrical and particularly electronics engineers have much fewer problems of materials availability. Several factors or driving forces need to be considered in the materials selection process (Charles *et al.*, 1989).

The performance of the material must meet specific requirements. It is necessary to match the task, which the component or device may have to perform, with the material resources. It is important to consider the whole range of service requirements that are likely to arise, such as the mechanical loads and loading regimes, hardness, rigidity, flexibility and particularly weight, in vehicle design, as well as a range of physical properties. These can then be surveyed and matched with the properties and characteristics of suitable materials. The requirements of both properties and processing are often needed in various combinations for particular applications. Electrical, thermal or heat resistance may be linked with resistance to both wear and corrosion, in order to improve reliability and to increase the life span of the product. These and other requirements can now be explored and tested with computer software packages, particularly with plastics and polymers, so that comparisons can be made. In this way, the materials choice can be narrowed and a suitable selection becomes possible (Institute of Materials, 1995, Cebon *et al.*, 1994).

Quality and styling requirements may be considered as an extension of performance requirements. Factors such as noise and vibration could cause significant fretting failures, which may be significant. The aesthetic features of surface finish, static build up, colour, texture, feel and smell, such as for leather seats and wood veneers, which have a marketing and sales dimension in vehicle design, are also controlled by materials selection and processing.

The method and scale of manufacture of the component or product are as significant, in the materials selection process, as the consideration of the in-service behaviour requirements. These processing factors are important in order to achieve the maximum effect with economy, precision and a high standard of finish. Thus, materials selection must take into account not only the in-service behaviour but also the influence, advantages and limitations, of the manufacturing process. For example, a car body panel may be made from timber, steel, aluminium or a GRP composite. Not only will the inherent properties of these materials differ but their fabrication into panels will involve different routes (Kalpakjian, 1991).

Materials are available from suppliers in many regular or standardized forms. These include wire, round and square bar, film, sheet and plate, angles and other extruded sections, granules, chips and pellets and finally, viscous fluids. These forms come in standard, preferred sizes, which have been established by practice and demand. Standardization, affecting both quantity and size, is now applied to the specification of most types of material and component. Non-standard sizes and quantities increase costs. Information on the structure, properties and behaviour of incoming materials and components will still require quality assurance, to ensure that the specifications are being met. It is common practice for larger companies, such as vehicle assembly plants, to purchase stock materials and components and then subject them to a quality audit. The assembly companies, such as Ford, Rover, Nissan, commonly known as the Original Equipment Manufacturer, OEM, are now reducing the size of their own design and development teams for work on new products, such as structural sub-assemblies, seats and body panelling. This work is now done in co-operation with their first-tier suppliers, who themselves cascade co-operative work down to third and fourth tier suppliers, such as raw materials manufacturers. Such simultaneous engineering down the product supply chain allows the OEM's to concentrate on the problems of final product manufacture, such as the vehicle itself, which will need to satisfy all the customers' requirements.

Economics and commercial factors play a vital part in vehicle engineering design. The selling price of a component or product is made up of a number of parts, such as the costs of raw materials, manufacture, marketing, transportation, installation, maintenance and profit. Keeping the materials and manufacturing costs low will either maximize the profit or ensure sales at a realistic market price. However, the component specification must still be met using the correct materials and manufacturing methods. For similar vehicle parts, such as tyres, the specification can vary widely, leading to the use of different materials and methods of manufacture. Tyres may be used for family saloons, sports and racing cars, vans and lorries, farm tractors and earth moving vehicles. These applications use both high and low cost materials, together with hand crafted and mass-production techniques.

Legislation requirements will influence the materials selection for a vehicle component. Health and safety factors govern such items as fuel tank integrity, windscreen vision, carbon and nitrogen oxides exhaust emissions, asbestos in friction materials and solvent/water based paints. Disposal methods, the cost of landfill and the economic necessity of recyclability now need serious consideration by the designer. Whilst the recycling of single material components is relatively easy, such as polypropylene copolymer bumpers, the recycling of multi-material products such as the starter battery, is a more complex affair. Both are being done at this time but the challenge is to actually design for recycling as well as for manufacture and behaviour in service. The life cycle analysis and total energy usage for a vehicle component, throughout its service and its re-entry into the vehicle, unfolds some very interesting problems.

All of these materials selection factors are a challenge to the materials technologist. The experience gained from the testing and working of materials helps to reinforce the ability to make successful materials selection and design decisions for a particular component. The final materials choice is often a compromise. In some cases, functional demands will dominate, whilst in others, cost or legislation may prove to be the main factors. It is only when all the information is collated that satisfactory materials decisions can be made. There is rarely a single materials selection solution.

2.5 Component materials case studies

2.5.1 Metals and alloys

These materials have the wide range of mechanical and physical properties of strength, stiffness and ductility, which are required for most vehicle bodywork and component parts. Consequently, metals and alloys are generally selected by designers for such engineering applications. Only for special properties, such as low density, high thermal and electrical resistivity or low wear rate, are plastics and polymers or ceramics and glasses considered for selection. More recently, the useful combination of properties, offered by composites of these groups of materials, is being realized (Ohring, 1995).

The current problem with construction steels, in meeting the selection criteria for the automotive industry, is their high density, in comparison to aluminium alloys and GRP composites. The total energy costs of owning a car is about 10% to build and 90% to run over its lifetime. Thus, whilst building costs can be reduced, reducing the vehicle weight would be easier. Traditionally, the car weight would be made up of about 70% steel and 15% cast iron, 4% rubbers and elastomers and the balance made up of glass, non-ferrous metals, plastics and other polymers. Thus, steels and cast irons were the obvious materials for review, either for improvement to higher strength steels or replacement with lighter materials such as aluminium, polymers and composites. This potentially severe competition led to the unification of major steel industries, world-wide, in order to develop 'lightweighting' steels via the ULSAB project (Ultra Lightweight Steel Auto Body). In principle, this was done by using higher strength, ductile steels, so that thinner, sheets and sections, could be used, to reduce body weight by 25% overall, down to about 200 kg, whilst optimizing structural performance and crash management. Porsche Engineering Services collaborated in the design and build of the ULSAB body. The materials costs and the manufacturing methods had to match those of mild steel and the body had to be recyclable. The main targets for action were suspension arms, engine mounting assemblies and chassis members. The manufacturing problems, particularly the fusion and resistance spot welding requirements in vehicle construction, with such steels have been resolved (Walker *et al.*, 1995). In addition, new manufacturing processes were developed as part of the project, such as hydroforming, which has the potential to make in one stage, components which were previously made from several parts and joined. The ULSAB project vehicle reduced the number of body parts used from 200 to about 150. The hydroforming process is more expensive but weight savings are possible as no weld flanges are required and since there is no welding, thinner sections can be used. Also stiffness can be maintained due to the elimination of spot weld joint discontinuities. Examples can be now found as sub-frames on the Ford Mondeo and Vauxhall Vectra, as well as the initial ULSAB project's, side roof rails from tube and roof panels from sheet. However, many designers and engineers still need convincing of hydroforming capabilities, since again the design must include the requirements of this process as well as the in-service behaviour of the component.

Following on from the success of the ULSAB project steel companies initiated the ULSAC project (Ultra Lightweight Steel Auto Closures). This focused on four main closure panels; doors, bonnets, boot-lids and tailgates. Again Porsche were contracted to provide the engineering management. The lessons learnt from the ULSAB project were implemented and weight savings of between 20 and 30% were achieved with all four parts, with no increase in costs compared with current steel closures of similar sizes and geometry.

Figure 2.4 Tony Shute driving a Lotus Elise at the Materials on the Move Conference

Stainless steels are used where the higher costs can be justified by the need for improved performance, which these materials offer, particularly in terms of corrosion resistance and operational economy. In addition to the common 18% chromium/8% nickel alloys for corrosion resistance, other alloying elements, such as molybdenum, titanium and niobium are added to improve formability and to avoid weld decay. However, whilst stainless steel railcar bodies have been used for many years in Scandinavia, due to the long term, life cycle cost advantages, in the UK and most other European countries, railcar bodies are now constructed from aluminium.

The automobile equivalent of the 'advanced aerospace materials development' is the development work in producing racing cars and sports cars. These are not experimental cars. They have a real job to do, which is to win races of various classifications. The success or experience gained is then used to develop mass-produced cars and their components. One such car was the Lotus Elise, produced by Lotus Engineering, following the end of the Lotus Elan production when its engine was no longer available (Shute, 1997) see Figure 2.4. The Elise was developed within two years, despite manpower and financial constraints, with the help of materials and component suppliers as well as other car manufacturers. Some established, specialist suppliers, such as Ciba-Geigy and Norsk-Hydro were enthusiastic to gain experience of the mass-production vehicle business. The Lotus objective was to produce a performance car by reducing weight rather than by increasing engine power. This was done by examining every component for its essential nature, using a back-to-basics approach. If the component was not essential then it was removed, so reducing weight and cost and also time of manufacture/ assembly. The chassis design of two torsion boxes, front and rear, joined by two longitudinal beams, made use of aluminium alloys. These were connected by overlap joints using adhesives rather than welding. Any critical or attachment points were supported by mechanical fixtures. Adhesives enabled thinner sections to be used since welding requires a minimum section thickness. Standard aluminium extrusions were used for a variety of components such as pedals, uprights and anti-roll bars, which significantly reduced the numbers of dies and tools and hence manufacturing costs (Litchfield, 1995). Following the lead of replacing the cast iron engine block with aluminium alloy, the brake discs were also reviewed and the traditional cast iron was replaced by a metal matrix composite, MMC. The lowest cost route is to cast aluminium alloy previously reinforced with particulate silicon carbide. An alternative process makes use of powder technology, where alumina powder is sintered to shape and then infiltrated with aluminium, to achieve minimum porosity. As with previous Lotus cars, polymer/fibre composites have continued in use for the bodywork and also for the front box section. The Elise is now on the market to earn income for its sponsors with a mass of 750 kg, reduced from the more common 1000 to 1200 kg, thereby achieving improved economy and performance. Lotus continue to search for further weight saving technology in such items as windscreens, wheels, tyres, drive shafts, gearboxes, exhaust systems, batteries and so on. By careful selection of materials, such as magnesium alloys and carbon fibre reinforced composites, it is estimated that a further 100 kg could be trimmed off the present vehicle weight. However, the automotive industry is familiar with and also understands steel and aluminium, particularly with respect to fatigue, so that new materials in modern designs face tough opposition, unless they offer something special or unique.

As with any other materials selection and design developments, it should be remembered that the costs of change and modifications at the design stage are significantly less than those during manufacture.

Following the experience and success of using aluminium alloys for the Lotus Elise, the aluminium industry and major car manufacturers, such as Audi have put considerable investments into the use of these lightweight materials. Aluminium reduces the vehicle weight by 35–40%, part for part. However, straight steel replacement is too expensive so that a space frame concept was developed, as in the Audi A8. The number of castings and extrusions were, thereby, reduced to 100 in contrast to 300 for a traditional steel body. 6000 Series aluminium alloys were selected which had satisfactory yield strengths but low stiffness, so that wall thickness was increased in critical areas. Aluminium alloys also lend themselves to stretch forming and hydroforming for close fitting parts such as body wings and panels, as in the Landrover Freelander, where aluminium bumper parts and crash cans are also used to avoid low speed impact damage to the body shell.

2.5.2 Plastics and polymers

The weight of plastics and other polymeric materials in motor cars currently stands at about 250 kg, some 25% of the total vehicle mass. The overall weight will continue to rise but the percent weight will probably remain the same due to an increase in the demand for greater safety and comfort. Without plastics the overall weight would increase significantly. European vehicle manufacturers tend to lead their competitors in the USA and the Far East in this respect and it is predicted that this component weight will increase to about 300 kg by the year 2002. Polymeric components are numerous in vehicle construction but can be classified into five major groups as follows: plastics for body parts, interior trim, instrument panels and headlamps; polymer foams for seats, padded safety components for interiors, sound insulation, wings and side panelling; surface coatings, adhesives and sealants, for seam finishing and corrosion protection; textile fibres for interior trim and carpeting; and finally natural and synthetic rubber for tyres, engine mountings, gaskets, drive belts and hoses. These groups do not all carry a load-bearing function but they do help to minimize the almost inevitable weight gains, as manufacturers develop new vehicle models with the added complexity of safety designs, demanded by legislation.

Of particular interest is the petrol tank, the front and rear bumpers, the battery case, the internal fascia panel and external lighting. Petrol tanks on saloon cars from Nissan are now constructed from a nylon/polyethylene sandwich using the blow moulding process. The spout is subsequently welded on (Watson, 1988). In this way complex shapes can be made to fit the design of the under-car. Polypropylene compounds are commonly used to manufacture bumpers, which have the required impact performance. This material is also used for battery cases. Nylon, polyethylene and polypropylene components, in addition to their weight saving, can also be recycled directly into other components or used as core materials in sandwich structures with virgin materials used for the skin. The current target for reuse and recycling is set by the European parliament at 80% at present rising to 85% by 2005. Fascia boards are also sandwich structures based on PVC skins with a polyurethane foam inner core. PVC is more difficult to recycle at this time due to the legislation covering halogen compounds. Lighting systems now make wide use of plastics materials. Thermosets are used for thermal stability as headlight reflectors, which are lacquered before aluminium coating. Glass is still widely used as the headlight lens but there is a trend towards polycarbonate. This needs to be hard lacquered with acrylic or polyurethane to avoid chemical and UV degradation. Rear lights present less of a problem, with lens made from acrylic and housings and reflectors made from ABS.

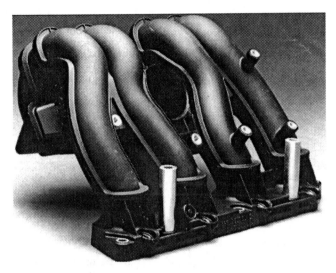

Figure 2.5 A typical engine manifold manufactured from Bayer glass filled nylon 6, for Mercedes Benz

Of engineering interest are plastics used in the engine compartment. In addition to the savings in weight and cost, they offer resistance to wear and corrosion as well as good electrical, thermal and insulation properties. Design opportunities are made possible by modern injection moulding technology together with the subsequent recycling advantages. Applications can be divided into four groups corresponding to the type of medium the components come into contact with; hot air, hot coolant, hot lubricant and fuel systems. Heat resistant nylons with glass fibre reinforcement, operating up to 130 °C with 35% glass fibre, play a prominent part in these applications such as intake manifolds (see Figure 2.5). Replacing cast aluminium with such nylons can save 4–5 kg. Whilst this weight saving is not so great as with bumpers or body panels, it is accompanied by better aspiration for the engine which will give a better performance. This comes from smoother wall surfaces and lower heat transfer to the incoming air. The manifolds are manufactured by the lost-core, injection process as single mouldings. Alternatively, injection moulded parts can be assembled by friction welding. Reinforced nylons are also used for cooling fans, toothed-belt pulleys, radiator tanks, rocker covers, water pump impellers, bearing cages, chain tensioners and fuel injector housings, see Figure 2.3.

Pneumatic tyres have been in use for about 100 years and without them there would be a much reduced transport system since they form an essential part of the vehicle suspension system. Tyres must establish and maintain contact between the vehicle and the road surface and, thereby, preserve life. The nominal area of contact of a car tyre with the road, the contact patch, is only about the size of the human hand. Therefore, the balance of materials selection, manufacturing method and tread design is a compromise, in order to achieve good wet road holding, low rolling resistance and good wear rates. Road handling, comfort and noise within the car is, in part, due to the tyre geometry and tread pattern as well as the texture of the road surface and the design of other car components (Williams *et al.*, 1995). Continuous materials developments are taking place, such as replacing the carbon black filler/reinforcement with silica and the steel wire reinforcement with aramid fibre. The re-treadability of commercial vehicle tyres is of continued interest as are the recycling and disposal of car tyres. As with thermosetting polymers, vulcanized rubbers cannot be directly recycled as polymers but the technology exists to de-polymerize such materials followed by re-polymerization. However, the economic drive for such technology is not yet in place in Europe or elsewhere.

2.5.3 Glass and ceramics

Perhaps the most critical parts of the motor vehicle, in terms of loading, temperature, and fatigue, as well as chemical activity, are in the engine and exhaust systems. Components such as the spark plug, piston, cylinder, cam shafts, valve parts, sealing products and the more recent catalytic converter all play an important role in engine performance. Unique solutions using ceramics have been developed for these components to satisfy various criteria, not the least of which are materials and manufacturing costs, which pose severe engineering problems (Kingery, 1986).

Glass still dominates the windscreen and window market primarily due to low cost and unique transparency, together with its social and legislative support. Nevertheless, the weights of such components could be reduced by 50% by the use and acceptance of impact resistant polymer laminates. Modern ceramic materials have some excellent engineering properties, such as high strength, good thermal shock resistance, low coefficients of expansion and good thermal and chemical resistance. However, they have only modest fracture toughness for moving parts in engines, despite some exciting technical ceramic developments and they have high manufacturing costs, for automotive applications. Thus, apart from the well-established alumina, spark plus insulators, which are not primary moving parts, ceramics have restricted use as components around the engine, at present, apart from tappet shims and turbo-charger impeller fans. These consist of silicon nitride, giving good wear, friction and noise characteristics. However, ceramics as coatings on fracture tough and refractory metals and also in composite materials with metals and polymers have maximized their potential advantages and minimized their limitations, as illustrated in the final section of this materials chapter.

2.5.4 Composite materials

Polymer composites
These materials have the advantage over steels of being lightweight and have been under investigation for vehicle components since the 1930s, starting with the phenolic-based fascia panels in Ford motor cars. Such materials were already established and used as for such items as battery cases, distributor caps and other electrical components. However, it was not until the 1950s that composite behaviour was better understood. The 1970s saw the development of new reinforcement and matrix materials, together with the evolution of new and innovative manufacturing methods.

Body parts using fibre-reinforced polymers are used by most vehicle manufacturers, for doors, tail-gates, rear spoilers and roofs. Dough moulding compounds, DMC, and pre-formed sheet moulding compounds, SMC, can be fabricated using low-cost tooling to make low volume parts economically. A significant development in this field is the one-piece roof for the Ford Transit van, using a 50 kg moulding with a coloured polyester or polyurethane skin, manufactured in 20 minutes compared with the hand lay-up process which previously took 3 hours. Composite body parts are not without criticism. Safety and crashworthiness remain under investigation, as is the recycling situation with thermosetting resins. The latter problem is being addressed by the use of long fibre reinforced thermoplastics. In addition, there still remains the general problem of painting and finishing. European paint plants traditionally work at 200 °C, so that only steel and aluminium can pass through without damage. The new Ford process eliminates this painting problem by using pre-coloured resin.

Other load-bearing parts have included the leaf springs on the Rover Sherpa van, which have since been replaced by hydraulic suspension on all four wheels. Similarly, drive shafts made from filament wound glass fibre composites with nylon outer sheaths have been eliminated with front wheel drive vehicles, another illustration of the effects of competition and design development. Friction materials, used as clutch plates, brake pads and shoes are also fibre-based composites, originally using asbestos fibres, in a phenolic/elastomer matrix. Asbestos has now been replaced, for health reasons, by other mineral fibres such as rockwool and vermiculite, and by aramid polymer fibres such as Kevlar. Such composition changes have had significant effects on the behaviour of friction materials and their performance will also be affected by the change from cast iron to the metal matrix composites in brake discs, used in critical applications such as racing and sports cars.

Metal matrix composites
As indicated above, ceramic materials and components have fracture toughness values, which are too low for most automotive components. However, they are finding much wider application as coatings, particularly when toughened with more ductile metals, in metal matrix composites, MMCs. A critical application is the engine piston, particularly the piston crown in diesel engines. The piston operates at temperatures of at least 350 °C, frequencies of 100 Hz (6000 rpm), with speeds from 0 mph to 60 mph, generating 1000 G. Aluminium/silicon alloys of eutectic composition for easy gravity casting are used for the body of the piston. 5% copper is added to the aluminium/silicon alloy for higher temperature applications. Such alloys are strong at working temperatures with low coefficients of expansion. The piston weight has been reduced over the years, although this has been largely due to advanced designs rather than improved materials. The piston crown and valve seat areas are now plasma coated with MMC, consisting of an aluminium alloy reinforced with alumina/silica whiskers. Whilst these coatings are not generally used in petrol engines, due to pre-ignition problems and knocking, they find wide application in diesel engines, which can now work at higher more efficient temperatures (Somiya *et al.*, 1989).

Piston rings act as springy seals to prevent gas and power leakage between the combustion space and the oil carrying crank-case. Combustion gases must not escape downwards and oil must not find its way into the combustion chamber where it would burn incompletely giving added emission problems. The rings must be wear resistant but must not wear the cylinder liner. They are coated with an MMC, such as a hard chrome plating reinforced with ultra-fine alumina particles located in the chrome micro-cracks. Similarly, engine bearings such as the main crank-shaft and big-end bearings make use of MMC coatings, usually alumina based, which greatly improve wear performance.

The use of MMC materials to replace cast iron in brake discs has been mentioned previously, with reference to the Lotus Elise. Such discs are commonly used on Formula One cars and Grand Prix motor cycles and all the European car manufacturers are evaluating MMCs for brake discs and drums, for use on new models. There are two main types of MMC, based on powder or fibre, but only powder based materials, PMMC, are financially viable at this time. Molten metal mixing is the lowest cost route currently available for PMMC, giving materials prices of about £8/kg. This produces an ingot, usually aluminium or magnesium based, via standard foundry techniques, where temperature control is essential to avoid chemical reaction between matrix and particles. The ingot can then be processed by rolling, extrusion, forging and drawing. Al–Si alloys, with more than 7% silicon, can be used with up to 30% silicon carbide, reinforcing particulates, which must be uniformly distributed. It should also be remembered that new and compatible brake pad materials must be developed in parallel with MMC discs. A related PMMC transport application is in brake discs for the railway industry, where up to 10 tonnes per train set can be saved compared with the use of cast iron. An interesting fact is that the largest single use of PMMC in Europe, at present, is as tyre studs in Finland to replace tungsten carbide in order to reduce weight.

Catalytic converters or auto-catalysts play a vital role in the modern motor vehicle. They perform under extreme conditions of chemical, thermal and rate constraints as well as requiring significant mechanical properties. Environmentally, air quality is affected by the products of photochemical reactions between the vehicle exhaust emissions of hydrocarbons and nitrogen oxides and the atmospheric oxygen and sunlight radiation. The problem became particularly acute in the Los Angeles basin of the USA, where the unusual atmospheric conditions meant that the photochemical reaction products remained at ground level. By 1975, legislation meant that nearly all vehicles in the USA were fitted with auto-catalysts and unleaded petrol became widely available, since lead contaminates the catalyst metals rendering them ineffective. Similar legislation is now world-wide, with the ultimate aim of zero emissions from motor vehicles.

There are some interesting materials problems in auto-catalyst design and manufacture. Petrol is not completely converted to carbon dioxide and water during combustion although lean burn engines are evolving. Hydrocarbons, carbon monoxide and nitrogen oxides remain in the exhaust gases, and even if these were completely converted, the carbon dioxide would still increase the so-called 'greenhouse effect'. The catalyst converts about 90% of the exhaust emissions to water, carbon dioxide and nitrogen, using the rare-earth metals, platinum, palladium and rhodium, stabilized with nickel, barium, lanthanum and/or zirconium, to prevent particulate sintering, which would cause loss of catalyst activity. This expensive, rare-earth group of catalyst metals must operate both at low initial temperatures as well as working at 1000 °C. The catalysts are carried by an alumina based coating, which is efficient and stable. This complex alumina system is used to coat an extruded ceramic monolith-honeycomb structure, manufactured from the mineral Cordierite. This structure consists of a series of internal channels or tubes, some 400 per square inch, producing the large surface area required for very rapid chemical reactions. The reactive surface area is increased by the nature of the alumina coating system, with the rare-earth catalyst metal particles having a mean diameter of 10 nanometres. It is estimated that the final reactive area is equivalent to that of three football pitches. The monolith honeycomb is finally enclosed in a stainless steel can, with an intumescent interlayer of ceramic mat or stainless steel mesh. This allows for differential thermal expansion and protects against mechanical vibration. Finally, the auto-catalyst is integrated with the rest of the vehicle exhaust system. Auto-catalysts for diesel engines are not yet established since they need to deal with the additional exhaust components of particulate carbon or soot and the attached lubricating oil and sulphur compounds, which not only poison human beings but also poison the rare-earth catalyst metals, so reducing their effectiveness. Diesel fuel exhaust problems will be resolved at some cost.

In conclusion, the development of modern materials lies at the heart of engineering design. In order to make the best use of the many available materials, the vehicle design engineer must have a fundamental understanding of the complex inter-relationships between the structures and compositions of materials and their properties and behaviour in service, together with a realisation of the effects of processing and fabrication on such relationships. The effects of the environment, legislation, economics and evolution will superimpose themselves on this purely rational, scientific approach to vehicle design. Hence, the presence of numerous other important and interesting chapters in this book.

2.6 References and further reading

Adler, U. (1993). *Automotive Handbook*. Bosch (ISBN 1-56091-372-X).

Ashby, M.F. and Jones, D.R.H. (1980/85). *Engineering Materials, I/II*. Pergamon (ISBN 0-08-012139-6 and ISBN 0-08-032531-9).

Atkinson, A.J. and Young, R.J. (1985). *Fracture Behaviour of Polymers*. Elsevier (ISBN 0-85334-7294-6).

Bolton, W. (1989). *Production Technology*. Butterworth-Heinemann (ISBN 0-434-90186-3).

BS 970, Part 3, 1991: 'Bright Bar for General Engineering Purposes'.

Brydson, J.A. (1985). *Plastics Materials*. Butterworth-Heinemann (ISBN 0-7506-1864-7).

Callister, W.D. (1987). *Materials Science and Engineering*. John Wiley (ISBN 0-471-13459-X).

Cebon, D. and Ashby, M.F. (1994). *Cambridge Materials Selector (Software)*. Granta Design Ltd.

Charles, J.A. *et al.* (1989). *Selection and Use of Engineering Materials*. Butterworth-Heinemann (ISBN 0-7506-1549-4).

Cottrell, A. (1985). *Introduction to Metallurgy*. Edward Arnold (ISBN 0-7506-1549-4).

Doremus, A. (1991). *Glass Science*. John Wiley (ISBN 0-471-89174-6).

Hall, C. (1989). *Polymer Materials*. Macmillan (ISBN 0-333-46397-X).

Harrison, A. (1997). *Advanced Materials and Process Development*. Ford Motor Company, Basildon, Essex, UK.

Institute of Materials, Materials Information Service, 1995, London, SW1Y 5DB

Kalpakjian, S. (1991). *Manufacturing Processes for Engineering Materials*. Addison-Wesley (ISBN 0-201-11690-1).

Kennedy, J. (1993). Adhesives in the Automotive Industry. *Materials World*, December 1993.

Kingery, W.D. (1986). *High Technology Ceramics*. American Ceramics Society (ISBN 0-916094-88-X).

Lenoe, E. *et al.* (1983) *Ceramics for High Performance Applications III* (ISBN 0-306-40736-1).

Litchfield, A. (1995). *The Aluminium Car*. Aluminium Extruders Association, Birmingham, B15 1TN

Mayer, R.M. (1993). *Design with Reinforced Plastics*. Design Council (ISBN 0-85072-294-2).

Mills, N.J.(1986). *Plastics*. Edward Arnold (ISBN 0-7131-3565-4).

Montgomery, D.C. (1991). *Design and Analysis of Experiments*. John Wiley (ISBN 0-471-52994-X).

Ohring, M. (1995). *Engineering Materials Science*. Academic Press (ISBN 0-12-524995-0).

Powell, P.C. (1983). *Engineering with Polymers*. Chapman & Hall (ISBN 0-412-24160-9).

Sheldon, R.P. (1982). *Composite Polymeric Materials*. Applied Science (ISBN 0-85334-129-X).

Shute, A. (1997). *Lotus Engineering*. Hethel, Norwich, UK.

Smallman, R.E. (1985). *Modern Physical Metallurgy*. Butterworth-Heinemann.

Smith, W.F. (1993). *Fundamentals of Materials Science and Engineering*. McGraw-Hill (ISBN 0-07-059202-0).

Somiya, S., Mitomo, M. and Yoshimura, M. (1989). *Silicon Nitride*. Elsevier (ISBN 1 85166 329 0).

Timmings, R. and May, T. (1990). *Mechanical Engineer's Pocket Handbook*. Newnes-Butterworth-Heinemann (ISBN 0-7506-0919-2).

Transformation Toughening: Part 4 – Fabrication, Fracture and Strength of Alumina-Zirconia Composites. *J. Mat. Sci*, **17** (1982).

Turner, S. (1983). *Mechanical Testing of Plastics*. Longman (ISBN 0-7114-5785-9).

Walker, E. and Lowe, K.(1995). 'Ultralight Auto Bodies', *Materials World*, December 1995.

Watson, M.N. (1988). *Joining Plastics in Production*. Welding Institute (ISBN 0-8530-0202-9).

West, J.M. (1986). *Basic Corrosion and Oxidation*. Ellis Horwood-John Wiley (ISBN 0-85312-997-5).

Williams, A.R. and Evans, M. (1995). 'Tyre Technology', *Materials World*, December 1995.

Further reading

The literature on engineering materials is wide and varied, dealing with metals, polymers, ceramics and composites, either as single subject text books or more general texts covering all of these materials. The books are set at a variety of undergraduate and M.Sc. levels. Whilst specific books on materials in vehicle design are more limited, this topic is extensively covered by journal and conference publications.

Of particular interest are the following texts, although other comparable books are available, as indicated above, which may be more suited to the needs of some engineering students and other research engineers.

1. Ashby, M.F. and Jones, DRH (1980/1985). *Engineering Materials, I/II.*
    These two volumes are set at undergraduate level and provide a survey of each of the engineering materials with extensive illustrations and basic data.
2. Cottrell, A. (1985). *Introduction to Metallurgy.*
    This is a standard text for undergraduate metallugists and metallurgical engineers covering structure, processing, property relationships.
3. Brydson, J.A. (1995). *Plastics Materials.*
    This is a detailed reference book for plastics and other polymeric materials dealing not only with structure, processing and property relationships but also with polymer manufacture, design and commercial topics.
4. Kingery, W.D. (1986). *High Technology Ceramics.*
    This is a more recent collection of specific ceramics topics based on a standard ceramic text, comparable with those of the Cottrel and Brydson.
5. Sheldon, R.P. (19982). *Composite Polymeric Materials.*
    This is a general composites book, set at post-graduate level, based on the author's experience in polymer physics.

3. The manufacturing challenge for automotive designers

P.G. Leaney, PhD, MSc, BSc, CEng, MIMechE
R. Marshall, MEng, AMIEE

The aim of this chapter is to:

- Present a case that competitive vehicle development requires 'design to manufacture' to be driven as a single process;
- Provide an indication of how that may be achieved;
- Illustrate the challenge and opportunities that designers should seek in exploiting manufacture as a competitive weapon;
- Provide an insight into manufacturing analysis for design purposes.

3.1 Introduction

> If we have a tradition it is this: everything can always be done better than it is being done (Ford, 1922).

The premise of this chapter is that effective manufacture can be better exploited as a competitive weapon by any automotive manufacturer. This premise is built on the accepted fact that once the product is designed then at least 70% of the cost is already committed (Figure 3.1). Thus

Figure 3.1 Influence of design

the first and most important stage in setting up effective manufacturing operations is to get the product design right. The case to be made is that manufacture can be used as a competitive weapon and should not be viewed simply as providing constraints to the designer. Manufacture should not be seen as limiting what the designer can do but rather to enable the designer to realize his product, provide a service to the consumer and make money.

> It is not the employer who pays wages. He only handles the money. It is the product that pays the wages and it is the management that arranges the production so that the product may pay the wages (Ford, 1922).

Where has manufacture been used as an effective competitive weapon? The Toyota Production System (TPS), and the concepts of just-in-time (JIT), have often been cited in recent years. An analysis of JIT (Ohno, 1988; Monden, 1997) and lean production (Womack *et al.*, 1990) allows generic lessons to be learnt that can apply to product development, and this is provided in Section 3.2.

However, the creator of the TPS, Mr Taiichi Ohno, when asked what had inspired his thinking has been quoted by Norman Bodek (President of Productivity Inc) in 1988 as saying he learned it all from Henry Ford. Ohno (1988) himself pays tribute to Henry Ford as the creator of the automobile production system that has since undergone many changes. However, even Henry Ford was inspired by precedent. His most highly acclaimed achievement was to introduce the moving line to assembly operations.

> Along about April 1st, 1913 we first tried the experiment of an assembly line. I believe that this was the first moving line ever installed. The idea came in a general way from the overhead trolley that the Chicago packers use in dressing beef (Ford, 1922).

More detailed study of Henry Ford (Ford, 1922; Ford, 1926) reveals many ideas that relate to the tenets of this chapter. These ideas include those that underpin JIT/lean production (JIT/ LP), continuous process improvement, design for manufacture and assembly, and so on. This should come as no surprise since Henry Ford was not only the chief architect of the product; he was also the chief architect of the manufacturing facilities and their operation. He personally drove design to manufacture as a single process. Henry Ford was an innovator of his time and built his success on the solid engineering but simplicity of his designs and on the low unit cost, but inflexibility, of mass production.

> Any customer can have a car painted any colour that he wants as long as it is black (Ford, 1922).

Since then, of course, the product technology and the manufacturing process technology has become very complex. Complexity has grown because of technological developments in response to customer demand, especially evident in electronics and software. Increasingly, stringent legislative requirements also need to be met governing safety and environmental impact. On top of all that automotive manufacturers now seek the efficiency of mass production whilst producing the product variety demanded by a maturing and an increasingly discerning market. In dealing with these pressures it is clear that product engineers concentrate on the product and manufacturing/ process engineers on developing processes and installing facilities. It is not surprising that design and manufacture has become compartmentalized and separated. This is not in the Ford tradition. Although Ford compartmentalized work by breaking it down to its simplest elements, the aim was always to produce efficient workflow and throughput. He always strived to maximize the work (added value) content with the minimum of human effort. This approach required a scientific study of work (Taylor, 1914) and included the extensive and carefully designed use of machinery.

This chapter is structured to address those who design, develop and manufacture automobiles with the aim of emphasizing the particular challenges/opportunities in seeking competitive advantage by driving 'design to manufacture' as a single business process. This is done by briefly reviewing the lessons of JIT/LP as applied to product development, see section 3.2. Section 3.3 develops the argument by leading on to the modern day concept of IPPD (integrated product and process development) which is presented as extending and encompassing the field of concurrent engineering consistent with the tenets of systems engineering. In doing so section 3.3 provides a rationale for the use of some methods and tools and section 3.4 draws out the mechanics of such in support of manufacturing analysis for IPPD implementation. Section 3.5 provides an insight into the vast range of processes with particular instances and examples where process developments have enhanced the product's design.

3.2 Lean product development and lean production

The 1980s have seen great strides in the reorganization of production around the JIT/LP philosophy but only very recently has it been recognized that effective product realization requires flexibility and leanness (i.e. agility) across the whole process of 'design to manufacture'; achieving this is one of the main challenges for the new millennium.

A key part of the new product development process is served by techniques that feed forward, into the design stage, the relevant information about the potential downstream consequences of decisions made early on in the product's design process. The downstream consequences include the product's manufacture and its subsequent use by customers in the market place. Early design decisions therefore ripple throughout the whole business organization. Defining quality as meeting (and exceeding) the needs and expectations of customers gives us a way to start to address decisions about the product's design in relation to the customers' perception.

The implication of a product's design on its subsequent manufacture is dependent on realizing, and providing some foresight to, the expanding responsibility of the manufacturing function in relation to the success of the business as a whole. It should, therefore, not only be concerned with efficient production (i.e. cost, manufactured quality and speed of response) but also with the product's development and the ability to engineer quality into the design. For this reason there is a general acceptance for the need to develop strong structural bridges between design and manufacturing (e.g. Clark and Fujimoto, 1991; Nevins *et al.*, 1989; Prasad, 1996; Corbett *et al.*, 1991; Huang, 1996). 'Design for manufacture' and concurrent design of the product and its manufacturing processes/facilities (collectively referred to as DFM) are key concepts in this regard and the rest of this chapter endeavours to draw this out within the context of a total view of design to manufacture.

This section concerns the JIT/LP approach to production (i.e. centred on throughput and speed of response) that allows us to draw out the lessons that could be applicable to product development. Before developing the lean production lessons it is necessary to identify and distinguish the 'production cycle' on the one hand and the 'development cycle' on the other.

In simple terms the production cycle represents the lead time taken between customers placing an order and receiving the goods where those goods were manufactured to their order and not simply taken from stock. The development cycle represents the time taken between identifying a market need and producing a new product to meet that need (i.e. product introduction), see Figure 3.2.

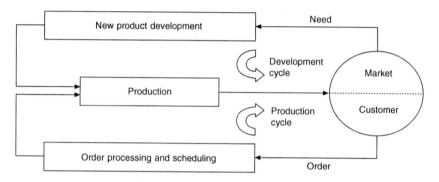

Figure 3.2 Production cycle and development cycle

Bower and Hout (1988) provide a prescription for boosting competitive power based on fast cycle capability. This capability must, ultimately, address not only the production cycle but also the development cycle. It is done by designing an organization (manufacturing organization or engineering organization) that performs without bottlenecks, delays, errors or high inventories. There is an analogy worth drawing out here. Whereas the production organization is seen to manipulate and process materials into components and products, the engineering organization is seen to manipulate and process ideas and information relating to the product's specification and emerging design.

Fast cycle time can be seen as a management paradigm. Compressing time reinforces and supports what capable managers are already trying to do. Fast cycle time causes costs to drop because production materials collect less overhead. Customer service improves due to shortened lead times. Quality is higher because you cannot speed up production or development unless everything is done 'right first time'. This section is, therefore, not used to define the JIT/LP philosophy but rather to draw out some key points with respect to its application to the production cycle and its potential relevance to the development cycle. What has come to be referred to as JIT/LP originally evolved from Toyota's aim to achieve efficiencies of a flow line while producing small batches. The three main elements underlying their strategic approach were (see Sugimori *et al.*, 1977; Ohno, 1988; Womack *et al.*, 1990):

- The right material at the right place at the right time.
- Continuous process improvement (Kaizen).
- Respect for the worker.

To save money the company decided to establish a production method that required as little stock and WIP (work in progress) as possible. It was therefore important to avoid making things that had not been ordered. At the same time a random sequence of orders are to be supported.

Many techniques and concepts have evolved to promote the growing relevance of the JIT/LP philosophy to the production situation. Some examples include SMED (set-up reduction), SPC (process control), Poka-Yoke (mistake proofing), 'zero defects' and 'right first time' (see Robinson, 1990).

The core of production activities is illustrated in Figure 3.3 alongside the development or design core which is discussed in the next section. The production core is derived from Figure 3.2 where PRODUCTION has expanded into SUPPLY, FABRICATE, ASSEMBLE and DISPATCH. This represents the basis of nearly all production models. Inspection and test do not appear explicitly as these are regarded integral with supply, fabricate and assemble to be consistent with the JIT/LP 'right first time' approach to production. Monitoring and control of production should be seen to envelop, and serve, the production core activities.

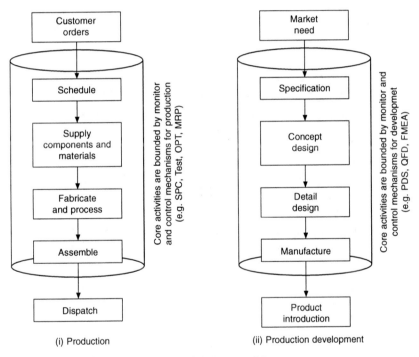

Figure 3.3 Core models

Implementing the JIT/LP philosophy in a company inevitability leads to conflicts with the way things are already done. For example, experience shows us that JIT/LP concerns the whole system and rules emerge such as 'the sum of local optimum is not equal to the optimum of the whole'. This can conflict with the classical method of return on investment (ROI) as a management technique for justifying (or otherwise) expenditure. Seeking high machine or process utilization for ROI justification may well fly in the face of successful JIT/LP production. ROI is now used to judge investment throughout companies in a piecemeal fashion without judging whether the technique works for the overall good of the company.

The preceding discussion, on the JIT/LP philosophy and the production cycle, leads us to identify a number of points that are relevant to the product development cycle.

Point 1. In the three main elements underlying Toyota's strategic approach to production if the word 'material' is replaced by 'information' then these elements are equally applicable to product development:

- The right information at the right place at the right time. Done by providing mechanisms for continual updating in small patches of information, i.e. not collecting all information together then throwing it 'over the wall' to the next downstream function, Figure 3.4.
- Continuous improvement via a formalized team-based organization and structured communication.
- Respect for the worker (e.g. white collar professional and not blue collar manual as before) by moving away from hierarchical working relationships in functional groups to product or project centred groups within a teamwork culture that is based on openness and trust.

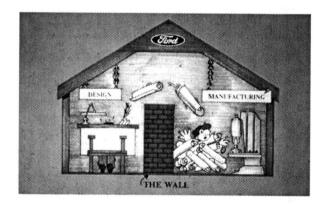

Figure 3.4 'Over the Wall', Historically the Way of Doing Business, courtesy of Munro and Associates, 1989

Point 2. A key feature of the JIT/LP approach to production is that the production system (supply, fabricate, assemble and dispatch) contains minimum WIP so that lead time through the system asymptotically approaches the processing time of value added operations only (i.e. no time is spent in buffer stocks). A system with no buffers has no tolerance to errors or delays. Minimal, or non existent, buffers thus promotes the importance of 'zero defects' and 'right first time'. To achieve this certain techniques have been developed and refined over the years. Some examples include SPC (i.e. closely monitor and control the process allowing for the natural variation of the manufacturing process but detecting trends to be corrected before defects are made) and Poka-Yoke (i.e. mistake proofing at the point of production and thus at the point of potential error introduction) and Process Improvement (i.e. waste elimination by identifying and eliminating non value added operations). Taguchi experimentation techniques also help identify the important controllable parameters to be addressed in process control. The aim, therefore, is to allow the designer some insight into the downstream consequences of his or her decisions. Listed below are some such consequences and some example techniques (see also sections 3.3 and 3.4):

- product and process function predictions (e.g. FMEA);
- market perception and acceptance (e.g. QFD);
- estimates of fabrication and assembly costs (e.g. DFMA).

It is worth noting at this stage that many of the techniques just mentioned are team driven.

Point 3. The final point is that the JIT/LP philosophy embraces a total system view (that includes a partnership with suppliers) requiring a strategic 'whole system' or holistic perception by management. Conflicts are seen to arise in changing from conventional manufacture to JIT/LP manufacture, as discussed with ROI for example, and all such conflicts must be faced and optimal trade-offs achieved. Expediency might dictate a bottom-up implementation in an incremental fashion but it is important that such piecewise implementation fits into a top-down strategic framework devised and supported by senior management.

The thrust of the three points above is not in the advocacy of any particular design or development methodology. It is unlikely that any such methodology will exist that allows all relevant constituencies to have their say, much less get everything they want. The purpose is therefore, to develop a strategic approach that builds bridges between the production cycle and the development cycle so that the design and manufacture of new or improved products can be achieved speedily and appropriately in response to market need. This is the subject of the next section.

3.3 Design to manufacture as a single process and IPPD

The aim of this section is to review the meaning of fast cycle capability in relation to the development cycle. The following discussion will lead to the view that the underlying principles of the JIT/LP philosophy, touched on in the previous section, are consistent with the aims of systems engineering (SE) and successful policy management when applied to product development. The starting point is the recognition of a simplified model for the total design activity as illustrated in Figure 3.3. The design core activity model illustrated here is based on that of Pugh (1990). This model does not deny the iterative nature of design but it does identify the general precedence of activities. In Pugh's model the core activities are, or should be, carried out within the mantle of a well defined, but evolving, product design specification (PDS). In the context of automotive engineering the total vehicle design specification is emphasized as representing a critical technical and managerial control mechanism.

The particular problem in designing and making new or improved products (vehicles) is how to break down and execute the engineering work into manageable portions so that it all fits together well in a total vehicle system to optimally meet custom needs. The nature of the car development/realization process is complex (Whitney, 1995) but is characterized in Figure 3.5. The process falls into roughly three phases: concept design, product design, and process or factory design, and largely follows the design core precedence of Pugh (1990). Each phase comes to an end with major decisions regarding styling or engineering feasibility, but a great deal of intercommunication between the phases is necessary. The circulating arrows in Figure 3.5 indicates the ongoing discussions and revisions that are typically necessary in order to ensure that the design is feasible and meets performance, manufacturability and cost requirements. As the design begins to gel the factory processes are designed and equipment is ordered. The major segments of the factory are powertrain, body shop and final assembly. Often an existing engine and transmission are used so preparation of their factories is a separate process. However, the functional chimneys of business organizations implies a 'divide to conquer' mentality. This worked well for Henry Ford who developed the techniques of mass production for his assembly line by breaking tasks down. Since then, however, even the Ford Motor Company has modified its approach in the light of the JIT/LP approach which advocates an emphasis on throughput rather than utilization and on shop floor teamwork in tackling more broadly defined work tasks. These developments come out of the re-evaluation of the 'system' or 'process' being addressed. The concepts underlying 'continuous process improvement' and 'business process re-engineering' (see Hammer and Champy, 1993) are providing the necessary reorientation in business thinking. The underlying concept of systems thinking is giving perceptive insight for seeking improvements. A re-evaluation of the role of systems engineering with a process perspective leads to broader opportunities.

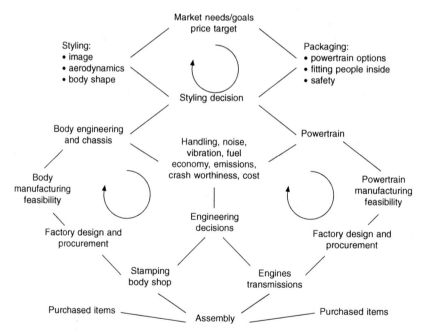

Figure 3.5 Outline of the car development process (Whitney, 1995)

In Japan extensive attention is paid to the design of processes. The Japanese tend to study and improve the process through team co-operation and consensus (Whitney, 1992a). This practice follows Galbraith's (1974) theory of contingency that there is no one best way to organize. The best organization depends upon the tasks' uncertainty and their mutual dependency. The difficult and unpredictable nature of the tasks make the 'product and process design jobs' everybody's job.

In Europe and the US more attention is being directed towards the development of concepts and tools for integrated design and manufacture (e.g. Prasad, 1996, DOD, 1996). With emphasis now shifting from 'product development practices' to 'integrated systems engineering' means that the fields of concurrent engineering and systems engineering are moving together. The aim of such a shift is to (Leaney, 1995):

- better integrate the formal methods (such as FMEA, QFD, DFMA, requirements capture and analysis) into the design process;
- better management of technical requirements versus business requirements versus customer requirements;
- better methodology and methods for negotiating and resolving design conflicts/trade-off decisions.

The discipline of systems engineering (SE) is relevant here where the word system means an ordered array of components or ideas to perform a function.

Systems engineering requires that the product or vehicle realization process be viewed as a systems-centred problem as opposed to a component centred problem. In a traditional component centred philosophy (driven by the division of labour in dealing with complexity – like developing a vehicle) not enough attention is given to interfaces and composite performance. The parts, components, and sub-systems all respond in an orchestrated way in providing the product's functions. It is this orchestration that is addressed by systems engineering. However, it is important to remember that the word system relates not only to the product itself but also to the manufacturing system(s) (Hitomi, 1996) as well as the management/technological system(s) that coordinate and direct the engineering effort that goes into designing and making physical artefacts to meet customer needs. The product/process/people model provides the framework for a total view of product development and project management (Sleath, 1998, Andersen *et al.*, 1995). In addition it opens the way for the application of systems engineering tools typically used by the aerospace sector for the engineering of other complex products such as automobiles (Loureiro, 1998; Shumaker and Thomas, 1998; Percivall, 1992). The key elements in all of this is to direct effort (people and finance) to best effect and to enable the management of change of the product and its enabling processes.

It is the authors' contention that there is significant utility in seeking competitive advantage through a fuller exploitation of manufacturing capability in the product's design. This is the manufacturing challenge for automotive designers and engineers. Practical guidance might be sought in studying the efforts of the US Department of Defense in promoting integrated product and process development, IPPD (DOD, 1996; Shumaker and Thomas, 1998). The DOD are reputed to be the greatest purchasing authority in the world and their impact on the technology and processes of new product/system development is immense. The DOD have mandatory procedures for major defence acquisition programmes. At the heart of this is IPPD.

IPPD is a management technique that simultaneously integrates all essential acquisition activities through the use of multidisciplinary teams to optimize the design, manufacturing and supportability processes. IPPD facilitates meeting cost and performance objectives from product concept through production, including field support.

The key tenets of IPPD are:

1. Customer focus.
2. Concurrent development of products and processes.
3. Early and continuous life cycle planning.
4. Maximize flexibility for optimization and use of contractor unique approaches.
5. Encourage robust design and improved process capability.
6. Event-driven scheduling.
7. Multidisciplinary teamwork.
8. Empowerment.
9. Seamless management tools.
10. Proactive identification and management risk.

For the automotive sector a continuing challenge relates to the use of design and engineering methods to create the concept and details of vehicles. The particular challenge is to know how to move on from engineering specialists who are organized into departments with functional specializations. Already a new pattern is emerging where a mix of platform team and functional organization is used in conjunction with a variety of strategies for reusing results from past or ongoing projects. Concurrent transfer permits faster introduction of cars with more recent design elements than sequential or modification, and it costs less than a complete new design. It is increasingly typical for critical or highly engineered elements like engines or bodies to be developed uniquely by platform team members whereas design of less unique elements like exhaust systems or trim is shared across more and more designs and is provided by functional organizations or, increasingly, first tier suppliers. However, a particular need remains for the development of a product realization infrastructure covering design and manufacturing (Whitney, 1995) and this is the kind of thing that IPPD seeks to address. IPPD should be developed as a core competence. The term 'core competence' is often used to call attention to capabilities that companies feel they really need to have in-house. Discussion of this topic was given a boost by Prahalad and Hamel (1990) who said, 'Core competencies are the collective learning in the organization, especially how to co-ordinate diverse production skills and integrate multiple streams of technologies.'

By way of example, the Ford Motor Company is developing its infrastructure in response to many of these, and other, issues. It has aligned itself as a global company under their Ford 2000 initiative. In addition, the company has identified its five core business processes and outside that they have set up their automotive component wing as an independent operation called Visteon. In this way Visteon moves from being a captive manufacturer to a contract manufacturer attracting business from whomever and seeking to develop its own core competencies.

Two of the five core competencies of the Ford Motor Company are the Ford Product Development System (FPDS) and the Ford Production System (FPS). FPDS is the new approach to the planning, design, development and manufacture of Ford's vehicles. It is characterized by being based on a systems engineering foundation; process driven; disciplined; effective on reusability; requirements driven (voice of the customer focused); and endeavours to include the structured involvement of manufacturing in the development process. FPS is a Ford manufacturing operating package used by the whole of Ford Automotive Operations across the world. It was designed to embrace all the best operating strategies in manufacturing today. The Ford vision statement for FPS is 'A lean, flexible and disciplined common production system that is defined by a set of principles and processes that employs groups of capable and empowered people who are learning and working safely together to produce and deliver products that consistently exceed customers expectations in quality, cost and speed of delivery'. FPS focuses on reducing waste, increasing equipment utilization and reducing inventory. FPS is based on five principles: effective work groups; zero waste and zero defects; alignment of capacity to market demand; optimized production throughput; total cost. Each FPS principle his a measurable to determine the extent to which each manufacturing plant performs.

Despite such laudable examples there continues to be a wider imperative to front load effort and resource in the development cycle. This is aptly illustrated by Hayes *et al.* (1988) who found that preproject planning and concept evaluation in the very early phases has just as powerful an impact on project performance, yet top managers typically pay least attention to those phases where influence is greatest. Figure 3.6 is based on the experience of one automotive company and this characteristic is all too common.

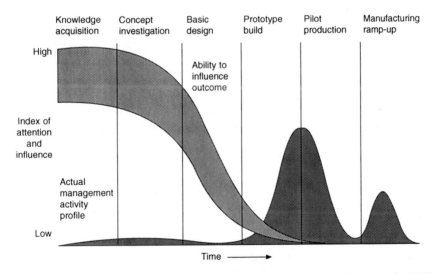

Figure 3.6 Timing and impact of management attention and influence (Hayes *et al.*, 1988)

It might seem to make sense that most management attention is absorbed at the time of major expense in the downstream phases such as setting up manufacturing facilities. The danger here is that the organization uses the project as a vehicle for developing its strategy rather than vice versa.

Promoting the idea of front loading effort and resource early in the development cycle depends on the development of effective concept evaluation techniques, and a contribution to this will be made by providing the designer with some early insight into the downstream consequences of his/her decisions. This means providing manufacturing input well before the detailing stage and we may identify this as a significant role for a 'design to value/cost' approach as part of the wider IPPD concept. Techniques that facilitate front loading effort in the product development cycle include DFMA (e.g. Boothroyd *et al.*, 1994), QFD (e.g. Clausing, 1994) and value engineering techniques (e.g. SAE, 1997). Engineering to costs, affordability and value are all key quantitative measures that can be applied at the concept design stage and embody 'whole life cost' information. The whole life cost of a product is defined as the total cost of acquisition, ownership and disposal. It is in this context that whole life cost applies not only to manufacturing aspects but also to quality of product performance, service and warranty. The designer must have access to effective costing methodologies and an effective tool kit of design rules to extract costs from a business organization that will influence, with confidence, the preferred concept designs. One major problem lies with the traditional accounting methods which lumps overheads as a factor of direct costs (direct labour, material and direct manufacturing costs); this distorts the perception required. Shumaker and Thomas (1998) argue that research into cost and affordability issues is essential. However, a number of techniques and methods are already established, which are supporting the tenets of IPPD. The mechanics of some of these are outlined in the next section.

3.4 Manufacturing analysis, tools and methods

The previous sections of this chapter developed the concept that manufacturing provides an opportunity to vehicle designers. A number of effective tools and techniques are available to product engineers that can be utilized to provide structured approaches to developing products optimized for manufacture and to provide some simple metrics upon which strategic decisions can be made with respect to manufacturability. This section examines some of the methods for attaining these opportunities and facilitation of design to manufacture as a single process.

3.4.1 Design for manufacture and assembly

Design for manufacture and assembly is a key facilitator of design and manufacturing integration. Through the use of some simple rules and additional numerical evaluation products may be effectively and efficiently examined for their ease of manufacture and assembly. Design for manufacture and assembly techniques are an engineering responsibility that provide a total product view. As such they must be applied early on in the development process before resource is committed to any one design and thus costly production problems avoided.

Three well known 'design for assembly' techniques are those of Boothroyd–Dewhurst and Lucas design for assembly (DFA) and Hitachi assemblability evaluation method (AEM) (Leaney, 1996a). These techniques are evaluative methods that analyse the cost of assembly of designs at an early stage in the design process, and use their own synthetic data to provide guidelines and metrics to improve the assemblability of the design (Leaney *et al.*, 1993).

The Boothroyd–Dewhurst DFA evaluation centres on establishing the cost of handling and inserting component parts. The process can be applied to manual or automated assembly, which is further subdivided into high speed dedicated or robotic. Regardless of the assembly system, parts of the assembly are evaluated in terms of ease of handling, ease of insertion and an investigation for parts reduction. The opportunity for this reduction is found by examining each part in turn and identifying whether each exists as a separate part for fundamental reasons. The fundamental reasons are (Boothroyd and Dewhurst, 1989):

1. During operation of the product, does the part move relative to all other parts already assembled? Only gross motion should be considered – small motions that can be accommodated by elastic hinges, for example, are not sufficient for a positive answer.
2. Must the part be of a different material or be isolated from all other parts already assembled? Only fundamental reasons concerned with material properties are acceptable.
3. Must the part be separate from all those already assembled, because otherwise necessary assembly or disassembly of other separate parts would be impossible?

The process of challenging the existence of each component in a product is key to efficient assembly. Products that consist of the minimum number of parts are not only enhanced for assembly but also provide knock-on benefits through reduced stock holding and inventory, reduced manufacturing or sourcing costs, and increased reliability.

In addition to DFA analyses, design for manufacture (DFM) analyses are used to aid in the detail design of parts for manufacture. DFM tools such as design for machining and design for sheet metalworking have been developed by the Boothroyd–Dewhurst partnership to address specific processes and the design of parts suited to those processes (Boothroyd, Dewhurst and Knight, 1994).

Since the early implementations of DFMA tools, steps have been taken to provide a more integrated approach covering a greater portion of the product life cycle. Boothroyd–Dewhurst have developed a number of Windows-based tools and Lucas DFA has been incorporated into an integrated suite called TeamSET (Tibbetts, 1995).

The tools are specific implementations of a basic set of guidelines for DFA which are aimed at raising the awareness of engineering to the importance of assembly. The generic guidelines (Leaney and Wittenberg, 1992) are:

1. Reduce the part count and types
2. Modularize the design
3. Strive to eliminate adjustments
4. Design parts for ease of feeding or handling
5. Design parts to be self aligning and self locating
6. Ensure adequate access and unrestricted vision
7. Design parts so they cannot be installed incorrectly
8. Use efficient fastening or fixing techniques
9. Minimize handling and reorientation
10. Utilize gravity
11. Maximize part symmetry
12. Strive for detail design that facilitates assembly

3.4.2 Quality function deployment

An integrated development process can be facilitated and enhanced through multifunctional techniques that span the activities of the product life cycle. Such techniques not only ensure the ability to trace key concerns throughout development but also provide a common and integrating approach to engineering and manufacture. One such technique is quality function deployment (QFD). QFD enables a development team to specify clearly the customer's wants and needs, and then to evaluate each proposed product or service capability systematically in terms of its impact on meeting those needs (Cohen, 1995).

The QFD process involves mapping customer requirements onto specific design features and manufacturing processes through a series of matrices. QFD can be employed at two levels. The first of this is to translate requirements of one functional group into the supporting requirements of a downstream functional group, and the second is a comprehensive organizational mechanism for planning and control of new product development (Rosenthal and Tatikonda, 1992). A localized application typically involves the first of these matrices (Figure 3.7). This matrix has the most general structure and is often called the house of quality (HOQ). Typically applications of QFD are limited to the HOQ, however, QFD can play a greater role as a linking mechanism throughout product development through the use of subsequent matrices.

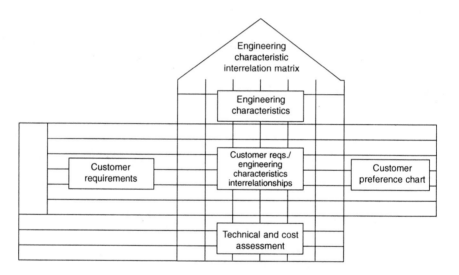

Figure 3.7 The House of Quality Matrix

After the house of quality matrix a number of additional matrices may be used to deploy the customer requirements through to production planning. Cohen (1995) presents the Clausing 'four-phase model' (Figure 3.8), that mirrors the process of design and manufacture. The ability of QFD to be deployed in this manner makes it unique among formal methods in its ability to span life cycle processes.

A good practical overview of both the benefits and pitfalls of QFD is given by Hasen (1989) who reports on experiences at Ford's body and chassis engineering. Some benefits include: provides a systematic approach in addressing the customer's wants and acts as a driver for other techniques such as FMEA, Taguchi, SPC; moves changes upstream where they are more

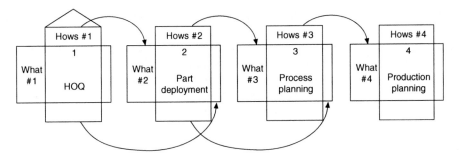

Figure 3.8 The four phase QFD model

economically accomplished; provides a valuable company record for the next product cycle; promotes teamwork and shared responsibility. Hasen reports that their initial experiences with QFD have allowed them to subsequently tailor the system towards their particular requirements and embody it in the business system.

3.4.3 Design for dimensional control

Design for dimensional control (DDC) refers to the total product dimensional control discipline which recognizes and manages variation during design, manufacture and assembly. It aims to meet customer quality expectations for appearance and function without the need for 'finesse' by shop floor operatives in manufacturing and assembly operations. DDC embodies a range of tools and techniques and also embodies an imperative for management to provide the appropriate organization of engineering effort that is consistent with the tenets of IPPD.

Major elements of production costs come from the failure to understand design for dimensional variation. This variation results in irreversible tooling and design decisions that forever plague manufacturing and product support. The aim of DDC is not to eliminate dimensional variation, but rather to manage it. The successful management of variation provides the following benefits:

- Easier manufacture and assembly
- Improved fit and finish
- Reduced need for shop floor 'finesse'
- Less work in progress
- Reduced cycle time
- Reduced complexity
- Increased consistency and reliability
- Improved ability for maintenance and repair

Robustness can be defined as a product insensitive to variations. DDC is the application of robustness thinking to dimensional variation. The approach is to seek the best overall economic solution to achieving control of dimensional variation through appropriate product design in conjunction with process design and process operation such that the resulting variation does not give rise to any concerns or symptoms through manufacture and assembly, test, and product operation (Leaney, 1996b).

DDC is built upon dimensioning and tolerancing (D+T) standards. However, traditional D+T practice was to 'define the result you want, not how to get it' in a one-way communication process between product engineering and manufacture. DDC now provides the mechanism to use the language of D+T and to close the feedback loop from manufacturing back into design. DDC is ultimately a combination of related processes within a framework aimed at robust design. The framework addresses the control of dimensional variation through initially designing for assembly and minimizing any inherent variation. Once optimized for assembly, variation is then controlled and managed through the use of assembly tolerance analysis, where the accumulation of tolerances stemming from component design and manufacture and assembly processes and procedures is analysed. A number of tools such as Variation System Analysis (VSA) and Valisys (Tecnomatix) exist to aid this process, providing a computer aided environment for the management of variation in assembly.

In addition to tolerance analysis the management of variation is achieved through the application of best practice guidelines for D+T, locating, measuring, and the consideration of manufacturing requirements up front in the development process.

3.4.4 Value engineering/analysis

Value engineering is a team-based evaluative technique which assigns a value to a product. The process attempts to enhance the value of the product by increasing its functional capability, for the same or lower cost. Or inversely, reducing the cost whilst maintaining the same functional capability. The goal is to eliminate unnecessary features and functions by optimising the value-to-cost ratio. This process thus provides a simple but structured approach to optimizing designs for both the customer and the manufacturer (SAE, 1997).

Care must be taken with the understanding of value as it is heavily dependent on the circumstances in which it is measured (Fox, 1993). This value can be divided into two components: a use, or functional, value and an esteem value. The use value reflects how the product satisfies the user's needs, and the esteem value is a measure of the desirability of the product. The two values are investigated analytically by a team of experts based on a preliminary design (Cross, 1994).

The process of value engineering consists of five phases: information, function, speculation, evaluation, and implementation. These phases span the following activities; information gathering and defining the function of the product and its constituent components, assigning a value to each component, generating and assessing alternatives, and finally implementing the proposed solutions.

3.4.5 Failure modes and effects analysis

Product failures through design or manufacturing faults are costly both in monetary terms and in the customer's perception of the product and manufacturer. Therefore a multifunctional approach to product system analysis done in a timely manner provides a valuable guard against the introduction of poor products.

Failure modes and effects analysis (FMEA) is a structured approach to the identification and evaluation through a risk priority number (rpn) of possible modes of failure in a product or process design. Failure is taken in its broadest sense, not as a catastrophic breakdown but as a consequence of not meeting a customer's requirements. The aim is to anticipate and design out all possible failures before they occur, removing the cost to manufacture, warranties, and customer satisfaction (see Figure 3.9).

Part	Function	Potential failure mode	Potential effects of failure	Severity	Potential causes of failure	Occur-rence	How will potential failure be detected?	Detec-tion	rpn	Actions
Tube	provide grip	hole gets blocked	vacuum on ink stops flow	7	debris ingress into hole	3	Check clearance of hole	5	105	enlarge hole or remove cap
Ink	provide writing medium	incorrect viscosity	high flow	4	too much solvent	2	QC on ink supply	4	32	introduce more rigid QC
Ink	provide writing medium	incorrect viscosity	low flow	4	too little solvent	2	QC on ink supply	3	24	no action required
Ball and seat	meter ink supply	incorrect fit	ball detached	8	total failure	2	Inspection checks	3	32	
Ball and seat	meter ink supply	incorrect fit	ball loose	6	blotchy writing	3	Sampling checks	6	108	introduce in process checks
Plug	close tube	wrong size	falls out	4	Moulding process not in control	2	no current checks or tests	8	64	eliminate part or control process variation

Figure 3.9 Product failure mode and effects analysis table (Fox, 1993)

3.4.6 Quality engineering

Dr Genichi Taguchi is possibly the most well known advocate of quality engineering (QE), so much so that Taguchi methods are often synonymous with QE. According to Taguchi (Taguchi, 1993) quality engineering pertains to the evaluation and improvement of the robustness of products, tolerance specifications, the design of engineering management processes, and the evaluation of the economic loss caused by the functional variation of products.

Taguchi (1993) defines quality as the amount of functional variation of products plus all possible negative effects, such as environmental damages and operational costs. Taguchi evaluates quality through a quality loss function (Figure 3.10). The quality loss function is expressed as the square of the deviation of an objective characteristic from its target, assuming the target to be the desire to meet customer satisfaction, any deviation from that value will mean a level of reduced satisfaction for the customer. Furthermore the greater the deviation, the greater the dissatisfaction to the customer.

In practice the Taguchi approach to quality engineering provides an analytical tool for

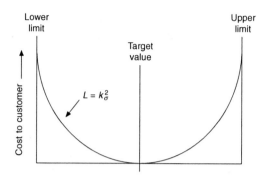

Figure 3.10 The quality loss function

designers to develop new products that can perform the desired functions, while keeping production costs below those for competitive products. The concept also highlights that it is not acceptable to just keep the parameter within the set limits, but that it is necessary to keep as close as possible to the nominal or target value.

3.4.7 Quality system 9000

Quality system (QS) 9000 was developed by Chrysler, Ford and General Motors as a united approach to the issue of supplier quality systems. The goal of QS 9000 is the development of fundamental quality systems that provide for continuous improvement, emphasizing defect prevention and the reduction of variation and waste in the supply chain (QS 9000, 1995). QS 9000 defines the fundamental quality system expectations of Chrysler, Ford, General Motors, and other subscribing companies for internal and external suppliers of production and service parts and materials.

These companies are committed to working with suppliers to ensure customer satisfaction, beginning with conformance to quality requirements, and continuing with reduction of variation and waste to benefit the final customer, the supply base, and themselves. The quality system itself is a harmonisation of Chrysler's Supplier Quality Assurance Manual, Ford's Q-101 Quality System Standard, General Motors' NAO Targets for Excellence, and ISO 9000 Section 4. It provides a number of rules and guidelines for quality requirements throughout the product life cycle, requiring a number of documenting procedures and the use of D + T, QFD, DFMA, VE, Taguchi, FMEA, and other CAD and CAE tools.

QS 9000 also incorporates an assessment against the 23 requirements from which the list below is taken. The assessment takes the form of a number of questions against each requirement that are graded between 'failure to meet the requirement' to 'effective meeting of the requirement with a marked improvement over the past 12 months that is meaningful to the customer'. The scores are then tallied and manipulated to provide a ranking;

- Management responsibility;
- Design control;
- Purchasing;

- Process control;
- Inspection and test;
- Handling, storage, packaging and preservation;
- Internal quality audits;
- Training;
- Servicing;

3.4.8 Group technology and cellular manufacture

The process of standardization and rationalization can provide a number of advantages to the design to products through the ability to reuse previously designed features, components, subassemblies or modules. In addition these standardized elements reduce manufacturing and assembly cost and may be used to structure manufacturing operations in an efficient manner.

Group technology (GT) is a method of manufacturing piece parts by classification of these parts into groups and subsequently applying to each group similar technological operations. On the shop floor GT facilitates the grouping of machine tools and other facilities around components that have similar processing characteristics. These groups then simplify manufacturing planning, flow of work, minimize set up times and component lead times. Though GT is aimed toward the efficiency of manufacture, in design GT promotes standardization, reduces design duplication, reduces the number of parts needing to be held in stock, part numbers and the associated documentation. GT also allows easy part data retrieval and reduces the development lead time.

However, GT was initially restricted to maintaining functional layout of machines whilst improving machine productivity. As GT has developed, a different term has been used to represent a broader interpretation that expands upon process-based groups including the formation of groups around products and people (Alford, 1994). This broader view is termed cellular manufacture, the distinction is not always clear. Burbidge (1994) suggests that these groups complete all the parts or assemblies they manufacture. The group machines are laid out together in a designated area and are manned by their own team of operators.

3.4.9 Flexible and agile manufacture

Manufacturing flexibility is an essential part of addressing the market pressures for increased variety, reduced lead times and improved quality. Corrêa and Slack (1996) highlight the benefits of manufacturing flexibility, particularly the change in competitive strategy from economies of scale to economies of scope. However, care must be taken when dealing with manufacturing flexibility as the term has no agreed definition; in fact there are a number of flexibilities that are subsumed within the general concept. Possibly the best generic definition of flexibility is the ability to respond effectively to changing circumstances (Nilsson and Nordahl, 1995), or the ability to cope with the uncertainty of change effectively and efficiently (Tincknell and Radcliffe, 1996). Specific types of manufacturing flexibility include:

- Volume/mix flexibility – to accept a change in production volumes or a range of products;
- Product changeover flexibility – to changeover to the production of a new product;
- Operational flexibility – to absorb changes to the product during its working life;
- Routing flexibility – to manufacture or assemble along alternative routes;

- Machine flexibility – to perform various tasks on a variety of parts;
- Location flexibility – to move the production of a particular product to different factories.

Manufacturing flexibility relies upon manufacturing strategy and the implementation of flexible facilities and working practice, but equally the responsibility of design and engineering functions to provide a product that is sympathetic to flexibility. This includes the consideration of DFMA, part commonization, product modularity, and an up-front loading of effort. Manufacturing flexibility is a collection of product and process design concepts, aimed at ensuring the competitive edge of a manufacturer (Barnett, Leaney and Matke, 1996). Issues for flexibility are:

- Typically flexible systems will have greater short term cost, but will realize greater long term savings. However, care must be taken as flexibility cannot be achieved indefinitely;
- Flexible systems will typically be more complex both in design and in operation;
- Flexible systems must be given time for adaptation, thus decreasing the time available for the actual operation for a given cycle time;
- Flexible systems can be developed to accept changes in capacity, but this will affect the size of the facilities and often require the inclusion of redundancy.

Agile manufacturing is a concept that has gained momentum in enabling rapid response to market needs. It aims to provide the flexibility of response with the efficiency of lean production, not only in the manufacturing environment but throughout the whole organization. Gould (1997) defines the agile approach as 'the ability of an enterprise to thrive in an environment of rapid and unpredictable change, and draws comparison between this goal and those of other initiatives such as mass customization, the fractal factory, holonic manufacturing, and holonic enterprise.' Booth (1995) suggests that the path to agile manufacturing (Figure 3.11) is a combination of process integration to reduce lead time, and flexibility in minimizing the costs of complexity associated with variety. He also proposes three aspects to the change to agile manufacture; the organization, people's working methods, and information systems. Owen and Kruse (1997) group these into internal and external agility. Internal agility is the ability to respond rapidly to change by localized changes to the product or processes. External agility

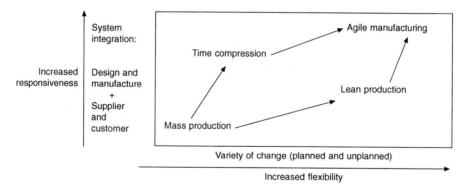

Figure 3.11 The path to agile manufacturing (Booth, 1995)

covers the organizational approach through the extended enterprise, companies focusing on their core competencies and forming strategic partnerships with suppliers to address change.

The final consideration is the product. As the product design is important in flexibility so it affects the concept of agile manufacturing. Appropriate consideration of design techniques and product architecture can facilitate agility by the provision of modular products, products that allow the introduction of variety later on in the manufacturing process and reusable design.

3.4.10 Modularity

Modularity has a rather unfortunate legacy in that many companies and engineers believe, incorrectly, that they understand what modularity means and that they already utilize a form of modular product architecture. In addition modularity is often seen purely as a process of decomposition or demarcation of product architecture into subassemblies (Whitney, 1992b). Modules have a number of characteristics that provide fundamental differences between them and convenient groups of components in a subassembly:

- Modules are co-operative subsystems that form a product, manufacturing system, business, etc;
- Modules have their main functional interactions within rather than between modules;
- Modules have one or more well-defined functions that can be tested in isolation from the system and are a composite of the components of the module;
- Modules are independent and self-contained and may be combined and configured with similar units to achieve a different overall outcome.

Modularity is typically utilized for its ability to rationalize variety through the partitioning of product functions (Pahl and Beitz, 1996; Smith and Reinertsen, 1991) and allow for flexibility of application. This advantage has been applied widely; throughout the electronics industry for computer manufacture, within the automotive industry on the Max Spider (Weernink, 1989) and the Renault Modus (Figure 3.12 – Smith, 1995), and within the aerospace industry on the

Figure 3.12 The Renault Modus

Joint Strike Fighter, a highly common modular range of aircraft for airforce, marine, and navy use (JSF, 1997). However, variety is only one aspect of product modularity. One of the key elements of modularity is its fresh approach to meeting the requirements of effective new product introduction.

The use of a modular approach such as Holonic Product Design (Marshall, 1998b) to product development has been shown to provide a number of advantages to both design and manufacture (Marshall, 1998a):

- Modularity provides product variety to the customer. However, variety can be offered efficiently through a limited number of modules and the use of common modules. Variety can also be introduced without unnecessary reengineering, in reduced timescales and at lower cost;
- Modularity allows customers to control variety, providing flexibility in operation and in support through improved serviceability and upgrade;
- Modularity presents an opportunity to manage process complexity and combine teams with the modules for which they are responsible. Requirements for modules to integrate together then encourages integration across teams and presents a greater system for efficient and effective product development;
- Modularity addresses product complexity through decomposition of systems, partitioning of functions, analysis of interactions and modular assembly. The result is greater reliability, service, and upgrade;
- Modularity allows more efficient and effective manufacture and assembly. Part standardization addresses quality, economies of scale and improved supplier relations. Processes can be structured around the product, modules assembled in parallel, testing done on individual modules, variety introduced late and thus orders rapidly fulfilled;
- Modularity also provides structure to the application of other related processes such as DFA, value engineering and group technology.

3.5 Materials processing and technology

The increasing customer and legislative pressure upon vehicle manufacturers constantly challenges the use of traditionally accepted design concepts, materials and manufacturing processes. The strive for improved performance, responsiveness, power, reliability and economy has forced vehicle manufacturers to look toward the use of advanced materials and the processing of materials in new and improved ways. In addition, process technology has been heavily investigated to actually realize previously unfeasible product design concepts. This section introduces a comprehensive process taxonomy and further details a selection of its process technologies and example case studies.

3.5.1 A manufacturing process taxonomy

The following taxonomy shows the wealth of processes available to the product engineer (Bralla, 1986; DeGarmo, Black and Kohser, 1990; Kalpakjian, 1991). Many of these processes offer unique characteristics that enable specific requirements to be met and designs realized. It is important for the product engineer to be aware of such processes in order to provide the minimum of constraints upon the concept design phase. The following section will highlight a number of these processes and provide examples of their use within the automotive industry.

Casting processes

Expendable mould, multiple use pattern
  Green sand/Dry sand casting
  Sodium silicate – CO$_2$ moulding
  Shell casting
  V – process
  Eff - set process
  Plaster mould
  Ceramic mould
  The Shaw process
  Expendable graphite moulding
  Rubber mould casting
Expendable mould and pattern
  Full mould (evaporative pattern)
  Investment casting

Permanent mould
  Gravity die casting
  Pressure die casting
  Squeeze casting
  Slush casting
  Centrifugal casting
  Continuous casting
  Electromagnetic casting
  Rotational moulding
  Injection moulding
  Reaction injection moulding
  Compression moulding
  Monomer casting/contact moulding

Material removal processes

Electromachining
  Electrochemical machining
  Electrical discharge machining
  Electrolytic hole machining
  Laser beam machining
  Ultrasonic machining
  Plasma arc machining
  Electron beam machining
Fluid processes
  Fluid or water jet machining
  Abrasive water jet machining
  Abrasive flow machining
  Hydrodynamic machining
Mechanical machining
  Single point
    Turning
    Planing

Multiple point
  Drilling
  Tapping
  Boring
  Reaming
  Sawing
  Filing
  Knurling
  Broaching
  Milling
  Facing
  Routing
Abrasives
  Grinding
  Abrasive machining
  Honing
  Lapping

Surface processes

Physical surface treatment
  Shot peening
  Shot blasting
  Polishing
  Abrasive cleaning
  Tumbling
  Wire brushing
  Belt sanding
  Electropolishing

Coating
  Electroplating
  Anodising
  Blackening
  Metal spraying
  Thermal spraying
  Plasma spraying
  Powder coating
  Organic/inorganic coating

Chemical surface treatment
Carburizing
Nitriding
Induction hardening
Alkaline cleaning
Solvent cleaning
Ultrasonic cleaning

UV curable coating
Chemical vapour deposition
Sputtering
Painting
Carbon film deposition

Joining processes
Fusion welding
Oxyfuel welding
Arc welding
Manual metal (MMA)
Tungsten Inert Gas (TIG)
Metal Inert Gas (MIG)
Resistance welding
Laser welding
Induction welding
Electron beam welding
Ion beam welding
Ultrasonic welding
Plasma arc welding
Electro slag welding
Electro gas welding

Solid state welding
Forge welding
Friction welding
Diffusion bonding
Brazing and soldering
Adhesive bonding
Mechanical joining
Fasteners
Bending
Crimping

Forming processes
Rolling
Forging
Open die hammer forging
Impression die drop forging
Press forging
Automatic hot forging
Upset forging
Swaging
Roll forging
Rotary forging
Orbital forging
Extrusion
Spinning
Bending
Shearing
Blanking
Tailored blanks
Pressing

Piercing
Stretching
Drawing
Deep or shell forming
Ironing
Superplastic forming
Embossing
Hydroforming
Explosive forming
Vacuum forming
Blow moulding
Composites weaving
Composites layup
Powder processing
Slip casting
Powder metal injection moulding
Pressing and sintering
Isostatic pressing

Treatment processes
Heat treatment
Annealing
Precipitation hardening
Stress relieving

Hot working
Cold working
Shot peening
Shot blasting

3.5.2 Hydroforming

Hydroforming was developed in the late 1970s in the United States. The process consists of placing pre-bent steel tubing into a die of the component to be formed. High pressure fluid (typically 500 bar) is then used to form the tube into the exact shape of the component. The benefits of this process include (Christiansen, 1997):

- reduced number of parts;
- reduced number of forming and welding/joining operations;
- more uniform strain distribution;
- reduced springback;
- reduced number of secondary operations (punching etc);
- reduced part weight;
- improved structural strength and stiffness;
- reduced dimensional variation;
- reduced scrap.

However, hydroforming also has a number of drawbacks. The hydroforming dies are relatively slow and costly to produce. Such dies are dedicated to the particular car model thus requiring a set of dies per vehicle and the need to have new dies if the part is modified in any way. In addition, die tolerances have to be extremely high, the process cycle time is relatively high and care has to be taken with the pre-bending of the steel tubes to be formed. Hydroforming has seen use within the automotive industry upon components such as Buick Park Avenue side roof rails, BMW 5000 rear axles, Mercedes Benz exhaust manifolds and GM Corvette lower rails, roof bow and instrument panel beam.

3.5.3 Tailored blanks

Tailored blanks are sheets of steel typically laser welded together prior to stamping or forming (Figure 3.13). Developed by Audi in the mid-1980s, the tailoring process includes the ability to combine different grade, thickness, strength and coating of material in one single blank and

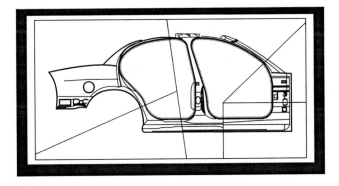

Figure 3.13 Single piece tailored blank body side outer panel

thus provide the capability to engineer the properties of the material and thus the finished component for optimum performance (Barrett, 1997). The advantages of using tailored blanks include:

- reduced number of parts and associated tooling;
- reduced tool/die costs;
- reduced weight;
- reduced processing times;
- economy with smaller production runs;
- improved quality;
- improved dimensional stability.

Tailored blanks currently see widespread use within the automotive industry at Ford for inner floor side members, liftgate inners, and pillar and fender reinforcements, at Rover for door inner panels, longitudinal members and wheel arches, Volkswagen for B and C pillars, and BMW in a number of components for its 5 and 7 series (Barrett, 1997).

3.5.4 The Cosworth process

The Cosworth process was developed to meet the need for consistent high quality aluminium alloy castings for the automotive industry. To meet their needs Cosworth established a research and development programme to study the fundamental problems in moulding, melting, casting and heat treatment of aluminium alloy castings (Clegg, 1991).

The reasoning behind this move was that commercial quality aluminium alloy castings produced by more traditional methods such as sand or gravity die casting contained porosity which impairs their metallurgical integrity and thus reduces the attainable level of mechanical properties. A second problem is the inherent inaccuracy of castings produced from moulds made of the commonly used silica sand.

To address these problems Cosworth developed a complete process to produce castings for the Cosworth engine in 1978. The process was extended by the opening of a new foundry in 1984 (McCombe, 1986). Since then the process has earned Cosworth the Queen's award for Technological Achievement and take-up by major motor manufacturers such as the Ford Motor Company who use the process in plants such as the Windsor Ontario plant (McCombe, 1990). The advantages of the Cosworth process include (Clegg, 1991):

- exceptional high strength and ductility;
- considerable weight saving, allowing improved design through lighter and more robust components;
- dimensional accuracy which can exploit modern manufacturing processes;
- components which have total commitment to value engineering;
- castings which are free of porosity and inclusions;
- flexibility to meet changing markets.

These are also met by economical production through comparatively inexpensive tooling, high metal yields, and minimum machining required. Due to the quality of the castings produced by this process, in addition to automotive components they are typically used in the aerospace and defence industries. Some typical components include: high performance cylinder heads for racing engines, engine blocks, transmission cases, gas turbine front end components, and flight refuelling manifolds.

3.5.5 Adhesive bonding

Adhesive bonding is attractive to a wide range of industries as it enables flexibility in material, selection of components, product design and component manufacture through to final assembly. Such factors can produce a significant advantage and savings over conventional joining methods such as welding or riveting. The specific advantages of adhesive bonding are (BASA, 1993):

- Most materials can be joined to themselves or different materials;
- Bonded joints are stiffer than spot welded or mechanically fastened joints;
- Adhesives allow a reduction in part count and therefore associated savings in cost and weight;
- Materials to be joined can be thinner;
- Bonded assemblies have a smooth clean finish;
- The high temperature effects of welding are removed;
- Components are not weakened by keyways, holes, slots, etc;
- Sound deadening and vibration damping is improved.

However, there are a number of additional disadvantages to adhesive bonding:

- Joints are essentially permanent;
- Adhesives are temperature sensitive with some becoming brittle at low temperatures and most having a maximum operating temperature of around 150°C, although 250°C is possible;
- Time is required for hardening;
- Surfaces must be clean;
- Quality control of the bond can be difficult.

The most important aspect of adhesive bonding is the selection of the appropriate adhesive for the application. Many adhesives exist for a multitude of applications and exist in three distinct categories: thermoplastics, elastomers, and thermosets.

Thermoplastic adhesive types include vinyl co-polymers, saturated polyesters, polyacrylates and polysulphides. They are typically used for bonding wood, glass, rubber, metal and paper products. Elastomeric adhesives are composed of both and natural and synthetic rubbers and are used for bonding flexible materials to rigid materials. Thermosetting adhesives include epoxies, polyurethanes, amino-phenol resins, isocyanates, and the silicones. They are all transformed into tough heat-resistant solids by the addition of a catalyst or the application of heat and are used for structural bonding of metallic parts.

3.5.6 Rapid prototyping

Rapid prototyping is a process for producing physical prototype parts from a computer aided design through the use of 'layer manufacturing'. Through a CAD system, the design is sliced into thousands of cross sections which can then be transformed into a physical prototype through a process of layering a polymer or paper in accordance with the cross sections. The actual process of producing the layered prototype can be performed through a number of technologies:

- Stereolithography (SLA) – a UV laser is used to cure the surface of a liquid resin;
- Fused deposition modelling (FDM) – molten polymer is extruded to form each layer;
- Selective laser sintering (SLS) – a laser is used to sinter a polymer powder;
- Laminate object manufacture (LOM) – a laser is used to cut paper layers that are then bonded together.

These processes provide a range of cost, quality, throughput, and material. However, all share the benefit of requiring no tooling so that prototype parts can be produced directly from CAD, reducing cost and lead times. Parts can be efficiently produced to aid design verification, styling, prototype test, tooling production, and as a valuable communication aid within product development (Chee-Kai and Kah-Fai, 1998).

Rapid prototyping is not suitable for every application and account must be taken of the inherent disadvantages of rapid prototyping processes:

- All parts exhibit a stepped z axis construction;
- Minimum wall thickness restricted to 0.5 mm;
- Parts must include integral supports;
- Generation of the appropriate CAD file can be problematic;
- Trapped volumes within the parts.

3.5.7 Manufacturing technology case examples

K-series Rover engine
The Rover K-series engine has been designed in a unique layered construction. Rather than separate bolts and studs the five layers; oil, rail, main bearing ladder, cylinder head and cam carrier are held together in compression by ten long through bolts. In order for the design to function there cannot be any movement between layers, thus ruling out the use of conventional cork and fibre gaskets. To provide the metal-to-metal seal Rover utilize an anaerobic liquid gasket which is screen printed onto each layer directly (BASA, 1993).

Water pump housing
Figure 3.14 shows two water pump housings. The left hand side shows the traditional sand cast housing. The right hand side shows a pressure die cast replacement that is produced in two parts and adhesive bonded together. The die cast housing presents a greater initial investment but provides an increased production rate, improved quality and a reduced cost of £1.75 per housing. In addition the die-cast housing is more flexible as it allows the tube to be bonded in any direction the designer requires without need for further tooling.

Renault prop-shaft
Originally designed for the Peugeot 205 1985 Paris to Dakar rally car a carbon composite prop

Figure 3.14 Water pump housing

shaft was bonded to a steel end fitting. The design is now widely used, including the Renault Espace, and provides greatly reduced weight and improved sound deadening over the conventional steel shaft.

Rapid prototyping
Using rapid prototyping (RP) technologies as presented earlier a number of case studies can be presented. Ford Motor Co. utilized stereolithography to produce a two part injection mould tool for an ABS wiper motor cover for their explorer. The process reduced the lead time from 17 weeks to 6 weeks and reduced costs from $33 000 to $18 000.

The Rover group also used RP to produce patterns for outside shape and resin core boxes for the internal tracts of inlet manifolds. Lead time was reduced from 8 weeks to 3 weeks and a saving of £3210 (Sterne, 1996).

Renault ring gear
Previously the Renault 9 ring gear and drive hub assembly consisted of three components that were bolted together in order to sustain high dynamic torque loading. However, in an attempt to reduce weight and manufacturing costs, designers looked for alternatives. A shrink-fit design proved unable to meet the required dynamic torque loading, thus the assembly was enhanced through the use of an anaerobic adhesive (Figure 3.15). The use of the simplified design, manufacture and assembly process reduced weight by 15% and saved a cost of £1.10 per assembly. The success of the redesign has seen its adoption by Citroen and Rover and its evaluation by other gearbox manufacturers.

Apticote – surface engineering
Surface engineering covers a broad range of applications, an example of which is the Apticote ceramic coating. Cylinder bores or engine blocks are typically protected with a special liner that provides the hard wearing and oleofilic (oil retaining) surface required for such extreme conditions. However, the Apticote coating has been developed to be applied directly to the cylinder bore without the need for a liner. Originally produced for racing engines and now used

Figure 3.15 Renault 9 ring gear

in high-end road cars, the ceramic coating can be applied to a wide range of base materials including aluminium, steel, cast iron, metal matrix composites and hyperutectic alloys. In addition to the removal of the need for a cylinder liner, the coating provides a much finer finish whilst still maintaining an oleofilic nature. The finer finish reduces wear and pressure on piston rings, oil consumption and friction resulting in improved efficiency and working life.

Lexus sound deadening laminate panels
Similar to tailored blanks, Figure 3.16 shows the structure of Lexus panels that are constructed from laminated steel to greatly reduce noise. Two sheets of steel are bonded together with a low modulus adhesive. The laminated panels can then be formed in the normal manner. This process requires little or no extra tooling and reduces the need for sound deadening material, saving both on extra components and cost.

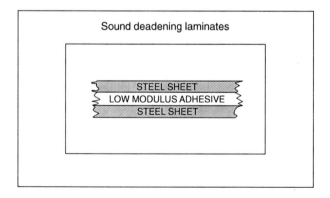

Figure 3.16 Lexus sound deadening panels

Mondeo – Presta camshaft assembly

Traditional camshafts are manufactured from cast iron which performs well, is relatively cheap but is also heavy and requires considerable machining which increases the cost and time of manufacture. To address these issues Presta have developed a composite camshaft design. Near-net shaped cams are broached with an internal spline profile. At each longitudinal cam location a hollow shaft is cold rolled to form a section of narrow annular ribs, matching the cam profile. The cam is then pressed over the ribbed area giving a positive locking by a combination of form fit and plastic deformation. This process reduces weight through the use of a hollow shaft and near net shape cam manufacture. In addition, the use of the welded cold drawn shaft requires no special machining. The process provides flexibility in material choice, arrangement of the components on the shaft for multi-valve technology, and torque capability, all in a simple, repeatable and controllable manner. This design has seen use in the 2.5 litre 24 valve Ford V6 Mondeo engine with the camshafts providing a 40% saving in weight and consequentially inertia (Matt *et al.*, 1995).

Aluminium intensive vehicle

Using Alcan's aluminium vehicle technology (AVT) Ford's North American concern has developed a fleet of 40 aluminium intensive vehicles (AIVs) (Broad, 1997). Based on the mid-sized Taurus/Mercury, the use of AVT has resulted in a bodyshell that is 47 percent (182 kg) lighter than its steel equivalent. With the use of AVT throughout the powertrain components the total weight saving is 318 kg over the steel Taurus.

The essential difference between the Ford AIV and other aluminium cars is the processes used in construction. Whilst most other aluminium vehicles utilize low volume technologies, Ford test vehicles were produced using conventional high volume processes in conjunction with Alcan's AVT structural bonding system. The use of bonding in conjunction with conventional spot welding technology maximizes the weight saving and also improves structural performance with increased bending and torsional stiffness, and increased fatigue resistance.

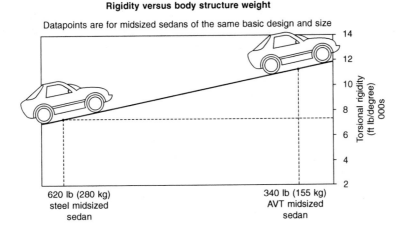

Figure 3.17 AVT vs. steel body structure comparison

Single piece wheel forming
Traditionally motor vehicle wheels are produced from a rolled and welded sheet of steel forming the rim of the wheel, with a disc of steel welded into the rim to form the hub. Figure 3.18 shows the cross section of a wheel formed from a single aluminium disc. The disc is spun and formed to produce a hub and integral rim reducing the weight of the wheel, reducing the number of components and eliminating the need for welding.

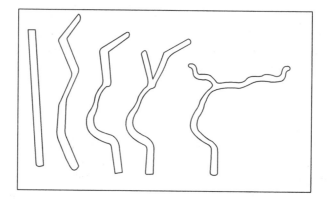

Figure 3.18 Single piece wheel section (courtesy of ASD Ltd., 1997)

3.6 Conclusions

* Competitive vehicle development depends on identifying market need and aligning the development activities across design and manufacture to meet those needs appropriately, without delays, errors or bottlenecks in the processing of information (e.g. the design) or processing of materials (e.g. production). This is the key to lean product development.
* Lean product development requires integrated product and process development, IPPD. It does this by adopting a systems engineering process used to translate operational needs and customer requirements through concurrent consideration of all life cycle needs including development, manufacturing, support and disposal.
* Once a product design is complete then, typically, 70% of the product cost is committed. The manufacturing challenge for automotive designers and engineers is to fully exploit manufacturing opportunities, through the product's design, to maximize value and minimize cost.
* Implementation of IPPD is supported through the use of methods and techniques that provide manufacturing analysis and requirements analysis. They are typically team driven and cut across the functional chimneys of traditional organizations. Examples outlined include DFMA, DDC, QFD, SE, Modularity, etc. Examples have also been provided where manufacturing process developments provide opportunities to the product designer.
* Further methods and techniques are sought, by practitioners and researchers alike, to enable front loading effort in the development cycle and the subsequent optimal deployment

of engineering effort with traceability to requirements and target outcomes. Particular emphasis is being placed on early estimation of life cycle value/cost estimating and requirements capture, analysis and deployment through the use of systems engineering toolsets.

3.7 Acronyms

AEM	Assemblability evaluation method	IPPD	Integrated product and process development
AIV	Aluminium intensive vehicle	JIT/LP	Just-in-time/lean production
AVT	Aluminium vehicle technology	HOQ	House of quality
CAD	Computer aided design	MRP	Manufacturing resource planning
CAE	Computer aided engineering	OPT	Optimized production technique
DDC	Design for dimensional control	PDS	Product design specification
DFA	Design for assembly	QE	Quality engineering
DFM	Design for manufacture	QFD	Quality function deployment
DFMA	Design for manufacture and assembly	ROI	Return on investment
		RP	Rapid prototyping
D+T	Dimensioning and tolerancing	SE	Systems engineering
FMEA	Failure modes and effects analysis	SMED	Single minute exchange of die
		SPC	Statistical process control
FPDS	Ford product development system	TPS	Toyota production system
		VE	Value engineering
FPS	Ford production system	VSA	Variation system analysis
GT	Group technology	WIP	Work-in-progress

3.8 References and further reading

Alford, H. (1994). Cellular manufacturing: the development of the idea and its application. *New Technology, Work and Employment*, 9(1), 3–18.

Andersen, E.S., Grude, K.V. and Haug, T. (1995). *Goal directed project management. Second edition.* Pub. Kogan-Page, London. ISBN 0-7494-1389-1.

Barnett, L., Leaney P.G. and Matke, J. (1996) Striving for manufacturing flexibility in body construction. *Automotive Manufacturing*. 55–62. ISSN 1357-9193.

Barrett, R. (1997). Strip mill products. *MBM-Metal Bulletin Monthly*, Oct/97, 39–43.

BASA. (1993). Product assembly with adhesives and sealants: a brief guide. BASA and Industrial Technology Magazine.

Booth, R. (1995). In the market. *Manufacturing Engineer*, 74(5), 236–239.

Boothroyd, G. and Dewhurst, P. (1989). *Product design for assembly*. Wakefield, RI: Boothroyd Dewhurst, Inc.

Boothroyd, G., Dewhurst P. and Knight, W.A. (1994). *Product design for manufacture and assembly*. New York: Marcel Dekker, Inc. ISBN 0-8247-9176-2.

Bower, J.L., and Hout, T.M (1988). Fast cycle capability for competitive power. *Harvard Business Review*. Nov-Dec, 110–118.

Bralla, J.B. (ed.) (1986). *Handbook of Product Design for Manufacturing – a Practical Guide to Low Cost Production*. McGraw-Hill. ISBN 0-07-007130-6.

Broad, A. (1997). Metals driven down the road. *MBM – Metal Bulletin Monthly*, August 1997.

Burbidge, J.L. (1994). Group technology and cellular production. *Advances in Manufacturing Technology VIII. Proceedings of the 10th National Conference on Manufacturing Research*, 5–7 Sept. London: Taylor & Francis Ltd, pp. 140–148.

Chee-Kai, C. and Kah-Fai, L. (1998). Rapid prototyping and manufacturing: the essential link between design and manufacturing. In: J. Usher, U. Roy and H. Parsaei, eds. *Integrated Product and Process Development*. New York: John Wiley and Sons, Inc. pp. 151–183, ISBN 0-471-15597-7.

Christiansen, S. (1997). Automakers shape up with hydroforming. *MBM – Metal Bulletin Monthly*, September 1997.

Clark, K.B., and Fujimoto, T. (1991). *Product development performance*. Harvard Business School Press, ISBN 0-87584-243-3.

Clausing, D. (1994). *Total quality development*. New York: ASME Press.

Clegg, A.J. (1991). *Precision casting processes*. Oxford: Pergamon Press.

Cohen, L. (1995). *Quality function deployment: how to make QFD work for you*. Reading, Mass: Addison-Wesley Publishing.

Corbett, J., Dooner, M., Meleka J. and Pym, C. (eds.) (1991). *Design for manufacture – strategies, principles and techniques*. Addison-Wesley. ISBN 0-201-41694-8.

Corrêa , H.L. and Slack, N. (1996). Framework to analyse flexibility and unplanned change in manufacturing systems. *Computer Integrated Manufacturing Systems*, 9(1), 57–64.

Cross, N. (1994). *Engineering Design Methods*. 2nd edn Chichester: John Wiley & Sons.

DeGarmo, E.P., Black J.T. and Kohser, R.A. (1990). *Materials and Processes in Manufacturing*. New York: Macmillan Publishing Company.

DOD, (1996). Department of Defense Regulation 5000.2–R. Mandatory procedures of major defense acquistion programs (MDAPs). 15 March, 1996.

Ford, Henry (1922). *My Life and My Work*, London: William Heinemann,

Ford, Henry (1926). *Today and Tomorrow*, London: William Heinemann. (Reprinted in 1998 by Productivity Press.)

Fox, J. (1993). *Quality through design: the Key to Successful Product Delivery*. Maidenhead: McGraw-Hill.

Galbraith, J.R. (1974). Organization design – an information processing view. *Interfaces*, **4**, 28–36.

Gould, P. (1997). What is agility? *Manufacturing Engineer*, 76(1), 28–31.

Hammer, M. and Champy, J. (1993). *Reengineering the Corporation – A Manifesto for Business Revolution*. New York: Harper Business.

Hasen, J.R. (1989). Quality function deployment (QFD) at Ford – body and chassis engineering. Proc. Auto. Tech. 89 Conf. Session on quality techniques C399/39. I. Mech. Engrs.

Hayes, R.H., Wheelwright S.C. and Clark, K.B. (1988). *Dynamic Manufacturing – Creating the Learning Organization*. The Free Press – Macmillan. ISBN 0-02-914211-3.

Hitomi, K. (1996) *Manufacturing Systems Engineering – a Unified Approach to Manufacturing Technology, Production Management, and Industrial Economics*. Second edition. Taylor & Francis. ISBN 0-7484-0324-8.

Huang, G.Q. (ed.) (1996). *Design for X – Concurrent Engineering Imperatives*. Chapman & Hall. ISBN 0-412-78750-4.

JSF. (1997). *Joint strike fighter program*. (URL: http://www.jast.mil/jsfmain.htm)

Kalpakjian, S. (1991). *Manufacturing Processes for Engineering Materials*. New York: Addison–Wesley Publishing Co.

Leaney, P.G. and Wittenberg, G. (1992). Design for assembling: the evaluation methods of Hitachi, Boothroyd and Lucas. *Assembly Automation*, 12(2), 8–17.

Leaney, P.G., Abdullah, A.S., Harris J. and Sleath, D. (1993). A case study in the DFA evaluation methods of Hitachi, Lucas and Boothroyd–Dewhurst. *Proceedings of the 1993 International Forum on Design for Manufacture and Assembly*, Newport, Rhode Island, USA. pp. 1–21.

Leaney, P.G. (1995). Design integration – setting the scene, Proceedings of IEE/IMechE Colloquium; *'Design integration:- people co-operating across the professions'*, IEE HQ, Savoy Place , London, 29 June, 1995, IEE Digest No. 1995/155, pp. 1–6, ISSN 0936-3308.

Leaney, P.G., (1996a). Case experience with Hitachi, Lucas and Boothroyd-Dewhurst DFA methods. In: G.Q Huang, ed. *Design for X - Concurrent Engineering Imperatives*. London: Chapman & Hall. pp. 41–71.

Leaney, P.G. (1996b). Design for dimensional control. In: Huang, G.Q ed. *Design for X – Concurrent Engineering Imperatives*. London: Chapman & Hall.

Loureiro, G. (1998). *A systems engineering and concurrent engineering framework for the integrated development of complex products*. PhD Thesis, Loughborough University.

Marshall, R. (1998a). *Design modularisation: a systems engineering based methodology for enhanced product realisation*. PhD Thesis, Loughborough University.

Marshall, R. (1998b). *Holonic product design (HPD): a workbook*. Dept. of Manufacturing Engineering, Loughborough University: IPPS Group Report No. 98/8.

Matt, L., Geiler M. and Hiscott, M. (1995). Assembled camshaft technology into volume production. *Proceedings of the IMechE AutoTech/95 Conference, Birmingham, November 1995*, pp. C498/3/094.

McCombe, C. (1986). Cosworth castings – right first time and every time. *Metals Industry News*, 3(3), 6–7.

McCombe, C., (1990). Enter Cosworth. *Foundry Trade Journal International*, 13(2), 3.

Monden, Y. (1997). *Toyota Production System: an Integrated Approach to Just-in-time.*, Engineering and Management Press, ISBN 0-89806-180-6.

Nevins, J.L., Whitney D.E. *et al.*, (eds.) (1989). *Concurrent Design of Products and Processes: a Strategy for the Next Generation in Manufacturing*. New York: McGraw-Hill, Inc.

Nilsson, C-H. and Nordahl, H. (1995). Making manufacturing flexibility operational Part 2: Distinctions and an example. *Integrated Manufacturing Systems*, 6(2), 4–10.

Ohno, T. (1988). *Toyota production system: beyond large scale production*. Translation of 1978 Japanese edition 'Toyota seisan hoshiki', Pub. Productivity Press, ISBN 0-915299-14-3.

Owen, D. and Kruse, G. (1997). Follow the customer. *Manuf. Engineer*, 76(2), 65–68.

Pahl, G. and Beitz, W. (1996). *Engineering Design: A Systematic Approach.* – 2nd edn London: Springer Verlag, Ltd.

Percivall, G.G. (1992). Systems engineering in the automotive industry. Proceedings of the 2nd Annual Conference, NCOSE, pp. 510–518.

Prahalad, C.V. and Hamel, G. (1990). The core competence of the corporation. *Harvard Business Review*, May–June, pp. 79–91.

Prasad, B. (1996). *Concurrent Engineering Fundamentals: Integrated Product and Process Organization*. New Jersey: Prentice-Hall Inc.

Pugh, S. (1990). *Total Design – Integrated Methods for Successful Product Engineering*. Addison-Wesley. ISBN 0-201-41639-5.

QS 9000 (1995). Quality system requirements QS-9000. West Thurrock, UK: Carwin Continuous Ltd.

Robinson, A. (ed.), (1990). *Modern approaches to manufacturing improvement*. American Technical Publishers Ltd. ISBN 0-915299-64-X.

Rosenthal, S.R. and Tatikonda, M.V. (1992). Competitive advantage through design tools and practices. In: Susman, G.I. ed. *Integrating Design and Manufacturing for Competitive Advantage*. New York: Oxford University Press. pp. 15–35.

SAE, (1997). *Value based decisions for automotive engineering*. Pub. SAE, Ref SP–1266. ISBN 1-56091-978-7 (Includes SAE papers 970760 to 970771).

Shumaker, G.C. and Thomas, R.E. (1998). Integrated processes in defense manufacturing, In: Usher, J. Roy, U. and Parsaei, H. eds. *Integrated Product and Process Development*. New York: John Wiley & Sons, Inc., pp. 281–308, ISBN 0-471-15597-7.

Sleath, D. (1998). *Competitive Product Development.* PhD Thesis. Loughborough University.

Smith, D.S. 1995. Tomorrow's world – details of the Renault Modus. *The Renault Magazine.* 131, 14.

Smith, P.G. and Reinertsen, D.G. (1991). *Developing Products in Half the Time.* New York: Van Nostrand Reinhold, Inc.

Sterne, D. (1996). Rapid prototyping and manufacturing in the automotive sector. In: Loughborough University Advanced Automotive Engineering: Advanced Manufacturing Technology Module 4, pp. 1–13.

Sugimori, Y. *et al.* (1977). The Toyota production system and Kanban system – materialisation of just-in-time and respect for human system. Proc. 4th Intl. Conf. on Production Research, Tokyo, Taylor & Francis, London, 1–12.

Taguchi, G. (1993). *Taguchi on Robust Technology Development: Bringing Quality Engineering Upstream.* New York: ASME Press.

Taylor, Frederick. W. (1914). *The Principles of Scientific Management.* New York/London: Harper.

Tibbetts, K. (1995). *An introduction to TeamSET.* Birmingham: CSC Manufacturing, Computer Sciences Ltd.

Tincknell, D.J. and Radcliffe, D.F. (1996). A generic model of manufacturing flexibility based on system control hierarchies. *International Journal of Production Research,* 34(1), 19–32.

Weernink, W.O. (1989). A modular spider. *Automation and Design,* Nov, 65–69.

Whitney, D.E. (1992a). State of the art in Japanese computer aided design methodologies for mechanical products – industrial practice and university research. *Scientific Information Bulletin.* US Navy Office of Naval Research, NaAVSO P-3580, vol. 17, no. 1 (Jan–March).

Whitney, D.E. (1992b). *Systematic Design of Modular Products at Telemechanique.* (URL:http://web.mit.edu/ctpid/www/Whitney/europe.html).

Whitney, D.E. (1995). CAD and product development in the US automobile industry.

Womack, J.P. Jones D.T. and Roos, D. (1990). *The Machine that Changed the World.* Rawson Associates – Macmillan Pub., ISBN 0-89256-350-8.

Further reading

Bralla 1986 (see above list for reference).
    A comprehensive handbook of manufacturing processes and materials. Provides clear and concise guidance for the design of products for efficient and effective manufacture.
SAE 1997.
    A collection of papers on value based techniques for automotive engineering. Provides insight into value lessons learned and techniques for matching the voice of the customer with the voice of the producer.
Smith and Reinertsen, 1991.
    An examination of techniques to address the rapid introduction of new products. Taking a management perspective and drawing upon a wealth of experience, the authors provide a framework for an integrated approach to product development.
Usher, J., Roy, U. and Parsaei, H., eds. 1998. Integrated Product and Process Developmen. New York: John Wiley and Sons, Inc. pp 151–183, ISBN 0-471-15597-7.
    Taking the DOD definition and providing a valuable insight into the tools, techniques, and philosophy of integrated product and process development. Provides practical experience of IPPD application and suggests the way forward from concurrent and systems engineering through the IPPD concept.

4. Body design: The styling process

Neil Birtley, FCSD

The aim of this chapter is to:
- Review the role of the stylist and aerodynamist;
- Give an overview of the design stages from concept to final design;
- Demonstrate the need for engineers and stylist to work in tandem.

4.1 Introduction

Very little has been written elsewhere on this subject, despite the regular appearance of stylists' sketches, and clay models in the specialist motoring press. The work of the automobile stylist remains little understood by many, including other industrial designers. Some interesting insights into the way stylists think, and the nature of their business, is given by Armi, 1988 where he states that (car styling) '. . . is amongst the least understood of the commercial arts'.

The operational procedures are mainly evolved from those created by Harley Earl, who set up the first purpose-built styling department at General Motors in 1927. The method includes creating designs on paper as 2D sketches, then converting them to full size 2D orthogonal illustrations. These are used to create templates, etc., to create full-size three-dimensional models, usually in clay, though wood and plaster are sometimes used.

The full size mock-up is then used as the basis for all body surface information required by the engineering department for structural design and tooling to be developed. The generation of concept sketches, their conversion into full size 'tape drawings' and renderings and the creation of scale or full size clay models are all part of the stylist's job. Whilst many of their skills are developed by training and experience, much of the decision making and interpretation is very intuitive. It is a fact in the development of styling techniques that the execution of sketches, whether for exteriors or interiors, follows an almost indefinable set of unwritten rules, with quite wide parameters.

The creation of appropriate forms, whether alluring, sophisticated, brutal, functional, taut or soft, to give a vehicle a visual appeal which will help to sell it, is the essence of the job and the part of it that the majority of stylists enjoy most. The detailing is important too, but this is seen only on closer inspection, after the basic shape has caught the eye. Good detail cannot save a bad design, but bad detail can ruin a good one. There are plenty of examples of both on today's roads.

This is confirmed by McKim when he states that there is a '. . . fundamental relationship between idea sketching, and imagination', and goes on to say that 'creativity seems to come from the unconscious'. A car stylist's skill is certainly in this category.

Sketches are produced and selected, via a series of presentations and discussions, to create further development. This achieves the progression to full size elevational views and onto the full size model. The creation of the clay model is a long, tedious and very costly process, carried out with the aid of teams of expert sculptors. These are the 'modellers' working under the supervision and direction of the stylists during the design development stages. The model, when finished and also at various earlier stages, is carefully measured, either manually or electronically, to supply information to the body engineers. It is this activity that was the focus of much of the original interfacing of CAD with the styling studios.

The entire styling process is subjected to various decision-making interventions, within the studio and by more senior management. The system devised by Harley Earl is the usual role model, but there are many variants of it, including the totally ad hoc approach common in vehicle design consultancies and smaller studios.

4.2 The studios, working environment and structure

There are conventional basic facilities that allow the styling process to be performed. Strict security is always imposed on studios to protect commercial confidentiality. Inside the styling building, studio space for designing and modelling the exteriors and interiors of vehicles is provided, plus the necessary workshop services and administration to support the activity.

The most usual layout is to have the building divided into two main areas by a wide corridor, with studios to one side and workshops to the other, Figure 4.1 with a viewing area or showroom at one end of the building, and an outdoor viewing courtyard attached, usually surrounded by a high wall or fence and tree screen to keep prying eyes and cameras at bay.

Figure 4.1 Typical styling studio layout

At one Detroit Technical Centre each model division has its own studio with interior and exterior sections, and the situation is similar at other large manufacturers. Massive studio complexes like these are not found outside America. On average the European and Japanese Studios (and the individual line American Studios), employ about 15 or so stylists and about 20–30 clay modellers. The smaller design consultancies may have as few as 3 or 4 stylists with proportionally smaller teams of modellers, who may well double as fabrication shop workers at other times.

Most stylists prefer to be close to the models. Ovens to soften the hard waxy clay for use will be positioned where they are easily vented and convenient for the modellers (Figure 4.1). In the exterior studios one or more modelling tracks, with calibrated surface plates on the floor, will be installed. The model will be placed on locating pins, and a large metal arched structure (referred to as a 'bridge') will run on guides either on or below the floor on either side of it. This bridge will have a sliding calibrated cross beam with adjustable measuring bars which can be slid in and out and up and down on this beam and its vertical supports. This enables the model to be accurately calibrated at any point on its surface. A simpler system using a single vertical post sliding on a track, with adjustable measuring bar, is also quite common Figure 4.2.

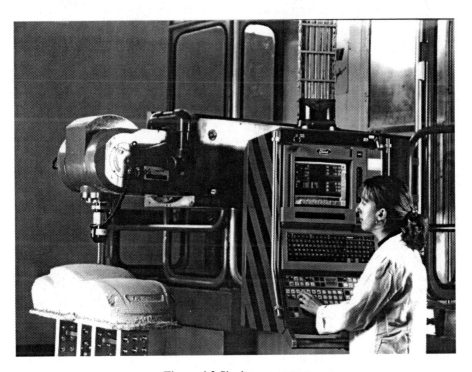

Figure 4.2 Single post system

Increasingly technology is being brought to bear in this area and various electronically controlled model recording, measuring and milling devices are being introduced (Figure 4.3) which are manually or electrically propelled on their tracks.

Figure 4.3 Electronic scan/mill system

Fabrication services, wood, metal and fibreglass workshops will support different aspects of the studio's work. This would include modelling armatures, ergonomic rigs, and fibreglass replicas. A paint booth for spraying models and other vehicles will usually be included. There is also usually a colour and trim section, who work alongside the other studios and devise the paint colour ranges, colours, patterns and weaves of trim fabrics. They may even have quite significant inputs into the whole interior design process, and work closely with textile, paint and vinyl manufacturers.

The chain of responsibility is usually quite short in studio management, where a Stylist will report directly to a Studio Head, who reports to the Senior Management or a Board. In some companies the Styling Department is under the control of Engineering, in others is a stand alone department responsible directly to Senior Company Management and working with Engineering or Product Planning.

The stylists receive information and guidance from engineering, and engineering ultimately receive information from styling on the shapes they must then enable to be manufactured. Differences may affect the slant on decisions on styling issues to a greater or lesser extent. There are no companies in which an engineering department works under the control of styling, but the styling voice can be very powerful, given the importance of its role in selling vehicles.

4.3 Product planning

The role of Product Planning is extremely important to the commercial success of the company. They work very closely with marketing and programme timing areas, who are frequently incorporated into the Product Planning Department. They have overall responsibility for analysing market research information, the performance of competitive companies, and their own sales together with the established product cycles and whether current tooling is life expired. Based on this information they formulate strategies for the replacement of existing models or the introduction of totally new ones.

Costing targets are established based on current manufacturing costs and the required profit objectives. Many competitors' vehicles are closely scrutinized, and even totally dismantled to establish their costs, construction methods and quality. Once the need to develop a new model or even facelift an existing one has been established, then the views of engineering and styling are obtained prior to setting out the product requirement 'brief' for the project.

The 'brief' is prepared in written form, consisting of a description of the vehicle required, model range, engines, options and variants, together with the cost and timing targets. More detailed specification lists, especially for the interiors may well be issued as the programme gets underway and changes in the original strategy may well be put forward.

Product Planning maintain an involvement for the duration of the project, being present at and part of all major decision making stages in styling and engineering. They also ensure that Marketing are preparing the groundwork upon which they will eventually present the vehicle to the dealers and the public.

4.4 Brainstorming

Once the product brief has been received by the styling department, copies are passed on to the studio manager who will call a meeting of all the stylists and the modelling supervisors to inform them of the requirements, dates, the lead times needed etc. If the studio is part of a design consultancy, the client's brief will be revealed at this point. The alternative avenues will be explored and theories and opinions exchanged. This activity is often referred to as 'brainstorming'; even the wildest ideas can be discussed, as nothing has even been suggested yet on paper. The use of this technique is of long proven value in many fields of design and elsewhere and is generally regarded as very necessary (Cross, 1994). At the end a few important 'bullet points' may be noted and kept as reference throughout the project.

4.5 The package

The 'package' is the industry's term for the three dimensional (3D) view, full sized orthogonal drawings which show the basic mechanical and ergonomic requirements of the intended vehicle. They are produced by an expert team of layout engineers and ergonomists, in consultation with product planning and styling. It is supplied to all styling and engineering areas involved. The original master drawing will be produced on translucent plastic 'vellum' with 100 mm grid lines on it called 'Ten Lines'. Typically the proposed location of the driver and passengers will be shown, in the agreed range of percentile mannequin outlines. Most probably from fifth percentile female to ninety-fifth percentile male are used for the range of sizes. This determines the necessary seat adjustments, sight lines and spatial requirements. Steering wheel, pedals, handbrake and gearshift positions will be shown, with ranges of movement where applicable. Luggage volume, fuel tank and critical clearances for engine and other components will be shown, as will the ideal position for the windscreen and side glass planes. Wheel location, suspension travel and all dimensional data will be included (Figure 4.4).

Figure 4.4 Package drawing

These drawings provide the basic dimensional controls or all other developments in all departments. The grid references becoming the basis of every subsequent engineering or workshop drawing produced. This enables any drawing to be directly compared or added to any other with accuracy. All departments must be informed of any change immediately, or the consequences could be colossal.

Once the studio has the package information they are able to produce their package-related sketches, and proceed with full sized tape drawing. In order to produce Typical Vehicle Package Drawing Information, the following will be included (all fully dimensioned):

(a)   Overall length, width and height
(b)   Wheelbase and sizes of wheels and tyres
(c)   Front and rear track (measured to the centre of the tread)
(d)   Front and rear overhang (measured from wheel centre)
(e)   Approach and departure angles
(f)   Engine and drive train outline and radiator location.
(g)   Fuel tank location
(h)   Windscreen location and angle
(i)   Legal parameters of all lights and signals
(j)   Minimum/maximum bumper heights and clearances
(k)   Suspension clearances
(l)   Fifth percentile female and ninety-fifth percentile male mannequins (all seats) with adjustment range of driver's seat. 'H' (hip) points should be shown
(m)   Eyelipse and vision angles (driver only) and obscuration zones
(n)   Driver reach zones and steering wheel location and adjustment
(o)   Heater box and Plenum
(p)   Headroom and Shoulder to bodywork and other passenger clearances
(q)   Eye to screen header rail dimension (does it clear belted head swing, including the air-bag?)
(r)   Pedal and gear lever positions.
(s)   Body opening lines for passenger entry and exit (window lines).
(t)   Luggage stack, load compartment requirements.
(u)   Basic body outline (may be preliminary guide on first package, prior to completion of design process).

4.6 Review of competition

Product planning will have provided a list of main competitors; the stylists may add to this. Package Engineering may well supply full package drawings of those vehicles considered to need close inspection plus lists of dimensions from their records.

Other display boards within this area contain photographs and pictures culled from magazines and brochures that the stylists find of relevance or of inspirational value. This helps to create the right 'atmosphere' in which to work. The images of the competition serve as a benchmark reference to what had already been done, and also may contain features perceived as desirable and worthy of inclusion in a new design. The company's own existing vehicle will be displayed alongside photographs of the competition, possibly also as actual vehicles in the studio as well as by photographs. These act as inspiration to the stylists.

4.7 Concept sketching and package related sketching

It is ideas that are being looked for, not refined designs. Up to 50 or so may be drawn in total. The sketches produced will be loose and fairly unconstrained. Accuracy at this early stage is not a prime requirement, a basic concept is. How elaborate the sketches are is very dependent on the individual stylist and the visual message that is being portrayed (Figure 4.5). The appreciation of the mechanical and ergonomic requirements is also a matter of intuition. The end result is almost always a slightly exaggerated or caricature image of the design in the stylist's mind. Each stylist will have their own preferences for media type. To many people stylists' sketches may seem incomplete; areas and lines may be left out if not needed to create the message.

Figure 4.5 Example of a concept sketch

McKim explains that when sketching the designer will sketch only what is of interest and will leave out that which is already understood or deemed irrelevant. The business of getting mental images down on paper is referred to as 'ideation' and quite a great deal has been written about this (McKim and Tovey). It involves a cycle of seeing, imagining and drawing.

In the opinion of many stylists their early mental images of vehicles and forms are very fleeting, they may 'see' a very real image, but it will fluctuate and change its form and detail as a myriad of different thoughts flood in (Armi, 1988). At the concept sketching phase the stylist will be making guesses at an eventual wished-for outcome in a way that disguises the high degree of analytical activity going on, almost subconsciously, that can only be based on experience and a wide knowledge of the subject. Only by interpreting the messages from the eye does the stylist know that the image is the one they intended. A few preferred themes will be agreed on for further development, or a particular feature will be highlighted and incorporated into future sketches (Figure 4.6).

The process may then be repeated, perhaps several times, until the team are happy they have a good selection of ideas to present to management. For this the concepts which will by now be

Figure 4.6 Informal selection of concepts

to a more consistent standard and on the same sized paper, usually A3 or A2, which are arranged neatly on display boards. The sketches shown at this stage may be still fairly unfettered by dimensional constraints or, have paid much more attention to the package information, perhaps using a photo reduction of a computer-produced perspective view or photographs of a similarly dimensioned vehicle as a sketch pad underlay. The level accuracy of these sketches will be quite good, but artistic license will still creep in.

Accurate sketches are required before the process can proceed to the next stage, the tape drawing. Immediately usable designs are needed, and the whole sketch programme will be based on tightly constrained underlays and will not go through the very loose phase. Where 'facelifts' of existing designs are concerned, the sketching will only be done on this basis, usually with a large photograph as an underlay, or by retouching a photograph.

The advantage is that creativity can run wild for a while, there is little imposed or self criticism for even quite significant inaccuracies of proportion, and size. The disadvantages are that only a fully attuned eye can properly interpret the actual intent and several more stages must be gone through before engineers or modellers can make use of the information.

Where sketching is controlled by package dimensions from the outset, by various means, the results are almost certainly much more immediately useful to others. The job of turning these illustrations into full size tapes requires less interpretation and judgement. It could be given to another individual to carry out, not the original artist.

If an accurate photo-like rendering is required, the illustrator must take great care to analyse the proposed surfaces and theoretical lighting sources very carefully. This will take a great deal of time, a week or more for the package drawing to be converted to a perspective that will be rigidly adhered to along with orthographic views.

This type of illustration is often required in consultancies, where clients wish to see a very accurate image, without the need or expense of a model. Such drawings can be readily digitized straight into an engineering CAD system and the first 3D real model to be seen could come from a multi-axis mill. Much more time is spent involved in the development of the model, and details like bumpers, lights, handles, wipers, add graphics. There is liaison with the engineers on feasibility issues, and an endless stream of other much more mundane tasks, all of which are important in their own way and require the expert eye and judgement of a stylist.

4.8 Full sized tape drawing

The sketch programme provided the basic aesthetic and proportional 'theme' that the studio and management agreed to proceed with. Firstly, a full sized package blueprint side elevation is thoroughly sketched, free from wrinkles and pinned to a display board, at a comfortable working height, where roof and door sills can both be easily reached. Important features and clearance points may be emphasized with a coloured pencil or some thin chart tape (typically the air cleaner, radiator and passenger head clearances) as the height of the bodywork over these areas can be of great aesthetic significance. Next a sheet of clear or translucent plastic film is tightly placed or stapled over the package print. The materials used are very simple, rolls of black crepe photographic adhesive tape ranging from 1 mm to 150 mm.

The sketch or sketches of the chosen design are placed on a display board. A length of tape will be placed along the ground line as shown on the package. This will give a visual base to the rest of the drawing. Using the stylist's usual mixture of learned skills and intuition and bearing in mind the clearance points, the lines of the sketch are translated into tape lines over the package (Figure 4.7).

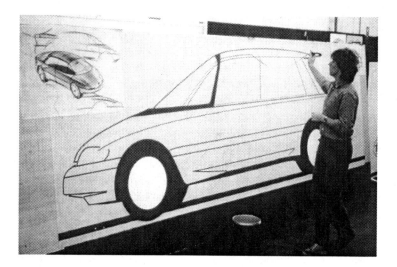

Figure 4.7 Translation of sketch into tape drawings

The technique is to lay another piece of tape next to the one wished to be moved so that its position is recorded until the original or a new piece is relaid. Only lines are visible at this stage there is no representation of form or section. Shadows and mock perspective via blacking out a portion of the side window area give an impression of solidity and being a true side elevation. Strips of white tape or cut out paper may be added to represent highlights, and wider black tape or dark paper to represent undercuts and shadows. Skilful application of thick and thin tapes for door shut lines can also enhance the impression of section or form.

Tyres may be blocked in using wide tape, paper, or card and could be illustrations, or photographs of actual wheels. Large taped up paper cutouts may sometimes be overlaid on the finished 'tape', to look at alternative treatments, perhaps for a roof pillar, screen angle, or front end silhouette. etc. These are simply referred to as 'overlays'. The tape drawing will be photographed with and without them as a record.

After management clearance the tapes may be enhanced by shading. This can be achieved by many methods; such as air brushes, or marker pens. Only well trained and expert stylists can do this. The end results, especially when photographed, can look very convincing but in reality there is no 3D information there which could be transferred. What exists is still a 2D representation, accompanied by sectional information. An impression of the size, profiles, proportions and graphic effect of glass areas has been gained together with some idea of how the surfaces might be.

The final tape review will be to a fairly large group of management, involving product planners, engineers, styling, costing and timing, and a verbal presentation will most probably be made by the studio manager.

4.9 Clay modelling

Clay modelling of interiors and exteriors is the most evident part of the conventional car styling activity whenever a studio is entered. The most time consuming part of the studio operation is the physical creation of the models. The clay that is used is almost exclusively confined to the car styling business, and was first used by Harley Earl.

Various formulations have been tried, but they are still wax based. The clay consists of sulphur, red-earth, various other filler powders, lanolin oil, antiseptic and buffering agents to protect the skin of the users, and is a reddish mid-brown in colour, though newer grey varieties are becoming more popular.

When warm it has the consistency of soft 'plasticene', but soon cools and hardens to a consistency where it can be carved with a variety of sculpting tools, many specifically designed and made for the job. It has the advantage over water-softened sculptors clay in that it does not dry out, crack or shrink. It can also be re-used several times. Before modelling can begin, an armature or substructure to support the clay must be built. The normal format is to have a metal frame or chassis to carry wheels on adjustable height axle supports to reach the floor and take the weight. The model will weigh about 3 tonnes when finished.

On this base there will be a boxed wooden structure, possibly covered with a rigid plastic foam, to within about 150 mm of the expected surface of the final model. If it is too close it will be cut off later (Figure 4.8).

The modellers liaise directly with the stylist. They will generally work upright, moving around the model, but where the work is on the lower part of the model, they sit on small box like stools with castors. Where door sills, or underpans are concerned they will frequently have to lie on the floor. The modellers are coordinated as a team and try to keep the model site as

Figure 4.8 Claying up armature to hardpoints

clear as possible to keep the end product constantly in view. They will begin to cover the armature with clay just about the time when the stylists are finalizing the tape drawings.

Wooden pegs will be driven in at all of the critical clearance points to ensure that these points are not infringed. The screen and side glass planes will also be put in according to the package information. Alternatively, an electronic mill can convert the tape drawing coordinates directly into a 3D shape via the 'ten lines'. These coordinates can also be used by the engineering team.

A block model with a longitudinal centre section cut into it is produced, and a few cross sections which correspond to the tape drawing, and the packaged glass planes and screen and other major features. The surfaces in between will be sculpted visually and with the aid of long plastic 'splines', usually 2–3 m long strips of perspex about 10 mm square section, to true up in between. The flexibility of the clay allows changes to continue until the stylists are finally satisfied. A design is usually only put onto one side of the model up to the centreline, and only 'balanced' over onto the other side when the stylists are satisfied.

Occasionally the stylists will use photographic tapes to highlight the clay model's feature lines. These will be used as a basis for experimental changes to gauge the opinions of the design team. Where a feature line or crease is to be preserved, a knife cut, guided by splines or a tape, will be made, and a strand of nylon fishing line inserted to create a harder edge that will not accidentally be erased.

The engineers will by now have been given a great deal of tentative surface information and will be discussing feasibility issues with the studio. Where changes become necessary they will be implemented on the model, the modellers again using the 'ten line' grid coordinates to translate the change from drawing to model. This process can to-and-fro for some time until the model is a long way from the initial design. Alternatively, where time is the essence, stylists can start designing in clay, with only the package and a few very quick sketches. The stylist will know the shape and surface when it finally arrives.

Where constant sections, such as on bumpers, or some body sides are concerned, a solid template, made to a drawing or tape of the required section will be made. This can be used to create a constant shape by 'dragging' the template over the model. An extrusion can be used for small constant sections, which can be stuck in place with the aid of white spirit or shellac.

Where large gently contoured surfaces, like roofs are concerned, 'sweeps' will be used to true up the surfaces – these are usually made of steel or aluminium, about 2 mm thick, 60 mm wide and 2000 mm long. A surface made from 'sweeps' can be quoted as such to engineering for their design package.

Where interior models are concerned the techniques are much the same, except that templates are smaller, and there will be much more detail and very few large simple surfaces. Detailing on models is carried out once all the major surfaces are resolved, the same routine of sketch, orthographic drawing and templates being used, or simply sketches.

When all the surfaces are as they should be, but left deliberately very slightly proud (may be 1/2 mm) it may be given a lightly 'combed' finish with a finely serrated surfacing 'slick'. This gives it a matt finish so that forms can be checked out visually, without the interference of any reflections, on shiny patches of clay. The surface is then 'slicked up' using flexible plastic or steel strips to hone it to a glass-smooth finish. Small adjustments to the surface using the slicks will correct all the highlights; this is a very skilled operation.

This is a time when quite major changes can occur, tapes will be positioned and repositioned, lines will be scribed, the modellers may be asked to make 'quickie' changes to the clay, that if kept will be properly modelled once the model is returned to the studio. It is not advisable to move a model too frequently, as cracks and distortions may occur.

Once the model is finished, the body surfaces will be covered with Dinoc (an adhesive backed paint film) to simulate paint, Figure 4.9. Any bright metal areas covered with carefully applied metal foil, and the window areas with black or graphite grey Dinoc. Sometimes a 'see-through' is made, with the roof supported on the pillars and some metal tubes, and with a perspex screen and side glass to create a more realistic impression. This is more costly and difficult to achieve, and it is difficult to change when problems occur, but enables direct comparisons to be made with alternatives and competitor vehicles.

Usually all the vehicles on show will be Dinoc'd (Figures 4.9 and 4.10) or painted the same colour, to avoid any adverse effect a different colour may have on design judgements. Metallic

Figure 4.9 Clay model being Dinoc'd

Figure 4.10 Presentation of full size clay model

silver is a favourite in many companies, as it shows form very well and 'flips', which means where surfaces change angle towards the light it contrasts very well. The whole process of modification, review by the stylists, re-Dinoc and presentation to management may take place several times at various intervals before the final approval is given.

These design reviews are performed at critical stages of the design process. They involve styling, engineering, marketing and management, which means that the models are presented with a certain amount of theatre. The model will be fully 'pointed off', this means that all of the dimensional and surface coordinates will be checked at each ten-line grid intersection, and more frequently in complex areas. It will then be passed to engineering who will then generate the final body surface line drawing.

A 'scan/mill' type of digitally controlled modelling 'bridge' is used to track over the surfaces and all of the coordinates will be stored and digitized into the engineering CAD systems, from which the body surface line drawings will be produced.

The body surface information will go to all body and tooling design areas where flanging, crowning, piercing, welding and all structural issues will be resolved. In some companies even the seats are handled in this way.

Some companies have a practice of making a 'prove-out' clay model, this means that a totally new clay model of the approved design is made in the studio from the body surface information once all the requisite engineering activity is completed. This is done to check that no visual problems have arisen due to digitization. The next opportunity to check if there are any body surface problems creeping into the manufacturing process is when the wood models of the die-making aids are assembled together as a complete body model, known as the 'cube', for final inspection. Stylists may well be involved in the final sign-off of 'the cube'. Once the model is fully measured, pointed or scanned a fibreglass mould may be taken from it, from which a fibreglass replica will be made. This is usually just an exterior shell mounted on a simple chassis with a set of wheels. It will be kept as a permanent physical record of the model.

At the same time, the work of the stylists and modellers on the project will be to assess the possibilities of model variants, such as estate cars, convertibles, coupés, or vans. All of these must be styled and modelled in much the same way as the original project, though the starting point is established.

Once the primary design model of the interior (usually the top of the range variant) is completed, all the series variants down to the 'fleet' model must be tackled and modelled. Very careful judgement and an ability to pull things together in the mind's eye is very necessary for continuity.

The development of the interior will take much longer than the exterior because of its complexity. Small clay models are frequently viewed, supported by detailed illustrations, showing their relationship to the rest of the interior. A common feature of interior clay models is the use of stylist's illustrations to represent instrument clusters, radios, door trim panels, etc. The final interior model is often a composite of many materials, such as clay, fibreglass mouldings, plaster casts, etc. Card and 'Foamcore' (card and styrofoam sandwich) models built up as sections, much like a balsa wood model aircraft, are built Figure 4.11.

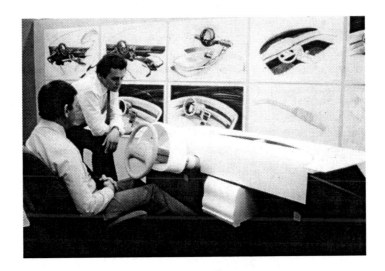

Figure 4.11 Example of Foamcore or card model

The great advantage of clay is its sheer flexibility as a medium. It is 'plastic' and can be modelled into any shape whatsoever; it can be hand sculpted, and honed to a metal-like or glass-like surface. It can also be textured to simulate vinyl or fabric using patterned rollers. If it is damaged it is easy to repair, and is also dimensionally very stable, as long as it has a sound armature. It can be Dinoc'd or painted, with either gloss enamels or water based paint. Fabrics and vinyls can be pinned or glued to it and finally it can be re-used.

It is seen as being advantageous by some managements to get to grips with 3D at a very early stage, using 1/4, 1/5 or sometimes 3/8 scale models, despite the scaling effects on the visual interpretations of the form.

It can be seen that the styling of a vehicle is a central part of the overall design of that vehicle. It affects all the engineering teams; from aerodynamics due to the shape, to suspension due to the confines of the space and the positioning of the windows and wheels. This means that effective vehicle design can only be achieved by efficient communication between all the teams involved. It also has the effect of controlling the duration of the design process as the clay model is the hub of all other processes.

4.10 2D systems

As with the 'conventional' process the concept design phase of a project will be less constrained and creativity will be to the forefront. Currently a variety of approaches are used even with a single organization, to constrain designers too much would most likely prove counter productive.

Some designers still prefer to use paper based sketches for the early 'ideation' phase, others will produce small 'thumb nail' sketches as a guide to putting their ideas onto 2D 'paintbox' systems via a digitizer pad.

Increasingly designers, particularly those who have gained some exposure to these systems during their early training, favour creating visuals directly on the 2D screen. Either way the end results will have no clue as to their method of creation.

The great advantage of using electronic sketch systems is not only the abandonment of paper pens, vast arrays of pencils, spirit markers and overflowing waste bins, but the ease with which images can be stored, printed, projected full size or transmitted to any location in the world. Either as they happen or at any subsequent time. This gives management great flexibility in their decision making process and allows much more interaction between design teams based in different countries. It must be remembered that the visual skills the designer needs to use this process are no different to those needed for conventional paper sketching. However considerable adaptation is needed to get to grips with the feel of the digitizer pad and electronic stylus, which, whilst used like a pencil, can represent any innovative tool called up. There is also the 'problem' that the pad is not in the same place as the screen where the images appear as they are drawn, unlike the paper pad we all understand. The majority of designers, however, seem to rapidly acclimatize.

4.11 3D systems

The next stage of the process is to transfer selected designs into the 3D modeller/surfacer systems. In general this means using the output of the 2D system as a guide to inputting, in exactly the same way as using paper based sketches. Systems do not exist which can directly take on board 2D non-dimensioned/non-orthographic (i.e. sketches or renderings) graphics and translate them into 3D models. They must essentially be re-created. These 3D wire frame, surface or solid modular systems are quite varied in their capabilities and will be chosen by companies to suit their needs and budgets. Most incorporate colour rendering systems which can deliver anything from simple surface shading to full paint, reflections, transparent glass, fabric textures and even surface dust or raindrops if required.

There are designers who have become quite adept at manipulating 3D systems as the initial stage of the design, skipping totally any early 2D, paper or computer based work. The disadvantage is that others must wait until the model is more complete to view the idea.

Another way, commonly used to input information into 3D systems is to scan or digitize a scale model, derived from earlier 2D work, as described in the conventional process.

Once data is resident in any 3D system it is then available to all who are on line to need it in the other areas of design and engineering. The fully simultaneous and interactive design process is then up and running, with models and later on tools being milled from the system whenever required.

4.12 References and further reading

Armi, C. Edson (1988). *The Art of American Car Design*. Penn. State Univ. Press.
Cross, N. (1994). *Engineering Design Methods, Strategies for Product Design*. J. Wiley & Sons.
McKim, R.H., Experiences in Visual Thinking.
Tovey, M., Vehicle Stylists' Creative Thinking.

Further reading

1. *Harley Earl and the Dream Machine*, Stephen Bailey.
   A quite well illustrated account of the career of Harley Earl, who became the first 'Chief Stylist' at General Motors and arguably the world's first 'Chief Stylist'. From being a jobbing 'customizer' in Hollywood he was invited to Detroit as a trial to help GM sales and stayed. From masking special bodies for film stars to mass produced cars in one step! Interestingly, he could not draw, but told others what he wanted or did it himself in 3D.
2. *The Car Programme*, Stephen Bailey, Conran, Victoria & Albert Museum, London.
   A quite informative booklet, produced by the V. & A., sponsored by Conran to accompany an exhibition on the design and launch of the original Ford Sierra. Various illustrations, studio photographs and a mock-up of a corner of a studio with a clay model were set up in the old museum 'Boiler House.' The first of the 'Boiler House Exhibitions at the V & A.' Not widely available at the time, but a good general account of the planning and styling to sign-off. Accurately reported much of Ford's input.
3. *Let's Call it Fiesta*, Edvard Seidler, Lausanne.
   A sparsely illustrated account of the market research, planning and styling of the first Ford Fiesta. A quite revolutionary car for Ford of Europe. Their first transverse front wheel drive car. A well written and very accurate account including much Ford information. The car sold beyond the company's wildest dreams, due to looks, performance and price.
4. *Chrome Dreams, Styling since 1895*. Paul C. Wilson.
   Mostly a picture book looking at cars with many American models ones from a visual viewpoint. There is no background or much technical information. The pictures are not always accurately described.

5. Body design: Aerodynamics

Robert Dominy, PhD, BSc(Hons), CEng, MIMechE

The aim of this chapter is to:
- Review the role of the stylist and aerodynamist;
- Review the basic aerodynamic concepts related to vehicles;
- Indicate the basic computations required aerodynamic design.

5.1 Introduction

Throughout the history of the motor car there have been individual vehicles that have demonstrated strong aerodynamic influence upon their design. Until recently their flowing lines were primarily a statement of style and fashion with little regard for the economic benefits. It was only rising fuel prices, triggered by the fuel crisis of the early 1970s, that provided a serious drive towards fuel-efficient aerodynamic design. The three primary influences upon fuel efficiency are the mass of the vehicle, the efficiency of the engine and the aerodynamic drag. Only the aerodynamic design will be considered in this section but it is important to recognize the interactions between all three since it is their combined actions and interactions that influence the dynamic stability and hence the safety of the vehicle.

5.2 Aerodynamic forces

Aerodynamic research initially focused upon drag reduction, but it soon became apparent that the lift and side forces were also of great significance in terms of vehicle stability. An unfortunate side effect of some of the low drag shapes developed during the early 1980s was reduced stability especially when driven in cross-wind conditions. Cross-wind effects are now routinely considered by designers but our understanding of the highly complex and often unsteady flows that are associated with the airflow over passenger cars remains sketchy. Experimental techniques and computational flow prediction methods still require substantial development if a sufficient understanding of the flow physics is to be achieved.

The aerodynamic forces and moments that act upon a vehicle are shown in coefficient form in Figure 5.1. The force and moment coefficients are defined respectively as

$$C_f = \frac{F}{\frac{1}{2}\rho v^2 A} \qquad C_m = \frac{M}{\frac{1}{2}\rho v^2 A l}$$

where F is force (lift, drag or side), M is a moment, ρ is air density, v is velocity, A is reference area and l is a reference length. Since the aerodynamic forces acting on a vehicle at any given speed are proportional to both the appropriate coefficient and to the reference area (usually frontal area) the product $C_f A$ is commonly used as the measure of aerodynamic performance, particularly for drag.

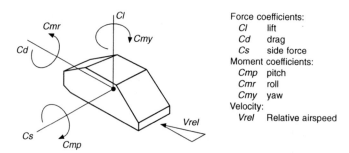

Force coefficients:
Cl   lift
Cd   drag
Cs   side force
Moment coefficients:
Cmp   pitch
Cmr   roll
Cmy   yaw
Velocity:
Vrel   Relative airspeed

Figure 5.1 Lift, drag, side force and moment axes

The forces may be considered to act along three, mutually perpendicular axes. Those forces are the drag, which is a measure of the aerodynamic force that resists the forward motion of the car, the lift which may act upwards or downwards; and the side force which only occurs in the event of a cross-wind or when the vehicle is in close proximity to another. The lift, drag and pitching moments are a measure of the tendency of those three forces to cause the car to rotate about some datum, usually the centre of gravity. The moment effect is most easily observed in cross-wind conditions when the effective aerodynamic side force acts forward of the centre of gravity, resulting in the vehicle tending to steer away from the wind. In extreme, gusting conditions the steering correction made by the driver can lead to a loss of control. Cross-wind effects will be considered further in section 5.5.

5.3 Drag

The drag force is most easily understood if it is broken down into five constituent elements. The most significant of the five in relation to road vehicles is the *form drag* or *pressure drag* which is the component that is most closely identified with the external shape of the vehicle. As a vehicle moves forward the motion of the air around it gives rise to pressures that vary over the entire body surface as shown in Figure 5.2a. If a small element of the surface area is considered

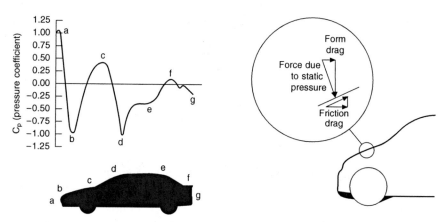

Figure 5.2 (a) Typical static pressure coefficient distribution; **(b)** The force acting on a surface element

then the force component acting along the axis of the car, the drag force, depends upon the magnitude of the pressure, the area of the element upon which it acts and the inclination of that surface element Figure (5.2b). Thus it is possible for two different designs, each having a similar frontal area, to have very different values of form drag.

As air flows across the surface of the car frictional forces are generated giving rise to the second drag component which is usually referred to as *surface drag* or *skin friction drag*. If the viscosity of air is considered to be almost constant the frictional forces at any point on the body surface depend upon the shear stresses generated in the boundary layer. The boundary layer is that layer of fluid close to the surface in which the air velocity changes from zero at the surface (relative to the vehicle) to its local maximum some distance from the surface. That maximum itself changes over the vehicle surface and it is directly related to the local pressure. Both the local velocity and the thickness and character of the boundary layer depend largely upon the size, shape and velocity of the vehicle.

A consequence of the constraints imposed by realistic passenger space and mechanical design requirements is the creation of a profile which in most situations is found to generate a force with a vertical component. That lift, whether positive (upwards) or negative induces changes in the character of the flow which themselves create an *induced drag* force.

Practical requirements are also largely responsible for the creation of another drag source which is commonly referred to as *excrescence drag*. This is a consequence of all those components that disturb the otherwise smooth surface of the vehicle and which generate energy absorbing eddies and turbulence. Obvious contributors include the wheels and wheel arches, wing mirrors, door handles, rain gutters and windscreen wiper blades but hidden features such as the exhaust system are also major drag sources.

Although some of these features individually create only small drag forces their summative effect can be to increase the overall drag by as much as 50%. Interactions between the main flow and the flows about external devices such as door mirrors can further add to the drag. This source is usually called *interference drag*.

The last of the major influences upon vehicle drag is that arising from the cooling of the engine, the cooling of other mechanical components such as the brakes and from cabin ventilation flows. Together these *internal drag* sources may typically contribute in excess of 10% of the overall drag (e.g. Emmelmann, 1982).

5.4 Drag reduction

Under the heading of drag reduction the designer is concerned not only with the magnitude of the force itself but also with a number of important and directly related topics. Firstly there are the effects of wind noise. Aerodynamic noise is closely associated with drag creation mechanisms which often exhibit discrete frequencies and which tend to arise where the air flow separates from the vehicle surface. Flow separation is most likely to occur around sharp corners such as those at the rear face of each wing mirror and around the 'A' pillar of a typical passenger car. Because of the close relationship between drag and noise generation it is not surprising that drag reduction programmes have a direct and generally beneficial effect upon wind noise. Such mutual benefits are not true of the second related concern, that of dynamic stability. The rounded shapes that have come to characterize modern, low drag designs are particularly sensitive to cross-winds both in terms of the side forces that are generated and the yawing moments. Stability concerns also relate to the lift forces and the changes in those forces that may arise under typical atmospheric wind conditions.

The broad requirements for low drag design have been long understood. Recent trends in vehicle design reflect the gradual and detailed refinements that have become possible both as a result of increased technical understanding and of the improved manufacturing methods that have enabled more complex shapes to be produced at an acceptable cost. The centre-line pressure distribution arising from the airflow over a typical three-box (saloon) vehicle has been shown in Figure 5.2a. A major drag source occurs at the very front of the car where the maximum pressure is recorded (Figure 5.2a, point 'a') and this provides the largest single contribution to the form drag. This high pressure, low velocity flow rapidly accelerates around the front, upper corner (b) before slowing again with equal rapidity. The slowing air may not have sufficient momentum to carry it along the body surface against the combined resistance of the pressure gradient and the viscous frictional forces resulting in separation from the body surface and the creation of a zone of re-circulating flow which is itself associated with energy loss and hence drag. The lowering and rounding of the sharp, front corner together with the reduction or elimination of the flat, forward facing surface at the very front of the car addresses both of these drag sources (Hucho, 1998). A second separation zone is observed at the base of the windscreen and here a practical solution to the problem is more difficult to achieve. The crucial influence upon this drag source is the screen rake. Research has clearly demonstrated the benefits of shallow screens but the raked angles desired for aerodynamic efficiency lead to problems not only of reduced cabin space and driver headroom but also to problems of internal, optical reflections from the screen and poor light transmission. Such problems can largely be overcome by the use of sophisticated optical coatings similar to those widely used on camera lenses but as yet there has been little use of such remedies by manufacturers. Figure 5.3 demonstrates the benefits that may be achieved by changing the bonnet slope and the screen rake (based on the data of Carr (1968)).

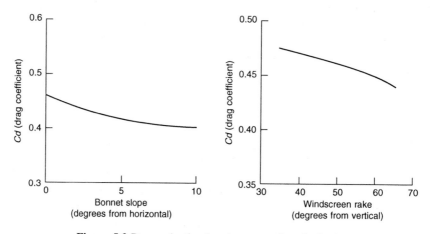

Figure 5.3 Drag reduction by changes to front body shape

There is further potential for flow separation at the screen/roof junction which similarly benefits from screen rake and increased corner radius to reduce the magnitude of the suction peak and the pressure gradients.

The airflow over the rear surfaces of the vehicle is more complex and the solutions required to minimize drag for practical shapes are less intuitive. In particular the essentially two-dimensional considerations that have been used to describe the air flow characteristics over the front of the vehicle are inadequate to describe the rear flows. Figure 5.4 demonstrates two alternative flow structures that may occur at the rear of the vehicle. The first Figure 5.4a occurs for 'squareback' shapes and is characterized by a large, low pressure wake. Here the airflow is unable to follow the body surface around the sharp, rear corners. The drag that is associated with such flows depends upon the cross-sectional area at the tail, the pressure acting upon the body surface and, to a lesser extent, upon energy that is absorbed by the creation of eddies. Both the magnitude of the pressure and the energy and frequency associated with the eddy creation are governed largely by the speed of the vehicle and the height and width of the tail. A very different flow structure arises if the rear surface slopes more gently as is the case for hatchback, fastback and most notchback shapes (Figure 5.4b). The centreline pressure distribution shown in Figure 5.2a shows that the surface air pressure over the rear of the car is significantly lower than that of the surroundings. Along the sides of the car the body curvature is much less and the pressures recorded here differ little from the ambient conditions. The low pressure over the upper surface draws the relatively higher pressure air along the sides of the car upwards and leads to the creation of intense, conical vortices at the 'C' pillars. These vortices increase the likelihood of the upper surface flow remaining attached to the surface even at backlight angles of over 30 degrees. Air is thus drawn down over the rear of the car resulting in a reacting force that has components in both the lift and the drag directions. The backlight angle has been shown to be absolutely critical for vehicles of this type (Ahmed *et al.*, 1984). Figure 5.5 demonstrates the change in the drag coefficient of a typical vehicle with changing backlight angle. As the angle increases from zero (typical squareback) towards 15 degrees there is initially a slight drag reduction as the effective base area is reduced. Further increase in backlight angle reverses this trend as the drag inducing influence of the upper surface pressures and trailing vortex creation increase. As 30° is approached the drag is observed to increase particularly rapidly as these effects become stronger until at approximately 30° the drag dramatically drops to a much lower value. This sudden drop corresponds to the backlight angle at which the upper surface flow is no longer able to remain attached around the increasingly sharp top, rear corner and the flow reverts to a structure more akin to that of the initial squareback. In the light of the reasonably good aerodynamic performance of the squareback shape it is not surprising that many recent, small hatchback designs have adopted the square profiles that maximize interior space with little aerodynamic penalty.

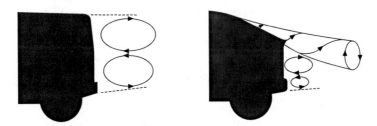

Figure 5.4(a) 'Squareback' large scale flow separation. **(b)** 'Hatchback/Fastback' vortex generation

The more traditional notchback or saloon form, not surprisingly, is influenced by all of the flow phenomena that have been discussed for the forms discussed above. As the overcar flow passes down the rear screen the conditions are similar to those of the hatchback and trailing; conical vortices may be created at or near the 'C' pillar. The inclination of the screen may be sufficient to cause the flow to separate from the rear window although in many cases the separation is followed by flow re-attachment along the boot lid. Research has shown that in this situation the critical angle is not that of the screen alone but the angle made between the rear corner of the roof and the tip of the boot (Nouzawa *et al.*, 1992). This suggests that the effect of the separation is to re-profile the rear surface to something approximating to a hatchback shape and consequently the variation in drag with this effective angle mimics that of a continuous, solid surfaced 'hatch'. It follows that to achieve the minimum drag condition that has been identified to correspond to a backlight angle of 15° (Figure 5.5) it is necessary to raise the boot lid, and this has been a very clear trend in the design of medium and large saloon cars (Figure 5.6). This has further benefits in terms of luggage space although rearward visibility is generally reduced. Rear end, boot-lid spoilers have a similar effect without the associated practical benefits. The base models produced by most manufacturers are usually designed to provide the best overall aerodynamic performance within the constraints imposed by other design considerations and the spoilers that feature on more upmarket models rarely provide further aerodynamic benefit.

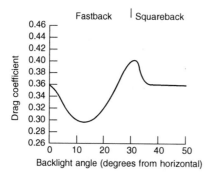

Figure 5.5 The influence of backlight angle on drag coefficient

Figure 5.6 High tail, low drag design

Attention must also be paid to the sides of the car. One of the most effective drag reduction techniques is the adoption of boat-tailing which reduces the effective cross-sectional area at the rear of the car and hence reduces the volume enclosed within the wake (Figure 5.7). In its most extreme configuration this results in the tail extending to a fine point, thus eliminating any wake flow, although the surface friction drag increases and the pressures over the extended surfaces may also contribute to the overall drag. Practical considerations prevent the adoption of such designs but it has long been known that the truncation of these tail forms results in little loss of aerodynamic efficiency (Hucho *et al.*, 1976).

Despite the efforts that have been made to smooth visible surfaces it is only recently that serious attempts have been made to smooth the underbody. The problems associated with

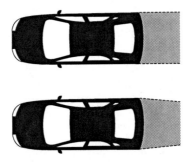

Figure 5.7 Boat tailing: reduced wake

underbody smoothing are considerable and numerous factors such as access for maintenance, clearance for suspension and wheel movement and the provision of air supplies for the cooling of the engine, brakes and exhaust must be given considerable weight in the design process. Just as the airflow at the extreme front and rear of the car were seen to be critical in relation to the overcar flow, so it is necessary to give comparable consideration to the air flow as it passes under the nose of the vehicle and as it leaves at the rear. It comes as a surprise to many to learn that the sometimes large air dams that are fitted to most production vehicles can actually reduce the overall drag forces acting on the car despite the apparent bluntness that they create. The air dam performs two useful functions. The first is to reduce the lift force acting on the front axle by reducing the pressure beneath the front of the car. This is achieved by restricting the flow beneath the nose which accelerates with a corresponding drop in pressure. For passenger cars a neutral or very slight negative lift is desirable to maintain stability without an excessive increase in the steering forces required at high speed. For high performance road cars it may be preferred to create significant aerodynamic downforce to increase the adhesion of the tyres. The side effects of aerodynamic downforce generation such as increased drag and extreme steering sensitivity are generally undesirable in a family car. Lowering the stagnation point by the use of air dams has also been shown in many cases to reduce the overall drag despite the generation of an additional pressure drag component.

The shaping of the floorpan at the rear of the car also offers the potential for reduced drag (Figure 5.8). As the flow diffuses (slows) along the length of the angled rear underbody the pressure rises, resulting in reduced form drag and also a reduced base area, although interactions between the overcar and undercar air flows can result in unexpected and sometimes detrimental effects. Such effects are hard to generalize and detailed experimental studies are currently required to determine the optimum geometry for individual vehicle designs, but typically it has been found that diffuser angles of approximately 15° seem to provide the greatest benefits (e.g. Howell, 1994).

5.5 Stability and cross-winds

The aerodynamic stability of passenger cars has been broadly addressed as two independent concerns. The first relates to the 'feel' of a car as it travels in a straight line at high speed and

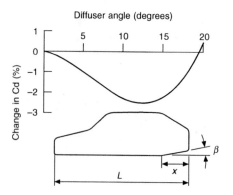

Figure 5.8 Rear, underbody diffusion

in calm conditions and to lane change manoevrability. The second concerns the effects of steady cross-winds and transient gusts that are associated with atmospheric conditions and which may be exaggerated by local topographical influences such as embankments and bridges.

The sources of straight line instability in calm conditions has proved to be one of the most difficult aerodynamic influences to identify. This is largely because of the complex interactions between the chassis dynamics and relatively small changes in the magnitude of lift forces and centre of pressure. Qualitative observations such as driver 'feel' and confidence have proved hard to quantify. New evidence suggests that stability and particularly lane change stability degrade with increases in the overall lift and with differences in lift between the front and rear axles (Hewell, 1998).

The influence of cross-winds is more easily quantifiable. Steady state cross-winds rarely present a safety hazard but their effect upon vehicle drag and wind noise is considerable. Most new vehicles will have been model-tested under yawed conditions in the wind tunnel at an early stage of their development but optimization for drag and wind noise is almost always based upon zero cross-wind assumptions. Some estimates suggest that the mean yaw angle experienced in the U.K. is approximately 5° and if that is correct then there is a strong case for optimizing the aerodynamic design for that condition.

The influence of transient cross-wind gusts such as those often experienced when passing bridge abutments, or when overtaking heavy vehicles in the presence of cross-winds is a phenomenon known to all drivers. To reduce the problems that are encountered by the driver under these conditions it is desirable to design the vehicle to minimize the side forces, yawing moments and yaw rates that occur as the vehicle is progressively and rapidly exposed to the cross wind. The low drag, rounded body shapes that have evolved in recent years can be particularly susceptible to cross-winds. Such designs are often associated with increased yaw sensitivity and commonly related changes of lift distribution under the influence of cross-winds can be particularly influential in terms of reduced vehicle stability. The influence of aerodynamics is likely to be further exaggerated by anticipated trends towards weight reduction in the search for improved fuel efficiency. Although methods for testing models under transient cross-wind conditions are under development, reliable data can, as yet, only be obtained by full scale testing of production and pre-production vehicles. At this late stage in the vehicle development programme the primary vehicle shape and tooling will have been defined so any remedial aerodynamic changes can only be achieved at very high cost or by the addition of secondary devices such as spoilers and mouldings; also an undesirable and costly solution. To evaluate the transient behaviour of a vehicle at a much earlier stage of its design it is necessary not only to develop model wind tunnel techniques to provide accurate and reliable data but most importantly to fully understand the flow mechanisms that give rise to the transient aerodynamic forces and moments. Initial results from recent developments in wind tunnel testing suggest that the side forces and yawing moments experienced in the true transient case exceed those that have been measured in steady state yaw tests (Docton, 1996).

5.6 Noise

Although some aerodynamic noise is created by ventilation flows through the cabin the most obtrusive noise is generally that created by the external flow around the vehicle. Considerable reductions have been made to cabin noise levels which may be attributed in part to improved air flows with reduced noise creation and also to improved sealing which has the effect both of reducing noise creation and insulating the occupants from the sound sources. Figure 5.9 provides an approximate comparison between the different noise sources (engine, tyres and aerodynamics) that have been recorded in a small car moving at 150 km/h (based on the data of Piatek, 1986). The creation of aerodynamic noise is mostly associated with turbulence at or the body surface and moves to reduce drag have inevitably provided the additional benefit of noise reduction. Although there is a noise associated with the essentially random turbulence that occurs within a turbulent boundary layer it is the sound associated with eddy creation at surface discontinuities that has both the greatest magnitude and also the most clearly defined (and annoying) frequencies. Improvements in rain gutter design and the positioning of windscreen wipers reflect some of the moves that have been made to reduce noise creation and improved manufacturing techniques and quality control have also resulted in major noise reduction as a consequence of improved panel fit. Protrusions such as wing mirrors and small surface radii such as at the 'A' pillar remain areas of particular concern because of their proximity to the driver and because of the

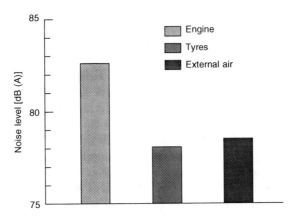

Figure 5.9 Noise sources (Piatek, 1986)

relatively poor sound insulation provided by windows. It has been demonstrated that it is the noise associated with vortex (eddy) creation that is the dominant aerodynamic noise source over almost the entire audible frequency range (Stapleford and Carr, 1971). One of the largest, single noise generators is the sun roof. Its large size results in low frequencies and large magnitudes and poorly designed units may even lead to discernible low frequency pressure pulsing in the cabin. Despite customer demand for low cabin noise there has been a parallel increase in the number of sun roofs that have been fitted to new cars. Open windows can create similar problems. Increased use of air conditioning is the best practical solution to this particular problem.

5.7 Underhood ventilation

The evidence from numerous researchers suggests that the engine cooling system is responsible for between 10% and 15% of the overall vehicle drag, so it is not surprising to note that considerable effort has been focused upon the optimization of these flows. Traditionally the cooling drag has been determined from wind tunnel drag measurements with and without the cooling intakes blanked-off. The results from those wind tunnel tests must be treated with caution since the closure of the intakes may alter the entire flow-field around a car. Underhood flow restrictions arising from the ever-increasing volume of ancillary equipment under the bonnet has further focused attention on cooling air flows, and this is now one of the primary applications for the developing use of computational flow simulation codes. Many of the sources of cooling drag are readily apparent such as the resistance created by the relatively dense radiator matrix and the drag associated with the tortuous flow through the engine bay. In general any smoothing of the flow path will reduce the drag, as will velocity reductions by diffusion upstream from the cooling system, although the implications of the latter upon the heat transfer must be considered. Less obvious but also significant is the interaction between the undercar flow and the cooling flow at its exit where high turbulence levels and flow separations may to occur. Careful design to control the cooling exit flow in terms of its speed and direction can reduce the drag associated with the merging flows but in general the aerodynamics are compromised to achieve the required cooling.

The potential for underhood drag reduction is greatest if the air flow can be controlled by the use of ducting to guide the air into and out from the radiator core. Approximate relationships between the slowing of the cooling airflow and the pressure loss coefficient, are widely described in the published texts (e.g. Barnard, 1996). The high blockage caused by the radiator core has the effect of dramatically reducing the air velocity through the radiator and thus much of the air that approaches the radiator spills around it. The relatively small mass flow that passes through the core can exhibit substantial non-uniformity which reduces the effectiveness of the cooling system. These problems can be much reduced if the flow is ducted into the radiator in such a way as to slow the flow in a controlled and efficient manner, and careful design of the degree of diffusion can greatly improve the efficiency of the cooling flow. Increasing the diffusion slows the air flowing through the radiator which reduces both the drag force and the heat transfer. Although the reduced heat transfer rate results in a requirement for a larger radiator core surface area, the drag reduction is proportionately greater than is the corresponding reduction in heat transfer. A low speed, large area core therefore creates less drag for a given heat transfer rate. Inevitably, compromises are necessary. The larger core adds weight and cost and the generally close proximity of the radiator to the intake leaves little scope for the use of long, idealized ducting. Too much diffusion will lead to flow separation within the intake which may result in severe flow non-uniformities across the face of the radiator. Gains are also available if the air is ducted away from the radiator in a similarly efficient manner, but in most cases the practical complexity of such a system and the requirement for a source of cooling air to the ancillaries has prevented such measures.

5.8 Cabin ventilation

Sealing between the body panels and particularly around the doors has achieved benefits in terms of noise reduction and aerodynamic drag, but the almost complete elimination of leakage flows has also led to changes in the design of passenger compartment ventilation. To achieve the required ventilation flow rates greater attention must be paid not only to the intake and exit locations but also to the velocity and path of the fresh air through the passenger compartment. The intake should be located in a zone of relatively high pressure and it should not be too close to the road surface where particulate and pollutant levels tend to be highest. The region immediately ahead of the windscreen adequately meets all of these requirements and is also conveniently located for air entry to the passenger compartment or air conditioning system. This location has been almost universally adopted. For the effective extraction of the ventilation air a zone of lower pressure should be sought. A location at the rear of the vehicle is usually selected and in many cases the air is directed through the parcel shelf and boot to exit through a controlled bleed in the boot seal. Increasing the pressure difference between the intake and exit provides the potential for high ventilation air flow rates but only at the expense of a flow rate that is sensitive to the velocity of the vehicle. This is particularly noticeable when the ventilation flow is heated and the temperature of the air changes with speed. A recent trend has been to use relatively low pressure differences coupled with a greater degree of fan assistance to provide a more controllable and consistent internal flow whether for simple ventilation systems or for increasingly popular air conditioning systems.

5.9 Wind tunnel testing

Very few new cars are now developed without a significant programme of wind tunnel testing. There are almost as many different wind tunnel configurations as there are wind tunnels and comparative tests have consistently shown that the forces and moments obtained from different facilities can differ quite considerably. However, most manufacturers use only one or two different wind tunnels and the most important requirement is for repeatability and correct comparative measurements when aerodynamic changes are made. During the early stages in the design and development process most testing is performed using small scale models where 1/4 scale is the most popular. The use of small models allows numerous design features to be tested in a cost effective manner with adequate accuracy.

For truly accurate simulation of the full scale flow it is necessary to achieve geometric and dynamic similarity. The latter requires the relative magnitudes of the inertia and viscous forces associated with the moving fluid to be modelled correctly and the ratio of those forces is given by a dimensionless parameter known as Reynolds number (Re):

$$Re = \frac{\rho u d}{\mu}$$

where ρ is the fluid (air) density, u is the relative wind speed, d is a characteristic dimension and μ is the dynamic viscosity of the fluid. For testing in air this expression tells us that the required wind speed is inversely proportional to the scale of the model but in practice the velocities required to achieve accuracy (using the correct Reynolds number) for small scale models are not practical, and Reynolds number similarity is rarely achieved. Fortunately, the Reynolds numbers achieved even for these small models are sufficiently high to create representative, largely turbulent vehicle surface boundary layers, and the failure to achieve Reynolds number matching rarely results in major errors in the character of the flow. The highest wind speeds at which models can be tested in any particular wind tunnel are more likely to be limited by the ground speed than by the air speed. The forward motion of a vehicle results not only in relative motion between the vehicle and the surrounding air but also between the vehicle and the ground. In the wind tunnel it is therefore necessary to move the ground plane at the same speed as the bulk air flow, and this is usually achieved by the use of a moving belt beneath the model. At high speeds problems such as belt tracking and heating may limit the maximum running speed, although moving ground plane technology has improved rapidly in recent years with the developments driven largely by the motor racing industry for whom 'ground effect' is particularly important. A considerable volume of literature is available relating to the influence of fixed and moving ground planes upon the accuracy of automotive wind tunnel measurements (for example Howell, 1994, Bearman *et al.*, 1988).

The use of larger models has benefits in terms of Reynolds number modelling and also facilitates the modelling of detailed features with greater accuracy, but their use also requires larger wind tunnels with correspondingly higher operating and model construction costs.

The forces acting upon a wind tunnel model are usually measured directly using a force balance which may be a mechanical device or one of the increasingly common strain gauge types. The latter has clear benefits in terms of electronic data collection and their accuracy is now comparable to mechanical devices. Electronic systems are also essential if unsteady forces are to be investigated. Lift, drag and pitching moment measurements are routinely measured and most modern force balances also measure side force, yawing moment and rolling moment. These latter three components relate to the forces that are experienced in cross-wind conditions.

Although direct force measurements provide essential data they generate only global information and provide little guidance as to the source of the measured changes or of the associated flow physics. That additional information requires detailed surface and wider flow-field measurements of pressure, velocity and flow direction if a more complete understanding is to be achieved. Such data are now becoming available even from transient flow studies (e.g. Ryan and Dominy, 1998), but the measurements that are necessary to obtain a detailed understanding of the flows remain surprisingly rare despite the availability of well-established measurement techniques.

5.10 Computational fluid dynamics

The greatest obstacle to the complete mapping of the flow-field by experimentation arises solely from time constraints. Recent developments in the numerical modelling of both external and internal flows now provide the engineer with a tool to provide a complete map of the flow field within a realistic timescale. Although the absolute accuracy of simulations is still questionable there is no doubt that, as a pointer to regions of interest in a particular flow, they have revolutionized experimental studies. The complexity of the flow around and through a complete vehicle is immensely intricate and despite the claims of some it is unlikely that within the next decade numerical simulations will achieve sufficient accuracy to replace wind tunnel testing as the primary tool for aerodynamic development.

The relationships between the pressure, viscous and momentum forces in a fluid flow are governed by the Navier–Stokes equations. For real flows these equations can only be solved analytically for simple cases for which many of the terms can be neglected. For complex, three-dimensional flows such as those associated with road vehicles it is necessary to achieve an approximate solution using numerical methods. Although different approaches may be adopted for the simulation there are aspects of the modelling that are common to all. Initially the entire flow field is divided into a very large number of cells. The boundaries of the flow field must be sufficiently far from the vehicle itself to prevent unrealistic constraints from being imposed upon the flow. From a pre-defined starting condition (e.g. a uniform flow velocity may be imposed far upstream from the model), the values of each of the relevant variables are determined for each cell. Using an iterative procedure those values are repeatedly re-calculated and updated until the governing equations are satisfied to an acceptable degree of accuracy. As a rule the accuracy of a simulation will be improved by reducing the volume of each cell although there are particular rules and constraints that must be followed near surfaces (e.g. Abbott and Basco (1989).

Unlike the aerospace industry, where aerodynamics is arguably the single-most important technology, automotive manufacturers rarely have sufficient resources to develop CFD codes for their own specific applications and in almost all cases commercially available codes are used. A danger of this approach is that users who are not fully conversant with the subtleties of the numerical simulation can overlook minor and sometimes major shortcomings in their predictions.

5.11 References and further reading

Abbott, M.B., and Basco, D.R. (1989). *Computational Fluid Dynamics: an Introduction for Engineers.* Longman. ISBN 0-582-01365-8.

Ahmed, S.R., Ramm, G and Faltin, G. (1984) Some salient features of the time-averaged ground vehicle wake, SAE International Congress and Exposition, Detroit. Paper no. 840300.

Barnard, R.H., (1996). *Road Vehicle Aerodynamic Design.* Longman, ISBN 0-582-24522-2.

Bearman, P.W., DeBeer, D., Hamidy, E. and Harvey, J.K. (1988). The effect of moving floor on wind-tunnel simulation of road vehicles, SAE International Congress and Exposition, Detroit. Paper no. 880245.

Carr, G.W., (1968). The aerodynamics of basic shapes for road vehicles, Part 2, Saloon car bodies. MIRA report no. 1968/9.

Docton, M.K.R. (1996). The simulation of transient cross winds on passenger vehicles, Ph.D. Thesis, University of Durham.

Emmelmann, H-J. (1982). Aerodynamic development and conflicting goals of subcompacts – outlined on the Opel Corsa. *International Symposium on Vehicle Aerodynamics*, Wolfsburg.

Howell, J. (1994). The influence of ground simulation on the aerodynamics of simple car shapes with an underfloor diffuser. Proc. RAeS Conference on Vehicle Aerodynamics, Loughborough.

Howell, J. (1998). The Influence of Aerodynamic Lift in Lane Change Manoevrability. Second M.I.R.A. Conference on Vehicle Aerodynamics, Coventry.

Hucho, W.H. (ed.) (1998). *Aerodynamics of Road Vehicles: from Fluid Mechanics to Vehicle Engineering*, 4th edition, S.A.E., ISBN 0-7680-0029-7.

Hucho,W.H., Janssen, L.J. and Emmelman, H.J. (1976). The optimization of body details – a method for reducing the aerodynamic drag of road vehicles. SAE International Congress and Exposition, Detroit. Paper no. 760185.

Nouzawa, T., Hiasa, K., Nakamura, T., Kawamoto, K. and Sato, H. (1992). Unsteady-wake analysis of the aerodynamic drag on a hatchback model with critical afterbody geometry. SAE SP-908, paper 920202.

Piatek, R. (1986). Operation, safety and comfort: in '*Aerodynamics of Road Vehicles*', Butterworth-Heinemann (ed. Hucho,W.H., 1986).

Ryan, A., and Dominy, R.G. (1998). The aerodynamic forces induced on a passenger vehicle in response to a transient cross-wind gust at a relative incidence of 30°. SAE International Congress and Exposition, Detroit. Paper no. 980392

Stapleford, W.R., and Carr, G.W. (1971). Aerodynamic noise in road vehicles, part 1: the relationship between aerodynamic noise and the nature of airflow, MIRA report no. 1971/2.

Recommended reading

Abbott, M.B., and Basco, D.R. (1989). *Computational Fluid Dynamics: an Introduction for Engineers.* Longman. ISBN 0-582-01365-8.

Barnard, R.H., (1996). *Road Vehicle Aerodynamic Design.* Longman, ISBN 0-582-24522-2.

Hucho, W.H., (ed.) (1998). *Aerodynamics of Road Vehicles: from Fluid Mechanics to Vehicle Engineering*, 4th edition, S.A.E., ISBN 0-7680-0029-7.

6. Chassis design and analysis

John Robertson, BScEng, CEng, MIMechE

The aim of this chapter is to:

- Introduce the loadings on vehicle structures;
- Introduce the different types of vehicle structure and their use;
- Indicate how these loadings can be analysed simply and with the use of computers;
- Suggest the requirements for sound vehicle structural design;
- Give examples of simple structural analysis which highlights the processes involved for vehicle structures.

6.1 Load case, introduction

The loads imposed on the chassis or body structure of a passenger car or light commercial vehicle due to normal running conditions are considered in this chapter. That is, the loads caused as the vehicle traverses uneven ground and as the driver performs various manoeuvres.

There are five basic load cases to consider.

1. Bending case
This is loading in a vertical plane, the x–z plane due to the weight of components distributed along the vehicle frame which cause bending about the y-axis, see Figure 6.1(a).

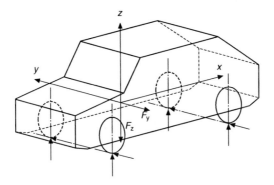

Figure 6.1(a) Vehicle bending case

2. Torsion case
The vehicle body is subjected to a moment applied at the axle centrelines by applying upward and downward loads at each axle in this case. These loads result in a twisting action or torsion moment about the longitudinal x-axis of the vehicle, see Figure 6.1(b).

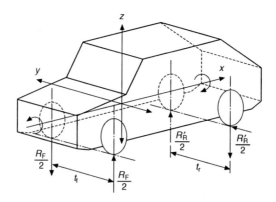

Figure 6.1(b) Vehicle torsion case

3. Combined bending and torsion

In practice, the torsion case cannot exist without bending as gravitational forces are always present. Therefore, the two cases must be considered together when representing a real situation, see Figure 6.1(c).

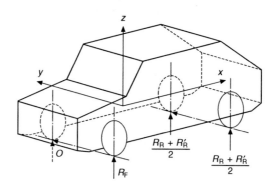

Figure 6.1(c) Vehicle combined bending and torsion

4. Lateral loading

This condition occurs when the vehicle is driven around a corner or when it slides against a kerb, i.e. loads along the y-axis, Figure 6.1(d).

6. Fore and aft loading

During acceleration and braking longitudinal forces are generated (along the x-axis). Traction and braking forces at the tyre to ground contact points are reacted by mass times acceleration (deceleration) inertia forces, see Figure 6.1(e).

The most important cases are those of 1 (bending), 2 (torsion) and 3 (bending and torsion) as these are paramount in determining a satisfactory structure (Pawlowski, 1964). The lateral

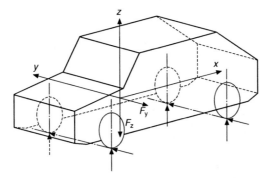

Figure 6.1(d) Vehicle lateral loading

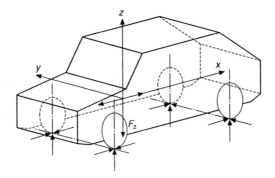

Figure 6.1(e) Vehicle fore/aft loading

loading and fore-aft loading cases require attention when designing the suspension mounting points to the structure but are less significant on the structure as a whole. Other localized loading conditions such as loads caused by door slamming, seat belt loads, etc., are not considered in this work.

6.1.1 Bending case

The bending conditions depend upon the weights of the major components of the vehicle and the payload. The first consideration is the static condition by determining the load distribution along the vehicle. The axle reaction loads are obtained by resolving forces and taking moments from the weights and positions of the components (i.e. the equations of statics). The structure can be treated as a two-dimensional beam as the vehicle is approximately symmetric about the longitudinal x-axis. A typical medium size passenger car load distribution is shown in Figure 6.2.

Examination of Figure 6.2 shows a typical list of the major components of the vehicle that are considered. The distributed loads are estimates of the weight per unit length for the body of the vehicle including trim details. The unsprung masses consisting of wheels, brake discs/drums and suspension links are of course not included as they do not impose loads on the structure.

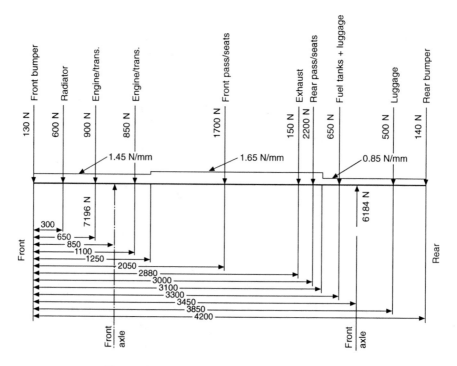

Figure 6.2 Typical passenger vehicle bending loads

From the load diagram Figure 6.2, the bending moment diagram and shear force diagram can be constructed in the normal way. Figures 6.3 and 6.4 show these diagrams constructed

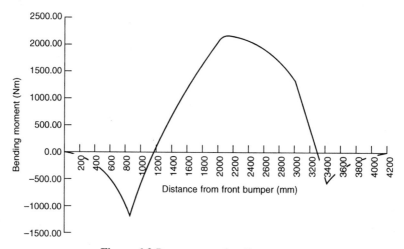

Figure 6.3 Passenger car bending moments

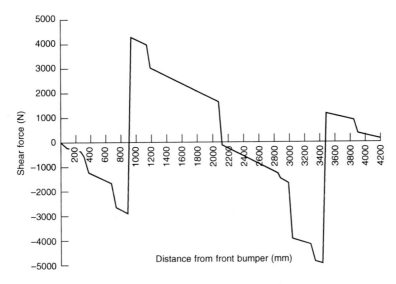

Figure 6.4 Passenger car shear force diagram

using a computer spreadsheet method. Values taken from these diagrams can be used to determine stress conditions on a chassis frame or on the side-frame of a passenger car.

The Dynamic Loading must be considered as the vehicle traverses uneven road surfaces. For example, the vehicle may pass over a hump-back bridge at such a speed that the wheels leave the ground. The resulting impact of the vehicle returning to earth is cushioned by the suspension system but inevitably causes a considerable increase in loading over the static condition. Experience gained by vehicle manufacturers indicates that the static loads should be increased by factors of 2.5 to 3.0 for road going vehicles. Off-road or cross-country vehicles may be designed with factors of 4 (Pawlowski, 1964).

6.1.2 Torsion case

The case of pure torsion can be considered simply as being applied at one axle line and reacted at the other axle. The condition of pure torsion cannot exist on its own because vertical loads always exist due to gravity, as mentioned in the introduction. However, for ease of calculation the pure torsion case is assumed.

The maximum torsion moment is based on the loads at the lighter loaded axle, and its value is the wheel load on that lighter loaded axle multiplied by the wheel track. See the following section (6.1.3) for further explanation. The loads at the wheels are then as shown in Figure 6.1(b).

The torsion moment $$\frac{R_F}{2}t_f = \frac{R_R}{2}t_r \qquad (6.1)$$

The front and rear track t_f and t_r respectively may be slightly different and the rear axle load R_R is usually smaller then R_F for a modern passenger car even when fully laden. In this situation R_R is the load on the rear axle for the fully laden case R_F will be less than the front axle load for the same fully laden condition.

Once again these loads are based on static reaction loads but dynamic factors in this case are typically 1.3 for road vehicles (Pawlowski, 1964). For trucks which often go off road 1.5 and for cross-country vehicles a factor of 1.8 may be used.

6.1.3 Combined bending and torsion

If the static loading cases of bending and torsion are combined the loading condition shown in Figure 6.1(c) will be achieved. This represents the situation arising if one wheel of the lighter loaded axle is raised on a bump of sufficient height to cause the other wheel on that axle to leave the ground. Pawlowski (1964) recommended that a maximum bump height of 200 mm should be considered as most cars have a suspension bump to rebound travel of 200 mm or less. The present writer considers the 200 mm bump height will lift the other wheel of the same axle off the ground. In this condition all the load of the lighter axle is applied to one wheel.

If this principle is applied to the vehicle described in Figure 6.2 and assuming the front track $t_f = 1450$ mm and rear track $t_r = 1400$ mm.

The load on the right wheel of the rear (lighter loaded) axle will be the total axle load $R_e = 6184$ N, the torque on the body 4328 N-m and R'_F is 5971 N.

Resultant wheel loads at the front axle are

Right wheel
$$R_{FTR} = \frac{R_F}{2} - \frac{R'_F}{2} = \frac{7196}{2} - \frac{5971}{2} = 613 \text{ N} \tag{6.2}$$

Left wheel
$$R_{FTL} = \frac{R_F}{2} - \frac{R'_F}{2} = \frac{7196}{2} + \frac{5971}{2} = 6583 \text{ N} \tag{6.3}$$

If the left front wheel had been lifted instead of the right rear wheel the same situation would have occurred, i.e. the left rear wheel load will reduce to zero before the right front wheel. Any further lifting of the left front wheel (or right rear wheel) will not increase the torque applied to the vehicle structure.

6.1.4 Lateral loading

When cornering, lateral loads are generated at the tyre to ground contact patches which are balanced by the centrifugal force $\dfrac{MV^2}{R}$ where M is the vehicle mass, V is the forward speed, R is the radius of the corner (see Figure 6.1(d)).

The worst possible condition occurs when the wheel reactions on the inside of the turn drop to zero, that is when the vehicle is about to roll over. In this condition the structure is subject to bending in the x–y plane. The condition approaching the roll-over is shown in Figure 6.5 and depends upon the height of the vehicle centre of gravity and the track. At this condition the resultant of the centrifugal force and the weight passes through the outside wheels contact patch (A)

$$\frac{MV^2}{R} h = Mg \frac{t}{2}$$

Therefore lateral acceleration $= \dfrac{V^2}{R} = \dfrac{gt}{2h}$ \hfill (6.4)

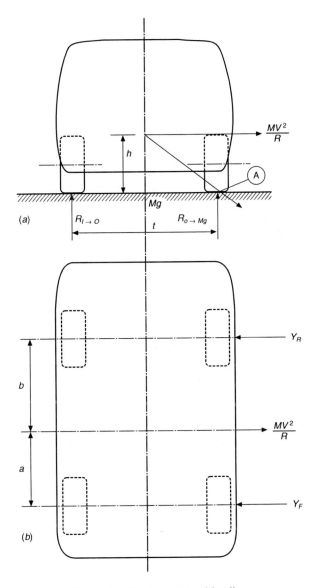

Figure 6.5 Maximum lateral loading

Therefore the lateral force at the centre of gravity $= \dfrac{MV^2}{R} = \dfrac{Mgt}{2h}$

The side forces at the front tyres $= Y_F = \dfrac{Mgt}{2h} \dfrac{b}{(a+b)}$ (6.5)

At the rear tyres $= Y_R = \dfrac{Mgt}{2h} \dfrac{a}{(a+b)}$ (6.6)

The structure can be considered now as a simply supported beam subject to lateral loading in the x–y plane through the centre of gravity. A more accurate model would consider distributed loads in a similar manner to that described in section 6.1.1 for bending in the x–z plane. Normal driving conditions never approach this situation because from equation 6.4 when h (the height of the centre of gravity from a road surface datum) for a modern car is typically 0.51 m and track of 1.45 m

$$\text{Lateral acceleration} = \frac{gt}{2h} = \frac{g1.45}{2*0.51} = 1.42g$$

That is, the lateral acceleration is 1.42 times gravitational acceleration. This does not occur as conventional road tyre side forces limit lateral acceleration to about 0.75g.

Kerb bumping may cause high loads and roll over in exceptional circumstances. The high lateral loads causing bending in the x–y plane are not critical as the width of the vehicle (or beam depth) easily provides sufficient bending strength and stiffness. Suspension mounting brackets must, however, be designed to withstand these high shock loads. For safety reasons these high lateral shock loads are usually assumed to be twice the static vertical load on the wheel.

6.1.5 Longitudinal loading

When the vehicle accelerates or decelerates, the mass times acceleration or inertia force is generated. As the centre of gravity of the vehicle is above the road surface the inertia force provides a load transfer from one axle to another. While accelerating, the weight is transferred from the front axle to the rear axle and vice versa for the braking or decelerating condition. To obtain a complete view of all the forces acting on the body the heights of the centres of gravity of all components will be required. These are often not known, therefore a plot of bending moments along the vehicle is not obtainable. A simplified model considering one inertia force generated at the vehicle centre of gravity can provide useful information about the local loading at the axle positions due to traction and braking forces.

Figure 6.6 shows the forces due to traction and braking for (a) front wheel drive acceleration (b) rear wheel drive acceleration and (c) braking.

For (a) front wheel drive, the reaction on the driving wheels is

$$R_F = \frac{Mg(L - a) - Mh\left(\dfrac{dV}{dt}\right)}{L} \tag{6.7}$$

For (b) rear wheel drive, the reaction on the driving wheels is

$$R_R = \frac{Mga + Mh\left(\dfrac{dV}{dt}\right)}{L} \tag{6.8}$$

For (c) the braking case, the reactions on the axles are

$$R_F = \frac{Mg(L - a) + Mh\left(\dfrac{dV}{dt}\right)}{L} \tag{6.9}$$

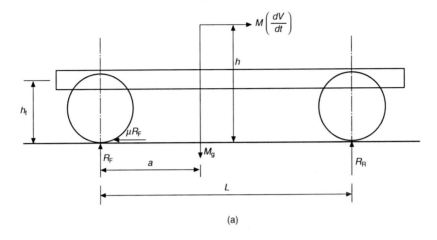

(a)

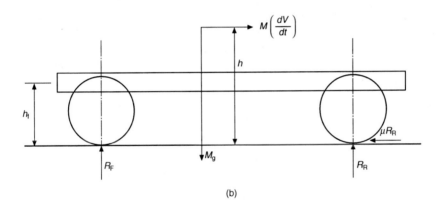

(b)

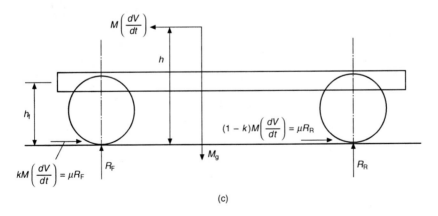

(c)

Figure 6.6 Load transfer due to acceleration, (a) front wheel drive; (b) rear wheel drive; (c) braking (deceleration)

$$R_R = \frac{Mga - Mh\left(\dfrac{dV}{dt}\right)}{L} \tag{6.10}$$

The limiting tractive and braking forces are controlled by the coefficient of adhesion between the tyres and road surface. These tractive and braking forces at the road surface apply additional bending to the vehicle structure through the suspension systems. Similarly the inertia force through the centre of gravity offset from the frame by $(h - h_f)$ applies an additional bending moment. Newcomb and Spurr (1966) provide details of brake proportioning and further information on axle and load transfer.

6.1.6 Asymmetric loading

This loading condition is illustrated in Figure 6.7(a) and occurs when one wheel strikes a raised object or drops into a hole that has a raised edge. The resulting loads are vertical and longitudinal applied at one corner of the vehicle. This condition results in a very complex loading on the vehicle structure. The magnitude of the force exerted on the wheel and hence through the suspension to the structure will depend on the vehicle speed, suspension stiffness, wheel mass, body mass, etc. As the shock force is only applied for a very short period of time it can be assumed that the wheel continues at a steady speed and therefore the shock force R_u acts through the wheel centre. The horizontal component will then be $R_{ux} = R_u \cos \alpha$ and the vertical

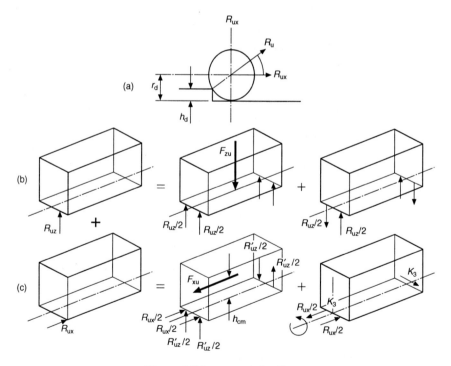

Figure 6.7 Asymmetric loading

component $R_{uz} = R_u \sin \alpha$. The angle α is approximately $\sin^{-1}(r_d - h_u)/r_d$ assuming the tyre does not deflect excessively. Note that the horizontal component will increase relative to the vertical for small radius wheels.

Consideration of the vertical load on its own causes an additional axle load, an inertia load through the vehicle centre of mass and a torsion moment on the vehicle structure (see Figure 6.7(b)). Similarly, considering the horizontal load on its own it can seen from Figure 6.7(c) that additional bending in the vertical plane (x–z) and a moment about the z axis are applied to the structure. Hence from the structural loading, this load can be analysed by the superposition of 4 load conditions.

6.1.7 Allowable stress

The load conditions discussed in sections 6.1.1 to 6.1.6 result in stresses occurring throughout the vehicle structure. It is important that under the worst load conditions that the stresses induced into the structure are kept to acceptable limits.

Consideration of the static loads factored by an appropriate amount should give a stress level certainly below the yield stress. For example, if the bending case for a road going passenger car is considered the maximum allowable stress should be limited as follows:

Stress due to static load × Dynamic Factor ≤ 2/3 × yield stress

This means that under the worst dynamic load condition the stress should not exceed 67% of the yield stress. Alternatively, the safety factor against yield is 1.5 for the worst possible load condition. A similar criterion is applied to other load conditions. This procedure is usually satisfactory for designing against fatigue failure, but fatigue investigations are necessary especially where stress concentrations occur at suspension mounting points.

6.1.8 Bending stiffness

The previous sections have considered loads and stresses and now there is a need to determine whether the structure is sufficiently strong. An equally important design requirement is to assess the structural stiffness; in fact, many designers consider stiffness is more important than strength. It is possible to design a structure which is sufficiently strong but yet unsatisfactory because it has insufficient stiffness. Designing for acceptable stiffness is therefore often more critical than designing for sufficient strength.

For the passenger car, the bending stiffness is determined by the acceptable limits of deflection of the side frame door apertures. If excessive deflections occur then the doors will not shut satisfactorily, i.e. the alignment of the door latches are such that doors cannot be opened or closed easily. Local stiffness of the floor is important for passenger acceptance. If the floor deflects under the passengers' feet it causes passenger insecurity. Floor panels are usually stiffened by swages pressed into the panels which give increased local second moments of area hence reducing deflections. These in turn reduce panel vibrations. A flat thin sheet metal panel will act like a drum skin vibrating at a frequency that depends on such factors as the size, the thickness and the edge restraint conditions of the panel. Some modern luxury cars now use sandwich material consisting of two thin panels separated by a honeycomb material which leads to a much quieter vehicle. Local stiffness needs to be increased at many other places within the structure, for example at the door, hood/bonnet, boot/trunk hinge mounting points, suspension attachment points, seat mounting points and mounting points for other major components. This is achieved by adding reinforcing plates and brackets in the body sections at hinge points, door latches, suspension pivot points, etc.

The acceptable deflections (or required stiffness) can be determined for some components of the structure. The acceptable misalignment of door latches will be determined by the design features of the latch. Other components such as the stiffness of floor panels are determined by the experience of the vehicle manufacturer or during the development of the vehicle.

6.1.9 Torsional stiffness

The acceptable torsional stiffness can be evaluated for specific criteria while for other criteria it is based on experience and development, as described in the previous section. A typical medium sized saloon (sedan) fully assembled can have a torsional stiffness of 8000 to 10 000 N-m/degree (Webb, 1984). That is when measured over the wheel-base of the vehicle. Experience shows this to be acceptable for a road going passenger car. If the stiffness is low, driver perception is that the front of the vehicle appears to shake with the front wing structures tending to move up and down.

Practical problems of doors failing to close properly occur when the vehicle is parked on uneven ground such as with one wheel on a kerb. A similar problem will occur if the jacking points (to permit wheel change) are positioned at the corners of the vehicle. The torsional stiffness is also influenced by the windscreen and back-light glass. Studies by Webb (1984) show a reduction in torsional stiffness with the glass removed of approximately 40%. Therefore the glass is subject to load and hence stress which again if excessive can cause the glass to crack. Open-top sports cars with no structural roof panel are likely to have poor torsional stiffness unless the underbody is reinforced. For these cars, the handling of the vehicle is extremely important and if the torsional stiffness is low, this has a detrimental effect on the handling characteristics. Therefore great care is taken to ensure that the torsional stiffness is adequate.

6.2 Chassis types, introduction

Now that the loading cases that are applied to a vehicle structure or chassis frame have been considered, the various types of structure that could be used can be investigated, and a design appraisal to assess their suitability for the loads imposed can be made. A range of structural types will be examined, and their effectiveness at resisting the various load conditions and how the passenger car structure can be developed.

6.2.1 Ladder frames

The early motor cars were constructed with a ladder frame structure on which was placed the body of the vehicle containing the passenger seats. The earliest designs often had no roof so the body did little to protect against the weather while later designs provided this protection with a roof, doors etc. Even so, the body did not contribute much to the vehicle structure. It was often made of wood with very low stiffness compared with the chassis frame. Therefore the high stiffness (in bending) ladder frame carried virtually all the bending and torsion loads.

The greatest advantage of the ladder frame is its adaptability to accommodate a large variety of body shapes and types. It is still widely used for light commercial vehicles such as pick-ups through to the heavy truck for this reason. Bodies ranging from flat platforms, box vans and tankers to detachable containers can all be easily attached to ladder frames. The ladder frame is so called because it resembles a ladder with two side rails and a number of cross beams.

Most designs are made with channel section side rails and either open or closed section cross beams (Figure 6.8). Good bending strength and stiffness for weight are obtained with deep beam side rails as the ratio of second moment of area/cross sectional area can be optimized. The flanges contribute to the large second moment of area and the whole flange areas carry high stress levels; therefore this is an efficient use of material (Figure 6.9). The open channel section provides easy access for attaching brackets and components. Attaching them to the web avoids holes in the highly stressed flanges (Figure 6.10). Another characteristic of the channel section is the shear centre being offset from the web (Roark, 1975). Local twisting of the side frames can be prevented by ensuring that brackets attaching components to them are as shown in Figure 6.10. Unfortunately, the Torsion Constant is very small and hence its torsional stiffness is low. If a ladder frame chassis is constructed with cross beams of channel section as well as the side frames the torsional stiffness of the whole is very low.

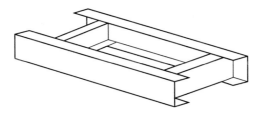

Figure 6.8 Basic ladder frame channel sections

The simple peripheral frame, which is the most simple ladder frame in Figure 6.11, the torsion in the cross members is reacted as bending in the side frames and the bending in the cross members reacted as torsion in the side frames. All members are loaded in torsion and due to their low torsion constants the frame has low torsional stiffness. If the open sections are replaced by closed box sections then the torsional stiffness is greatly improved. This is done on vehicles such as the Land Rover. However, the strength of the joints becomes critical as the maximum bending on all members occurs at the joints (see Figure 6.11) and the attachment of brackets becomes more complex (see Figure 6.12).

6.2.2 Cruciform frames

It is possible to design a frame to carry torsion loads where no element of the frame is subject to a torsion moment. The cruciform frame shown in Figure 6.13 is made of two straight beams and will only have bending loads applied to the beams. This type of frame has good torsional

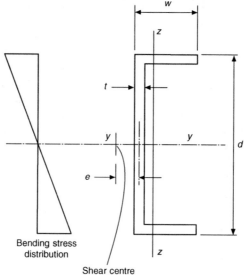

Second moment of area, $I_{yy} = 2\left\{ \dfrac{wt^2}{12} + wt\left(\dfrac{d-t}{2}\right)^2 \right\}$

Torsion constant, $J = \left\{ \dfrac{2wt^3}{3} + \left(\dfrac{(d-t)t^3}{3}\right) \right\}$

Shear centre offset, $e = \left\{ \left(\dfrac{(w-t)}{2}\right)^2 (d-t)^2 \right\} \dfrac{1}{4 I_{yy}}$

Figure 6.9 Channel section properties

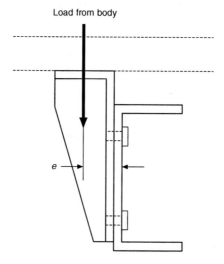

Figure 6.10 Attachment to channel section

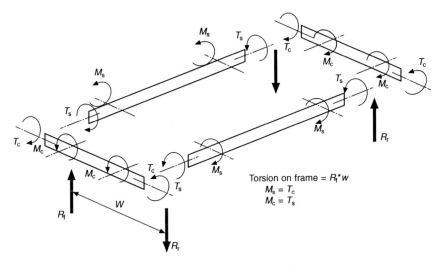

Figure 6.11 Torsion load on ring frame

Torsion on frame = $R_f^* w$

$M_s = T_c$

$M_c = T_s$

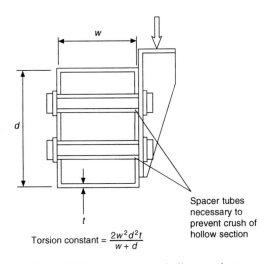

Torsion constant = $\dfrac{2w^2 d^2 t}{w + d}$

Spacer tubes necessary to prevent crush of hollow section

Figure 6.12 Attachments to hollow sections

stiffness provided the joint at the centre is satisfactorily designed. It should be noted that the maximum bending occurs at the joint hence joint design becomes critical.

Combining the properties of the cruciform frame with those of the ladder frame assists in obtaining both good bending and good torsional properties. The cross beams at the front and rear not only assist in carrying the torsion moment but assist in carrying the lateral loads from the suspension mounting points, see Figure 6.14.

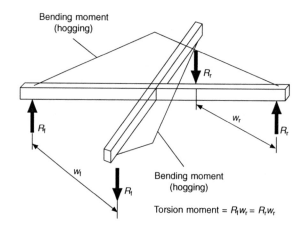

Figure 6.13 Cruciform frame – members subject to bending

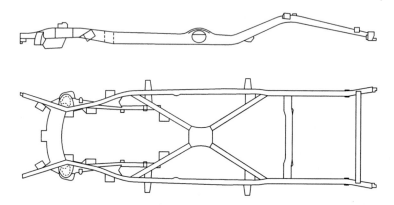

Figure 6.14 An early ladder frame with cruciform

6.2.3 Torque tube backbone frames

It was noted in Figure 6.12 that a closed box section has vastly improved torsional stiffness compared with an open section. This property has been exploited in the Lotus chassis shown in Figure 6.15 where the main backbone is a closed box section through which runs the drive shaft between the gearbox and the final drive unit. The splayed beams at the front and rear extend to the suspension mounting points while additional transverse members tie the suspension mounting points together resisting lateral loads. In this type of structure the backbone is subject to bending and torsion loads, the splayed beams to bending and the transverse members to compression or tension from the lateral loads from the suspension.

6.2.4 Space frames

The frames described in the previous three sections are all essentially 2-dimensional or at least their depth is very much less than their length and breadth. Adding depth to a frame considerably

Figure 6.15 A typical torsion tube frame

increases its bending strength and stiffness (i.e. truss type bridges). 3-dimensional Space frames have been used for specialist cars such as sports racing cars; an example is shown in Figure 6.16. This type of vehicle design can be used for low volume production with G.R.P. bodies.

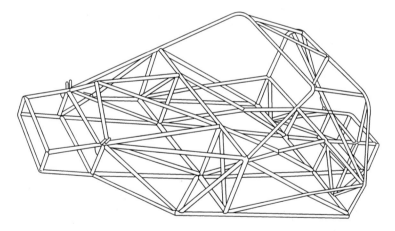

Figure 6.16 Space frame chassis

In this type of structure it is imperative to ensure all planes are fully triangulated so that the beam elements are essentially loaded in tension or compression. Due to the welded joints some bending and torsion restraints will occur at the joints, but to rely on these restraints will render

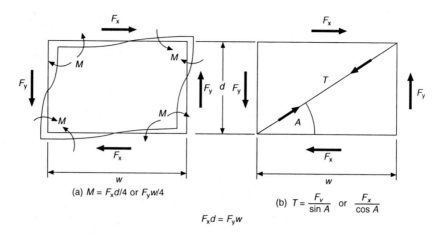

(a) $M = F_x d/4$ or $F_y w/4$

$F_x d = F_y w$

(b) $T = \dfrac{F_v}{\sin A}$ or $\dfrac{F_x}{\cos A}$

Figure 6.17 Ring frame and diagonal braced frame

the structure far less stiff. Consider the situation illustrated in Figure 6.17(a) and (b). At (a) the stiffness of an 'open' rectangular frame depends on the bending of the elements where at (b) the stiffness is provided by the diagonal member subject to direct tension or compression. For a practical structure 'open' apertures are necessary for the windscreen, back-light, access to the engine compartment, doors, etc. which can result in this type of structure being less torsionally stiff.

6.2.5 Integral structures

The modern mass-produced passenger car is almost exclusively produced with sheet steel pressings spot welded together to form an integral structure. This is a structure where the component parts provide both structural and other functions. The depth of a structure such as a space frame, can improve the stiffness and in the integral structure the whole side frame with its depth and the roof are made to contribute to the vehicle bending and torsional stiffness. A typical passenger car integral structure is shown in Figure 6.18.

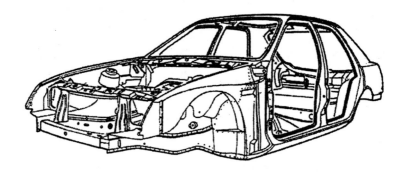

Figure 6.18 A typical passenger car integral structure

Such a structure is geometrically very complicated and the detailed stress distribution can only be determined by the use of Finite Element methods (see section 6.4). The structure can be described as a 'Redundant Structure' as some parts can be removed (i.e. redundant) and the structure will still carry the applied loads although with less efficiency or greater flexibility. The stress distribution within the structure is not only a function of the applied loads but also of the relative stiffnesses of the many components. The details of this analysis is beyond the scope of this chapter (Roark, 1975). The advantages of the Integral Structure are numerous. It is stiffer in bending and torsion, it is of lower weight than when using a chassis and separate body, it can be produced with lower cost, and it produces a quieter car for the passengers.

Section 6.3 describes a method for determining the main load paths through the integral structure for bending and torsion load cases. The function of the main structural members can be demonstrated with this method.

6.3 Structural analysis by simple structural surfaces method

There are many ways of modelling a vehicle structure for the purpose of determining loads and stresses within the structure. The most elementary form described as a beam has already been considered in Figure 6.2. Complex models are considered in section 6.4. Before these complex models are examined, it is useful to have intermediate models that help understanding of the main load paths within the structure.

One most useful method was developed by Pawlowski (1964) is called Simple Structural Surfaces. It is possible with this method to determine the loads on the main structural members of an integral structure. Although this type of structure is highly redundant it is possible by careful representation of the main elements in the structure to determine loads and hence stress by the simple equations of statics.

For example, a simple van structure can be represented as shown in Figure 6.19. This shows that the body structure is represented by 10 structural components or Simple Structural Surfaces. These are the roof, floor, 2 side-frames, front panel and windscreen frame, rear frame plus three floor cross-beams. Figure 6.19 shows the torsion load case previously described in Section 6.1.2 and the forces acting on each Simple Structural Surface. If the geometry of the vehicle and the axle loads are known, all the edge loads $K_1 \ldots K_n$, acting can be evaluated between the Simple Structural Surfaces. From these loads the sections for the window pillars, floor cross-beams, etc. can be evaluated to give acceptable levels of stress and deflection.

6.3.1 Definition of a Simple Structural Surface (SSS)

A Simple Structural Surface is 'rigid' in its own plane but 'flexible' out of plane. That is, it can carry loads in its plane (tension, compression, shear, bending) but loads normal to the plane and bending out of the plane are not possible. Figure 6.20 illustrates diagrammatically the principle of the Simple Structural Surface.

6.3.2 Simple Structural Surfaces representing a box van in torsion

Using the same basic model of Figure 6.19 the torsion load case can be considered in detail.

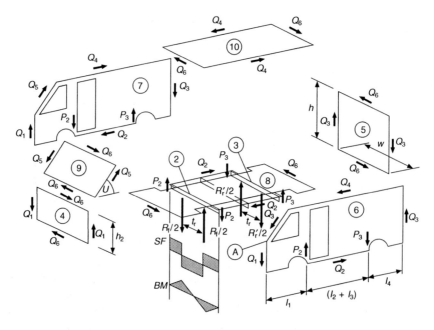

Figure 6.19 Simple structural surfaces representing a van structure

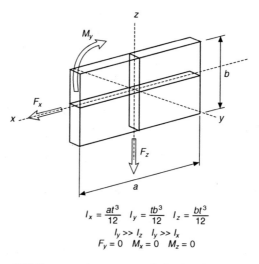

$$I_x = \frac{at^3}{12} \quad I_y = \frac{tb^3}{12} \quad I_z = \frac{bt^3}{12}$$

$$I_y \gg I_z \quad I_y \gg I_x$$

$$F_y = 0 \quad M_x = 0 \quad M_z = 0$$

Figure 6.20 Forces and moments only in the plane of an SSS

Taking the axle with the lightest load as explained in 6.1.2 equal and opposite loads (up and down) are applied to the front and rear cross beams (SSS-2 and SSS-3). In this example when fully laden a van front axle is most probably the lightest loaded axle so $\frac{R_f}{2}$ is taken as acting up and down on SSS-2.

The moment at the front cross-beam must be reacted by an equal moment at the rear crossbeam, therefore:

$$\frac{R_f}{2}*t_f = \frac{R'_r}{2}*t_r \tag{6.11}$$

R'_r will be less than the rear axle load and different from the axle load R_f if the rear track t_r is different to the front track t_f.

The equilibrium of the SSS-2 and SSS-3 can be obtained by taking moments, and as the values of R_f and R'_r are known the values of P_2 and P_3 are obtained

SSS-2 (Front cross-beam)

$$P_2 w \frac{R_f}{2} t_f = 0 \tag{6.12}$$

SSS-3 (Rear cross-beam)

$$P_3 w \frac{R'}{2} t_r = 0 \tag{6.13}$$

P_2 and P_3 will in fact be equal in magnitude because they both act at the width of the vehicle and the torque at front and rear must be equal.

Now consider the loads from the cross-beams acting on the left-hand sideframe (SSS-6).

Edge loads Q_1 to Q_5 will occur around the periphery of the sideframe applying an opposing moment to the moment applied by P_2 and P_3. The moment equation can be developed for SSS-6 by taking moments about A, the base of the windscreen pillar, see Figure 6.19.

$$P_3(\ell_1 + \ell_2 + \ell_3) - Q_3(\ell_1 + \ell_2 + \ell_3 + \ell_4) - Q_4(h_1 - h_2) - Q_2 h_2 - P_2 \ell_1 = 0 \tag{6.14}$$

Consider the equilibrium of SSS-4, 5, 8, 9, 10. These surfaces must all be held in equilibrium by complementary shear forces which balance the moments applied from the side-frames. The right-hand side-frame must of course be loaded exactly opposite to the left-hand side-frame.

SSS-4 (Front panel)

$$Q_6 h_2 - Q_1 w = 0 \tag{6.15}$$

SSS-5 (Rear door frame)

$$Q_6 h_1 - Q_3 w = 0 \tag{6.16}$$

SSS-8 (Floor panel)

$$Q_6(\ell_1 + \ell_2 + \ell_3 + \ell_4) - Q_2 w = 0 \tag{6.17}$$

SSS-9 (Windscreen frame)

$$Q_6(h_1 - h_2)/\sin \alpha - Q_5 w = 0 \tag{6.18}$$

SSS-10 (Roof)

$$Q_6 \ell_5 - Q_4 w = 0 \tag{6.19}$$

There are now six equations, 6.14 to 6.19, and six unknowns Q_1 to Q_6 so a solution can be obtained. By substituting for Q_2, Q_3 and Q_4 from equations 6.17, 6.16 and 6.19 into equation 6.14 an equation in one variable Q_6 is derived. Hence the value of Q_6 can be obtained and then the other unknowns using equations 6.15 to 6.19.

It should be noted that the roof, floor, front windscreen frame and rear door frame are all subject to complementary shear. The floor crossbeams are subject to bending moments and shear forces while the side-frames are also loaded in bending and shear. The centre cross beam SSS-1 has no loads applied in this case, but will be loaded in the bending case.

6.3.3 Box van structure in bending and torsion

As previously explained in Sections 6.1.2 and 6.1.3 the torsion case always is combined with bending so using the principle of superposition the load conditions from the two cases can be added to obtain the loads on individual members of the structure.

6.3.4 Simple Structural Surfaces representing a saloon car in bending

A passenger car structure such as for a saloon car is constructed with a more geometrically complex structure than a box van. However, it is still possible to model with Simple Structural Surfaces as shown in Figure 6.21. Detail models will vary according to mechanical components, especially the suspensions, see Figure 6.22. In this model the front suspension loads will be applied at the top of the front wing, as for a strut suspension, while the rear suspension loads are applied to the inner longitudinal member under the boot floor. This arrangement would

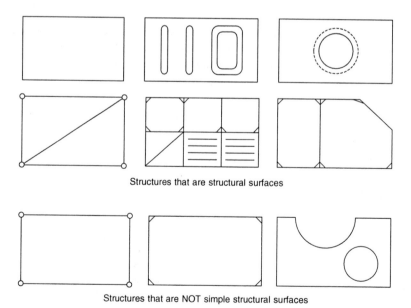

Structures that are structural surfaces

Structures that are NOT simple structural surfaces

Figure 6.21 Definition of Simple Structural Surfaces

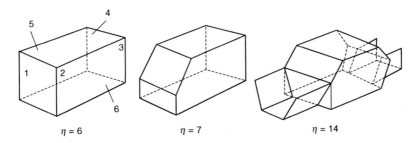

Figure 6.22 Vehicle structures represented by SSS

apply for a twist beam/trailing arm suspension. Other suspension types and body types (i.e. Hatchback) will require different SSSs to represent the structure.

The diagram, Figure 6.23, shows a half model for simplicity; the distribution of loads, again for simplicity, are limited to five loads plus the uniformly distributed load representing the body weight. The main loads are, F_{1z} = (radiator, bumper, battery)/2, F_{2z} = (engine)/2, F_{3z} = one front passenger and seat, F_{4z} = one rear passenger, seat, and half fuel tank, F_{5z} = (luggage)/2.

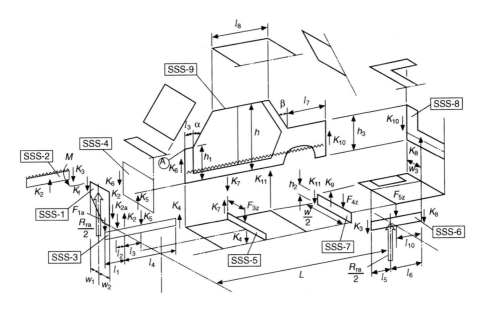

Figure 6.23 SSS model, saloon car – bending

The reactions at the front and rear axles are determined first. Take moments about the rear axle and obtain $R_{zf}/2$ (a half-model is being considered). Take moments about the front axle and obtain $R_{rz}/2$ and check by resolving vertically that vertical forces are in equilibrium. Then working through the Simple Structural Surfaces the edge forces can be obtained.

SSS-1 (Transverse SSS representing the strut tower)

Resolving Forces $\qquad K_1 + K_2 - \dfrac{R_{fz}}{2} = 0$ $\qquad\qquad$ (6.20)

Moments $\qquad K_1 = \dfrac{R_{fz}}{2} \dfrac{w_2}{(w_1 + w_2)}$ $\qquad\qquad$ (6.21)

SSS-2 (Upper front longitudinal)

Resolving Forces $\qquad K_1 - K_3 - u(\ell_1 + \ell_3) = 0$ $\qquad\qquad$ (6.22)

Moments $\qquad K_1\ell_3 - u\left(\dfrac{(\ell_1 + \ell_3)^2}{2}\right) - M = 0$ $\qquad$ (6.23)

SSS-3 (Lower front longitudinal)

Resolving Forces $\qquad F_{1z} + F_{2z} + K_5 - K_2 - K_4 = 0$ $\qquad\qquad$ (6.24)

SSS-4 (engine fire wall)

Resolving Forces (and by symmetry) $\quad K_5 - K_6 = 0$ $\qquad\qquad$ (6.25)

SSS-5 (Floor cross-beam (front))

Resolving Forces (and by symmetry) $\quad K_7 - K_4 - F_{3z} = 0$ $\qquad\qquad$ (6.26)

SSS-6 (Longitudinal under boot)

Resolving Forces $\qquad K_9 + K_8 - R_{rz}/2 + F_{5z} = 0$ $\qquad\qquad$ (6.27)

Moments $\qquad K_9 = (R_{rz}\ell_6/2 - F_{5z}\ell_{10})/(\ell_5 + \ell_6)$ $\qquad$ (6.28)

SSS-7 (Floor cross-beam (rear))

Resolving Forces (and by symmetry) $\quad K_9 - K_{11} - F_{4z} = 0$ $\qquad\qquad$ (6.29)

SSS-8 (Rear panel)

Resolving Forces (and by symmetry) $\quad K_{10} - K_8 = 0$ $\qquad\qquad$ (6.30)

There are now eleven equations, 6.20 to 6.30, and the eleven unknowns ($K_1 \dots K_{10}$, M) can now be evaluated. However, the equilibrium of the right-hand side-frame must be verified by resolving forces and moments.

SSS-9 (Right-hand side-frame)

Resolving Forces $\qquad K_6 - K_7 + K_{11} + K_{10} - u(L + \ell_6 - \ell_3) = 0$ $\qquad\qquad$ (6.31)

Moments about (A) $\qquad K_{10}(L + \ell_6 - \ell_3) + K_{11}(L - \ell_3 - \ell_5) - K_7 (\ell_4 - \ell_3)$

$\qquad\qquad - u(L + \ell_6 - \ell_3)^2\, 2 = 0$ $\qquad\qquad$ (6.32)

It should be noted that the SSSs 1 to 9 are subject to loads while the rear boot top frame, rear screen, roof, windscreen, floor panel and boot floor have no loads applied to them. This analysis shows that the side-frame carries the major loads and is the main structural member for determining the bending stiffness and strength of the car.

6.3.5 Simple Structural Surfaces representing a saloon car in torsion

Using the same model as for the previous section, the torsional load condition can be considered, see Figure 6.24. This shows a half-model for simplicity.

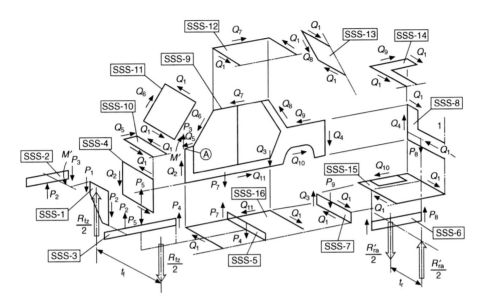

Figure 6.24 SSS model, saloon car – torsion

If the front axle load is assumed to be lighter than the rear then the maximum torque that can be applied to the structure is

$$\frac{R_{fz}}{2} * t_f = \frac{R'_{rz}}{2} * t_r \qquad (6.33)$$

The front axle load R_{fz} and the vehicle front and rear track t_f and t_r respectively are known, i.e. where the suspension mounting points are positioned, R'_{rz} can be obtained.
Now the equilibrium of SSSs can be considered, beginning at SSS-1.

SSS-1 (Representing the strut tower)

Resolving forces $\qquad\qquad P_1 + P_2 \frac{R_{fz}}{2} = 0 \qquad (6.34)$

Moments $\qquad\qquad P_1 = \frac{R_{fz}}{2} \frac{w_2}{(w_1 + w_2)} \qquad (6.35)$

Note: these are similar to equations 6.20 and 6.21.

SSS-1′ (Strut tower on left-hand side)
The loads on this SSS will be equal but opposite to the right-hand side.

SSS-2 (Upper front longitudinal)

Resolving forces $\qquad\qquad\qquad P_3 - P_1 = 0$ (6.36)

Moments $\qquad\qquad\qquad\qquad M - P_1\ell_3 = 0$ (6.37)

SSS-2′ (Upper front longitudinal on left-hand side)
This will have equal but opposite loading.

SSS-3 (Lower front longitudinal)

Resolving Forces $\qquad\qquad\quad P_2 + P_4 - P_5 = 0$ (6.38)

Moments $\qquad\qquad\qquad\qquad P_5 = P_2\ell_4(\ell_4 - \ell_3)$ (6.39)

SSS-3′ (Lower front longitudinal on left-hand side)
This will have equal but opposite loading.

SSS-5 (Floor cross beam (front))

Moments $\qquad\qquad\qquad\qquad P_4(t_f - 2w_2) - P_7w = 0$ (6.40)

SSS-6 (Longitudinal under boot floor)

Resolving Forces $\qquad\qquad\quad P_9 + P_8\dfrac{R'_{rz}}{2} = 0$ (6.41)

Moments $\qquad\qquad\qquad\qquad P_9 = \dfrac{R'_{rz}}{2}\ell_6/(\ell_6 + \ell_5)$ (6.42)

SSS-6′ (Longitudinal under boot floor on left-hand side)
This will have equal but opposite loading.

At this stage there are nine unknowns ($P_1 \ldots P_5, P_7 \ldots P_9, M'$) and nine equations, 6.34 to 6.42. These are not simultaneous equations and can be solved in sequence.

The remaining SSSs must be considered, which are all panels in shear that make a shear box.

SSS-4   (Engine fire wall)

Moments $\qquad\qquad\qquad\qquad P_5(t_f - 2w_2) - Q_1h_1 - Q_2w = 0$ (6.43)

SSS-7   (Floor cross beam (rear))

Moments $\qquad\qquad\qquad\qquad P_9t_r - Q_1h_2 - Q_3w = 0$ (6.44)

SSS-8   (Rear panel)

Moments $\qquad\qquad\qquad\qquad P_8t_r - Q_1(h_3 - h_2) - Q_4w = 0$ (6.45)

SSS-10 (Front parcel tray)

Moments $\qquad\qquad\qquad\qquad Q_1\ell_9 - Q_5w = 0$ (6.46)

SSS-11 (Windscreen frame)

Moments $\qquad\qquad\qquad\qquad Q_1(h - h_1)/\cos\alpha - Q_6w = 0$ (6.47)

SSS-12 (Roof panel)

      Moments $\qquad\qquad Q_1\ell_8 - Q_7w = 0 \qquad\qquad\qquad\qquad$ (6.48)

SSS-13 (Back-light frame)

      Moments $\qquad\qquad Q_1(h - h_3)/\cos\beta - Q_8w = 0 \qquad\qquad$ (6.49)

SSS-14 (Boot/Trunk top frame)

      Moments $\qquad\qquad Q_1\ell_7 - Q_9w = 0 \qquad\qquad\qquad\qquad$ (6.50)

SSS-15 (Boot/Trunk floor panel)

      Moments $\qquad\qquad Q_1(\ell_5 + \ell_6) - Q_{10}\,w = 0 \qquad\qquad$ (6.51)

SSS-16 (Main floor)

      Moments $\qquad\qquad Q_1(L - \ell_5 - \ell_3) - Q_{11}w = 0 \qquad\qquad$ (6.52)

Note: SSS-11 to SSS-16 are all in complementary shear.

SSS-9 (Side frame)
      Moments about A

$$Q_4(L + \ell_6 - \ell_3) + Q_3(L - \ell_5 - \ell_3) + P_7(\ell_4 + \ell_3) + M' + Q_6(\ell_9 \cos\alpha)$$
$$- Q_7(h - h_1) - Q_8 \cos\beta(L + \ell_6 - \ell_7 - \ell_3) - Q_8 \sin\beta(h_3 - h_1)$$
$$- Q_9(h_3 - h_1) - Q_{10}(h_1 - h_2) - Q_{11}(h_1) = 0 \qquad\qquad (6.53)$$

There are a further 11 simultaneous equations, 6.43 to 6.53, with 11 unknowns ($Q_1 \ldots Q_{11}$). These can all be solved using matrix methods but it should be noted that equations 6.43 to 6.52 can all be rearranged to give Q_2 to Q_{11} in terms of Q_1. By substituting the necessary terms in equation 6.53 this equation becomes an equation with only one unknown Q_1. Hence the value of Q_1 is obtained. By substituting for Q_1 in equations 6.43 to 6.52 the remaining unknowns are obtained.

Final checks should be made to ensure no arithmetic errors occur by three further procedures. First, resolve forces vertically on the side-frame

$$Q_2 + P_3 - P_7 - Q_3 - Q_4 - Q_6 \cos\alpha + Q_8 \cos\beta = 0 \qquad\qquad (6.54)$$

Second, resolve forces horizontally on the side-frame

$$Q_{11} + Q_{10} - Q_9 - Q_8 \sin\beta - Q_7 - Q_6 \sin\alpha - Q_5 = 0 \qquad\qquad (6.55)$$

Third, check that the shear flow on panels in complementary shear are equal.

Shear flow is the shear force per unit length and for panels subject only to complementary shear the shear stress and hence shear flow must be equal on each side. If this is applied to SSS-12, the roof panel then

$$q_1 = \frac{Q_1}{w} = q_7 = \frac{Q_7}{\ell_8} \qquad\qquad (6.56)$$

Similar checks should be made on SSS-10, 11, 13, 14, 15 and 16. Panels SSS-4,7 and 8 that are subject to other forces P_5, P_8, P_9 although subject to shear will not have equal shear flows due to these forces.

Examination of Figure 6.24 reveals that shear is applied to all the 'panels'. These include components such as the windscreen frame, the back-light frame, the boot/trunk top frame, the rear panel, the floor panel and the boot/trunk floor panel. It is imperative that these panels and frames are constructed to have good shear stiffness. Floor panels require swaging to prevent buckling while the roof curvature assists in preventing the same.

The windscreen frame and the back-light frame must be constructed with stiff corner joints to ensure that the shear is carried across the frame up to the roof, see Figure 6.17(a). A single poor frame stiffness will result in poor vehicle torsional stiffness. The windscreen frame and the back-light are stiffened by the glass which acts as a shear panel. Modern glazing methods result in the glass being bonded to the frame so that the glass is retained in frontal impacts. This bond also acts as a good shear connection and hence the glass is subject to shear stress. If the surrounding frame is insufficiently stiff then the glass can be over stressed and glass cracking can result.

Another observation is that the rear panel and boot/trunk top frame are subject to shear. In fact these two components are not very good SSSs because of the large discontinuity caused by the boot/trunk lid. This problem is often overcome by constructing the rear with a high sill or 'lift over' which makes poor access for loading baggage. Also the sides of the rear panels which house the rear lights are often made wide as are the sides of the boot top frame.

A better structure will incorporate a panel or cross brace in the plane of the rear seat back. This used to be a design feature but modern cars do not usually include this feature because customers require rear folding seats to permit accommodation of long objects protruding from the luggage area.

6.4 Computational methods

In the context of vehicle fundamentals, mention of computational methods for structural analysis must be made, as these methods are now fundamental in the vehicle design process. Structural analysis is now centred around the Finite Element Analysis method where the vehicle structure is divided into small elements. The equations of statics (and/or dynamics) plus the equations of stress analysis and elasticity for each element are solved simultaneously using matrix methods. This is not considered in this chapter, but the theoretical aspects of this method are contained within the many textbooks available on the subject (NEL, 1986).

The complexity of Finite Element models has increased enormously as engineers have attempted to model vehicles in greater detail. In this section, simple and complex models are described, and examples of how complex models can be sub-divided into more manageable problems are given.

Early models, which for initial investigations are still used today, use simple beam elements. The beam elements are chosen to represent the main structural members such as sills, window pillars, engine rails, and floor cross beams. Panels such as the floor, roof, and bulkheads can be represented by 'equivalent' beams that have stiffnesses equivalent to the shear panels. Figure 6.25 from Lotz (1991) shows an example of a beam element model of a cabriolet car. This model shows the use of 'equivalent' beams to represent the panel members.

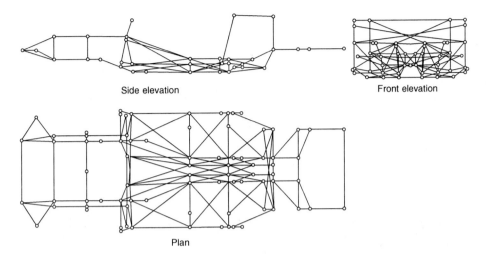

Side elevation

Front elevation

Plan

Figure 6.25 Basic beam model of Cabriolet body with equivalent beams representing panels (Lotz, 1991)

Later models use plate or shell elements that more accurately represent sheet metal components. Figure 6.26 from Kuo and Kelkar (1995) shows these to be quadrilateral or triangular elements and a complete model consists of thousands of these elements. The load input points for a vehicle of this complexity probably number 30 with components in three orthogonal directions. Hence, both the number of loads and the number of elements result in a very large data-set. This requires considerable preparation time and a long computer processing time.

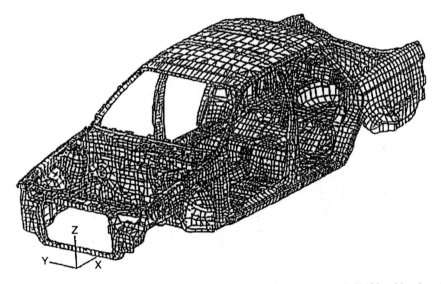

Figure 6.26 Complete body Finite Element Model, with 66310 elements, and 61420 grid points (Kuo and Kellar, 1995)

At the initial design stage it is not necessary nor possible to determine the stresses and deflections in such detail because the detail geometry of the components will not be known. Therefore the application of the Simple Structural Surfaces method, as described in Section 6.3, and the use of simple beam element programs can provide the designer with useful design tools. Results from these methods applied to the whole body can then be used for the loading applied to sub-assemblies modelled with Finite Element Methods using shell and plate elements.

Figure 6.25 Lotz (1991), shows a beam element model that evaluates the whole cabriolet vehicle structure, the results from which were used in investigating the rear seat pan sub-assembly shown in Figure 6.27.

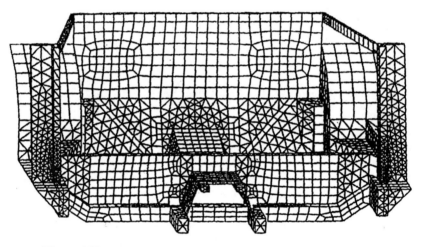

Figure 6.27 Finite Element Model of rear seat structure (Lotz, 1991)

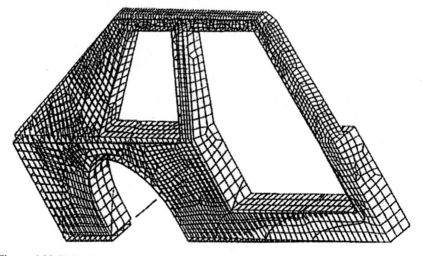

Figure 6.28 Finite Element Model of sideframe indicating displacements (Hansen, 1996)

An example of applying the Simple Structural Surfaces method is described by Hansen (1996) where the edge loads applied to the side-frame were obtained. These were then used to determine the stresses and deflections within the side-frame. Figure 6.28 shows the original and deflected shape of the side-frame.

6.5 Summary

This chapter has outlined the basic design load cases that are considered when designing a passenger car structure. These are the loads that are subject to the vehicle when traversing roads and other surfaces. Impact load conditions are not considered here. The Simple Structural Surfaces method is outlined and applied for the two main load cases of bending and torsion. Loads on individual members obtained from this method can be used to determine local stresses and deflections. The fundamental design of the structure and its main components can be established in this way. Application of Finite Element Methods can then be applied to the fundamental design to achieve improved details and greater structural efficiency.

6.6 References and further reading

Bastow, D. (1987). *Cars Suspension and Handling*. Pentech Press.
Hansen, R. (1996). A feasibility study of a composite vehicle structure, Cranfield University, MSc thesis.
Kuo, E.Y., and Kelkar, S.G. (1995). *Vehicle body structure durability analysis*. SAE Paper 951096.
Lotz, K.D. (1991). Finite element analysis of the torsional stiffness of a convertible car body, Cranfield University, MSc Thesis
NEL (1986). *A finite element primer*, National Agency for Finite Element Methods and Standards (NEL).
Newcomb, T.P., and Spurr, R.T. (1996). *Braking of Road Vehicles*, Chapman and Hall.
Pawlowski, J. (1964). *Vehicle Body Engineering*, Business Books.
    [This text demonstrates many simple methods for analysing vehicle structural designs and provides excellent background reading on the subject.]
Roark, R.J. (1975). *Formulas for Stress and Strain*, McGraw-Hill.
Webb, G.G. (1984). *Torsional stiffness of passenger cars*, C172/84, I.Mech.E.

7. Crashworthiness and its influence on vehicle design

Bryan Chinn, PhD

The aim of this chapter is to:
- Indicate the relationship between injury and accident type;
- Introduce the subject of vehicle crash dynamics;
- Demonstrate methods for vehicle and component design to reduce accident injury levels;
- Indicate the possibilities of active and passive safety.

7.1 Introduction

Crashworthiness was, for many years, seen by the automotive designer as something to be tolerated and necessitated only that the seat belts complied with the British Standard, and that the modest requirements of an impact at 48 km/h into a solid barrier were met, but over the last 10 to 15 years attitudes have changed. Various governments have campaigned to reduce the accident toll and many car manufacturers, led by Volvo, now promote safety as a sales feature. This has led to a fall in the accident and injury rate, particularly in the UK, notable as the road safety 'chart topper' within the European Union. However, in spite of the improvements, road accidents, a modern epidemic, is the most frequent cause of premature death and 47 000 occupants die in car crashes each year in the EU alone; this is almost equivalent to one Jumbo Jet dropping out of the sky every three days.

This chapter begins with a review of accident and injury mechanisms, an essential starting point if crashworthiness is to be properly understood and improved. The principal impact types are identified, as are the mechanisms of injury for the most vulnerable, head and chest, and most frequently injured body regions. Seat belts are the most effective, and airbags the most recent, injury reducing devices; the benefits and disadvantages of both are reviewed and related to injury savings and patterns.

Injuries vary by severity, location and combination, they can only be compared, case by case, by the use of a ranking system. The Abbreviated Injury Scale, 'AIS' (American Association for Automotive Medicine) was devised to do just this and is now used throughout the world. It assigns a value of from zero to six for every conceivable injury; zero is uninjured and six is not survivable. Of course, a casualty (an injured person) can have many injuries and, therefore, Maximum AIS, 'MAIS', was devised to identify the most severe injury sustained by one casualty.

Technologists tend to regard the computer and its attendant CAD and finite element packages, as the beginning and end of crashworthiness, but this is a mistake and will lead to a lack of understanding of the basic principles, with frustration a far more likely outcome than improved protection. Classical mechanics is the basic tool used to illustrate the general dynamics in both front and side collisions, Section 7.3, and this is followed in Section 7.4 by an examination of the effect of the vehicle crush characteristics in impacts into a rigid barrier and in collisions between two vehicles. Rigid barriers are used in legislative tests but accidents are frequently collisions between two vehicles: this section explains the importance of this difference and the consequences.

Structural collapse and the associated energy absorption and intrusion is fundamental to crashworthiness. Section 7.5 reviews the way in which these characteristics affect safety and what may be changed to bring about improvements. This is particularly difficult in side impacts where space is limited and this section concludes the chapter with an outline of the construction and testing of a novel side impact airbag system.

7.2 Accident and injury analysis

7.2.1 Injury by impact type

General classification
To assess the priority of any safety device it is necessary to consider which occupants are affected, the impact types that are relevant to the device, the injury severity distribution and the potential protection that can realistically be achieved.

The figures for fatalities given in Table 7.1 (Consumers' Association, 1993) show that frontal impacts are the most important, closely followed by side impacts and Thomas *et al.* (1992) agree with this.

Harms *et al.* (1987) presented a breakdown of injury sources. The results are for impact type and injury severity for different body areas, an analysis that indicates priorities. The results

Table 7.1 Consumers' Association 1993 figures for the fatal casualties in different impacts (Consumer's Association, 1993)

		Belted	Unbelted	All
Drivers	Frontal	61	50	56
	Side	30	29	31
	Rear	1	2	2
	Rollover	2	7	4
	Other	6	12	8
Front passengers	Frontal	45	42	42
	Side	44	27	42
	Rear	0	4	1
	Rollover	0	8	1
	Other	11	19	14
Rear passengers	Frontal	35	35	35
	Side	40	36	37
	Rear	10	4	45
	Rollover	5	15	14
	Other	10	11	

indicate that contact with the steering assembly is a substantial cause of injuries to the head and face for belted drivers whereas for the front seat passenger the seat belt is cited as the cause of a large proportion of the chest and abdomen injuries.

It is important to consider the likely cost-benefit of a safety system, for which an analysis of the frequency and severity of injuries is necessary. It may be better to save a small number of fatalities rather than a large number of slight injuries. The Consumers' Association Secondary Safety Rating System (Consumer's Association, 1993), as detailed in Table 7.3, allows the priority for injury prevention to be determined.

Table 7.2 Injury sources for belted occupants in frontal collisions for vulnerable body regions (AIS 2)

	Head/Face	Neck	Chest	Abdomen	Total
Steering wheel	159	3	30	10	202
Seat belt	0	0	180	9	189
Other vehicle	23	1	4	2	30
A-pillar	20	1	7	0	28
Windscreen and frame	32	0	0	0	32
Fascia	13	0	7	2	22
Other occupant	7	1	1	7	16
Bonnet	20	0	0	0	20
Roof	9	4	0	0	13
Glass	10	0	0	0	10
Front header	10	1	0	0	11
Own seat	1	0	6	0	7
Other	6	0	3	5	13

Side impacts
The proportion of side collisions in all injury accidents lies between 15% and 40% (Otte, 1982, 1990, Rouhana and Foster, 1985, Danner *et al.*, 1987, Ropohl, 1990). However, if only belted occupants are included then front impacts account for more than 50% and side impacts 20–25% of all injury accidents to car occupants (Niedere, Waltz and Weissnerb, 1980, Kallieris and Mattern, 1984, Morris *et al.*, 1995); furthermore if only serious and fatal injuries (AIS 3-6) are considered then the proportion of injuries attributable to side impacts increases by 50%. Multiple serious injuries to the head, chest, abdomen, and pelvis, are typical in side impacts (Gloynes *et al.*, 1989; Thomas *et al.*, 1987).

In Fildes and Vulcan's study (1990) of 150 side impacts, struck-side front seat occupants were injured mainly by interior vehicle structures. In general, far-side occupants were less frequently injured: 36% were due to contact with other occupants, 27% by contact with interior vehicle surfaces, 18% from seat belts and 18% by contact with the dashboard. Rouhana and Foster (1985) describe similar results. Fildes and Vulcan also found that the most frequently injured body regions were the abdomen 90%, the chest 70%, and the head and upper extremities 63%.

Contact sources in lateral impacts have been examined by Mackay *et al.* (1983). For sources of injury to fatally injured occupants >AIS 3, the most frequent contact source was an exterior object such as another vehicle, pole or tree 42%, the side-header 19% and via ejection 6%. In their survey in 1993 Miltner and Salwender (1995) found a clear correlation between the total injury severity and the energy equivalent speed (EES) for occupants seated on both sides. Within the critical range of 30–50 km/h for nearside occupants and 40–60 km/h for far-side occupants, the probability of severe injuries of MAIS 4–6 increased from approximately 20% to more than 90%. The conclusion is that far-side occupants are less endangered than those adjacent to the impact. Nevertheless, within the critical EES range of 40–59 km/h there is still the risk of severe injury for far-side occupants: 25% compared with 50% for the near side. Thus all seating positions must be considered when occupant protection measures are being developed.

7.2.2 Injury patterns and seat belt use

Frontal impacts
Throughout Europe the majority (more than 50%) of front seat car occupants wear a seat belt and in the UK this figure is as high as 92% (Dept. of Transport, TRL, 1996). The effectiveness of seat belts has been established beyond doubt. In the UK, Rutherford *et al.* (1985), following the 1983 introduction of the front seat belt law, found the number of patients taken to a hospital was reduced by 15%, the number of patients requiring admission to a hospital was reduced by 25% and the number of fatalities fell by nearly 26%.

Table 7.3 Consumers' Association Secondary Safety Rating system

Area	Weighting
Front impacts	
Steering wheel head/face impact	69
Steering wheel chest impact	51
Steering wheel and body shell intrusion	50
Driver and passenger leg impact area	42
Seat strength	38
Front belt design	38
Rear belt design	20
Header/pillar padding	18
Side impacts	
Side impact structure and padding – front	29
– rear	8
Roof rail padding	23
Rear impact	
Head restraints – front	34
– rear	5
Rollover/complex accidents	
Door locks	17
Fuel system	13
Front belt system	12
Rear belt system	1

Nevertheless, although seat belts have effected a substantial overall reduction in injury, the pattern has changed. Mackay *et al.* (1995) identified four categories of seat belt inadequacy and, in turn, have identified typical injuries:

- Head and face contact with the steering wheel is almost certain to occur in collisions of about 50 km/h in which the head will arc forward and downwards with a horizontal translation of some 60 cm to 70 cm; injuries are usually AIS 1 to 3. The suggested solution is an airbag, but this has been found to cause problems for out of position drivers.
- Rear loading from unrestrained occupants can cause injuries to correctly restrained front seat occupants although this problem has greatly diminished as a result of legislation that requires rear occupants to wear seat belts.
- Misuse of the seat belt is frequent with those who are overweight who tend to place the seat belt over the abdomen instead of low across the pelvis; the consequence is often severe abdominal injuries at relatively low impact speeds.
- The most frequent injuries caused by the seat belt are fractures to the ribs and sternum, particularly for the elderly.

7.2.3 Injury patterns and airbag use

Airbag implementation
In the early 1960s, faced with disappointing low seat belt use, work in the US turned to passive restraint systems. The intention was to protect car occupants without them having the need to take any action themselves, such as fastening a seat belt. The US regulation FMVSS 208 provides for protection in frontal impacts and requires passive protection of the front seat occupants and the industry has used airbags as the way to meet this requirement. Airbags are controlled by performance requirements specified as dummy criteria, which must be met in the standard 30 mph (48 km/h) full-frontal impact test without the use of seat belts. The requirement is that the criteria must be met without the use of seat belts, and this effectively controls the size of the air bag that is needed. American air bags are typically 70 litres in volume, are deployed at an impact speed of 16 km/h and inflate very rapidly. In Europe, where seat belt use is frequent, smaller bags specifically intended to complement the use of a seat belt are fitted. This smaller bag is known as a 'European' or face bag and is 30–45 litres for the driver's side and 60 litres for the passenger's side. They are designed to deploy at an impact speed of between 24 km/h and 32 km/h and inflate more slowly than American bags, typically within 50 ms.

Air bags: benefits and injuries
Dalmotas et al. (1996a) found that supplementary air bag systems significantly reduce (26.7%) the risk of severe head and facial injuries among belted drivers (collision severities at 40 km/h). However, these benefits are being negated by air bag-induced injuries, most notably to the face in moderate and low speed collisions, and to the upper extremities at all collision severities.

The safety benefits achieved at higher collision severities are being negated by the higher incidence of a bag induced injury in low and moderate collision severities. When seat belt use is very frequent, the vast majority of air bag deployments in low speed collisions serve no useful purpose. In such collisions, the injury outcome is either unchanged or adversely affected. While the majority of air bags related injuries are AIS 1 facial and AIS 1–3 upper extremity injuries, they can include AIS > 3 injuries to other body regions when the occupant is close to the deploying air bag.

Of particular concern are possible adverse air bag occupant interactions if the seat is located forward of the middle position. Evidence from Canadian case studies shows that the proximity of an occupant to the air bag module has a strong influence on the response of the neck and the chest (Melvin *et al.*, 1993). Dalmotas *et al.* (1996a) quotes a report by the NHTSA to the US Congress 1996 which shows that the current air bag systems are unlikely to reduce the risk of moderate injuries to belted or unbelted drivers. He claims that if the deployment threshold of airbags was increased then belted occupants would be much better protected.

Alternatively, the belted driver may be better protected if current airbags were less aggressive. Reducing the inflation rate and bag pressure in combination with seat belt improvements, such as belt pre-tensioners, is possibly a much better way of substantially improving protection in impacts where injuries of AIS 1–3 may be expected. This has been achieved in Europe without compromising protection in high speed collisions.

Manufacturers test an airbag only for a standard driving position, which does not account for the many positions in which occupants sit. It has been suggested that some drivers may be as close as 24 cm from the steering column hub. Lau *et al.* (1993) have shown in animal experiments, the rapid deployment of the airbag can generate complex biomechanical forces between the head neck and torso and within the chest and this result is important in the deployment to protect out of position occupants. Walter and James (1996) report on an airbag deployment that occurred when the driver was close to the steering wheel. The initial part of the expansion was restricted by the chest, which was then subjected to large forces generated when the bag began to inflate rapidly. This caused an increase in the upward expansion of the bag, and, in turn, produced substantial shearing forces on the skin of the neck, hyperextension of the neck, and increased forces on the chest.

It is possible that a steering wheel with an uninflated air bag will be stiffer than one optimized for face contact. Car design, therefore, needs to ensure that airbag installation in cars does not increase face and head injuries at low impact severities. Other analyses of crash data (Thomas *et al.*, 1994), have shown that head and face injuries represent only 30% of the economic cost of steering wheel injuries with more than 60% resulting from chest and abdomen injuries. The most effective type of airbag should reduce both head and torso injuries in conjunction with an effective European type of seat belt.

7.3 Vehicle impacts: general dynamics

7.3.1 Front impact

The accident review above has shown that the most frequent fatal and serious injury producing accident impact is a frontal collision; side impact is the next most frequent vehicle designers have, therefore, had a tendency to concentrate upon improvements to provide better protection in frontal impact not only because it is the most frequent but greater space available ahead of the passengers' compartment allows more scope for treatment. Nevertheless, frontal impacts remain of concern and it is important to consider the general dynamics of such collisions before studying the improvements in crashworthiness that may lead to reduced injury severity. Figure 7.1 illustrates a full head-on collision between two vehicles.

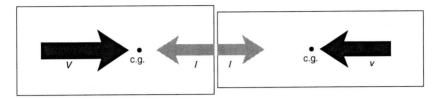

Figure 7.1 Head-on collision

The notation adopted (Macmillan, 1983) one is that upper case is used to denote vehicle one and lowercase for vehicle two. The subscript *1* is used for conditions immediately before the impact and two for conditions immediately after it. It is important to note that impulse and momentum are related as follows:

$$\text{Impulse} = \text{Force} \times \text{time} = \text{change in momentum}$$

During an impact (duration typically 100 to 200 ms) a variable force F acts between the two vehicles and, therefore, the linear impulse, I, is given by

$$\int_{t_1}^{t_2} F dt = I$$

When $t_1 = t_2$, then $I = I_2$. Considering the momentum of each vehicle gives

$$MV_1 - I = MV$$

$$mv_1 - I = mv$$

where v_1 and V_1 are initial velocities of the vehicles *towards* the point of impact, and v and V is their velocity at any time during the impact. This notation and sign convention was chosen because, by applying symmetry principles, it is possible to derive expressions for the velocities and other variables of the second vehicle when those related to the first vehicle have been determined.

The velocity during the impact is an important parameter which will be denoted by V_r where:

$$V_{r1} = V + v$$

Figure 7.2 is a curve of the integral with time of the acceleration of a vehicle during an impact at 48 km/h and represents a typical velocity curve.

It is evident from Figure 7.2 that the velocity follows a general trend of decreasing until it reaches zero and then becomes negative as the vehicle rebounds. Thus, the impact can be divided into two phases. In the first, from t_1 to t_0 the vehicle structures are compressed and distorted until V_r is reduced to zero and the vehicles are moving together; in the second phase some of the elastic strain energy in the vehicle structures is restored and the vehicles separate with a negative velocity $-V_{r2}$. During the first phase the impulse between the vehicles is I_0 and during the second phase it is $(I_2 - I_0)$.

The two equations contain three unknown quantities V, v, and V_r, so they are insufficient to determine the final velocities without additional information. It is known from Newton that the impulse $(I_2 - I_0)$ is proportional to V_r and the ratio is the coefficient of restitution e. Thus, we have

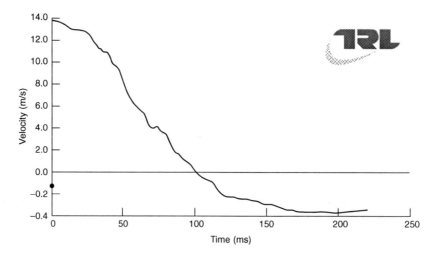

Figure 7.2 Velocity against time of a VW Polo at 48 km/h impacting a Ford Mondeo travelling at 48 km/h

$$I_2 - I_0 = eI_0 \text{ and}$$
$$\therefore \quad I_2 = (1 + e)I_0$$

Considering the time t_0, we can write $I = I_0$, $v = v_0$, and $V = V_0$ and this gives

$$V_{r2} = -eV_{r1}$$

If the instant t_2 is considered then:

$$mv_1 - I_2 = mv_2 \text{ and } MV_1 - I_2 = MV_2 \text{ which together with}$$
$$v_2 + V_2 = -e(v_1 - V_1)$$

can be rearranged to give the full expressions for the velocities for the general case as follows:

$$v_2 = v_1 - M(1 + e)(v_1 + V_1)/(m + M)$$
$$V_2 = V_1 - m(1 + e)(v_1 + V_2)/(m + M)$$

The totally inelastic collision, $e = 0$, represents a perfect energy absorber (total energy absorption), which is likely to be optimum for occupant protection and is the ideal towards which car designers may strive. In practice such ideals do not exist. It is useful, therefore, using the impact illustrated in Figure. 7.2 to put some practical figures into the equations: the mass m of the Polo was 972 kg and the mass M of the Mondeo was 1504 kg; both were travelling at 13.78 m/s. The final velocity of the Polo was –3.77m/s and that of the Mondeo was 3.30 m/s. Thus e, from the equations, was 0.4. The reader may wish to determine the consequences for a perfectly elastic and perfectly inelastic collision. A further simplification to consider is a stationary struck vehicle with $V_1 = 0$.
    Whence

$$v_2 = v_1(m - eM)/(m + M)$$

$$V_2 = - mv_1(1 + e)/(m + M)$$

and finally, an impact into a fixed rigid barrier where

$$V_1 = 0 \text{ and } M = 8$$

$$(1/M = 0) \quad \text{whence} \quad v_2 = - ev_1$$

These results may seem trivial but it can be shown that it is possible to apply the same technique to the solution of the more difficult general problem of the plane impact as illustrated in Figure 7.3. Consider the vehicle to the right of the axes' intersection to have an initial vector velocity $\underline{v}$ and an initial angular velocity $\dot{\theta}_1$. At the surface of impact there is a compressive impulse I (*Fdt*) together with a tangential impulse J. The tangential impulse is caused by a combination of friction and interlocking between the two surfaces. Let the coefficient of this interaction be λ thus $J = \lambda I$. The value of λ will almost certainly vary throughout the impact, but may be considered to have a definite value at t_2 the moment with which we are concerned.

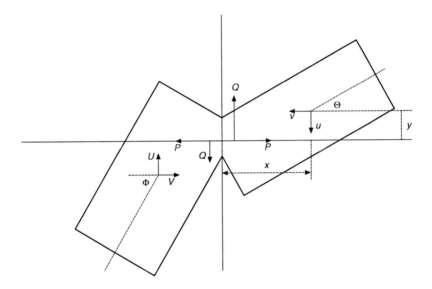

Figure 7.3 Plan of impact between two vehicles

If the vector velocity $\underline{v}$ makes an angle of θ with the normal to the impact surface then the components of velocity, normal (v) and tangential (u), are

$$v = \underline{v} \cos \theta$$

$$u = \underline{v} \sin \theta$$

also

$$\tan \theta = u/v \qquad\qquad (\theta = \tan^{-1} (u/v))$$

The momentum equations for one vehicle may now be written, and are for the linear momentum

$$I = mv_1 - mv$$

$$\lambda I = mu_1 - mu$$

for the angular momentum

$$Iy - \lambda Ix = mk2 - mk^2$$

Similar expressions for the other vehicle may be derived using V_1, U and ϕ and written by replacing lower case letters with upper case. The equations for the rates of approach of the two vehicles and of the relative motion of the impacting surfaces may be expressed as follows:

$$\text{Velocity along the } x \text{ axis } p = v + V - y - Y$$

$$\text{Velocity along the } y \text{ axis } q = u + U + \dot{\theta}x + \dot{\phi}X$$

The equation for restitution $I_2 = (1 + e)I_0$, as defined in 7.3.1 above, is still valid and may be used together with the previous eight equations to define the angular and resolved linear velocities of the two vehicles as follows:

Vehicle to the right of the axes' intersection

$$v_2 = v_1 - I_2/m$$

$$u_2 = u_1 - \lambda I_2/m$$

$$\dot{\theta}_2 = \dot{\theta}_1 + \frac{y - \lambda X}{mk^2} I_2$$

Vehicle to the left of the axes' intersection

$$V_2 = V_1 - I_2/M$$

$$U_2 = U_1 - \lambda I_2/M$$

$$\dot{\Phi}_2 = \dot{\Phi}_1 + \frac{Y - \lambda X}{M K^2} I_2$$

7.3.2 Side impact

The dynamics of side impact can clearly be considered in the same mathematical way as for frontal impact. However, because the vehicle structure between the occupant and the impact plane is so much smaller in a side impact than in a frontal impact, the assessment of injury potential for a given relative velocity and relative mass is much more closely related to the extent of intrusion and the velocity of the intruding component relative to that of the occupant. This complex problem is considered in Section 7.4.2.

7.4 Vehicle impacts: crush characteristics

7.4.1 Impact into a rigid barrier

Development of the equations
The analysis in Section 6.3 has shown that it is possible to predict the dynamic outcome of an

accident and to determine the amount of energy absorbed provided that the coefficient of restitution, denoted by *e*, for a particular impact is either known or can be reliably assumed. However, this analysis does not, and cannot, determine the value of *e*, nor describe how the energy loss is distributed in deforming the two vehicles, and nor can it be used to determine how much the vehicles are crushed.

Such questions may only be answered by an approach which examines the crushing and whilst computer analysis using finite element techniques provides details of the collapse of particular vehicle components or combinations of components it does not easily provide an insight into the overall behaviour of a vehicle during an impact.

Macmillan (1983) proposes an alternative approach based upon the results of many impact tests into barriers. The acceleration, velocity and displacement (crush) results of barrier impact tests tend to display similar characteristics. Figure 7.4 is a typical shape.

The acceleration curve has high frequency modulation caused by the erratic crumpling of the vehicle structure. The velocity and displacement curves are progressively smoother because of the filtering effect inherent in integration. However, these curves need to be idealized in order to examine the overall vehicle behaviour during an impact and hence, in turn, the effect of this behaviour on the vehicle occupants.

Macmillan (1983) stated, that what is needed is an analytical expression for the smoothed curves that satisfy the following criteria:

- It must be simple enough to be manipulated.
- It must satisfy the boundary conditions found in curves from impact tests.
- It must correlate well with known test cases and hence justify its use to predict the outcome over a range of unknown examples.
- It must be capable of representing the behaviour of vehicles with different crush characteristics with changes to a small number of variables.

The expression must also be applicable for all values of *e* from 0 to 1 and must satisfy the condition that

$$\frac{da}{dt} = 0 \text{ at } t = t_2$$

which ensures that an instantaneous rate of change of acceleration does not occur at the end of the impact.

Macmillan proposed the dimensionless equation for acceleration as follows:

$$\text{Acceleration } a = -\frac{cv_1}{t_2}\left(\frac{t}{t_2}\right)\left(1 - \frac{t}{t_2}\right)^\beta$$

where *c* is a dimensionless constant, to be determined, and *b* is a dimensionless index greater than unity.

Let $T = \frac{t}{t_2}$ and integrate which becomes

$$v = v_1 a_v(T)$$

where

$$\frac{a_v(T)}{c} = \frac{(1 - T)^{\beta+1}}{\beta + 1} - \frac{(1 - T)^{\beta+2}}{\beta + 2} - \frac{e}{c}$$

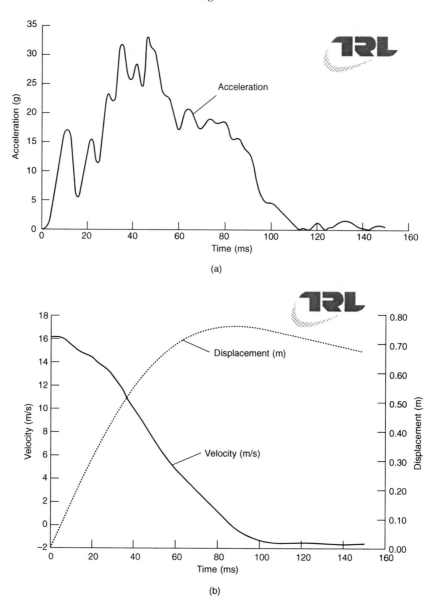

Figure 7.4 (a) Acceleration against time of an impact of a Vauxhall Cavalier at 58 km/h into a rigid barrier (b) Velocity and displacement against time of the impact shown in Figure 7.4(a)

substituting for $a_v(T)$ and integrating gives

$$S = v_1 t_2 a_s(T)$$

The parameter β_0 is called the *structure index* because soft nosed vehicles have small values of β_0 and hard nosed vehicles have larger values. A typical value for a medium sized car is $\beta_0 = 2$.

Two further parameters which characterize the crushing characteristics of a vehicle can now be defined. The first is the *crush modulus, C_m*, analogous to the modulus of elasticity. It is defined as the initial slope of the force displacement curve (F vs S) curve. It can be shown that:

$$C_m = \frac{mc}{t_2^2}$$

For a medium car C_m has a value typically of 1 – 1.5 kN/mm. Figure 7.5 shows a force deflection curve for a Vauxhall Cavalier into a rigid barrier and the slope of the initial part is the crush modulus.

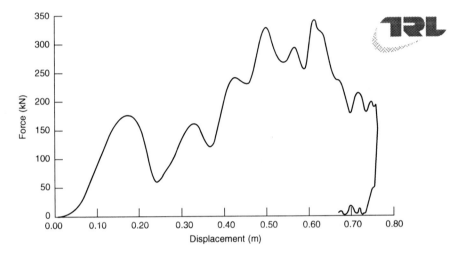

Figure 7.5 Force deflection curve for a Vauxhall Cavalier into a rigid barrier at 58 km/h

The crush modulus, $C_m = \dfrac{df}{ds} = \dfrac{mc}{t_2^2}$; the measured value = 1.24 kN/mm

The final vehicle impact parameter defines how severe an impact must be to cause the structure to collapse with plastic deformation and is called the Impact Severity Factor, K_s.

The severity of an impact can best be quantified by the magnitude of the mean force that is induced in the structure as follows:

$$\text{Mean force} = \frac{m(v_1 - v_2)}{t_2} = \frac{m(1 - e)v_1}{t_2}$$

When the mean force is small, the impact is almost elastic ($e \rightarrow 1$), and when it is large the deformation is almost plastic ($e \rightarrow 0$). Thus, e varies with mean force as shown in Figure 7.6. If the curve is assumed to be exponential then the equation may be written

$$K_s \ln\left(\frac{1}{e}\right) = \frac{m(1 + e)v_1}{t_2} \quad \text{where}$$

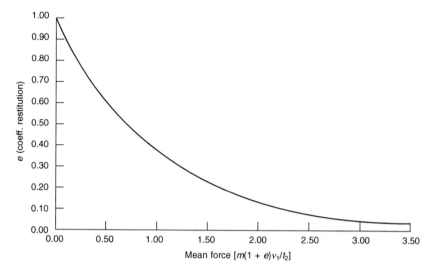

Figure 7.6 Mean force versus coefficient of restitution, *e*

$$a_k(e) = \frac{1 + e}{\ln\left(\dfrac{1}{e}\right)}$$

Macmillan (1983) gives a typical value of K_s for a medium car as 65 kN; for the Cavalier into the rigid barrier at 58 km/h it was 175 kN.

Special cases
Maximum force f_m: Maximum acceleration occurs when $t = t_m$ and $df/dt = 0$ and by differentiating, it can be shown that the maximum force is

$$f_m = \frac{mcv_1}{t_2} \, a_m \; (\beta)$$

Maximum dynamic displacement (crush) S_m: Maximum displacement occurs when $v = 0$ and is defined as follows

$$S_m = v_1 t_2 \, a_s \left(\frac{t_0}{t_2}\right)$$

7.4.2 Impacts between two vehicles

Frontal impacts
The above treatise has examined the process of crushing, in an impact with a fixed rigid barrier. It will be shown that this approach may be extended to the study of central impacts between two vehicles.

   It should be noted that *m* refers to the vehicle on the left and *M* to the vehicle on the right. These are used as subscripts in the following equations. Time is measured from the instant of

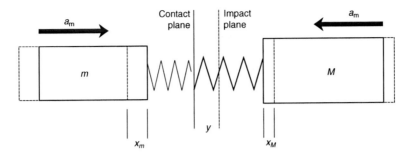

Figure 7.7 The displacement and acceleration of two vehicles during the crushing process of a central impact

contact, x_m is the movement of the vehicle of mass m, y is the movement of the plane of contact and f is the force between the vehicles at that plane. The deformation of the two vehicles is given by

$$s_m = x_m - y \quad \text{and} \quad s_M = x_M - y$$

The closing velocity is

$$V = \dot{s} = \dot{x}_m + \dot{x}_M$$

the closing acceleration is

$$\dot{V} = \ddot{x}_m + \ddot{x}_M = a_m + a_M = f\left(\frac{1}{m} + \frac{1}{M}\right)$$

and

$$f = \frac{mM}{m + M}\dot{p}$$

Using the method given in the previous section and writing C_c for the combined crash modulus, then this can be expressed in terms of the individual crash moduli as follows:

$$C_c = \frac{C_m C_M}{C_m + C_M}$$

This approach enables a designer to examine the performance of the design car when impacting various other cars assuming that certain facts for each car are known. However, the performance of a car in an impact will depend upon the crush characteristics of the colliding members of each vehicle and, therefore, particular criteria can be met only if the design is realised in practice.

Side impacts
The outcomes for the impact of two vehicles when expressed mathematically, is simply a function of the dynamics of the impact and the mass and crush characteristics of the colliding vehicles, as described above. However, in frontal impacts the occupant is some distance from the impact plane and the injury outcome is very dependent upon the crush characteristics of the vehicle front, which, in turn, may be modified to suit the requirements of restraint systems. In side impacts it is the extent and velocity of the intrusion when the occupant is struck that is the prime cause of injury; Neilson (1969) provides a very elegant explanation.

Figure 7.8 shows a car Q of mass M' struck in the side by a car P of mass M travelling at a velocity V. At a given time t the car P would have penetrated to A (x from the origin) if its front was rigid but in practice it reaches B (y from the origin).

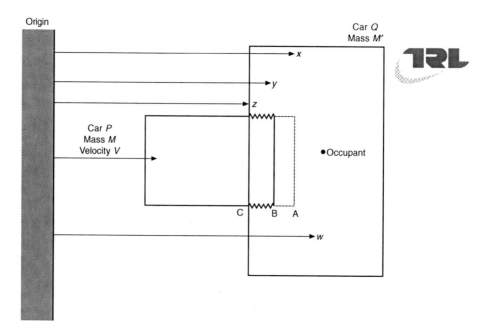

Figure 7.8 A representation of a side impact

The side of car Q is z from the origin and is penetrated $(y - z)$ and the occupant is at position w. Car P is braking with a force Mbg during the impact, while the side force between the wheels of car Q and the ground is Msg. The occupant of car Q is assumed to slide while restrained by a frictional force of coefficient F. The force R between the two cars during an impact is assumed to be constant.

The equations of motion are:

$$M\ddot{x} = - R - Bmg$$

$$M'\ddot{z} = R - SM'g$$

$$\ddot{w} = Fg$$

where at $t = 0$, $x = y = z = 0$, $w = w_0$, $\dot{x} = \dot{y} = V$, $\dot{z} = 0$, $\dot{w} = 0$

Then $\qquad\qquad x = Vt - (R/M + Bg)t^2/2 \quad$ and $\quad z = (R/M' - Sg)t^2/2$

The factor critical to injury potential is the velocity of the occupant relative to that side of the car being hit. Various different circumstances were considered by Neilson (1969), and two are examined here as follows:

The occupant remains in position and the front of car P is not crushed. Then the velocity at which the restrained occupant is struck by the side of his car is

$$\dot{y} - \dot{z} = (V^2 - aw_0)^{\frac{1}{2}} \text{ where } a = R/M + R/M' + Bg - Sg$$

If the occupant slides across the seat before hitting that part of his car being crushed, then his velocity at contact will be $(V^2 - 2\gamma w_0)^{\frac{1}{2}}$ where $\gamma = R/M + Bg + Fg$ and this applies until this velocity reaches $V\left(1 - \dfrac{\gamma}{\alpha}\right)$ when the cars have finished colliding and are moving together.

It is more convenient to express these velocities in terms of the depth X to which the impacted car is crushed rather than in terms of the crushing force R. This can be shown to be

$$X = \frac{V^2}{2(R/M + R/M')}$$

One factor that has a substantial effect on the outcome is the relative mass of the two vehicles. Figure 7.9 shows the velocity with which the occupant hits the intruding component for various relative vehicle masses of an impact at 48 km/h (30 mile/h). It is interesting to note that if the relative velocity, of an occupant into the car side, for cars of equal masses is considered then this velocity doubles, if the mass of the striking car is double the mass of the struck car.

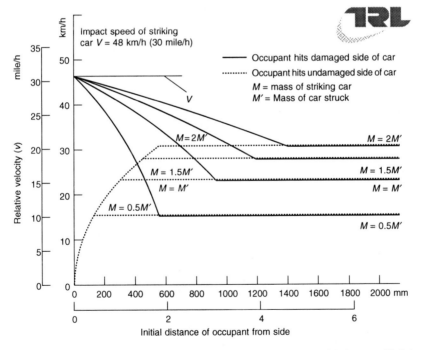

Figure 7.9 Effect of vehicle mass on occupant relative impact velocity in a side impact (Neilson, 1969)

It should also be noted that Figure 7.9 gives values of velocity for the occupant striking a damaged and undamaged part of the struck car. An undamaged part is an area that is adjacent to a damaged part but does not intrude. Neilson showed that if the occupant hits a deforming panel then the speed may be almost that of the striking vehicle but this speed will be between a half and zero of the striking vehicle if an undeforming panel is struck.

7.4.3 Derivation of a typical stress–strain curve

The performance of a seat belt is dependent not only upon the belt design but also on the collapse characteristics of the car. It is well known that the deceleration of a car upon impact is a function of successive loading, up to their ultimate stress, of the various vehicle parts close to the point of impact and the interaction of these parts with the next component. The structure behind the impact point remains largely intact but suffers a deceleration that varies with impact severity and from car to car. It would be convenient to be able to predict the pattern theoretically but even with modern finite element techniques this has proved very difficult. The following treatise (Neilson, 1963) is, therefore, based upon considering different likely deceleration patterns and examining the consequences.

A linear stress–strain relation leads to a quarter sine wave with a maximum deceleration occurring at the moment the car is brought to a rest. The equation is:

$$M\ddot{x} = \text{load} = -\bar{k}(x - X) \text{ and this corresponds to}$$

$$M\ddot{x} = -V\sqrt{\frac{K}{M}} \sin \sqrt{\frac{\bar{k}}{Mt}}$$

All impacts from different velocities V take the same time $\dfrac{\Pi}{2}\sqrt{\dfrac{M}{\bar{k}}}$, the deceleration distance is

$-V\sqrt{\dfrac{M}{k}}$ and the maximum deceleration is $-V\sqrt{\dfrac{M}{k}}$. Conversely if the deceleration time curve is linear with a constant force f, then $\ddot{x} = -tf$ and the stress–strain relation, which is found by

rearranging this to give $x = -V\dfrac{\ddot{x}}{f} + \dfrac{\ddot{x}^3}{6f^2}$ has an initial slope Mf/V, which increases with the

strain until at the maximum strain of $2V\sqrt{\dfrac{2fV}{3f}}$, the stress is $M\sqrt{2fV}$.

by rearranging this to give $x = -V\dfrac{\ddot{x}}{f} + \dfrac{(\ddot{x})^3}{6f^2}$

has an initial slope Mf/V, which increases with the strain until at the maximum strain of

$$2V\sqrt{\frac{2fV}{3f}} \text{ the stress is } M\sqrt{2fV}.$$

Experimental data shows that the deceleration tends to zero as the car comes to rest rather than at the end of the impact. This is because in spite of the extensive crushing of the vehicle front, there is usually some elastic energy that causes the car to rebound.

A vehicle designer will want to know the vehicle deceleration pattern for impacts into a given object at different speeds and thereby predict the outcome for a restrained occupant over a range of impact severities. It may be hoped that if the characteristics are measured in one impact they may then be predicted for impacts into the same obstacle but at different velocities. This is possible only if the vehicle always deforms according to the same stress–strain curve but unfortunately this is unlikely because of the complicated pattern of the collapse of the various parts of the car. However, it may be useful at certain stages of design to assume only one stress–strain curve.

Illustrated in Figure 7.10(b) is a deceleration-time curve for a half sine wave from V to zero. The corresponding stress–strain curve is also shown (Figure 7.10(a)) and, from that, the deceleration-time curves for impacts at $7V/8$, $3V/4$, $V/2$ and $V/4$ are computed. It is interesting to note that the maximum deceleration for the impact at velocity V, which occurs after half the deceleration is completed, is reached only by impact from velocities of greater than about $7V/8$. Rebound was assumed to be zero for all calculations.

7.4.4 The effect of crush on seat belt performance

The injuries caused in accidents by seat belts have been discussed above in Section 7.2 but it may be helpful to the designer to understand the link between seat belt characteristics and potential injury. It is well known that a relatively slight impact can lead to serious injury to unbelted occupants. This is often a head or chest injury and is caused usually by the occupant being thrown against the car interior and also by ejection.

Seat belt wearers, for the most part, are well protected against all but the most severe of impacts, but there is a limit to the effectiveness of the belt and this limit may be reached typically in four ways:

- Intrusion, caused by collapse or penetration of the occupant's compartment.
- Extension of the seat belt allowing the occupant to strike some part of the car; typically the steering wheel.
- Transmission of localized loads to the wearer through the webbing; most likely for passengers
- High deceleration in severe impacts may be sufficient to exceed injury thresholds, particularly for the chest.

Neilson (1963) discusses at great length mathematically the factors that influence the loads transmitted to the occupant from a seat belt and, in turn, how the loads are affected by the vehicle deceleration pattern.

7.5 Structural collapse and its influence upon safety

7.5.1 Frontal impacts

The interaction of the vehicle deceleration with a restrained occupant has been discussed above and it is clear that the deceleration pattern is critical to the outcome for the occupant. What is not so clear is how the vehicle deceleration may be influenced by structural design. Williams

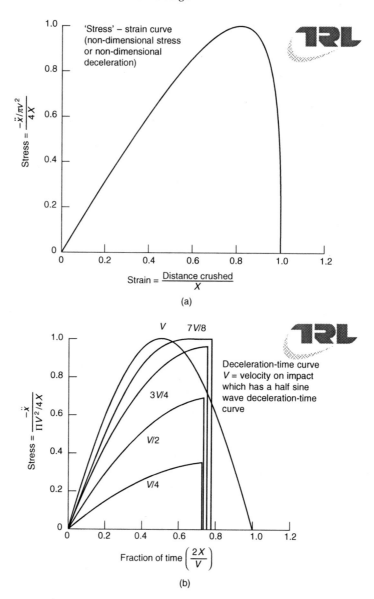

Figure 7.10 (a) Stress–strain curve; (b) Deceleration time curve for a half sine wave from V to zero

(1995) states that current vehicles with metal structures, absorb energy in many different ways in a frontal impact, none of which bear much resemblance to the controlled folding or inversion mechanisms of idealized metal tubes.

Figure 7.11 illustrates the basic structure of a modern saloon car and shows the integrated chassis and passenger compartment. The main longitudinal members will absorb much of the energy in a frontal collision and must also support the engine, suspension and subframe components.

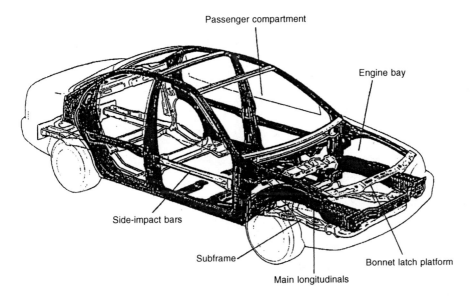

Figure 7.11 Basic structure of a modern saloon car (Williams, 1995)

That most of the energy is absorbed by the longitudinal members is the basis of conflict between legislation, which requires impact tests into a rigid barrier, and safety. In practice the most common accident is a collision where two cars collide head-on with a partial overlap, typically 40%. The resulting deceleration will depend upon the relative position of the longitudinal members, if they meet then the pulse will depend upon the collapse characteristics and the outcome for the occupant will be determined by the efficiency of the restraint system. If the longitudinal members do not meet then there is likely to be a substantial collapse of the relatively weak body panels and although the performance of the restraint system will still be vital to the protection of the head and chest, lower limb injuries from intrusion may occur that cannot be readily controlled by restraint systems.

However, for the frontal impact test, in Europe, although not in the US, the rigid barrier has been replaced by a deformable offset barrier that is more representative of an opposing vehicle. Thus it will be even more important for the car designer to understand the way in which the vehicle collapses and particularly so for the strong energy absorbing structural members. Such understanding is likely to be sought mainly from a combination of the use of lumped mass models constructed in computer packages such as MADYMO and finite element models using packages such as DYNA 3D. Lumped mass models are useful for large parametric studies and then the detailed behaviour of critical structural components can be analysed using finite element techniques.

It is beyond the scope of this chapter to discuss the use of packages such as MADYMO and DYNA 3D but it is important to understand the link between the deformation of components and the deceleration pattern. Figure 7.12 shows the force deflection plotted from an impact of a medium saloon car into a deformable barrier at 64 km/h. This, therefore, is a better representation of what happens in an accident than an impact into a rigid barrier.

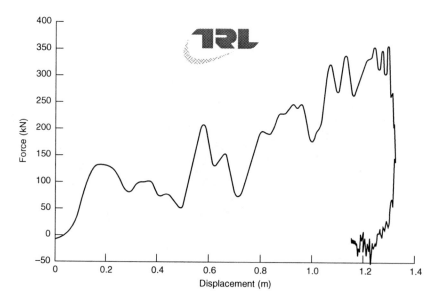

Figure 7.12 Force deflection curve for a Ford Sierra into a deformable barrier at 58 km/h

Macmillan (1983) likens the formation of this curve to the collapse under strain of a vertical strut, fixed at its lower end and constrained to move down at its upper end by a force F sufficient to cause vertical a movement denoted by S. As the displacement S is increased from zero, the force applied to the strut induces bending moments that cause it to deform as shown in Figure 7.13. The action continues until the bending moment M reaches a value denoted by M_p, which is the moment generated as the yield point is reached and the deformation becomes plastic. The buckling of a sheet of metal, compressed end-on, resembles that of the simple strut described. In an impact the crumpling of a vehicle front end comprises the simultaneous and linked formation of a large number of plastic hinges that determines the crush characteristics and hence the deceleration pattern. However, whilst the initial buckling may be largely that of sheet

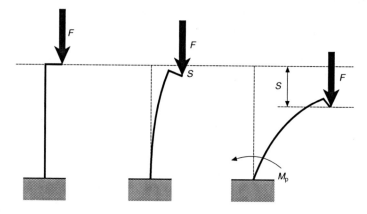

Figure 7.13 Collapse of a strut

metal, the deformation rapidly becomes a function of the collapse of stiff structural members usually of approximately rectangular cross section.

It is often possible to consider that the structure of the front end of a car comprises rectangular components in parallel and calculate the deceleration pattern based upon this assumption. However, this will give a value that is appropriate only if the component under consideration collapses and this assumes that the target is stiffer. In practice, and as stated above, this is unlikely and what will probably occur is that a strong member strikes a weaker member, which then collapses and determines the deceleration pulse. The overall outcome for both vehicles will be a pulse that depends upon the sequential collapse of components from the weakest to the strongest on whichever vehicle they happen to be and consequently the collapse is difficult to predict.

Vehicle manufacturers do not generally publish the results of impact tests on their vehicles, especially those that led to the development of a particular body structure. However, independent organizations such as the Transport Research Laboratory have, over the years, attempted to improve the crashworthiness of an existing vehicle to demonstrate examples of structural changes that would improve occupant protection. A good example of this is a modified Rover Metro designated 'ESV 87' and first exhibited in 1987. Figure 7.14 is taken from a paper by Hobbs *et al.* and illustrates how the main structure was changed in an attempt to improve the collapse characteristics.

The standard subframe was very strong and the substantial loads transmitted through to the passenger's compartment caused high peaks in the deceleration pattern and extensive intrusion. Modifications included four rectangular pre failed steel tubes incorporated into the subframe. However, although the structural changes proposed by Hobbs *et al.* (1987) greatly reduced intrusion on the driver's side, intrusion on the passenger side increased slightly and the seat belt loads greatly increased. Potential head injury was reduced for the passenger only.

Nevertheless, the test was an impact into a 30 degree barrier and was part of the research to determine an impact procedure to replace the frontal rigid barrier test that was considered to be not typical of accidents. Professor Lowne (TRL), as Chairman of the EEVC Impact Test Procedures Committees, has led the development of the frontal impact and side impact test procedures, both using a deformable barrier, that was introduced in Europe in 1998. It is confidently expected that these procedures will lead to greatly improved safety.

7.5.2 Side impacts

The injury study (7.2) showed that intrusion was always substantial in side impacts and in general the greater the intrusion the greater the injury severity. This indicates that structural reinforcement may help to reduce injury potential. Neilson (1963) showed that the velocity with which the door strikes the occupant is the most important factor and he investigated mathematically the effect of the vehicle relative mass and other factors upon this velocity (see 7.4.2). The analysis is extended here, to provide a more detailed examination of the injury mechanisms and the consequences of structural changes.

Research by Hobbs (1989) has shown that injury reduction was not as great as predicted with cars extensively reinforced. This research claimed that the door profile when striking the occupant was more relevant to the outcome than intrusion.

More recently, Håland (1994), in agreement with Neilson (1963), has shown that the two

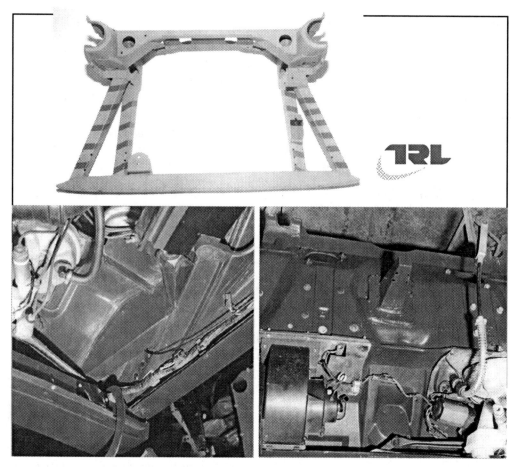

Figure 7.14 Engine subframe with front beam and energy absorbing tubes, box section triangulation and firewall energy absorber

factors mainly responsible for injuries in side impacts is the velocity of the inner door when it strikes the occupant and the time history of the inner wall during the period of contact with the occupant. The stiffness of the front of the striking car and the strength of the door and side structure of the struck car will determine the velocity of impact with the occupant. A very soft door structure will allow the striking vehicle to penetrate the struck vehicle, and hence the occupant, at a velocity only a little less than the impact velocity particularly if the striking car has a stiff front structure. Conversely, if the struck vehicle side structure is strongly reinforced and the bullet car has a weak front end then the occupant will be struck with a velocity approaching the 'momentum' velocity, which is about one half of the impacting car's velocity if both vehicles are of approximately equal mass (see Section 7.4.2). Thus, for a side impact of 48 km/h the door to occupant impact velocity can vary from about 13 m/s to down to about 7 m/s. Although, Håland suggests that to achieve 7 m/s the reinforcement would be so extensive as to be impracticable and that 9 m/s is achievable with a well-reinforced structure.

It is, of course, essential to fit padding to the inside of the door so that the force on the occupant is minimized but this will tend to cause the contact to be earlier in the impact and more energy will be transferred to the occupant. The effect of this energy transfer is a function of the stiffness of the padding but, very importantly, and, in turn, the criteria that are used to determine the potential chest injury will influence the measured benefits.

Acceleration-based criteria like TTI, thoracic trauma index, can be reduced by padding whereas deformation-based criteria like VC (viscous criteria) and chest deflection can increase because of prolonged occupant contact and an increase in energy transfer. This was verified by a series of side impact tests in the USA in which the standard padding was compared with the standard to which had been added 75 mm of 'medium stiff' padding to the inside of the door. The TTI figures were typically 35% lower in the padded cars whilst the maximum chest deflection was 35% greater (Wasko *et al.*, 1991).

The TTI is specified in the US side impact regulation but VC is used in the European side impact test procedure. Thus, to some extent the car designer is limited by the regulators. Moreover, it is easier to reduce the deflection and acceleration-based criterion, TTI, than to reduce the deformation and velocity-based criterion, VC. A typical rib acceleration time curve has two peaks. The first, and usually the largest peak, occurs during the initial contact with the padding and the second peak occurs when the padding is fully crushed. The initial gradient of the padding can be optimized to reduce TTI. To reduce the VC, the padding must be softer than the human torso to compensate for the prolonged contact; there will be a net reduction in chest deflection if the padding causes the spine to move from the door fast enough to compensate for the padding thickness. With careful choice of padding it is possible to reduce the chest deflection by an amount greater than the padding crush.

Passing the American requirements is possible with padding of stiffness of approximately 50–100 kN/m and a depth of 50 mm in the chest area (Deng, 1989). However, if the European requirement is to be met then the padding stiffness needs to be less than 80 kN/m and probably between 60–80 kN/m (Viano, 1987) but of greater depth than 50 mm thus creating design difficulties.

A greater depth of padding can be used below the armrest level in the door without infringing on the space needed for the occupant's arm. This is required to protect the pelvis and it has been shown (Pipkorn, 1992) that about 75 mm of padding is acceptable to most car designers and this depth of soft polyethylene type of foam of density 30–40 kg/cm that has a characteristic between constant stiffness and constant force, provides good protection at 50 km/h. This foam also has good energy absorbing properties of about 70% deformation. Thus, injuries in side impact can be reduced by the careful choice and positioning of protective padding, but currently, substantial protection can really only be achieved by the use of a combination of padding and airbags in conjunction with excellent quality seat belts that are now universal in Western countries and particularly so in Europe.

Airbag technology and hence the choice of systems is developing rapidly and the designer will, no doubt, choose a combination that best suits the intended vehicle. Nevertheless, the system pioneered by Håland and now marketed by Autoliv is an excellent approach and is discussed briefly below to provide the reader with an insight into the problems and solutions.

7.5.3 Side impact airbag systems

The Autoliv airbag and padding system
Håland and Pipkorn (1993) found that a combination of a side airbag, developed by Autoliv,

placed in the chest/abdominal area and thick soft padding in the pelvic/thigh area gave a considerable improvement in the protection of all body segments of the struck side occupant in car-to-car side impacts. Chest injury criteria, TTI and VC, were significantly lower with an 8-litre airbag compared with 50 mm thick padding in sled tests simulating a 50 km/h (30 mile/h) side impact into a well-reinforced car. In the latest version the side airbag has a volume of 12 litres and a length of about 450 mm to be able to protect occupants of different sizes, with the seat in the most rear to most forward position. Two small and very fast gas generators, of the same type used for pyrotechnical seat belt pretensioners but with a larger pyrotechnic charge (in total 4 g), are used. The bag must be fully inflated within 10–12 ms, while there still is about 100 mm clearance between the door inner wall and the occupant's chest. The bag inflation takes 7–8 ms with the type of gas generators used. This means that a sensor must trigger the system within 2–5 ms after the initial impact.

A recent study by the Accident Research Centre at the University of Birmingham showed that the optimum position for the location of the sensor was found to be the rear lower quadrant of the front door and is appropriate for almost 90% of the impacts. An undeformed part of the car will not start to move until 7–10 m/s after first car-to-car contact (EEVC, Friedel, 1988). The sensor must be located close to the outer surface of the car and must also be approximately in line with the occupant, because 80–90% of the life threatening injuries in struck side impacts are attributed to door intrusion close to the occupant (Hartemann *et al.*, 1976; Harms *et al.*, 1987).

Autoliv has chosen a pyrotechnical, non-electrical, sensor. The sensor is located in the lower rear part of the door, 30–40 mm inside the door outer skin, which will only trigger in the case of door intrusion with a risk of personal injuries and not in the case of, for example, parking damage or low impact speeds. The sensor element is a percussion cap that fires above a certain impact speed, typically 1.0–1.6 m/s and above a certain contact force, typically 1 kN. Within 1 m/s from sensor contact the flame from the percussion cap has been distributed to the two gas generators by means of shock tubes.

Evaluation of the Autoliv protective system
Håland (1993) conducted two series of tests on the Autoliv system. The first corresponding to a 48 km/h (30 mile/h) and the second to a 32 km/h (20 mile/h) car-to-car side impact. The chosen door velocity, in the first series of tests, was 9 m/s which represented a car with good car body reinforcement (Mellander *et al.*, 1989). The deceleration of the door was 20 g. The door test velocity in the second series of tests was chosen to be 6 m/s. Three basic configurations were tested:

1. Configuration 'A' was a reference door having 10 mm thick and stiff (80 kg/cu m) polyethylene padding to give a stiff door response. This covered a flat rigid door inner wall and the 'B'-pillar.
2. The 'B' configuration had a 50 mm thick chest padding and a 75 mm thick pelvis padding. The chosen material was polyethylene foam with open cells and a density of 30 kg/cu.m. The material was soft with a progressive characteristic (about 60 kN/m at an impact area of 175 sq. cm).
3. The 'C' configuration consisted of an 8-litre airbag for the chest and the same pelvis padding as in configuration 'B'. The airbag can be considered as soft hidden padding that

does not infringe the space for the occupant's arm. The characteristic of the bag is also progressive (30–60 km/h at an impact area of 175 sq. cm). The airbag was unventilated. Two small gas generators of the same type employed for pyrotechnical seat belt pretensioners were used. The pyrotechnic charge was 3.5 grams, and the air bag was inflated in about 8 ms.

The results of the three configurations are shown in Figure 7.15, from which it can be seen that potential injury to the chest, both TTI and VC, abdomen and pelvis were substantially reduced to below the human tolerance values when configuration 'C' was used, thus, the chest airbag configuration 'C' performed better than the chest padding configuration 'B'.

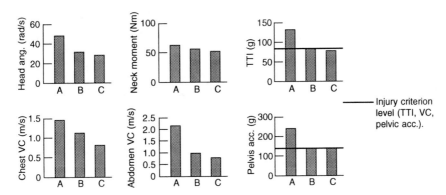

Figure 7.15 Håland's test results at 9 m/s (48 km/h (30 mile/h) side impact for the A, B and C configurations

Volvo Side Impact Protection System (SIPS)
Volvo has developed a system together with Autoliv in which the bag is inflated through an upholstery seam in the back of the seat. When fully inflated it covers the chest and the abdomen down to the armrest level. The contact area is the same as for the door bag. The performance of the seat-mounted side airbag is similar to the door-mounted bag when the bag has the same initial bag pressure and the same ventilation area. The sensor of the same type as evaluated in the study is not located in the door but at the side of the lower outer seat structure.

The current Volvo Side Impact Protection System permits a quick response of the inside of the door in a side impact. The sensor fires within 5 ms in a 50 km/h side impact and the bag is fully deployed within 12 ms from first car-to-car contact. The sensor and the seat mounted side airbag forms a self-contained system and the bag has an optimum position regardless of the size of the occupant, since the bag moves with the seat. Another advantage with a seat mounted side airbag is the less demanding environment compared with the door mounted equivalent. This side airbag, introduced during model year 1995, was the first to be installed in a mass production car.

Estimate of the benefits of a side impact protection system
Estimating the benefit of a safety device such as the side airbag is a complex prediction analysis. A simple approach by Håland, justified by his test results, is to assume that the side

airbag would reduce AIS 3+ chest injuries by one value on the AIS scale. If this is applied to the accident distribution curves and to all accident severities, then the results predict a saving of 25% of all AIS 3+ injuries currently occurring.

7.6 References and further reading

American Association for Automotive Medicine (AAAM) (1990). The Abbreviated Injury Scale. 1990 revision. AAAM, Des Plaines, IL 60018, USA.

Consumers' Association (1993). *The Secondary Safety Rating System for Cars*, London.

Dalmotas, D.J., Hurley R.M. and German, A. (1996a). Supplemental restraint systems: friend or foe to belted occupants? *40th Annual Proceedings of AAAM*, pp. 63–76.

Danner, M., Langwieder, K. and Hummel, T. (1987), Eleventh international technical conference on experimental safety vehicles, Washington DC, May 12–15, 1987 p. 201–11. IRRD 830208.

Deng, Y.C. (1989). The Importance of the test method in determining the effects of door padding in side impacts, 33rd Stapp Car Crash Conference, Washington, DC, Oct 4–6. SAE Publication P-227, 1989, pp. 79–85.

Department of Transport (1996). Road accidents in Great Britain 1995. HMSO. London.

EEVC, European Experimental Vehicle Committee. Chairman Prof. B. Friedal BAST, Brüderstraße 53, Postfach 100150, D-51401 Bergisch Gladbach.

Fildes, B. and Vulcan, A. (1990). Crash performance and occupant safety in passenger cars involved in side impacts. Proceedings of the IRCOBI Conference, Lyon, France.

Gloynes, P., Rattenbury, S. Wellor, R. and Lestina, A. (1989). Mechanisms and patterns of head injuries in fatal frontal and side impact crashes.

Håland, Y. (1994). Car-to-car side impacts. Doctoral thesis. Dept. of Injury Prevention, Chalmers University of Technology. Gothenburg, Sweden.

Håland, Y., Pipkorn B., (1993). A Parametric Study of a Side Airbag System to meet Deflection based Criteria. 1993 IRCOBI conference on the Biomechanices of Impacts, September 8–10, Eindhoven, 339–353.

Harms, P.L., Renouf, M., Thomas, P.D. and Bradford, M. (1987). Injuries to restrained car occupants: what are the outstanding problems? Proceedings of the 11th International ESV Conference, Washington, Washington: NHTSA.

Hartemann, F., Thomas, C., Foret-Bruno, J.Y., Henry, C., Fayon, A., Tarrière,C., Patel, A., 1976, 'Description of lateral impacts', Proc. of the 6th Int. Technical Conference on Experimental Safety Vehicles, Washington, D.C., pp. 541–563.

Hobbs, C.A., Lowne., R.W., Penoyre S. and Petty, S.P.F., 1987. Progress towards improving car occupant protection in frontal impacts. Proceedings of the 11th International ESV Conference, Washington, Washington: NHTSA.

Hobbs, C.A., 1989. The Influence of Car Structures and Padding on Side Impact Injuries. Proceedings of the 12th International ESV Conference, Göteborg, Sweden. Washington: NHTSA.

Kallieris, D., and Mattern, R. (1984). Relastbankeitsgrenzen und Verletzungsmechanik der angegurteten Fahrzeuginsassen beim Seitandfall. FAT-Schriftenveihe.

Lau, I.V., Horsch, J.D.,Viano, D.C. and Andrzejak, D.V., (1993). Mechanism of injury from airbag deployment loads. Accident analysis and prevention, pp. 25–29

Lau, I.V., Viano, D.C. (1984).The viscous criterion-bases and applications of an injury severity index for soft tissues. Proceedings of the 30th Stapp Car Crash Conference, SAE Technical Paper 840888, Warrendale, PA 1984.

Mackay, M. (1973). Vehicle safety legislation – its engineering and social implications. I.Mech.E.

Mackay, M. *et al.* (1995). Smart seat belts – what they offer. Automotive Passenger Safety. Selected papers from Autotech 95, 7–9 November 1995. paper 32/145. Suffolk: Mechanical Engineering Publications.

Macmillan, R.H. (1983), Dynamics of vehicle collisions. Publ: Interscience Enterprises. ISBN 0907776078, –25–00.

Mellander, H., Ivarsson, J., Korner, J., Nilsson, S., 1989. 'Side impact protection system — a description of the technical solutions and the statistical and experimental tools', Proceedings of the 12th Int.Technical Conference on Experimental Safety Vehicles, Göteborg, pp. 969–976.

Melvin, J., Horsch, J., McCleary, J., Wideman, L., Jensen, J. and Wolamin, M. (1993). Assessment of an airbag deployment load with the small Hybrid III dummy. SAE 933119.

Miltner, E., and Salwender, H.J. (1995). Influencing factors on the injury severity of restrained front seat occupants in car-to-car head-on collisions. *Accident Analysis and Prevention*, v27, n2. pp. 143–50.

Morris, A., Hassan, A., Mackay, M., Hill, J. (1995). 'Head injuries in lateral impact collisions', The 1993 Int. IRCOBI Conference on the Biomechanics of Impacts, Eindhoven, Sept. 8–10, 1993, pp. 41–55.

Niedere, P., Waltz, F. and Weissnerb, R. (1980). Verletzingsursachen beim Pkw-Inassen, Verletzumpsmmderung dorch moderne Sichenbeitseinrichtungen Unfallheilkunds.

Neilson, I.D., (1963). The dynamics of safety belt assemblies for motor vehicles. Road Research Laboratory note LN/303/IDN.

Neilson, I.D., (1969). Simple representations of car and unrestrained occupant impacts in road accidents. Road Research Laboratory report LR 249.

Otte, D. et al. (1982). Erhebungen am Unfallort. Unfall-und Sicherheitsforschung Strassenverkehr.

Pipkorn, B. (1992). 'Car to Car side impacts. Development of a two dimensional BioSid dummy', Department of Injury Prevention report R015, Chalmers University of Technology, Göteborg.

Ropohl, D. (1990). Die vechtsmedijmische Rekonstrucktion von Verkehrsurfallen. DAT – Schirftenveihe Technik, Markt, Sachverstandigenwesen. Band 5.

Rouhana, S., and Foster, M. (1985). Lateral impact – an analysis of the statistics in the NCSS. Proceedings of the 29th STAPP Car Crash Conference.

Rutherford, W.M., Greenfield, T.H.R.M., and Nelson, J.K. (1985). The medical effects of seat belt legislation in the UK. HMSO Research Report, No 13:ISBN 0113210396.

Thomas, C., Foret-Bruno, J.Y., Brutel, G., and Le Coz, J.Y., (1994). Front passenger protection: what specific requirements in frontal impact? Proceedings of the International IRCOBI Conference on the Biomechanics of Impacts, Lyon, September 1994. Pp. 205–16. Bron: IRCOBI.

Thomas, C., Henry, C., Hartemann, F., Patel, A., Got, C. (1987). 'Injury Pattern and Parameters to Assess Severity for Occupants Involved in Car-to-Car Lateral Impacts', 11th ESV conference, Washington, DC., 1987, 49–61.

Thomas, P., Bradford, M., and Ward, E.,(1992). Vehicle design for secondary safety. VTI Rapport 380a pp. 153–173. Linkoping, VTI.

Viano, D.C. (1987). 'Evaluation of the benefit of energy absorbing material in side impact protection', Proc. of the 31st Stapp Car Crash conference, SAE Technical Paper 872213, Warrendale, PA 1987.

Walter, D., and James, M. (1996). An unusual mechanism of airbag injury. *Injury*. v 27, n.7. pp. 523–4. Oxford: Elsevier Science.

Wasko R. J., Cambell, K., Henson, S.E. (1991). Results of MVMA full vehicle side impact tests on 1990 model year Pontiac 6000 vehicles using BioSid and Sid, 13th Int. Technical Conference on Experimental Safety Vehicles, Paris, Nov. 4–7, 1991, pp. 567–573.

Williams D.A. (1995). Angled compression of energy absorbing composite tubes. PhD Thesis, Cambridge University.

Reference description

The most noteworthy references here are Neilson (1963 and 1969), Macmillan (1983) and Håland (1994). Neilson was a brilliant applied mathematician who worked for the Transport and Road Research Laboratory from 1961 until he retired in 1988. The two reports contain an extensive theoretical analysis of the behaviour of a car and its occupants in front and side impacts together with an equally thorough examination of the behaviour of seat belts and how this is related to vehicle impact performance. Macmillan's treatise is very extensive and begins with an analysis of car behaviour using Newtonian principles; this is followed by a systematic examination of the collapse of the vehicle structure from simple considerations through to the development of complex formulae from which the impact performance of a vehicle may be predicted. The development of the formulae are linked to the summation of the results of very many full scale barrier impact tests. I have included Håland because it is a novel and thorough investigation of the use of airbags in side impact protection, it contains a very good description of accident studies with very many references and it discusses, very lucidly, the use and consequences of different thorax injury criteria.

Acknowledgements

The figures accompanied by the TRL Logo are reproduced by kind permission of Dr R. Kimber, Research Director, Transport Research Laboratory, Old Wokingham Road, Crowthorne, RG45 6AU. The aforementioned figures are the copyright of The Transport Research Laboratory, and may not be copied or reproduced unless permission is given in writing by Dr R. Kimber.

The co-operation of DETR (formerly DOT) Vehicle Standards and Engineering Division who funded the TRL research described in this chapter is gratefully acknowledged.

8. Noise vibration and harshness

Brian Hall, PhD, BScEng, CEng, MIMechE

The aim of this chapter is to:

- Introduce the basic concepts and importance of vibration theory to vehicle design;
- Consider the role of the designer in vibration control;
- Demonstrate methods for the control of vibration to help the elimination of noise and harshness;
- Indicate methods by which the designer can control vibration and noise to create an equitable driving environment.

8.1 Introduction

Noise, vibration and harshness (NVH) have become increasingly important factors in vehicle design as a result of the quest for increased refinement. Vibration has always been an important issue closely related to reliability and quality, while noise is of increasing importance to vehicle users and environmentalists. Harshness, which is related to the quality and transient nature of vibration and noise, is also strongly linked to vehicle refinement.

Controlling vibration and noise in vehicles poses a severe challenge to the designer because unlike many machine systems, motor vehicles have several sources of vibration and noise which are interrelated and speed dependent. In recent years, the trend has been towards lighter vehicle constructions and higher engine speeds to meet the requirements for improved fuel consumption and engine performance. This has tended to increase the potential for noise and vibration, posing many new problems for automotive engineers. These developments have also coincided with a reduction in the time to market for new vehicles and created an increased dependency on computer-aided design and analysis with less time spent on prototype testing.

This accelerated development of new and highly refined vehicles is dependent on accurate dynamic analysis of vehicles and their subsystems and calls for refined mathematical modelling and analytical techniques. While NVH analysis has in recent years been aided by developments in finite element and multi-body systems analysis software, there is still an underlying need to apply basic vibration and noise principles in vehicle design.

There are many excellent texts dealing with vibration and noise, but few are devoted to automotive applications. It is therefore the objective of this chapter to address some important noise and vibration issues arising in vehicle design. It is assumed that the reader has some previous knowledge of noise and vibration theory, since space only permits a brief review of the fundamentals.

8.2 Review of vibration fundamentals

We begin this section by reviewing some vibration fundamentals in an automotive context and proceed to summarize the characteristics and response behaviours of vibrating systems. For further details of the fundamental pricinciples the reader is recommended to read some of the well established texts such as Timoshenko *et al.* (1974), Meirovitz (1986), Rao (1995) and Dimaragonas *et al.* (1992).

8.2.1 Basic concepts

Vibration arises from a disturbance applied to a flexible structure or component. Common sources of vibration in vehicles are road and off-road inputs to suspensions, rotating and reciprocating unbalance in engines, fluctuating gas loads on crankshafts, gear manufacturing errors and tooth loading effects in transmissions, generation of fluctuating dynamic forces in constant velocity joints and inertia and elasto-dynamic effects in engine valve trains.

Vibration sources are characterized by their time and frequency domain characteristics. In automotive engineering, most vibration sources produce continuous disturbances as distinct from shocks and short duration transients encountered in some machine systems. They can therefore be categorized principally as either periodic or random disturbances. The former are the easiest to define and originate from the power unit, ancillaries or transmission, while random disturbances arise from terrain inputs to wheels.

The simplest form of periodic disturbance is harmonic and might typically be produced by rotor unbalance. In the time domain this is represented by a sinusoid and in the frequency domain by a single line spectrum. It might be noted that a full representation in the frequency domain requires both amplitude and phase information. This is important when the disturbance includes several frequency components each of which may be phased differently to one another. Typical of these are the general periodic disturbances produced by reciprocating unbalance and crankshaft torque.

In the case of random disturbances, it is not possible to predict the precise level of the disturbance at any given time and hence it is not possible to express such disturbances as continuous functions in the time domain – only statistical representations are possible. From the vibration point of view, the frequency content of a random signal is very important. For example the frequency spectrum of a road input to a vehicle is a function of the spatial random profile of the road and the speed of the vehicle (see Chapter 10). For a given set of conditions this results in a large (theoretical infinite) number of frequency components distributed over a wide band of frequencies and is commonly represented by its power spectrum (Newlands, 1975).

All mass-elastic systems have natural frequencies, i.e. frequencies at which the system naturally wants to vibrate. For a given (linear) system these frequencies are constant and are related only to the mass and stiffness distribution. They are not dependent on excitation applied to the system provided that the system can be classified as linear. Non-linear effects (which often arise in automotive systems) are beyond the scope of this chapter. The interested reader should consult one of the specialised texts such as Thomsen (1997).

An arbitrary short duration disturbance applied to the system tends to excite all the system's natural frequencies simultaneously. Most systems have a very large number of natural frequencies, but normally only a few of the lower order ones are of interest because the higher ones are more highly damped. At each natural frequency a system vibrates in a particular way, depicted by the relative amplitude and phase at various locations. This is called the *mode of vibration*.

Lightly damped structures can produce high levels of vibration from low level sources if frequency components in the disturbance are close to one of the system's natural frequencies. This means that well designed and manufactured sub-systems, which produce low level disturbing forces, can still create problems when assembled on a vehicle. In order to avoid these problems at the design stage it is necessary to model the system accurately and analyse its response to anticipated disturbances.

The general approach to vibration analysis is to:

(a) develop a mathematical model of the system and formulate the equations of motion
(b) analyse the free vibration characteristics (natural frequencies and modes)
(c) analyse the forced vibration response to prescribed disturbances and
(d) investigate methods for controlling undesirable vibration levels if they arise.

8.2.2 Mathematical models

These provide the basis of all vibration studies at the design stage. The aim is to represent the dynamics of a system by one or more differential equations. It is possible to represent the distributed mass and elasticity of some very simple components such as uniform shafts and plates by partial differential equations. This is called the *distributed-parameter* approach. However, it is not generally possible to represent typical engineering systems (which tend to be more complicated) in this way. Hence the approach normally adopted is to model a system by a set of discrete mass, elastic and damping elements, resulting in one or more ordinary differential equations. This is called the *lumped-parameter* approach. Masses are concentrated at discrete points and are connected together by massless elastic and damping elements. The number of elements used dictates the accuracy of the model – the aim being to have just sufficient elements to ensure that an adequate number of natural vibration modes and frequencies can be determined while avoiding unnecessary computing effort.

Figures 8.1(a) and (b) compares the distributed and lumped parameter approaches for modelling a uniform beam undergoing free lateral vibration. In Figure 8.1(a) the displacement at a particular point is described in terms of a variable which is a continuous function of position and time, e.g. $y(x, t)$ and leads to a second order partial differential equation (Meirovitz, 1986). In Figure 8.1(b) the mass of the beam is broken up into a series of equal masses connected together by massless beam elements. The beam stiffnesses are calculated from a knowledge of the flexural rigidity of the beam. The displacements of the masses are represented by a finite number n of generalized coordinates (also called degrees of freedom) which are a function of time only, e.g. $y_1(t)$, $y_2(t)$, $\ldots y_n(t)$. Hence the number of generalized coordinates is equal to the number of degrees of freedom (DOF) of the model. Thus in general an n-DOF system will (a) be described by n-second order differential equations and (b) have n-natural frequencies and modes.

Thus it is apparent that the building blocks of discrete models are mass, spring and damping elements. In basic mathematical models these elements are two dimensional, representing translational and rotational motion. The dynamics of each element can be represented by a constituent equation (see Rao, 1995 for a table of elements and equations) and the dynamics of

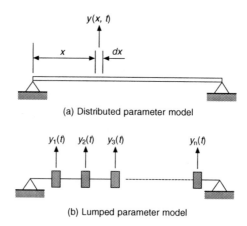

(a) Distributed parameter model

(b) Lumped parameter model

Figure 8.1 Types of vibration

the interconnected set of elements can be found either by applying Newton's second law to each of the mass elements or by use of an energy method e.g. Lagrange equations (Meirovitz,1986). In general the topology of the assembled model bears close similarity to the real system.

The system model
The following suspension model illustrates the use of modelling elements. A more detailed analysis of this model is developed later in this chapter.

Suspension systems have a variety of geometries and can be modelled in many different ways depending on the objectives of the analysis. One of the simplest forms of suspension model which is used for conceptual studies is the quarter vehicle model associated with one wheel-station (Figure 8.2). This type of model normally has two degrees of freedom comprising a sprung mass m_s (a proportion of body mass) and an unsprung mass m_u (incorporating a proportion of suspension components plus wheel, axle and brake). The suspension stiffness k_s and damping c are represented by linear elements in this simplified model. In reality suspension motion together with the characteristics of the spring and damper are all non-linear (see Chapter

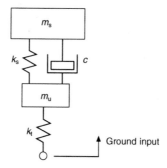

Figure 8.2 Quarter vehicle suspension model

10), however for small suspension motions linear representation is generally acceptable. The model is completed by including the tyre stiffness k_t. The input to the model is determined by the ground profile characteristics and the speed of the vehicle. It is convenient to measure horizontal motion of the ground relative to the vehicle, in which case the model is constrained to move within a vertical plane while the ground moves vertically at the bottom of the tyre spring.

8.2.3 Formulating the equations of motion

The first step in the formulation of the equations of motion is to assign a set of generalized coordinates (a minimum set of independent coordinates) to the model which describes the general motion. For the relatively simple multi-body systems discussed in this chapter, the equations of motion can then be determined from a set of free-body diagrams (FBDs) of the masses. The equation of motion can then be determined by applying Newton's second law to each free-body. For cases where the geometry of the model is complicated, the equations can be formulated more elegantly by energy methods (Ginsberg, 1988).

As an example of the FBD approach, consider the suspension model in the previous section. This 2DOF example (which leads to two second order differential equations) can be expressed in matrix form and is typical of linear multi-degree of freedom systems in general. In this case two generalized coordinates y_2 and y_1 are required to represent the displacements of the sprung and unsprung masses respectively and y_0 represents the ground input. If it is assumed that there is no separation between tyre and ground, the annotated model and FBDs are then as shown in Figure 8.3. Note that if deflections are measured from the static mean positions, there is no need to show the gravity force acting on the mass and the mean force in the spring – they are equal and opposite. Defining the displacements positive upwards results in the corresponding velocities and accelerations being positive upwards.

Applying Newton's second law to each of the two masses gives:

$$k_t(y_0 - y_1) - k_s(y_1 - y_2) - c(\dot{y}_1 - \dot{y}_2) = m_u \ddot{y}_1 \qquad (8.1a)$$

$$k_s(y_1 - y_2) + c(\dot{y}_1 - \dot{y}_2) = m_s \ddot{y}_2 \qquad (8.1b)$$

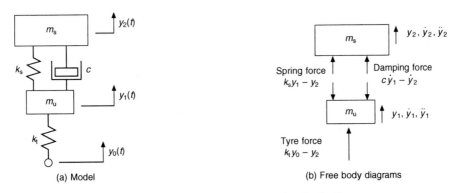

(a) Model   (b) Free body diagrams

Figure 8.3 The quarter vehicle model and free-body diagrams

These equations can be re-arranged and written in matrix form as:

$$\begin{bmatrix} m_u & 0 \\ 0 & m_s \end{bmatrix} \begin{Bmatrix} \ddot{y}_1 \\ \ddot{y}_2 \end{Bmatrix} + \begin{bmatrix} c & -c \\ -c & c \end{bmatrix} \begin{Bmatrix} \dot{y}_1 \\ \dot{y}_2 \end{Bmatrix} + \begin{bmatrix} (k_1 + k_s) & -k_s \\ -k_s & k_s \end{bmatrix} \begin{Bmatrix} y_1 \\ y_2 \end{Bmatrix} = \begin{Bmatrix} k_t y \\ 0 \end{Bmatrix} \tag{8.2}$$

Equation 8.2 is of a form typical of linear multi-degree of freedom (MDOF) vibrating systems. Such equations can be written more concisely as:

$$[M]\{\ddot{x}\} + [C]\{\dot{x}\} + [K]\{x\} = \{F(t)\} \tag{8.3}$$

where $[M]$, $[C]$ and $[K]$ are the mass/inertia, damping and stiffness matrices, $\{x\}$, $\{\dot{x}\}$ and $\{\ddot{x}\}$ are displacement, velocity and acceleration vectors, $\{F(t)\}$ is the excitation vector.

This general form is also applicable to rotational systems and will be used in a later application.

8.2.4 System characteristics and response

Single degree of freedom systems
Despite its limitations for accurately modelling most automotive systems, a knowledge of SDOF behaviour provides a basic understanding of more complex systems. The important features related to the classic SDOF model shown in Figure 8.4 are:

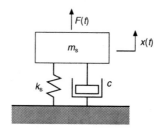

Figure 8.4 Classic SDOF model

1. The equation of motion is given by:

$$m\ddot{x} + c\dot{x} + kx = F(t) \tag{8.4}$$

2. The characteristics of the system are obtained from the free-vibration behaviour, i.e. when $F(t) = 0$.
3. If $F(t) = 0$ and $c = 0$, we have the equation for simple harmonic motion which has the solution:

$$x = X \cos(\omega_n t - \phi), \tag{8.5}$$

where X and ϕ are arbitrary constants (determined from the conditions at $t = 0$) and ω_n is the undamped natural frequency given by:

$$\omega_n = \sqrt{\frac{k}{m}} \tag{8.6}$$

i.e. the system vibrates at this frequency with amplitude X.

4. If damping is included there are two possible free vibration characteristics. If disturbed, the mass either returns to its equilibrium position with or without oscillation (termed underdamped and overdamped respectively). In the former case the oscillations are of progressively reducing amplitude. The characteristics are determined by the relative magnitude of c in relation to m and k. When c is such as to be on the boundary between the two characteristics it is said to critically damped. Then $c = c_c = 2\sqrt{mk}$.

The level of damping in a SDOF system is often described in terms of the damping ratio ζ defined as:

$$\zeta = \frac{c}{c_c} \tag{8.7}$$

The values of ζ for underdamped, critically damped and overdamped are thus < 1, 1 and > 1 respectively.

5. For an underdamped system ($\zeta < 1$) the solution for free vibration is:

$$x = Xe^{-\zeta\omega_n t} \cos(\omega_d t - \phi) \tag{8.8}$$

where X and ϕ are arbitrary constants and ω_d is the damped natural frequency given by:

$$\omega_d = \omega_n \sqrt{1 - \zeta^2} \tag{8.9}$$

6. If $F(t) \neq 0$ the solution to the equation of motion is made up from two components: the Complementary Function (CF) and the Particular Integral (PI). The CF is identical to the free vibration solution (i.e. equation 8.8 for $\zeta < 1$) and quickly dies away for realistic levels of damping to leave $x = $ PI.

7. If $F(t) = F_0 \sin \omega t$, the steady-state response of the mass (after the CF has become zero) is given by:

$$x = A(\omega) \sin [\omega t - \alpha(\omega)] \tag{8.10}$$

where $A(\omega)$ is the steady-state amplitude and $\alpha(\omega)$ is the phase lag – both are dependent on ω. Note: the steady-state response is at the excitation frequency ω.

8. It may be shown that

$$A(\omega) = F_0 \, | H(\omega) | = F_0 \left| \frac{1}{(k - m\omega^2) + (c\omega)i} \right| = \frac{F_0}{\sqrt{(k - m\omega^2)^2 + (c\omega)^2}} \tag{8.11}$$

$H(\omega)$ which is complex, is called the frequency response function. It relates the input (excitation) and output (response) in the frequency domain.

9. The amplitude response can be presented conveniently in dimensionless form in terms of the dynamic magnifier $D = kA/F_0$ and frequency ratio $r = w/\omega_n$. It may be shown (Rao, 1995) that

$$D = \frac{1}{\sqrt{\left[1 - \left(\dfrac{\omega}{\omega_n} \right)^2 \right]^2 + \left(2\zeta \dfrac{\omega}{\omega_n} \right)^2}} \tag{8.12}$$

Figure 8.5 shows D plotted for two different values of ζ.

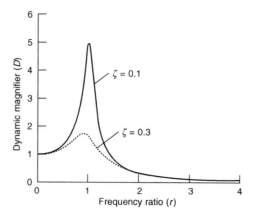

Figure 8.5 Amplitude response of a SDOF system

10. It follows from Figure 8.5 that:
    (a) maximum response amplitude occurs at resonance when $\omega \approx \omega_n$.
    (b) the amplitude is strongly influenced by the level of damping in the system when $\omega \approx \omega_n$. When ω and ω_n are appreciably dissimilar, damping has very little effect on response amplitude. This is an important point when considering the use of damping to control vibration levels.

Multi-degree of freedom systems
Realistic modelling of most forms of automotive vibration requires the use of MDOF models. These have two or more degrees of freedom and lead to two or more equations of motion which can be written in matrix form as shown in equation 8.3. The characteristics of such systems can be determined by considering the free vibration behaviour. This requires the excitation vector in equation 8.3 to be set to zero giving

$$[M]\{\ddot{x}\} + [C]\{\dot{x}\} + [K]\{x\} = \{0\} \tag{8.13}$$

Generally equation 8.13 represents a set of coupled differential equations.

(a) Negligible damping
One can begin to understand the characteristics of MDOF systems best by neglecting damping, i.e. setting $[C] = [0]$. This gives

$$[M]\{\ddot{x}\} + [K]\{x\} = \{0\} \tag{8.14}$$

Assuming solutions of the form $\{x\} = \{A\}\, e^{st}$ leads to a set of homogeneous equations

$$([M]s^2 + [K])\{A\} = \{0\} \tag{8.15}$$

The non-trivial solution of these is the characteristic equation (or frequency determinant)

$$| [M]s^2 + [K] | = 0 \tag{8.16}$$

This leads to a set of real roots, typically

$$s_i^2 = -\lambda_i = -\omega_i^2 \tag{8.17}$$

where λ_i is the *i*-th eigenvalue and ω_i the *i*-th natural frequency. For a system having more than two degrees of freedom it is necessary to find these by numerical methods or by using mathematical software (e.g. MathCAD, 1995). This is most easily accomplished by forming the eigenvalue equation (from equation 8.17).

$$\lambda[M]\{u\} = [K]\{u\} \tag{8.18}$$

Corresponding to each eigenvalue λ_i, there is an eigenvector $\{u\}_i$, which relates the relative amplitudes at each of the degrees of freedom, i.e. they describe a normal (or natural) mode of vibration. The eigenvectors can be found by substituting each eigenvalue in turn back into equation 8.15 or more directly using mathematical software. Vibration in the *i*-th mode then can be described by

$$\{x\}_i = \{u\}_i \, A_i \sin(\omega_i t + \alpha_i) \tag{8.19}$$

If the system is given a short duration disturbance at some arbitrary position all modes of vibration will tend to be excited and the ensuing motion will be a combination of these, i.e.

$$\{x\} = \sum_{i=1}^{n} \{x\}_i = [u]\{q(t)\} \tag{8.20}$$

where:

$$[u] = [\{u\}_1 \{u\}_2 \dots \{u\}_n]$$

is called the modal matrix and

$$\{q(t)\} = \begin{cases} A_1 \sin(\omega_1 t + \alpha_1) \\ \vdots \\ A_n \sin(\omega_n t + \alpha_n) \end{cases}$$

is a vector of *modal (principal) coordinates*.

Equation 8.20 represents a linear transformation between generalized coordinates $\{x\}$ and modal coordinates $\{q\}$.

The eigenvectors have a special property called orthogonality, such that when the product of eigenvectors $\{u\}_i$ and $\{u\}_j$ is formed with either the mass or stiffness matrices the result is zero, i.e. $\{u\}_i^T [M] \{u_j\} = 0$ and $\{u\}_i^T [K]\{u_j\} = 0$, provided that $i \neq j$,

If $i = j$ the result is $\{u\}_i^T [M] \{u_i\} = M_{ii}$ and $\{u\}_i^T [K]\{u_i\} = K_{ii}$ where M_{ii} and K_{ii} are called the *modal mass* and *modal stiffness* respectively.

The orthogonality property can be used to uncouple the equations of motion and express the equations in modal coordinates. This facilitates the solution of the general forced vibration problem and highlights the contribution which the excitation makes to each of the modes.

Replacing $\{x\}$ with $[u]\{q\}$ in equation 8.3, premultiplying by $[u]^T$ and applying the orthogonality condition gives:

$$\text{diag}[M]\{\ddot{q}\} + \text{diag}[K]\{q\} = [u]^T \{F(t)\} \tag{8.21}$$

where diag[M] and diag[K] are diagonal matrices containing modal mass and stiffness elements respectively. The *i*-th equation (of the set of *n*) is of the form

$$M_{ii}\ddot{q}_i + K_{ii}q_i = \{u\}_i^T \{F(t)\} \tag{8.22}$$

Hence a set of uncoupled equations is formed, each of which is similar to that for forced vibration of a SDOF system. When the set of solutions $\{q\}$ have been obtained they can be transposed back into generalized coordinates using the transformation $\{x\} = [u]\{q\}$.

(b) Viscous damping
In (stable) lightly damped systems the frequency determinant is

$$| [M]s^2 + [C]s + [K] | = 0 \tag{8.23}$$

For an *n*-DOF system this produces *n* complex conjugate roots having negative real parts providing information about the frequency and damping associated with each mode of vibration. There is a possibility that some of the roots will be equal or have zero real and/or imaginary parts. The latter being the case if rigid body motion is possible.

Except in those cases where damping has been deliberately added, damping in automotive systems is so low as to have a negligible effect on the natural frequencies and modes of vibration. However, damping must be considered in the analysis if the response of the system is required for a relatively short period of time in comparison to the natural periods of the system or when one or more components of a periodic excitation is at or near to one of the system's natural frequencies.

(c) Forced-damped vibration (harmonic excitation)
Since most of the excitations in automotive systems are of a periodic nature, this aspect of vibration analysis is of great important to us. In general, it is possible (using Fourier series) to decompose any periodic signal into a set of harmonic components having differing amplitudes and frequencies. After determining the response to each component, it is then possible to determine the overall response by adding the individual responses together using the principle of superposition provided the system is linear. This approach allows us to develop the solution in the frequency domain relating the response at location *i* to the excitation at location *j* via a frequency response function.

Many of the features of harmonic response analysis of SDOF systems extend to MDOF ones, e.g.:

(1) When subjected to harmonic excitation an MDOF system vibrates at the same frequency as the excitation.
(2) The displacement amplitudes at each of the degrees of freedom are dependent on the frequency of excitation and
(3) The dynamic displacement at each DOF lags behind the excitation.

In MDOF systems the excitation can be applied simultaneously at any of the DOFs. For a linear system the response at any of the DOFs is the sum of the responses due to each excitation force.

Consider a MDOF system with viscous damping subjected to a set of harmonic excitations $\{F(t)\}$ = $\{F\}f(t) = \{F\} \sin (\omega t)$ applied at the DOFs. Equation 8.3 can be written

$$[M]\{\ddot{x}\} + [C]\{\dot{x}\} + [K]\{x\} = \{F\}f(t) \tag{8.24}$$

By taking Laplace transforms of both sides with zero initial conditions, replacing s with $i\omega$ (Schwarzenbach *et al.*, 1984), and pre-multiply both sides by $(-\omega^2[M] + i\omega[C] + K)^{-1}$ results in

$$\{H_x(\omega)\} = [H(\omega)] \{F\} \tag{8.25}$$

$\{H_x(\omega)\}$ is a vector of frequency responses at the DOFs and $[H(\omega)]$ is a matrix of frequency response functions such that H_{ij} = frequency response at i due to unit amplitude excitation at j.

$$\{H(\omega)\} = \begin{bmatrix} H_{11} & H_{12} & \cdots & H_{1n} \\ H_{21} & H_{22} & \cdots & H_{2n} \\ \vdots & \vdots & \vdots & \vdots \\ H_{n1} & \cdots & \cdots & H_{nn} \end{bmatrix}, \tag{8.26}$$

Thus the frequency response at DOF i is

$$H_{xi}(\omega) = H_{i1} F_1 + H_{i2} F_2 + \ldots H_{in} F_i = \sum_{j=1}^{n} H_{ij} F_j \tag{8.27}$$

i.e. is made up of contributions from excitations at the various DOFs in the systems.

The frequency response functions (and hence the frequency responses) are complex if damping is included in the analysis. Amplitude and phase response at each DOF is given by taking the modulus and argument of the elements of $\{H_x(w)\}$ respectively.

(d) Forced damped vibration (random excitation)
Random excitation arises particularly from terrain inputs and is important in the analysis and design of vehicle suspensions. This form of excitation is non-deterministic in that its instantaneous value cannot be predicted at some time in the future. There are, however, some properties of random functions which can be described statistically. The mean or mean square value can be determined by averaging and the frequency content can be determined from methods based on the Fourier transform (Newlands, 1975).

The effect of random excitation on suspension design and analysis will be discussed in more detail in the chapter on Suspension Systems and Components in Chapter 10.

8.3 Vibration control

While it is recognized that the ideal form of vibration control is 'control at source', there is a limit to which this can be performed. For example the most dominant on-board source of vibration in motor vehicles is the engine. Here, engine firing and reciprocating unbalance combine to produce a complex source of vibration which varies with engine operating conditions. Reciprocating unbalance arises at each cylinder because of the fluctuating inertia force associated with the mass at each piston. This force acts along the cylinder axis and in multi-cylinder engines gives rise to a shaking force and moment acting on the engine block. By carefully arranging the relative crank positions on the crank shaft it is possible to reduce these forces and moments significantly, but because the forces contain a number of higher harmonic components the unbalance effect can never be completely removed. This topic will be discussed in more detail later in Section 8.3.5.

Another important source of on-board vibration is due to the unbalance of rotating parts.

While these may meet balancing standards it should be appreciated that there is no such thing as 'perfect balance'. Hence, small amounts of allowable residual unbalance are there to cause unwanted levels of vibration.

It follows that even when the best practices are followed there will always be some unwanted sources of vibration present. It is then necessary to minimize the effect of these on driver and passengers. In this section we review some of the ways in which this can be achieved.

8.3.1 Vibration isolation

This is a way of localizing the vibration to the vicinity of the source, thereby preventing its transmission to other parts of a (vehicle body) structure where it may result in the generation of noise. It can be achieved by the use of either passive or controllable vibration isolators. Passive isolators range from simple rubber in shear or combinations of shear and compression components to quite sophisticated hydro-elastic elements. The simpler forms of isolator are a cost-effective solution for sources with a limited range of operating conditions (amplitudes and frequencies). For situations where the source of vibration produces a range of operating conditions (as in i.c. engines) it is necessary to consider the use of hydro-elastic or controllable isolators. Some of these devices are discussed below. In all cases it is essential to understand the basic principles to achieve the best results.

The basic principles for selecting the appropriate isolator can be illustrated with reference to the SDOF model shown in Figure 8.6 (a more detailed treatment can be found in Snowdon, 1968). This represents a machine of mass m, subjected to a harmonic excitation arising from rotating unbalance $m_e r$, supported on elastomeric mounts having a complex stiffness k^* described by

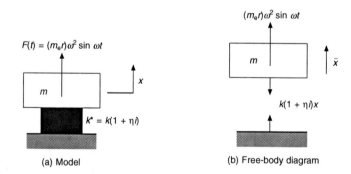

(a) Model   (b) Free-body diagram

Figure 8.6 SDOF vibration isolation model and free-body diagram

$k^* = k(1 + \eta i)$, where k is the dynamic stiffness and η the loss factor.

The effectiveness of the isolation (a function of frequency ω) can be defined by the transmissibility

$$T(\omega) = \frac{P}{F_0} \qquad (8.28)$$

where P = the amplitude of the force transmitted to the foundation and F_0 is the amplitude of the excitation force due to unbalance. Applying Newton's second law to the FBD gives:

$$m\ddot{x} + k(1 + \eta i)x = m_e r \omega^2 \sin \omega t = F_0 f(t) \qquad (8.29)$$

and the force transmitted to the foundation is

$$P(t) = k(1 + \eta i) x(t) \qquad (8.30)$$

Employing the approach outlined in Section 8.2.4 for harmonic excitation it can be shown that:

$$T(\omega) = k \sqrt{\frac{1 + \eta^2}{(k - m\omega^2)^2 + (k\eta)^2}} \qquad (8.31)$$

To appreciate how $T(\omega)$ varies with frequency it is helpful to show equation 8.31 in dimensionless form by dividing numerator and demoninator by k we then have

$$T(\omega) = \sqrt{\frac{1 + \eta^2}{\left[1 - \left(\dfrac{\omega}{\omega_n} \right)^2 \right]^2 + \eta^2}} \qquad (8.32)$$

where $\omega_n = \sqrt{\dfrac{k}{m}}$ is the natural frequency of the mass on its isolators.

For low damping elastomers η is of the order of 0.05 (Snowdon, 1968). The variation of transmissibility with frequency ratio $r = \dfrac{\omega}{\omega_n}$ for $\eta = 0.05$ is shown in Figure 8.7.

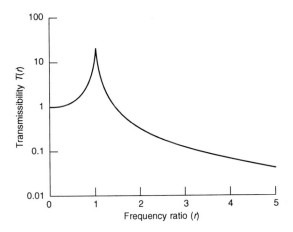

Figure 8.7 Transmissibility of an elastomeric isolator

For the isolators to be effective the transmissibility must be less than unity, i.e. P must be less than F_0. From Figure 8.7 it is clear that frequency ratios from 0 to approximately 1.4, P is greater than F_0 and therefore the isolators **magnify** the force to the foundation in this range.

Importantly around resonance ($r = 1$) the isolators magnify the force to the foundation approximately 20 times emphasizing the danger of not doing some simple analysis. As the value of r increases beyond 1.4 the isolators become increasingly effective. However, for very large values of r it is possible to induce wave-effects in isolators (Snowdon, 1968). These are due to local resonances in the distributed mass and elasticity of the isolator material and produce additional resonance peaks in the transmissibility curve (and reduced isolator performance) at certain frequencies of excitation.

An alternative way of describing isolator effectiveness is to use the term isolator efficiency. This is defined as

$$E_{iso} = [1 - T(\omega)]\ 100 \tag{8.33}$$

and expressed as a percentage.

8.3.2 Tuned absorbers

Vibration absorbers are useful for reducing vibration levels in those systems in which an excitation frequency is close to or coincides with a natural frequency of the system. Such absorbers consist of a spring-mass sub-system which is added to the original system. In effect, energy is transferred from the original system to the absorber mass which can vibrate with significant amplitude depending on the amount of damping contained in the absorber sub-system. These devices are particularly effective for reducing large amplitude oscillations in the original system but add another degree of freedom to the overall system, producing new natural frequencies above and below the original natural frequency. These result in resonant amplitudes (which can be controlled with an appropriate choice of damping in the absorber), permitting them to be used for variable speed applications.

The principles of undamped and damped tuned absorbers can be understood by outlining first the analysis of the damped absorber and then treating the undamped absorber as a special case of this. Assume the problem system has a SDOF with harmonic excitation as shown in Figure 8.8(a). By adding an absorber to the original system, the two-DOF system shown in Figure 8.8(b) is formed.

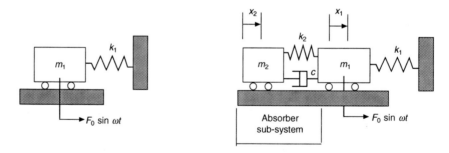

Figure 8.8 Models for analysing the tuned absorber. (a) Original system; (b) original system with absorber

The natural frequency of the original system is $\omega_1 = \sqrt{\dfrac{k_1}{m_1}}$ and at resonance $\omega = \omega_1$. By drawing the FBDs for the two mass system, it can be shown that the equations of motion in matrix form are:

$$\begin{bmatrix} m_1 & 0 \\ 0 & m_2 \end{bmatrix}\begin{Bmatrix} \ddot{x}_1 \\ \ddot{x}_2 \end{Bmatrix} + \begin{bmatrix} c & -c \\ -c & c \end{bmatrix}\begin{Bmatrix} \dot{x}_1 \\ \dot{x}_2 \end{Bmatrix} + \begin{bmatrix} (k_1 + k_2) & -k \\ -k_2 & k_2 \end{bmatrix}\begin{Bmatrix} x_1 \\ x_2 \end{Bmatrix} = \begin{Bmatrix} F_0 \\ 0 \end{Bmatrix}\sin \omega t \quad (8.34)$$

Following the method outlined in Section 8.2.4 it can be shown that the (complex) steady state responses $\overline{X}_1$ and $\overline{X}_2$ of the two masses are given by:

$$\begin{Bmatrix} \overline{X}_1 \\ \overline{X}_2 \end{Bmatrix} = \begin{bmatrix} ((k_1 + k_2 - m_1\omega^2) + c\omega i) & -k_2 \\ -k_2 & ((k_2 - m_2\omega^2) + c\omega i) \end{bmatrix}^{-1}\begin{Bmatrix} F_0 \\ 0 \end{Bmatrix}$$

This produces the amplitudes of vibration

$$X_1 = F_0\sqrt{\dfrac{(k_2 - m_2\omega^2)^2 + (c\omega)^2}{[(k_2 - m_2\omega^2)(k_1 + k_2 - m_1\omega^2) - k_2^2]^2 + (c\omega)^2(k_1 - m_1\omega^2 - m_2\omega^2)^2}} \quad (8.35)$$

and

$$X_2 = \dfrac{F_0 k_2}{\sqrt{[(k_2 - m_2\omega^2)(k_1 + k_2 - m_1\omega^2) - k_2^2]^2 + (c\omega)^2(k_1 - m_1\omega^2 - m_2\omega^2)^2}} \quad (8.36)$$

The undamped tuned absorber (c = 0)
In this case the amplitudes are given by:

$$X_1 = \dfrac{F_0(k_2 - m_2\omega^2)}{\Delta(\omega)}$$

$$(8.37,\ 8.38)$$

$$X_2 = \dfrac{F_0 k_2}{\Delta(\omega)}$$

where

$$\Delta(\omega) = (k_2 - m_2\omega^2)(k_1 + k_2 - m_1\omega^2) - k_2^2 9 \quad (8.39)$$

In order to make X_1 zero, $k_2 - m_2\omega^2$ must be zero and hence $\omega = \sqrt{\dfrac{k_2}{m_2}} = \omega_2$ the natural frequency of the absorber sub-system. It follows that for this case $\omega_1^2 = \omega_2^2 = \dfrac{k_1}{m_1} = \dfrac{k_2}{m_2}$ and hence the natural frequencies of the two sub-systems must be the same.

Resonance of the complete 2-DOF system (i.e. when X_1 and X_2 tend to infinity) occurs when ω coincides with the system's natural frequencies Ω_1 and Ω_2. This occurs when $\Delta(\omega) = 0$, representing the characteristic equation for the 2-DOF system.

In designing an untuned absorber it is necessary to consider the magnitude of the absorber mass in relation the original system mass m_1. In general the larger the mass ratio $\mu = m_2/m_1$, the more widely separated are the frequencies Ω_1 and Ω_2 and the wider is the range of frequencies at which the system can operate without exciting resonance.

The general response of the complete system is best described in terms of dimensionless amplitudes and frequencies ratios. Denoting the dimensionless amplitudes of m_1 and m_2 as $A_1 = \dfrac{k_1}{F_0} X_1$ and $A_2 = \dfrac{k_1}{F_0} X_2$ together with the frequency ratio as $r = \dfrac{\omega}{\omega_1}$ enables equations 8.37 and 8.38 to be written as:

$$A_1 = \frac{1 - r^2}{(1 - r^2)(1 + \mu - r^2) - \mu} \tag{7.40}$$

$$A_2 = \frac{1}{(1 - r^2)(1 + \mu - r^2) - \mu} \tag{7.41}$$

This enables the amplitude responses to be plotted for various values of μ. Figure 8.9 shows plots for $\mu = 0.2$. In Figure 8.9(a) A_{10} is the amplitude response of the original system.

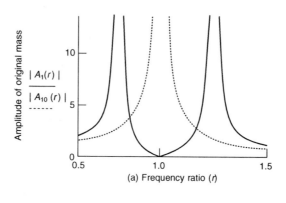

(a) Frequency ratio (r)

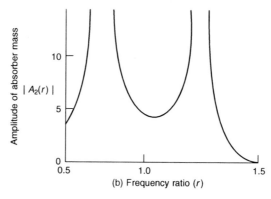

(b) Frequency ratio (r)

Figure 8.9 Frequency responses of a system fitted with an untuned absorber ($m = 0.2$)

The damped tuned absorber
The undamped absorber transfers energy from the original system to the absorber sub-system resulting in large amplitudes of vibration of the absorber mass. This can lead to the possibility of fatigue failure in the absorber spring. To overcome this problem it is necessary in practice to add some damping to the absorber. This also allows a wider operating range and limits the

resonant amplitudes in the region of the two natural frequencies. In this case the system is represented by the model in Figure 8.8(b) and the amplitude responses are given by equations 8.35 and 8.36. When $c = \infty$ the two masses in the system are effectively locked together resulting in a new undamped SDOF system having a natural frequency $\Omega_n = \sqrt{k_1/(m_1 + m_2)}$. When the response for this case is superimposed on that for $c = 0$, the response curves intersect at two points P and Q. This is shown in Figure 8.10. It can be shown (Dimaragonas *et al.*, 1992) that when $0 \leq c < \infty$ the amplitude response A_1 passes through P and Q as shown in Figure 8.10 for the case $\omega_1 = \omega_2$ and $\mu = 0.2$, illustrating how the resonant response amplitudes are limited. By careful optimization of the parameters (Snowdon, 1968) it is possible to minimize the resonant amplitudes. Unlike the case of the undamped absorber, it is not possible to reduce the amplitude of mass m_1 to zero at the original natural frequency of the system. Thus some of the effectiveness of an absorber is lost at this frequency when damping is introduced.

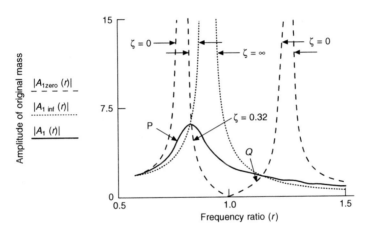

Figure 8.10 Response resulting from a damped absorber ($m = 0.2$)

8.3.3 Untuned viscous dampers

While tuned absorbers are tuned to a particular system resonance, an untuned viscous damper is a device designed to generally increase damping in a system and thereby reduce resonant amplitudes across a wide range of frequencies. These devices consist of an inertia (seismic) mass which is coupled to the original system via some form of damping medium, usually silicone fluid.

These devices are commonly used to limit torsional oscillations in crankshafts which have a number of natural frequencies and are subjected to a wide range of excitation frequencies. The torsional damper consists of a free rotating disc mounted on bearings inside the casing of the damper which is filled with silcone fluid. The casing of the damper is attached to the crankshaft at the opposite end to the flywheel. The principles of vibration control can be studied by assuming that the mass-elastic model of the crank shaft can be simplified down to a single rotating mass supported on a shaft fixed at one end with the mass subjected to a harmonic excitation torque $T_0 \sin \omega t$. An application of this will be discussed in more detail in Section 8.3.5. The resulting model is shown in Figure 8.11(a) where I_1 is the inertia of the mass about the shaft axis and K is the torsional stiffness of the shaft. Adding the damper of inertia I_2 and damping coefficient C results in the 2-DOF system shown in Figure 8.11(b). It is assumed here that the masses of the fluid and damper casing are negligible. θ_1 and θ_2 are the angular position of the two masses.

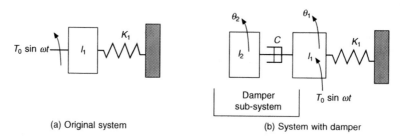

(a) Original system   (b) System with damper

Figure 8.11 Models for analysing the untuned viscous damper

The equations of motion are

$$I_1 \ddot{\theta}_1 + C \dot{\theta}_1 + K \theta_1 - C \dot{\theta}_2 = T_0 \sin \omega t$$

$$(8.42, 8.43)$$

$$I_2 \ddot{\theta}_2 - C \dot{\theta}_1 + C \dot{\theta}_2 = 0$$

Comparing with equations 8.34 and 8.35 it follows that the amplitude of vibration of the mass in the original system is

$$\theta_1 = T_0 \sqrt{\frac{(I_2 \omega^2)^2 + (C\omega)^2}{[(I_2 \omega^2)(K_1 - I_1 \omega^2)]^2 + (C\omega)^2(K_1 - I_1 \omega^2 - I_2 \omega^2)^2}} \qquad (8.44)$$

Using the following notation:

undamped natural frequency of the original system, $\omega_n = \sqrt{\dfrac{K}{I_1}}$, damping ratio, $\zeta = \dfrac{C}{2\sqrt{I_1 K}}$,

inertia ratio, $\mu = \dfrac{I_2}{I_1}$, dimensionless amplitude of I_1, $A_1 = \dfrac{K\theta_1}{T_0}$ and frequency ratio $r = \dfrac{\omega}{\omega_n}$,

it can be shown that

$$A_1 = \sqrt{\frac{(\mu r)^2 + 4\zeta^2}{(\mu r)^2(1 - r^2)^2 + 4\zeta^2[\mu r^2 - (1 - r^2)]^2}} \qquad (8.45)$$

A_1 is thus a function r, μ and ζ. For a given value of ζ the response will exhibit a single peak similar to that for a damped SDOF system. The extreme values of damping are $\zeta = 0$ and ∞.

When $\zeta = 0$ the system response is that of the original SDOF system having a natural frequency ω_n, and when $\zeta = \infty$, both masses move together as one and the undamped natural frequency is $\sqrt{K/(I_1 + I_2)}$. When A_1 is plotted on the same axes for these two extreme cases and for a given μ, the curves intersect at a point P. It can be shown (Thomson, 1989) that the curves for other values of damping also pass through P. These features are illustrated in Figure 8.12.

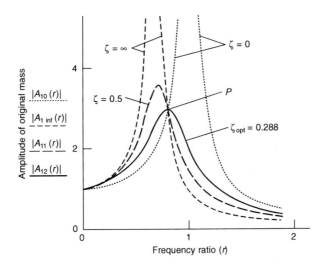

Figure 8.12 Response of an untuned viscous damper ($m = 1.0$)

Clearly the optimum value of damping is the one which has its maximum value at P. It can be shown (Thomson, 1989) that this is given by:

$$\zeta_{opt} = \frac{\mu}{\sqrt{2(1 + \mu)(2 + \mu)}} \tag{8.46}$$

In general the minimum peak amplitude decreases as μ and ζ increases up to the value given by equation 8.46.

8.3.4 Damping treatments

Damping treatments (in the form of high damping polymers) can be used to limit resonant response amplitudes in structures and are particularly effective for flexural vibration of panels and beams. In an automotive context they are used extensively to limit the resonant responses of body panels and bulkheads.

Because of the poor structural strength of high damping polymers it is necessary to either bond them to the surface of load-bearing elements or to incorporate them into load-bearing elements by sandwich construction. These forms of damping are termed *unconstrained-* and *constrained-layer damping* respectively, with the latter being by far the most effect way of deploying this type of structural damping treatment. Flexing of the load-bearing element produces shearing effects in the damping layer and thus vibrational energy is converted into heat and dissipated. Shear properties of polymer materials are generally temperature and frequency dependent. Furthermore their use in the form of constrained layer damping poses problems with bending and forming in manufacturing operations (Beards, 1996).

8.3.5 Applications

Two practical examples will be presented, both are of considerable importance in automotive vibration control. The first of these is concerned with isolation of the engine from the vehicle structure and the other relates to the control of torsional oscillation amplitudes in engine crankshafts.

In order to understand how the above principles relate to these two examples it is first of all necessary to understand the processes by which the engine vibration is generated. The first part of this sub-section therefore outlines how the excitation forces and moments arise in single cylinder and multi-cylinder i.c. engines.

Dynamic forces generated by i.c. engines

(a) Single cylinder engines (Shigley *et al.*, 1980; Norton, 1992).
The dynamic forces exerted on single cylinder engines arise from the reaction torque at the crankshaft and a shaking force acting along the line of stroke. Both are cyclic and related to engine speed. The torque on the crankshaft consists of a component generated by the gas force in the engine cylinder and on the inertia force component, associated with the acceleration of the connecting rod and piston assembly. The shaking force is attributed solely to the inertia forces. The component of torque arising from the gas forces has a fundamental frequency equal to running speed for a two-stroke engine and at half engine speed for a four-stroke engine, with higher order harmonics (called engine orders). The kinematics of reciprocating engine mechanisms results in inertia torque and shaking force components which have fundamental components at engine speed with higher order harmonics. For four-stroke engines this leads to half order components since an engine cycle is half the engine speed.

The fluctuating reaction torque and shaking force produce translational and rotational vibration of the engine and calls for careful mounting to minimize transmission of vibration to the vehicle body.

(b) Multi-cylinder engines
Since the gas torque from each cylinder is basically a pulse occurring during the expansion stroke it is essential that for a multi-cylinder engine the torque contributions from each cylinder are evenly spaced to give a regular train of pulses. This dictates that the crank throws in the order of firing must have a regular angular spacing. For two-stroke and four-stroke engines this is $360/n$ and $720/n$ respectively, where n is the number of cylinders.

Also, in a multi-cylinder engine, the shaking forces act along the line of each cylinder

(Figure 8.13) producing a shaking moment on the engine about an axis perpendicular to the plane containing the cylinders. For the example shown this is given by:

$$\vec{M}_{\text{Total}} = \sum_{i=1}^{n} \vec{z}_i \otimes \vec{F}_i \qquad (8.47)$$

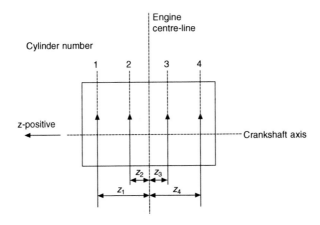

Figure 8.13 Axial location of cylinders relative to centre-line

By careful arrangement of the firing order and relative angular position of the cranks it is possible to cancel out some of the shaking force and shaking moment components. For example for a four cylinder four-stroke engine with equally spaced cylinders numbered from 1 at one end to 4 at the other, the relative crank spacings are at 180 degree intervals and possible firing orders are 1-3-4-2 and 1-2-4-3. It can be shown (Norton, 1992) with the latter firing order, that the first and third order inertia torques are balanced, there are 2nd, 4th and 6th order shaking force components and a second order shaking moment component (the 1st, 4th and 6th being balanced). Norton (1992) gives some general equations which can be used to check the state of balance of multi-cylinder engines.

Engine isolation
The variable operating conditions of automotive engines presents a complex problem in vibration isolation for the engineer. In addition to the fluctuating torque at the crankshaft and the shaking forces and moments identified in the previous section there are additional dynamic inertial loads arising from vehicle manoeuvring and terrain inputs to the wheels.

The primary components of engine vibration at idling are integer multiples of engine speed and the dominant component in four cylinder engines occurs at twice the engine speed and results from the combustion pulses. Idle speeds for four cylinder engines are typically in the range from 8–20 Hz producing dominant frequency components in the range from 16–40 Hz. Since the primary bending mode of passenger cars can be less than 20 Hz it is obvious that it is easy to excite body resonance at idle if engine isolation is not carefully designed.

When supported on its vibration isolators the powertrain (engine and gearbox) can be treated as a rigid body having six degrees of freedom (three translational and three rotational). In order

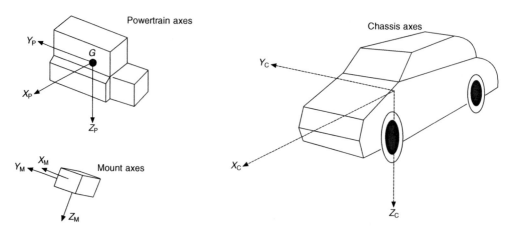

Figure 8.14 Reference axes for analysing the dynamics of a powertrain-mount system

to uncouple these motions the isolators should lie in planes aligned with the principal inertial axes passing through the centre of gravity of the unit. The problem facing the chassis mounting engineer is to select a set of appropriate mounts and position them in such a way as to isolate the chassis from the above excitations and restrain the engine against excessive movement due to the engine torque. In general, engine mounts require complex characteristics to meet the above demands. Some of the important features of these will be discussed below.

(a) Formulation of the vibration equations
In order to formulate the problem for dynamic analysis one must consider the position and orientation of the engine relative to a set of chassis-based coordinates together with the orthogonal characteristics of each mount, its location and orientation. Typical sets of axes are shown in Figure 8.15. In the static equilibrium position the mass centre G of the powertrain coincides with the origin of the chassis fixed axis system. In the following development of the equations of motion it will be assumed that the displacements of the powertrain are small.

Denoting the position of the i-th mount attachment point relative to the powertrain in the equilibrium position as $\vec{r}_{i0}$ (in chassis coordinates) and the translation and rotation of the powertrain relative to the chassis axes as $\vec{r}_G$ and $\vec{\Theta}$, the deflection of the mount is given by

$$\vec{d}_{ci} = \vec{r}_i - \vec{r}_{i0},\tag{8.48}$$

where

$$\vec{r}_i = \vec{r}_G + \vec{\Theta} \otimes \vec{\rho}_i\tag{8.49}$$

and $\vec{\rho}_i$ is the location of the mount attachment point relative to the powertrain axes.

If the coordinate transformation relating the orientation of the mount axes to those of the chassis is $[T_{mc}]$, then the mount deflection in terms of mount coordinates is

$$\{d_m\}_i = [T_{mc}]_i \{d_c\}_i\tag{8.50}$$

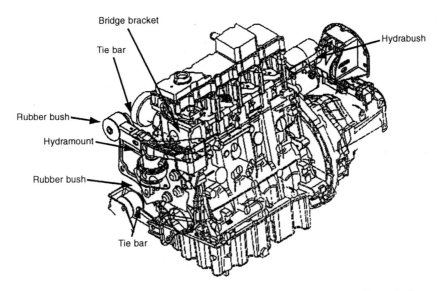

Bridge bracket

Tie bar

Hydrabush

Rubber bush

Hydramount

Rubber bush

Tie bar

Figure 8.15 A torque axis engine mounting system (courtesy of Rover Group Ltd)

The 'elastic' potential energy for a set of mounts typically having orthogonal complex stiffness components represented by the diagonal stiffness matrix $[k_m]_i$ is

$$V = \tfrac{1}{2} \sum_{i=1}^{n} \{d_m\}_i^T [k_m]_i \{d_m\}_i \tag{8.51}$$

in terms of mount displacements. Using the transformation in equation 8.50 the equivalent equation in terms of chassis coordinates is

$$V = \tfrac{1}{2} \sum_{i=1}^{n} \{d_c\}_i^T [k_c]_i \{d_c\}_i \tag{8.52}$$

where $[k_c]_i = [T_{mc}]_i^T [k_m]_i [T_{mc}]_i$ is a non-diagonal matrix.

The kinetic energy of the system is given by

$$T = \tfrac{1}{2} m \dot{\vec{r}}_G \cdot \dot{\vec{r}}_G + \tfrac{1}{2} \dot{\vec{\Theta}} \cdot \vec{h}_G \tag{8.53}$$

where $\vec{h}_G$ is the angular momentum of the powertrain about G, the components of which are given by $\{h_G\} = [M]\{\dot{\Theta}\}$.

$[M]$, the symmetrical mass/inertia matrix, is of the form:

$$[M] = \begin{bmatrix} m & 0 & 0 & 0 & 0 & 0 \\ 0 & m & 0 & 0 & 0 & 0 \\ 0 & 0 & m & 0 & 0 & 0 \\ 0 & 0 & 0 & I_{xx} & -I_{xy} & -I_{xz} \\ 0 & 0 & 0 & -I_{yx} & I_{yy} & -I_{yz} \\ 0 & 0 & 0 & -I_{zx} & -I_{zy} & I_{zz} \end{bmatrix} \tag{8.54}$$

where m is the mass of the powertrain, elements I_{xx}, I_{yy}, I_{zz} are moments of inertia and I_{xy} etc are products of inertia about the powertrain axes. The equations of motion can then be derived from the linearized form of Lagrange's equations (Meirovitz, 1986):

$$\frac{d}{dt}\left(\frac{\partial T}{\partial \dot{q}_k}\right) + \frac{\partial V}{\partial q_k} = Q_k, \; k = 1 \ldots 6 \tag{8.55}$$

where q_k are the generalized coordinates (components of $\bar{r}_G$ and $\bar{\Theta}$ and Q_k are a set of generalized forces derived from the excitations imposed on the powertrain). The result is a set of six equations in matrix form:

$$[M]\{\ddot{q}\} + [K]\{q\} = \{Q\} \tag{8.56}$$

These are a set of equations coupled in both the mass and stiffness matrices, the elements of the latter being complex (containing dynamic stiffness and loss factor values). The equations can be solved for harmonic inputs using the techniques described earlier.

The dynamic forces transmitted to the chassis can be determined for a particular mount configuration by solving the above equations and translating the dynamic displacements $\{q\}$ on the powertrain into mount deflections $\{d_c\}$ in chassis coordinates and transposing into mount coordinates with equations 8.50. The dynamic force components transmitted to the chassis at the i-th isolator is then given by

$$\{F_m\}_i = [k_m]_i\{d_m\}_i. \tag{8.57}$$

(b) Mount requirements and types
The requirements for engine mounts are:

   (i) a low spring rate and high damping during idling and
  (ii) a high spring rate and low damping for high speeds, manoeuvring and when traversing rough terrains.

The following types of mount attempt to meet these conflicting requirements:

   (i) Simple rubber engine mounts
      These are the least costly and least effective forms of mount and clearly do not meet all the conflicting requirements listed above. They do not provide the high levels of damping required at idling speeds.
  (ii) Hydro-elastic mounts
      These generally contain two elastic reservoirs filled with a hydraulic fluid. Some also contain a gas filled reservoir. This type of mount exploits the feature of mass-augmented dynamic damping which is a form of tuned vibration absorber. In operation there is relative motion across the damper which produces flexure of the rubber component and transfer of fluid between chambers, thereby inducing a change in mount transmissibility. This type of mount has become common in recent years and some examples are given in the literature (Kim *et al.*, 1992; Muller *et al.*, 1996).
 (iii) Semi-active mounts
      The operation of these mounts is dependent on modifying the magnitude of the forces

transmitted through coupling devices. They may be implemented via low-bandwidth low-power actuators which are suited to open-loop control. Some forms of hydraulic semi-active (adaptive) mount use low powered actuators to induce changes in mount properties by modifying the hydraulic parameters within the mount. The actuators may then be on–off (adaptive) or continously variable (semi-active) types. Considerable effort has been devoted to this type of technology in recent years. Some examples are given in Morishita *et al.* (1992) and Kim *et al.* (1993).

(iv) Active mounts
This type of mount requires control of both the magnitude and direction of the actuator force used to adjust the coupling device. High-speed actuators and sensors require having an operating bandwidth to match the frequency spectrum of the disturbance. Power consumption is generally high in order to satisfy the response criteria. Active vibration control is typically implemented by closed-loop control. An example of active engine mount modelling and performance is given by Miller *et al.* (1995).

(c) A typical example
A practical implementation of an engine mounting system for a four-cylinder diesel engine is shown in Figure 8.16. It comprises two mass carrying mounts (one a hydramount, the other a hydrabush, both passive mounts) and two torque reacting tie bars. The hydramount is linked to the power unit by an aluminium bridge bracket. Both tie bars have a small bush at the power unit end and a large bush at the body end. The lower tie bar has its power unit end carried by a bracket attached to the sump and its body end attached to a subframe which also carries the vehicle suspension. The vertical stiffnesses of the mass carriers have very little effect on the torque performance of the system and can therefore be tuned for ride. The function of the hydramounts is of course to improve ride. The tie bar fore and aft rates do not affect ride and can be tuned for the torque loading.

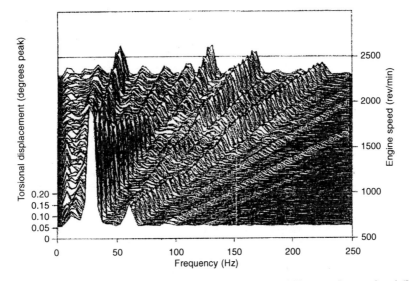

Figure 8.16 Waterfall plot for a multi-cylinder engine (courtesy of Simpson International (UK) Ltd)

Crankshaft damping

The torsional dynamics of crankshafts are dependent on the distribution of their mass and elasticity (defining the modal characteristics) and the excitations arising from the torque/cylinder discussed earlier in this section. Because the torque contains a number of harmonic components and engine speed is variable there is a tendency to excite a large number of torsional resonances as illustrated by the waterfall plot (torsional amplitude plotted as a function frequency for a range of engine speeds) in Figure 8.17.

Casing

Seismic mass

Figure 8.17 Typical crankshaft damper (courtesy Simpson International (UK) Ltd)

Because crankshafts are lightly damped in torsion, the resulting resonant amplitudes can be large, resulting in high cyclic stresses which tend to cause fatigue failure. One solution to this problem is to introduce some torsional damping into the system, with emphasis on damping the fundamental mode of vibration because this tends to be subjected to the highest torsional amplitudes. A typical crankshaft damper is shown in Figure 8.17 and a simplified selection procedure is outlined as follows.

(a) Determination of the mass-elastic model

The case of a 6-cylinder in-line diesel engine is taken as an example. The objective is to determine the torsional stiffnesses and disc inertias in the system model shown in Figure 8.18(a).

The disc inertias I_1 to I_6 are each made up from the moments of inertias of the crankshaft elements (webs and pins etc) about the crankshaft axis and the inertia equivalents for the connecting rod and piston components associated with each cylinder (Shigley *et al.* 1980). The moments of inertia of the crankshaft elements can be determined from a draughting package and torsional stiffnesses between cylinders K_1 to K_5 from a finite element model.

(b) Modal analysis of the mass-elastic model

Having determined the numerical data for the mass-elastic model in Figure 8.18(a), it is then possible to formulate the mass and stiffness matrices, $[M]$ and $[K]$ respectively – in our example

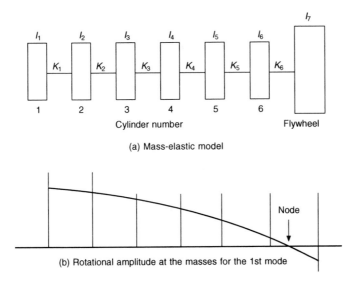

Figure 8.18 Mass-elastic model and first torsion mode of a six cylinder engine

these are 7×7 matrices. It should be noted that the mass-elastic model in this case is a so-called *free-free* system since it is not anchored to ground. This will result in a rigid-body mode in the eigenvalue solution identified as a vector of seven equal numerical values. The corresponding eigenvalue will be zero. Of particular interest is the first torsional mode which may be denoted by its frequency ω_1 and mode shape $\{u\}_1$, shown in Figure 8.18(b).

Biasing the selection of the damper towards the first mode of vibration, it is possible to replace the mass-elastic model in Figure 8.18(a) with a single DOF equivalent and when the damper is added to this model the two-DOF model in Figure 8.19 results. The ground fixing in this model coincides with the node position determined from the first mode of vibration and the damper casing mass is assumed negligible. The inertia of the damper casing can be incorporated into the eigenvalue analysis if data is available. The SDOF equivalent inertia of the crankshaft I_e, can be determined by equating the maximum kinetic energy of the SDOF system to that in the first mode of the mass-elastic system to the left of the node. This results in

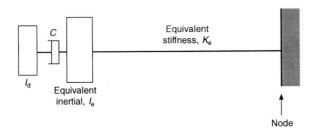

Figure 8.19 Equivalent model based on first mode of the mass-elastic model

$$I_e = \sum_{i=1}^{6}\left(\frac{I_i u_i}{u_1}\right)^2 \tag{8.58}$$

The equivalent torsional stiffness of the SDOF model K_e, is then detemined from $\omega_1 = \sqrt{\dfrac{K_e}{I_e}}$

or $K_e = I_e\omega_1^2$. The theory in Section 8.3.3 can then be applied to determine an optimum damping rate C_{opt} for a given damper inertia I_d.

8.4 Fundamentals of acoustics

An understanding of acoustic fundamentals is essential in controlling noise and interpreting noise criteria. This section outlines some of the basic principles in sound propagation.

8.4.1 General sound propagation

Sound is transmitted from the source to the receiver by an elastic medium called the path. In an automotive context this is the surrounding air or the vehicle body structure, giving rise to the term *structure-borne sound*.

The simplest form of sound propagation occurs when a small sphere pulsates harmonically in free space (away from any bounding surfaces). The vibrating surface of the sphere causes the air molecules in contact with it to vibrate and this vibration is transmitted radially outwards to adjoining air molecules. This produces a propagating (travelling) wave which has a characteristic velocity c, the velocity of sound in air. At some arbitrary point on the path, the air undergoes pressure fluctuations which are superimposed on the ambient pressure. A sound source vibrating at a frequency f, produces sound at this frequency. Taking a snapshot of the instantaneous pressure and traversing away from the source, the variation of pressure with distance is also sinusoidal. The distance between pressure peaks is constant and known as the *wavelength* λ. This is related to c and f by the equation:

$$\lambda = \frac{c}{f} \tag{8.59}$$

From this equation it is seen that as f increases λ decreases. In the audible range from 20 Hz to 20 kHz; the wavelength correspondingly varies from 17 m to 17 mm.

8.4.2 Plane wave propagation

The fundamentals of wave motion are most easily understood by considering the propagation of a plane wave (having a flat wavefront perpendicular to the direction of propagation). Denoting the elastic deformation as ξ at some distance x from a fixed datum and combining the continuity and momentum equations for the element with the gas law leads to (Kim *et al.*, 1993) the one dimensional wave equation

$$\frac{\partial^2 \xi}{\partial t^2} = c^2 \frac{\partial^2 \xi}{\partial x^2} \tag{8.60}$$

where the progation velocity

$$c = \sqrt{\frac{p\gamma}{\rho}} = \sqrt{\gamma RT} \qquad (8.61)$$

and the notation is as follows: p = the ambient pressure, ρ = corresponding density of the medium, γ = ratio of specific heats for air, R = universal gas constant and T = absolute temperature. For air at 20°C the magnitude of c is 343 m/s.

The general solution to equation 8.60 for harmonic waves is:

$$\zeta = Ae^{(\omega t - kx)} + Be^{(\omega t + kx)} \qquad (8.62)$$

The first term on the right hand side represents the incident wave (travelling away from the source) while the second term represents the reflected wave (travelling in the opposite direction). The wavenumber $k = \dfrac{\omega}{c}$ or $k = \dfrac{2\pi f}{c} = \dfrac{2\pi}{\lambda}$ and hence is defined as the number of acoustic wavelengths in 2π. c varies considerably for fluid and solid materials. Some typical values are shown in Table 8.1.

Table 8.1 Velocity of wave propagation in various media

Medium	c, m/s
Air at 1 bar and 20°C	343
Mild steel	5050
Aluminium	5000
Vulcanized rubber	1269
Water at 15°C	1440

Specific acoustic impedance, z
The impedance which a propagating medium offers to the flow of acoustic energy is called the *acoustic impedance*. It is defined as the ratio of acoustic pressure p to the velocity of propagation u. It can be shown (Reynolds, 1981) that

$$z = \frac{P}{u} = \rho c \qquad (8.63)$$

and for normal temperature and pressure (101.3 kPa and 20°C) is equal to 415 rayls (Ns/m³).

Acoustic intensity, I
This is defined as the time averaged rate of transport of acoustic energy by a wave per unit area normal to the wavefront. It is given by Reynolds (1981);

$$I = \frac{P_{rms}^2}{\rho c} \qquad (8.64)$$

where p_{rms} is the rms pressure fluctuation. For a harmonic wave $p_{rms} = \dfrac{\hat{p}}{\sqrt{2}}$ and then

$$I = \frac{\hat{p}^2}{2\rho c} \tag{8.65}$$

where $\hat{p}$ is the peak pressure.

8.4.3 Spherical wave propagation; acoustic near and far fields

Spherical waves more closely approximate true source waves, but approximate to plane waves at large distances from a source. It may be shown (Reynolds, 1981) that the wave equation in spherical coordinates is

$$\frac{\partial^2 (rp)}{\partial t^2} = c^2 \frac{\partial^2 (rp)}{\partial r^2} \tag{8.66}$$

The general solution for an incident wave only (no reflection) is

$$p = \frac{1}{r} A e^{(\omega t - kx)} \tag{8.67}$$

when equation 8.67 is used in conjunction with the definition for acoustic impedance it may be shown that

$$z = \rho c \frac{(kr)^2}{1 + (kr)^2} + i(\rho c) \frac{kr}{1 + (kr)^2} \tag{8.68}$$

At large distances from the source ($kr \gg 1 orr \gg \lambda/2\pi$), $z \to \rho c$. Then pressure and particle velocity are in phase and we are in what is called the *acoustic far field* where spherical wavefronts approximate to those of plane waves and pressure and velocity are in phase.

At distances close to the source ($kr \ll 1 orr \ll \lambda/2\pi$), $z \to i\rho ckr$. Then pressure and velocity are 90° out of phase and we are in what is called the *acoustic near field*.

The transition from near to far field is in reality a gradual one, but is normally assumed to take place in the vicinity of $\lambda/2 \pi$. For a harmonic wave at 1 kHz ($\lambda \approx 1$ kHz), $r = 50$ mm; while for at 20 Hz; $r = 2.5$ m. The far field/near field transition has important implications for microphone positioning in sound level measurements.

8.4.4 Reference quantities

Certain reference quantities are used for sound emission measurements. For sound transmission in air the reference rms pressure is taken to be $p_{ref} = 20$ μPa corresponding approximately to the threshold of hearing at the reference frequency of 1 kHz. With a reference impedance $z_{ref} = (\rho c)_{ref} = 400$ rayls; the reference intensity $I_{ref} = 10^{-12}$ W/m^2 from equation 7.64. Since the acoustic power is the intensity times the spherical area, for a reference area $A_{ref} = 1$ m^2 the reference sound power is 10^{-12} W.

8.4.5 Acoustic quantities expressed in decibel form

The human ear is capable of detecting acoustical quantities over a very wide range, e.g. pressure variations from 20 μPa to 100 Pa. Hence there is a need to represent acoustical data in a convenient form. This is achieved by using the decibel scale. The quantity of interest x is expressed in the form $10 \log_{10} (x/x_{ref})$, where both x and x_{ref} have units of power. Using the above reference quantities the sound power level (L_w), the sound intensity level (L_I) and the sound pressure level (L_p) are as follows:

$$L_W = 10 \log_{10} \left(\frac{W}{W_{ref}} \right), \text{ dB} \tag{8.69}$$

$$L_I = 10 \log_{10} \left(\frac{I}{I_{ref}} \right), \text{ dB} \tag{8.70}$$

$$L_p = 10 \log_{10} \left(\frac{p}{p_{ref}} \right)^2 = 20 \log \left(\frac{p}{p_{ref}} \right), \text{ dB} \tag{8.71}$$

When it is noted that the threshold of hearing corresponds to a sound pressure level of 0 dB it may be shown (Reynolds, 1981) that for normal temperature and pressure (101.3 kPa and 20 °C). L_W, L_I and L_p are related as follows:

$$L_I = L_p - 0.16 \text{ dB} \tag{8.72}$$

$$L_W = L_p + 20 \log_{10} \left(\frac{r}{r_{ref}} \right) - 0.16 \text{dB} \tag{8.73}$$

where $r_{ref} = 0.282$ m.

From equation 8.72 it is seen that L_p and L_I are approximately equal numerically, while equation 8.73 is useful for determining the sound power level of an acoustic source from sound pressure level measurements in free field conditions. For these whole field radiation conditions the reduction in sound pressure level with doubling of distance from the source is 6 dB.

8.4.6 Combined effects of sound sources

It is often necessary to determine the sound pressure level of two or more uncorrelated sound sources when the level for each source is known. This can be achieved by using the equation.

$$L_{p,total} \approx L_{I,total} = 10 \log_{10} (\Sigma \, 10^{0.1 L_{pi}}) \text{dB} \tag{8.74}$$

8.4.7 Effects of reflecting surfaces on sound propagation

When an incident wave strikes a reflecting surface the wave is reflected backwards towards the source. In the vicinity of the reflecting surface the incident and reflected waves interact to produce what is known as a *reverberant field*. The extent of the depth of this field towards the sound source is dependent on the absorptive properties of the reflecting surface. A typical interaction between incident and reflected waves is shown in Figure 8.20 and sound pressure level variations as function of distance r from the sound source are as shown in Figure 8.21.

Practical situations arise where the sound source is positioned close to a hard reflecting surface. Four idealized cases are (a) whole space radiation – when there are no reflecting surfaces, i.e. the source is in free space, (b) half-space radiation – when the source is positioned

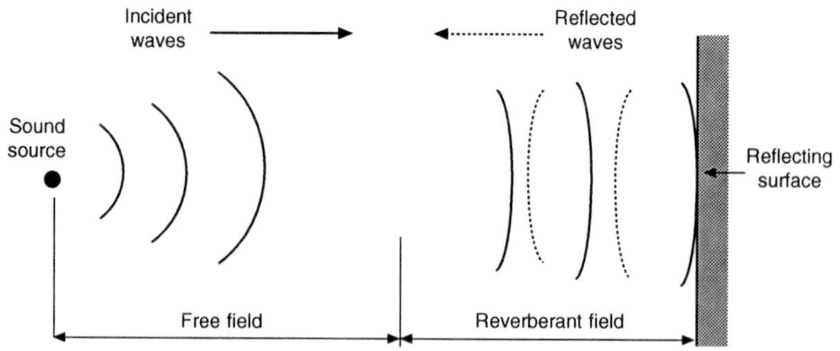

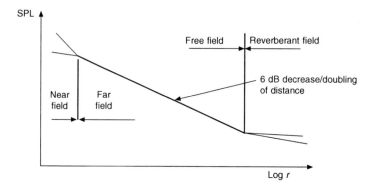

Figure 8.20 Interaction of incident and reflected waves

Figure 8.21 Sound pressure level as a function of distance from a simple spherical source

at the centre of a flat hard (reflecting) surface, (c) quarter space radiation – when the source is positioned at the intersection of two flat hard surfaces which are perpendicular to one another and (d) eighth space radiation – when the source is positioned at the intersection of three flat perpendicular hard surfaces. In each case there is an increase in acoustic intensity. The effect can be described by the *directivity index* DI in terms of the *directivity factor Q*.

$$DI = 10 \log_{10} Q \text{ dB} \tag{8.75}$$

where $Q = 1$ (DI = 0, dB) for a whole space, $Q = 2$ (DI = 3 dB) for a half space, $Q = 4$ (DI = 6 dB) for a quarter space and $Q = 8$ (DI = 9 dB) for an eighth space.

When evaluating sound power level (PWL) of a source from sound pressure level (SPL) measurements with the above source locations, equation 8.73 can be modified to

$$L_W = L_p + 20 \log_{10} \left(\frac{r}{r_{\text{ref}}} \right) - DI - 0.16 \text{ dB} \tag{8.76}$$

8.5 Human response to sound

The human ear is a delicate and sophisticated device for detecting and amplifying sound (Reynolds, 1981). It consists of an outer ear, a middle ear containing an amplifying device (the ossicles) and an inner ear containing the cochlea. This small snail-shaped element contains lymph and a coiled membrane to which are connected thousands of very sensitive hair endings of varying thickness. These respond to different frequencies, converting the sound stimulus into nerve impulses which are transmitted to the brain. A certain threshold level is required to stimulate the nerve cells, while over-stimulation can lead to temporary or permanent deafness. This latter effect was recognized in the early 70s as a cause of industrial deafness resulting in a number of regulations to protect workers.

The audible range for a healthy young person lies within the envelope shown in Figure 8.22. The frequency range extends from 20 Hz to 20 kHz and the SPL extends from the threshold of hearing at the lowest boundary to the threshold of feeling (pain) at the highest. It is observed that the sound pressure level at the upper and lower boundaries vary markedly with frequency. Typically at 1 kHz the range of sound pressure levels is from 0 to 130 dB. The shape of the curves for sounds of increasing loudness are generally similar to that for the threshold of hearing. It follows therefore that the human ear is most sensitive between 500 Hz and 5 kHz and is insensitive to sounds below 100 Hz.

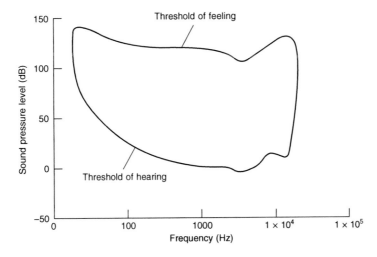

Figure 8.22 The audible range

8.6 Sound measurement

Automotive noise measurement is required for a variety of purposes dictating the need for a range of measuring equipment. In development work there is a requirement for measuring continuous noise levels such as that from drivetrains and their ancillaries, there are requirements for component noise testing for sound power, frequency analysis and source identification. For type approval there are requirements for assessing whole vehicle noise. Controlled test environments are also required to ensure that tests are repeatable and not weather dependent. This calls for special acoustical test chambers such as the anechoic chambers used to simulate free-field environments.

8.6.1 Instrumentation requirements

Sound level meters
The most basic instrument for sound measurement is a sound level meter comprising a microphone, r.m.s. detector with fast and slow time constants and an A-weighting network to enable measurements to be made which relate to human response to noise, leading to so called A-weighted noise levels L_{pA}, expressed in dB(A). Because of the frequency sensitivity of the human ear, the A-weighting network has the form shown in Figure 8.23. This emphasises the frequencies in 500 Hz to 5 kHz range and produces increasing attenuation below 100 Hz. (B.S. 5969, 1981) specifies four quality grades of sound level meter ranging Type 0 laboratory reference standards to Type 3 industrial grade meters. For development work Type 1 instruments are generally recommended. If the instrument is required for measurement of transient noise it should also be equipped with a peak hold facility.

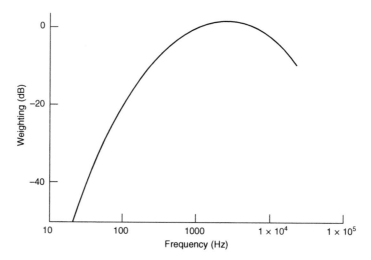

Figure 8.23 The A-weighting curve

Because sound levels are rarely constant (e.g. noise resulting from changes in engine speed), there is a need to average levels over prescribed intervals of time. This type of measurement leads to an equivalent noise level such as $L_{Aeq,T}$, which is related to noise deafness and annoyance criteria. This indicates the A-weighted noise level has been averaged over a measurement period T to give a level having the same energy content as a constant sound of the same numerical value. Mathematically this can be written:

$$L_{Aeq,T} = 10 \log_{10} \left(\frac{1}{T} \int_0^T \left(\frac{p_A(t)}{p_{ref}} \right) dt \right) \qquad (8.77)$$

An integrating type of sound level meter is required for this type of measurement.

Another requirement of sound level meters is to determine the level which has been exceeded for a prescibed portion of the measurement time, L_N, e.g. L_{90} represents the A-weighted level exceeded for 90% of the measurement period and is used in environmental background noise measurements associated with traffic noise.

Frequency analysers

Since the frequency spectrum of noise is closely related to the origins of its production, frequency analysis is a powerful tool for identifying noise sources and enables the effectiveness of noise control measures to be assessed.

The simplest frequency analysers split the frequency range into a set of octave bands having the following standardized centre frequencies: 31.5, 63, 125, 250, 500, 1k, 2k, 4k, 8k and 16k; Hz. These filters have a constant percentage bandwidth implying that the bandwidth increases with centre frequency giving increasingly poor discrimination at high frequencies. This can be improved with the use of third octave analysis. A number of product noise and environmental standards require the use of octave and third octave analysis and much of the performance data for noise control products is expressed in terms of octave band centre frequencies.

For serious noise control investigations, narrow band frequency analysers are a necessity. Instead of switching sequentially through a set of filters, the signal in a narrow band analyser is presented simultaneously to the inputs of all filters in the analysis range. The signal processing is done digitally. The outputs are updated many times per second and are fed to continuous display devices such as VDUs and/or downloaded to computers.

Sound intensity analysers

Sound intensity analysers allow sound power measurements to be made in-situ in the presence of background noise, i.e. they do not require special noise testing installations. They also allow noise source identification from sound intensity mapping.

A typical sound intensity probe consists of two closely spaced pressure microphones which measure the sound pressure and pressure gradient between the two microphones. Signal processing converts these measurements into sound intensity values in a sound intensity analyser.

8.7 Automotive noise criteria

As a result of the ever increasing numbers of vehicles on the roads of developed countries, the level of road traffic noise has continued to grow alarmingly. This is in spite of regulations imposed by governments and the significant reductions in noise levels which have already been achieved by new vehicles. The quest for quieter vehicles coupled with good design of new roads is set to go on in an attempt to drive overall noise levels down. Vehicle manufacturers are being faced with increasingly stringent noise regulations for new vehicles. The current limits for drive-by noise of new vehicles are being harmonized within the EEC and are intended to be progressively reduced into the foreseeable future.

8.7.1 Drive-by noise tests (ISO 362, 1981)

The procedure is to accelerate the vehicle in a prescribed way and in a prescribed gear past a microphone set up at a height of 1.2 m above a hard reflecting surface and 7.5 m from the path of the vehicle. The test area is required to be flat, have a low background noise level and not be influenced by reflecting obstructions, bystanders, tyre noise and wind noise. The test site should be as shown in Figure 8.24 and the vehicle should follow path A–B. When the vehicle reaches A the throttle should be opened fully and maintained in this position until the rear of the vehicle reaches B. A minimum of two measurements should be made on each side of the vehicle. In addition to the results of the measurements, vehicle details such as loading, rating, capacity and engine speeds should be reported.

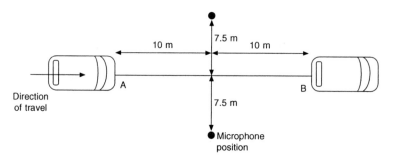

Figure 8.24 Drive-by test site and measurement locations

8.7.2 Noise from stationary vehicles

Since exhaust noise is one of the major sources of vehicle noise and vehicles spend a significant amount of time stationary in traffic queues, noise measurements are often taken from stationary vehicles in the vicinity of the exhaust silencer (ISO 5130, 1978). Measurements are carried out with the engine running at 75% of the speed at which it develops maximum power. When the engine speed has become constant the throttle is quickly released to the idling position while the A-weighted SPL is measured. For these measurements, the exhaust outlet and microphone are in the same horizontal plane with the microphone 500 mm from the exhaust outlet and at an axis of 45° to it. The background noise level is also measured and the maximum difference between the vehicle noise and the background noise is then compared with vehicle's specified noise level.

This noise test is currently being adopted by a number of European countries to check exhaust silencer performance as a part of routine vehicle testing.

8.7.3 Interior noise in vehicles

There are no legal requirements for assessing interior noise in vehicles. Because of the necessity for subjective assessment it has long been the practice to have this work performed by a team of experienced assessors. This has its disadvantages in a development programme where there is a need to quantify the essential characteristics and relate these to noise sources and transmission paths. To aid this process a number of ad hoc criteria have been developed by different manufacturers for specific types of noise. For example a modified form of Articulation Index, AI (Greaves *et al.* 1988) which is designed to quantify the intelligibility of conversation has been used. The audible range between 200 Hz and 16 kHz is split into sixteen third octave bands. The SPL is measured in each band and the articulation index A_i for the i-th band is determined from the equation:

$$A_i = \frac{W_f (A_0 - \text{SPL})}{A_0 - A_{100}} \qquad (8.78)$$

where A_0 = SPL for zero intelligibility

$\qquad A_{100}$ = SPL for 100% intelligibility

$\qquad W_f$ = weighting factor for each third octave band

The overall (single number) AI is then determined by adding together the sixteen individual indices.

Diesel powered vehicles pose very difficult problems for the NVH engineer. In particular the noise resulting from cold idling conditions contains periodic high frequency large amplitude sound pressure variations. These pulses of sound vary from cylinder to cylinder and from firing stroke to firing stroke leading to an irregularity which is subjectively very annoying.

An analyser has been developed by one manufacturer (Russell *et al.*, 1988) which is designed specifically to assess this problem. The analyser is capable of measuring the impulsive content (based on the kurtosis, the 4th statistical moment) of the sound pressure variation and irregularity of the sound pressure variation by measuring the standard deviation of the low pass filtered amplitude of the diesel knock pulses.

8.8 Automotive noise sources and control techniques

The frequency composition of a sound is one of its most identifiable features. When the sound occurs at a single frequency it is call a *pure tone*. However, the great majority of sounds are far more complicated than this, having frequency components distributed across the audible range. Because there are numerous sources of automotive noise, most of which are cyclic and related to engine speed, the result is what is called *broad band noise* containing a number of dominant frequency components related to engine speed. The frequency characteristics of this sound are represented by its frequency spectrum in a similar way to those for vibration sources.

8.8.1 Engine noise

Engine noise originates from both the combustion process and mechanical forces associated with engine dynamics. The combustion process produces large pressure fluctuations in each cylinder giving rise to high dynamic gas loadings and other mechanical forces such as piston slap. These forces combined with the dynamic forces from inertia and unbalance effects (which are generally dependent on engine configuration and speed) produce the excitations applied to the engine structure. The resulting vibration produces noise radiation from the various surfaces of the engine.

Noise control at source therefore has to be concerned with controlling the extent of cylinder pressure variations (combustion noise) and choice of engine configuration (dynamic effects). Both of these options tend to conflict with the need for small high-speed fuel efficient engines.

In the case of diesel engines there is evidence (Pettitt *et al.*, 1988) that combustion force reduction can be achieved by controlling the rate of pressure rise in cylinders. This requires careful attention to the design of combustion bowls and selection of turbocharger and fuel injection options. The mechanical noise associated with piston slap can be reduced by careful selection of gudgeon pin offset and minimizing piston mass.

Ranking of engine noise components indicates (Pettitt *et al.*, 1988) that most engine noise is radiated from the larger more flexible surfaces such as sumps, timing case covers, crankshaft pulleys and induction manifolds. It makes sense therefore to isolate these components from the vibration generated in engine blocks using specially designed seals and isolating studs. This enables the isolation of components such as sumps and intake manifolds from high frequency excitation components. Noise shields can also be effective in attenuating radiated noise from components such as timing case covers and the side walls of engine blocks. The shields are generally made from laminated steel (see Section 8.3.4) or thermoset plastic material designed to cover the radiating surface and are isolated from it by flexible spacers. The high internal damping of laminated steel can also be used to produce other effective noise reduced components such as cylinder head covers. Noise from crankshaft pulleys can be reduced either by using spoked pulleys or fitting a torsional vibration damped pulley.

8.8.2 Transmission noise

Gear noise rises with speed at a rate of 6 to 8 dB with a doubling of speed, while measurements have shown that gear noise increases at a rate of 2.5 to 4 dB for a doubling of power transmitted (Hand, 1982).

In an ideal pair of gears running at constant speed, power will be transmitted smoothly without vibration and noise. In practice, however, tooth errors occur (both in profile and spacing) and in some cases shaft eccentricities exist. If a single tooth is damaged or incorrectly cut a fundamental component of vibration is generated at shaft speed f_{ss}. If the shaft is misaligned or a gear or bearing is not concentric vibration (and noise) is generated at tooth meshing frequency f_{tm} with sidebands f_{s1} and f_{s2} given by:

$$f_{s1}, f_{s2} = f_{tm} \pm f_{ss} \qquad (8.79)$$

For a wheel having N teeth rotating at n rev/min; the tooth meshing frequency f_{mf} is:

$$f_{tm} = \frac{nN}{60}, \text{ Hz} \qquad (8.80)$$

Furthermore gear teeth are elastic and bend slightly under load. This results in the unloaded teeth on the *driving* gear being slightly ahead of their theoretical rigid-body positions and the unloaded teeth on the *driven* gear being slightly behind their theoretical positions. Thus when contact is made between the teeth on driving and driven wheels there is an abrupt transfer of load momentarily accelerating the driven gear and decelerating the driving gear. This leads to noise generation at tooth meshing frequency. Considerable effort has been devoted in recent years to correcting standard tooth profiles to account for tooth elasticity effects, but because gear teeth are subjected to variable loading, it is impossible to correct for all eventualities.

8.8.3 Intake and exhaust noise

Intake noise is generated by the periodic interruption of airflow through the inlet valves in an engine, thus creating pressure pulsations in the inlet manifold. This noise is transmitted via the air cleaner and radiates from the intake duct. This form of noise is sensitive to increases in engine load and can result in noise level increases of 10 to 15 dB from no-load to full load operation. When a turbocharger is fitted, noise from its compressor is also radiated from the intake duct. Turbocharger noise is characterized by a pure tone at blade passing frequency together with higher harmonics. Typical frequencies are from 2 to 4 kHz.

Exhaust noise is produced by the periodic and sudden release of gases as exhaust valves open and close. Its magnitude and characteristics vary considerably with engine types, valve configurations and timing. The fundamental frequency components are related to the engine firing frequency, which for a four-stroke engine is given (in Hz) by:

$$f = \frac{\text{enginespeed (rpm)}}{60} \times \frac{\text{number of cylinders}}{2} \tag{8.81}$$

Levels of exhaust noise vary significantly with engine loading. From no-load to full-load operation these are typically 15 dB. Turbocharging not only reduces engine radiated noise by smoothing combustion, but also reduces exhaust noise.

Attenuation of noise at engine intakes and exhausts calls for devices which minimize the flow of sound waves while allowing gases to flow through them relatively unimpeded. Such devices are effectively acoustic filters. The operational principles of intake and exhaust silencers ('mufflers' as they are called in the USA) can be divided into two types, dissipative and reactive. In practice, silencers are often a combination of both types.

Dissipative silencers contain absorptive material which physically absorbs acoustic energy from the gas flow. In construction, this type of silencer is a single chamber device through which passes a perforated pipe carrying the gas flow. The chamber surrounding the pipe is filled with sound absorbing material (normally long-fibre mineral wool) which produces attenuation across a very broad band of frequencies above approximately 500 Hz. The degree of attenuation is generally dependent on the thickness and grade of the absorbing material, the length of the silencer and its wall thickness. Figure 8.25 shows the cross-section through a typical dissipative exhaust silencer with venturi tube (see later paragraph).

Reactive silencers operate on the principle that when the sound in a pipe or duct encounters a discontinuity in the cross-section, some of the acoustic energy is reflected back towards the sound source thereby creating destructive interference. This is an effective means of attenuating low frequency noise over a limited range of frequencies. The effectiveness of this technique can be extended by having several expansion chambers within the same casing connected together by pipes of varying lengths and diameters (Figure 8.26). Silencers of this type increase the exhaust back-pressure and result in some power loss.

Intake noise attenuation is generally incorporated into the air filter and is achieved by designing the filter to act as a reactive silencer based on the Helmholtz resonator principle. For

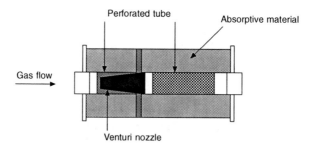

Figure 8.25 Two chamber dissipative silencer with venturi

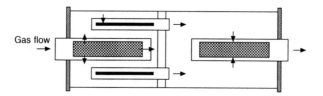

Figure 8.26 Two-chamber reactive silencer

an intake system (Figure 8.27) comprising an intake venturi pipe of mean cross-sectional area *A* and length *L* together with a filter volume *V*, the resonance frequency is given by:

$$f = \frac{c}{2\pi}\sqrt{\frac{A}{LV}} \qquad (8.82)$$

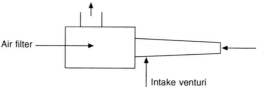

Figure 8.27 Air cleaner and venturi tube

where *c* is the velocity of sound in air. This type of design produces a low frequency resonance (a negative attenuation) but an increasing attenuation at higher frequencies. This can, however, be offset by high frequency resonances within the intake venturi (Peat *et al.*, 1990).

The exhaust systems of present-day vehicles are required to perform the dual task of reducing both the exhaust gas pollutants and exhaust noise. Catalytic converters are fitted immediately downstream from exhaust manifolds to ensure that they quickly achieve operating temperature and thus become quickly effective in urban driving. In addition to acting as exhaust gas scrubbers, catalytic converters also have an acoustic attenuation effect resulting from the gas flow through narrow ceramic pipes. This attenuation is produced by both interference and dissipation.

Additionally, silencers are positioned in the exhaust system downstream from the catalytic converter. These are required specifically to smooth exhaust gas pulsations and make them as inaudible as possible. The silencers and their pipework form an acoustically resonant system which is tuned to avoid exciting bodywork resonances which would aid transmission of structure borne noise. For this latter reason, it is common for silencers to have a double skin and insulating layer which also provides thermal insulation. Exhaust systems need to be isolated from vehicle bodywork to avoid transmission of structure-borne sound and for this reason are suspended from the underbody of the vehicle by flexible suspension elements. There is also a risk that the noise emitted from the tailpipe can cause body resonances if the exhaust is not properly tuned.

The following are some of the devices used to overcome specific silencer tuning problems.

(a) The Helmholtz resonator – a through-flow resonator which amplifies sound at its resonant frequency, but attenuates it outside this range.
(b) Circumferential pipe perforations – create many small sound sources resulting in a broadband filtering effect due to increased local turbulence.
(c) Venturi nozzles – designed to have flow velocities below the speed of sound they are used to attenuate low frequency sound.

Computer software is being increasingly used in the design and analysis of intake and exhaust silencer systems. One example of this is LAMPS (Loughborough and MIRA program for Silencers) (Peat *et al.*, 1990).

8.8.4 Aerodynamic noise (Barnard, 1996)

Aerodynamic noise is due principally to pressure fluctuations associated with turbulence and vorticity. For road vehicles this can be broken down into three noise generating components: the boundary layer distributed over the vehicle body, edge effects and vortex shedding at various locations on the vehicle body and also at cooling fans.

Boundary layer noise tends to be random in character and is spread over a broad band of frequencies. Noise levels due to boundary layer effects are not normally troublesone and the higher frequency components in the spectrum can easily by attenuated with absorbent materials inside body panels.

Edge noise is produced by flow separating from sharp corners and edges on the body structure. As the flow separates from an edge it rolls up into large vortices which also break up into smaller vortices. It is the intermittent formation and collapse of these vortices which leads to the narrow band characteristics associated with edge separation. The noise level associated with edge noise is generally higher than boundary layer noise and has a more defined band of frequencies. This band of frequencies is a function of vehicle speed such that changes in speed can be observed by the change in noise signature. It is possible to reduce edge noise by minimising protrusions from the body surface, making the body surface smooth and continuous and ensuring that gaps around apertures such as doors are well sealed. There is also a strong tendency for vortices to be produced at the pillars supporting the front windscreen. These vortices extend rearwards and envelope the front sidelights which tend to have a low resistance to sound transmission. There is generally very little which can be done to improve this problem since a change of profile to a well rounded contour while improving the aerodynamics is generally unacceptable from a vision point of view. Edge noise also arises at protuberances such as wingmirrors and at wheel trims. In these cases there is generally scope for improvements to the profiles without impairing their function.

Vortex shedding occurs when an airflow strikes a bluff-body producing a periodic stream of vortices downstream. This results in the production of pure tones – subjectively the most annoying of all sounds. The frequency f of the vortices are related to the air speed U and depth d of the bluff body by the equation

$$f = \frac{SU}{d},$$
(8.83)

where S is the Strouhal number. Typically $S = 0.2$ for a long thin rod, so that for a vehicle fitted with a roof-rack having 10 mm diameter bars (facing the air flow) and travelling at 113 km/h (70 mph), the vortices shed a frequency of 640 Hz, i.e. in the frequency range where the human ear is most sensitive. A much quieter form of roof carrier is the enclosed pod shaped design which tends to avoid the production of vortices.

The cooling fan is also a source of noise. In this case the fan blades shed helical trailing vortices which result in periodic pressure fluctuations when they strike downstream obstacles. To overcome this problem fan rotors are made with unevenly spaced blades and with an odd numbers of blades. The use of thermostatically controlled electrically driven fans now ensures that fan noise does not increase with engine speed, as was the case for earlier belt-driven designs.

Noise from internal air flows designed for ventilation and occupant comfort is becoming increasingly important as overall cabin noise is reduced. It is essential that in modern vehicles inlet and outlet apertures are carefully sited and designed to ensure that they do not themselves generate noise and that noise from the engine compartment is not carried into the cabin space by the ventilating air.

8.8.5 Tyre noise

As engine noise is progressively reduced and with the advent of electric vehicles, tyre noise is emerging as a serious problem. Tests have shown that tyre noise can be broken down into two components (Walker *et al.*, 1988) tread pattern excited noise and road surface excited noise. While the problem of road surface excited noise is the province of highway engineers, that of tread pattern excited noise clearly belongs to the automotive engineer. Tyre designers are concerned with reducing tyre noise at source while chassis engineers are concerned with reducing the transmission of noise from the tyre contact patch to the vehicle interior. The mechanism of tyre noise generation is due to an energy release when a small block of tread is released from the trailing edge of the tyre footprint and returns to its undeformed position. There is also a contribution from the opposite effect at the leading edge of the footprint.

With a uniform tyre block pattern, tonal noise (at a single frequency with harmonics when the wheel rotates at a constant speed) is generated. To overcome this problem tyre designers have produced block pitch sequences which re-distribute the acoustic energy over a wider band of frequencies. When tread patterns are taken into account there is a need to analyse the effect of the individual impulses produced across the width of the tyre. Computer software has been developed (Membretti, 1988) to aid this aspect of tread evaluation at the design stage. Models of tyres which take account of their structural dynamic characteristics and the air contained within them are also used at the design stage.

8.8.6 Brake noise

Despite sustained theoretical and experimental efforts over many years, the mechanism of noise generation in disc and drum brakes is still not fully understood. The problem of brake noise is one of the most common reasons for warranty claims on new vehicles with market research evidence suggesting that as many as 26% of owners of one year old medium sized cars complain of brake noise problems.

The intractable nature of the problem arises from the complex assemblage of components in which shoes or pads are held in contact with either a drum or disc under hydraulic and friction loading. A dynamically unstable brake system results in vibration of the brake components and noise is generated by components of significant surface area such as brake drums and discs. Progress towards a better understanding of the noise generating mechanisms has been aided by experimental investigations (e.g. Fieldhouse *et al.*, 1996) and mathematical models (e.g. Nishiwaki, 1991) have been put forward to aid the design of quieter brakes.

In the absence of comprehensive brake models, a number of noise 'fixes' have been implemented to cure specific problems. For low frequency drum brake noise these have included the addition of either a single mass or a combined mass and a visco-elastic layer applied at anti-nodes of the drum backplate (Fieldhouse *et al.*, 1996). At higher frequencies other solutions are necessary such as a redistribution of drum mass to eliminate some of the specific backplate diametral modes.

8.9 General noise control principles

8.9.1 Sound in enclosures (vehicle interiors)

For small enclosures of regular shape (e.g. cubes, cylinders) it is possible to determine the sound field in precise mathematical terms. There are in theory an infinite number of natural frequencies and modes and the situation is analogous to the vibration of homogeneous elastic solids of similar shapes. Broadband sources appropriately positioned in such enclosures are capable of exciting standing wave patterns resulting in SPLs which are very position sensitive.

The complex shapes encountered in vehicle interiors means that such analytical techniques are not applicable and the sound field tends to be diffuse. The sound pressure levels tend to vary much less throughout the vehicle interior compared to the standing wave behaviour described in the previous paragraph. In general there are a number of sound sources emitting noise into vehicle interiors and these can produce discrete components which are superimposed on a lower level of broadband noise.

8.9.2 Sound energy absorption

Absorption is one of the most important factors affecting the acoustic environment in enclosures.

Increasing the average absorption of internal surfaces is a relatively inexpensive way of reducing sound levels in enclosures and is effective in vehicle interiors.

The *absorption coefficient* α, is defined as the ratio of sound energy absorbed by a surface to the sound energy incident on it. Its value is dependent on the angle of incidence and since in an enclosure all angles of incidence are possible, the values of α are averaged for a wide range of angles. Absorption is also dependent on frequency and published data normally quotes values at the standard octave band centre frequencies.

For an enclosure having a number of different internal surface materials, the average absorption coefficient α can be determined (for n surfaces) from:

$$\alpha = \frac{\sum_{i=1}^{n} S_i \alpha_i}{\sum_{i=1}^{n} S_i}, \tag{8.84}$$

where S = surface area. In general, the materials having high levels of absorption are porous materials in which movement of air molecules is restricted by the flow resistance of the material. Typical absorption coefficients can be found in (Beranek, 1971).

8.9.3 Sound transmission through barriers

One of the principal noise transmission paths in a vehicle is through the bulkhead separating the cabin space from the engine compartment. The bulkhead can be considered to be a sound barrier. The effectiveness of sound barriers is normally quoted in terms of *transmission loss* TL. This is the ratio of the incident sound energy to that of the transmitted sound energy, expressed in dB. For a thin homogeneous barrier and random incidence (in the range from 0 to 72 degrees) it can be shown (Wilson, 1989) that the field incidence transmission loss is given by:

$$TL = 20 \log_{10} (fm) - 47 \text{ dB} \tag{8.85}$$

where f is the frequency of the sound in Hz and m is the mass per unit area of the barrier in kg/m^2.

Equation 8.85 applies to what is called the mass-controlled frequency region in which the transmission loss increases by 6 dB per octave increase in frequency, while doubling the barrier thickness or density increases transmission loss by 6 dB at a given frequency. It is evident from this that an effective means of increasing transmission loss is to use a high density material for acoustic barriers.

Sound transmission through barriers is governed at low frequencies by panel bending stiffness and panel resonances; these tend to reduce their low frequency effectiveness. Barriers can be considered to be mass-controlled above twice their lowest natural frequency, but below a critical frequency f_c. This frequency is related to the ability of sound in barriers to be transmitted as bending waves and occurs when the wavelength of the incident wave coincides with the bending wavelength λ_B. The lowest frequency at which this can occur is when the incident sound grazes the surface of the barrier and is given by:

$$f_c = \frac{c}{\lambda_B} \tag{8.86}$$

In practice the range of incidence angles varies from zero to something less than 90 degrees which means that the decrease in transmission loss associated with coincidence occurs at a frequency somewhat higher than the value given by equation 8.86. The effectiveness of barriers is also sharply reduced by even the smallest apertures and can pose problems in noise isolation when electrical trunking and pipework is required to run between the engine to cabin compartments.

In practice layered materials consisting of a dense core and surface layers of absorptive material can perform the dual role of providing sound absorption with a high transmission loss.

8.10 References and further reading

Barnard, R.H. (1996). *Road Vehicle Aerodynamic Design: An Introduction.* Longman.
Beards, C.F. (1996). *Structural Vibration: Analysis and Damping.* Arnold.
Beranek, L.L. (1971). *Noise and Vibration Control.* McGraw-Hill.
British Standards Institution, (1981). *B.S. 5969: Specification for Sound Level Meters.* BSI.
Dimaragonas, A.D., and Haddad, S. (1992). *Vibration for Engineers,* Prentice-Hall International.
Fieldhouse, J.D. and Newcomb, P. (1996). Double pulsed holography used to investigate noisy brakes. *Optics and Lasers in Engineering,* 25, (6), pp. 455–494, Elsevier Applied Science.
Ginsberg, J.H. (1988). *Advanced Engineering Dynamics,* Harper and Row.
Greaves, J.R.A., and Sherwin, C. (1988). Wind noise evaluation using a closed circuit climatic tunnel. *Proceedings of International Conference on Advances in the Control and Refinement of Vehicle Noise,* Paper No C18/88, IMechE.
Hand, R.F. (1982). Accessory noise control, Chapter in *Noise Control in Internal Combustion Engines.* D.E. Baxa, ed. Wiley-Interscience.
International Standards Organization. (1978). *ISO 5130, Acoustics – Measurement of Noise Emitted by Stationary Road Vehicles – Survey Method,* ISO.
International Standards Organization. (1981). *ISO 362 Measurement of Noise Emitted by Accelerating Motor Vehicles.* ISO.
Kim, G., and Singh, R. (1992). Resonance isolation and shock control chacteristics of automotive nonlinear hydraulic engine mounts, *Proceedings of ASME Winter Annual Meeting – Transportation Systems,* DSC. Vol. 44.
Kim, G. and Singh, R. (1993). A broadband adaptive hydraulic mount system, *Proceedings of ASME Winter Annual Meeting – Advanced Automotive Technologies,* DSC. Vol. 52.
MathCAD PLUS 6.0, (1995). MathSoft Inc, Massachusetts.
Meirovitz, L. (1986). *Elements of Vibration Analysis,* 2nd edn; McGraw-Hill.
Membretti, F.N. (1988). Tyre noise simulation at computer. *Proceedings of International Conference in Advances in the Control and Refinement of Vehicle Noise,* Paper No C35/88, IMechE.
Miller, L., Ahmadian, M., Nobles, C.M and Swanson, D.A. (1995). Modelling and performance of an experimental active vibration isolator, *Journal of Vibration and Acoustics,* Vol. 117, pp. 272–278, Trans ASME.
Morishita, S., and Mitsui, J. (1992). An electronically controlled engine mount using electro-rheological fluid, *Proceedings of SAE,* Technical Paper 922290.
Muller, M., Eckel, H.G., Leibach, M. and Bors, W. (1996). Reduction of noise and vibration in vehicles by an appropriate engine mount system and active absorbers, *Advances in Component Designs for Noise and Vibration Control,* Paper No 960185 in SP-1147, SAE.
Newlands, D.E. (1975). *Random Vibration and Spectral Analysis,* Longman.
Nishiwaki, M. (1991). Generalised theory of brake squeal, *Proceedings of IMechE Autotech Seminars,* Paper No. C427/11/001, Autotech, Birmingham, NEC.
Norton, R.L. (1992). *Design of Machinery,* McGraw-Hill.
Peat, K.S; Callow, G.D. and Bannister (1990). Improving the acoustic performance of an intake system.

Proceedings of International Conference on Quiet Revolutions – Powertrain and Vehicle Noise Refinement, Paper No. C420/021, IMechE.

Pettitt, R.A., and Towch, B.W. (1988). Noise reduction of a four litre direct injection diesel engine. *Proceedings of International Conference on Advances in the Control and Refinement of Vehicle Noise,* Paper No. C22/88, IMechE.

Rao, S.S. (1995). *Mechanical Vibrations,* 3rd edn, Addison-Wesley.

Reynolds, D.D. (1981). *Engineering Principles of Acoustics: Noise and Vibration Control.* Allyn and Bacon.

Russell, M.F; Worley, S.A and Young, C.D. (1988). An analyser to estimate the subjective reaction to diesel engine noise. *Proceedings of International Conference on Advances in the Control and Refinement of Vehicle Noise,* Paper No 30/88, IMechE.

Schwarzenbach, J. and Gill, K.F. (1984). *System Modelling and Control.* Arnold.

Shigley, J.E. and Uicker, J.J. (1980). *Theory of Machines and Mechanisms.*

Snowdon, J.C. (1968). *Vibration and Shock in Damped Mechanical Systems,* Wiley.

Thomsen, J.J. (1997). *Vibrations and Stability,* McGraw-Hill.

Thomson, W.T. (1989). *Theory of Vibration with Applications,* 3rd edn, Unwin Hyman.

Timoshenko, S., Young, D.H. and Weaver, W. (1974). *Vibration Problems in Engineering* (4th edn), Wiley.

Walker, J.C., and Evans, D.I. (1988). The effect of vehicle/tyre/road interaction on external and internal vehicle noise, *Proceedings of International Conference on Advances in the Control and Refinement of Vehicle Noise,* Paper No. 26/88, IMechE.

Wilson, C.E. (1989). *Noise Control: Measurement, Analysis and Control of Sound and Vibration,* Harper and Row.

9. Occupant accommodation: an ergonomics approach

J. Mark Porter, PhD, FErgS, EurErg
C. Samantha Porter, PhD, FErgS, CPsychol

The aim of this chapter is to:

- Highlight the need for ergonomic factors to be linked into the design process from concept onwards;
- Indicate common misconceptions when attempting to design vehicles for a population;
- Demonstrate methods by which human factors can be analysed and incorporated into the design process;
- Suggest by example how this design process can be achieved and an approach for the future.

9.1 Introduction

It is essential that the ergonomics input to a vehicle or a product takes place throughout the design process but nowhere is it more important than at the concept and early development stages of design. Basic ergonomics criteria such as the adoption of healthy and efficient postures for the range of future users need to be satisfied at a very early stage as there is usually only limited scope for modifications later in the process without incurring serious financial consequences.

However, automotive design has always been preoccupied with styling and engineering function. As a consequence, these aspects of the vehicle design typically take precedence over the ergonomics aspects of the interior, leaving those responsible for the well-being of the occupants with a difficult, if not impossible, task. For example, the seat designer is given the brief to ensure that potential owners or users of the vehicle proclaim it to be comfortable when sitting in the showroom and during a long journey. Many factors may contrive to prevent this, including:

- the potential owners and users may not have been accurately predicted in terms of their body size and proportions or their functional ability (e.g. the strength to operate controls; range of joint mobility to get in and out, to reach controls or to twist to see when reversing).
- the seat designer has little input to the 'design' of the driving package. As a result, the users may suffer poor postures, chronic discomfort and serious health problems such as herniated lumbar intervertebral discs (slipped discs) and unnecessary injury during crashes.

- the development of 'digital mock-ups' using CAD has reduced the number of traditional full size mock-ups that are prepared during a vehicle's development. These full size mock-ups were made for purposes of visualization and assessing manufacturing issues. However, they also had the tremendous benefit that they could be used for potential occupants to experience the driving package and accommodation throughout the vehicle's development. This possibility is now more limited and usually performed at a later stage than before.

The authors believe that the quality of accommodation provided by a vehicle is an extremely important issue, and this has been made the focus of this chapter. The quality of accommodation dictates what percentage of the potential user population will be able to fit in the vehicle, with adequate clearances around the body. Furthermore, prospective drivers must be able to see the displays and road environment whilst operating the various controls and maintaining a range of healthy postures. If any individual driver cannot achieve these basic, but essential, criteria then many will decide to purchase or use a more suitable alternative vehicle. The success or failure of a vehicle can be measured in many ways, but the single most important measure to the manufacturer is its percentage of the market share. It is inexplicable how commitment to the aesthetic of early concept designs decisions, such as the height and profile of the roof line dictating the internal headroom, can be allowed to 'design out' 10–20% of male drivers. Typically, this compromise is often made by default, rather than by decision, because the design team was not aware of the variety of methods and tools to help predict and quantify these issues. One of the purposes of this chapter is, therefore, to present information on these methods and tools. It is important to remember that it is much more expensive to attract an additional 10–20% of sales to a production vehicle with poor levels of accommodation than it is to prevent the 'designing out' of this percentage of potential purchasers during the design of the vehicle.

The traditional procedure adopted by many manufacturers in the development of a new vehicle was sequential and can be simply described as designing 'outside-in'. The exterior styling being considered first followed by fitting the engineering within this volume (Tovey, 1992). An alternative approach to designing a new vehicle is to design 'inside-out' (Porter & Porter, 1998). This approach would promote a clearer focus on the occupant issues inside the vehicle. For example, the size, number and age of the future occupants, together with details of their preferred postures, sight lines and reach envelopes, would help define the volume that they will require in the vehicle, not forgetting the space required for possessions such as shopping, golf clubs, pushchairs and pets. The control and display interfaces would then be designed around these people with a knowledge of the range of their hand, foot and eye locations. The exterior of the vehicle then needs to accommodate the people and the engineering.

Successful design has a lot to do with working within the imposed constraints (typically time, cost, and legislation) and achieving optimum compromises wherever practicable. Such compromises will not be achievable if the basic ergonomics issues affecting accommodation and comfort are not established from the very outset.

The chapter commences with a discussion of commonly held fallacies concerning ergonomics and its role in design. These fallacies need to be exposed as such and it is hoped that many readers will be able to provide their own examples of where they had similar misconceptions. This is followed by a description of how ergonomists operate in the vehicle industry, and how and when they communicate with designers and engineers during the design and development of a vehicle. The next section presents practical details on predictive and evaluative methods and tools to help promote high levels of occupant accommodation, ranging from anthropometry-based tools such as manikins and human modelling CAD systems to road trials and owner questionnaires. The 'inside-out' theme is then illustrated by two case studies conducted by the Vehicle Ergonomics Group, SAMMIE CAD and Coventry School of Art and Design. The increasing importance of this approach is supported by a discussion of future trends and the chapter concludes with suggestions for further reading.

9.2 Eight fundamental fallacies

Until fairly recently, many automotive companies did not have a formalized structure to identify and deal with ergonomics issues in the design and development of their vehicles. The reasons for this are likely to include several misconceptions about the use and value of ergonomics, as illustrated by the eight fundamental fallacies described below. The first five fallacies are taken from Pheasant (1996) and the last three are from Porter and Porter (1997). The examples have been taken from the authors' own experiences.

1. *The design is satisfactory for me – it will, therefore, be satisfactory for everybody else.*
In this case the resulting design is likely to accommodate the designers and senior management. The Vehicle Ergonomics Group has conducted several large surveys of car owners on behalf of major manufacturers. The results of one such survey showed that half of the male drivers complained that the seat was too narrow in the sports model. The reason for this apparent mistake was that the Chairman had taken considerable interest in this model and the seat was designed with his needs in mind. When he was measured he was found to have a 50th percentile (i.e. average) male hip breadth, which explained precisely why 50% of male owners had a problem: their hips were wider than the Chairman's.

2. *This design is satisfactory for the average person – it will, therefore, be satisfactory for everybody else.*
Unfortunately for automotive designers, there is no such thing as an average man or woman, or child for that matter. This was demonstrated in a classic study by Daniels (1952). 4000 flying crew were measured on a variety of dimensions and each 'average' category was calculated very crudely as the arithmetic mean plus and minus 0.5 standard deviations. The analysis showed that none of the crew was 'average' on 10 dimensions, with less than 2% being 'average' for 4 dimensions. This has been historically known as the 'fallacy of the average man' as he or she does not exist.

Consequently, human variability is often a problem in design. The statistical mean of a sample of dimensions or weights gives absolutely no information whatsoever concerning the variation in that sample. The stature of two drivers can be measured and their mean value calculated. Neither are likely to have exactly this mean stature and it is possible that their statures will lie many centimetres above and below this fictional 'average' stature. If the doorway were designed using this mean stature, then one of the two drivers will have been 'designed-out'; a completely unacceptable 50% success rate. However, from an engineering viewpoint the mean is a very useful predictor of the dimension and weight of each item on the production line because their variability has been designed out by careful selection of materials and methods of manufacture. A 100% success rate is expected for fitting a CD-ROM into a computer because they are all made to an identical specification. This approach does not work for people and, to sell as many vehicles as possible, it is essential to understand human variability. The ergonomics methods to describe and cope with variability are not well known to designers and engineers and this has lead to poor communication and, in many cases, poor design.

An example of poor design is the air bag fitted to the steering wheel in cars in USA and Canada. Small female drivers have been advised to de-activate their driver bags following a series of deaths arising from the explosive force of the bag, not the impact of the car. The bags were designed to accommodate 50th percentile male crash dummies and this decision put small females at risk because they sit much closer to the steering wheel. Consequently, they can hit the air bag whilst it is still expanding. Average or 50th percentile values should never be used for specifying clearances, minimum reach distances or maximum control forces.

Variability in a product's characteristics is typically designed-out by the manufacturing process in order for the product to be successful. Variability is 'designed-in' genetically for humans and any product which does not accommodate this variability results in people being designed-out or suffering in some way.

3. *The variability of human beings is so great that it cannot possibly be catered for in any design – but since people are wonderfully adaptable it doesn't matter anyway.*
People are adaptable but the research has shown conclusively that they are not adaptable enough. The authors have conducted several studies to examine the relationship between exposure to driving a car and reports of discomfort and sickness absence from work. The results clearly show that low back troubles are the major complaint experienced by high-mileage (<25000 miles/year) drivers (around 60–80% of whom experience low back discomfort frequently).

In the survey of 600 members of the British general public (Gyi and Porter, 1995) it was found that the mean number of days ever absent from work with low back trouble was 22.4 days for high mileage car drivers, who drove for more than 25 000 miles in the last 12 months, compared with only 3.3 days for low mileage car drivers who drove for less than 5000 miles. For those who drive a car as part of their job the mean number of days ever absent with low back trouble was 51.3 days for those who drove for more than 20 hours a week compared to 8.1 days for those who drove for less than 10 hours a week. This amounts to a six-fold increase in sickness absence.

Several other studies have identified that prolonged car driving is clearly related to the occurrence of low back pain. It has been shown in cross-sectional surveys of US males (Frymoyer *et al.*, 1983; Damkot *et al.*, 1984), British males (Walsh *et al.*, 1989) and French commercial travellers (Pietri *et al.*, 1992). A case-control study of US adults (Kelsey and Hardy, 1975) has also shown that men who have ever had a job where they spent half or more of their working day driving were nearly three times as likely to develop an acute herniated lumbar disc (i.e. a 'slipped' disc).

This link between a high exposure to driving and an increased prevalence of low back trouble has been established with respect to criteria from discomfort, to sickness absence arising from low back trouble through to the degeneration of the lumbar intervertebral discs. Interestingly, several of these studies have reported that the relationship is strongest when the exposure to driving is greater than half of the working day (Porter and Gyi, 1995).

Porter et al. (1992) found that there were fewer reports of low back trouble or sickness absence for those cars that offered high levels of adjustability within the driving package and when an automatic gearbox was provided (the benefit of the automatic gearbox arising from the reduction of both postural constraints and postural fixity). This demonstrates that design can indeed cater for human variability with good effect.

4. *Ergonomics is expensive and since products are actually purchased on appearance and styling, ergonomics considerations may conveniently be ignored.*
Ergonomics can be expensive if it is an after-thought, brought in to rectify a problem that was only detected late in the development process. Making something the right shape and size need not be more expensive than making it the wrong shape and size. The ergonomics must be considered early on and this requires a pro-active contribution from the ergonomist. Human modelling CAD systems can provide the necessary support for this function in the earliest stages of design (see Section 9.4.6 below).

The public are increasingly aware of consumer issues such as safety, usability and health. The authors consider that some cars should display health warnings in their sales literature. For example, following the development work on the Fiat Punto seats (Porter, 1995), it was found that the percentage of drivers reporting discomfort in the Punto at the end of the 150 minute trials were 5%, 5%, 15% and 5% in the upper, mid and low back and the buttocks respectively. One of its competitors, which is a very popular car in Europe, performed much worse with 20%, 35%, 40% and 30% of drivers reporting discomfort in the body areas listed above. It is likely that many owners of this car have developed back trouble since acquiring the car. The legal issues involved have already attracted attention. The first named author acted as an expert witness in a recent case in the UK where the court decided that a high mileage salesman had suffered his slipped disc as a consequence of his poor accommodation in his company car.

5. *Ergonomics is an excellent idea. I always design things with ergonomics in mind, but I do it intuitively and rely on common sense so I don't need tables of data.*
Achieving optimum compromises using solely intuition is unlikely to be consistently successful, particularly when the vehicle designers' and engineers' personal biases and preferences are not fully in accord with those of the wide variety of customers. Female drivers experience problems such as the steering wheel being too close when the seat is adjusted forwards to easily reach the pedals. Similarly, pedal designs may not take account of the fact that many women would like to drive with high-heeled shoes; seat belts do not adjust sufficiently, causing chafing over the neck and/or chest; and control knobs are designed without consideration of long fingernails (Thompson, 1995). To quantify these problems, a recent study of a small family sized car revealed that 42% of females considered the pedals to be uncomfortable and 25% of females complained of the seat belt position (Petherick and Porter, 1996). Data such as these provide clearly essential feedback to demonstrate the need for iterative design modifications until the design is deemed satisfactory.

6. *The design is not satisfactory for me – it will, therefore, be unsatisfactory for everybody else (a variation on fallacy 1 above).*
A typical complaint of long legged male motoring journalists is that the seat cushion does not offer sufficient thigh support. Their review of the car subsequently states that the seat is too

short in the cushion. It is not clear that any extra support would actually be more comfortable for the journalist but a longer seat cushion may be disastrous for the great majority of other drivers with shorter legs. Either they would find it difficult or impossible to operate the pedals or they would slide forward and slouch as a consequence, leading to back pain.

7. Percentiles are a very clear and simple way to present and use information concerning body size.

The concept of percentile values is very easy to understand (see Section 9.4.2 if you are not familiar with this concept), but the fallacy arises because it is assumed that the usage of such data is equally easy. Even ergonomists, who should know better, fall into the trap of referring to mythical people such as a 5th percentile female or a 95th percentile male. Anthropometric dimensions are poorly correlated which means that people of the same stature can have markedly different leg lengths, arm lengths, back lengths and so on.

Percentiles are univariate and only refer to one dimension at a time. A percentile value should never be used without obtaining details of the age range, nationality and occupational groups included in the original survey data. The date of the survey is important too, due to the issue of secular growth where the mean growth rate over generations in North America and the UK is continuing to run at approximately 10 mm per decade.

It is mathematically impossible to have a person or manikin who is 5th or 95th percentile in all component vertical dimensions and in overall stature as well (McConville, 1978). Unfortunately, it is possible to construct a 50th percentile manikin without such problems – it will be internally consistent because only the statistical means for segment lengths can be added together to equal the mean overall length. It is unfortunate because such manikins abound, presumably because they are so easy to make, adding fictional support to fallacy 2 above.

It is likely that only a few manufacturers offer an 'in–out' adjustable steering wheel because the use of 5th percentile female and 95th percentile male manikins (however, these may be constructed, see above comment) suggests that one fixed position will suit all drivers. A more sophisticated analysis would need to consider a tall male driver with long legs but short arms (for his stature) and a short female driver with short legs and long arms (for her stature). Such an analysis convincingly demonstrates the need for 'in–out' adjustment of the steering wheel to comfortably accommodate all drivers.

8. Designing from 5th percentile female to 95th percentile male dimensions will accommodate 95% of people.

This will be true if only one dimension is relevant to the design solution (i.e. univariate accommodation such as standing headroom). However, some vehicle workstations require simultaneous accommodation on a large number of dimensions (i.e. multivariate accommodation). This is especially true of fighter aircraft where the pilot is strapped into the seat, preventing forward leaning for controls out of easy reach, or slumping because of limited headroom. Because there is a poor correlation between body dimensions it follows that those males who are designed out because of limited headroom (5% of males in theory for a large random sample) will not necessarily be the same 5% who are designed out for having arms that are too long or the 5% with legs that are too long, hips too broad and so on. Similarly, those females who are designed out because they have legs, arms, sitting eye heights, etc. that are too small will not just constitute 5% of the females.

A classic study by Roebuck (1995) demonstrated the seriousness of this problem when examining air crew selection standards. If the cockpit was designed using anthropometric data ranging from 5th–95th percentile male dimensions (so the fallacy would specify accommodation for 90% of males), they found that this would result in nearly 50% of the air crew being designed out, rather than the 10% some might expect. Thankfully, the situation is less critical in the automotive industry where head-up displays, ejector seats and the frequent exposure to very high g-forces do not need to be considered. On the other hand, it is more than likely that poor vision from a vehicle, due to a driver's sitting eye height being too low/high or too far forward/rearwards, will have contributed to many accidents over the years. Similarly, a poor fit of the seat belt and contact of the knees with the dashboard will add to the severity of any accident sustained.

9.3 Ergonomics in the automotive industry

Traditionally, in the automotive design process most of the ergonomics input was provided at the end of the styling process. Very crude occupant data would have been used to derive the 'hard points' in the package (see Section 9.4.2) with which the stylists must comply. However, a number of factors have contrived to improve this situation dramatically over the past few years. Most manufacturers are now producing well designed, reliable motor cars and ergonomics is one of the arenas in which they are now competing, and looking for a market advantage. User-centred design is on the increase in every area of product design and the purchasing population is becoming more aware of ergonomics issues and much more discriminating about the products they buy. Five years ago some of the major automotive manufacturers did not employ their own in-house ergonomists, but now almost all of them do. However, there are still large parts of the automotive industry, such as smaller design-houses and other suppliers, who do not employ ergonomists and whose ergonomics knowledge is extremely limited. Increasingly, ergonomics practice, tools and techniques are being employed in a huge variety of ways throughout the industry, as a recent informal survey revealed. For the purpose of this chapter, ergonomists in all the automotive manufacturers in this country and the design director of a smaller design house were contacted, and interviewed about the ergonomics practice, tools and techniques used in their company. In general, there is a move towards an 'inside-out' approach to automotive design.

Most ergonomists are no longer associated with 'package engineering' but are part of the design function, reporting to local design chiefs, or in some cases European design chiefs. Senior management consider the ergonomist's contribution to be fundamental to the design process. Many large automotive manufacturers employ qualified ergonomists in teams of between 2 and 5. In those larger companies where there are no ergonomists, the ergonomics responsibility still resides with the engineers, and the thoroughness with which the designing-in of human consideration is carried out is entirely dependent upon the skill set of the engineer. In the smaller design houses, it is the designers who take on the ergonomics responsibility and again the quality is dependent upon the experience of the individual. When working for a large manufacturer, some of the ergonomics of the vehicle will be specified by the manufacturer but when working for smaller companies they may be entirely responsible for the ergonomics of the vehicle.

In almost all cases where there are ergonomists they are involved 'pre-program' i.e. they have an input to the definition of the vehicle and the prioritization of its characteristics before the design program begins. Alternatively, they are involved from the beginning of the concept design stage, particularly in foreign-owned companies where pre-program may take place at the parent company. Once the design program begins the already defined ergonomics targets and deliverables must be met. Failure to achieve a target or a deliverable means the program can be stopped.

Where ergonomics has become a formal part of the design process, ergonomists are responsible for all aspects of ergonomics and where there are separate departments (e.g. seating) they are expected to liaise with the department, raise awareness of ergonomics issues and 'red flag' problems. In some cases they also have a role in the teaching of ergonomics to other disciplines involved in the design process. In all cases, they have input to the design of displays, controls, logic of operation of displays and controls, packaging, seating, ingress/egress, and access to luggage areas. Where ergonomics compromises are made they are informed ones and their extent of the impact is understood. At the other end of the spectrum, where there are no ergonomists, there are no formal procedures for considering the ergonomics aspects of the vehicle and understanding of the issues can be very limited. Where designers are responsible for ergonomics they most often still design by one or more of the fallacies described above.

The tools and techniques used by ergonomists in the automotive industry are those described in this chapter. They are using the same data sources and CAD tools. They all run user trials (see 9.4.8 for details) to evaluate their products; in some cases using only company staff as subjects, in other cases with members of the general public. They have good links with universities and are involved in research projects with them. Where there are no ergonomists, the data sources used tend to be out of date and user evaluations are rarely conducted.

Ergonomists have had to find methods of communicating ergonomics information to those who need to use it. In general, ergonomics information is numeric and dry in its presentation. This style may well be appropriate for ergonomists and engineers, but there is good evidence that it is not appropriate for the majority of designers (Porter and Porter, 1997). Designers are trained to communicate visually and in all cases the ergonomists contacted have had to learn to communicate in this way. For example: translating numeric data into visual data; producing CAD plots to show reach zones which are fed directly into the software that the designers are using; making marks directly onto clay models to communicate design recommendations; and using all the multi-media tools which are commonplace in design presentations and the visual exploration of design ideas.

9.4 Ergonomics methods and tools to promote occupant accommodation

This section describes a variety of methods and tools that vehicle ergonomists employ when defining or evaluating occupant accommodation.

9.4.1 Standards, guidelines and recommendations

There are a wide range of standards, guidelines and recommendations available in many areas of ergonomics that are pertinent to automotive design. They are extremely useful in the early stages of concept design as they are readily available and quick and easy to use.

Standards, and their associated procedures, have been developed in areas of ergonomics that have been intensively researched, e.g. occupant packaging. The Society of Automotive Engineers (SAE) in the US has been particularly active in the generation of such Standards, and many of these form part of the legislation which covers automotive design. The most relevant to the ergonomist when considering occupant packaging are:

SAE J826	H-point (ISO 6549)
SAE J1100	Seating reference point
SAE J1100	H-point travel path
SAE J1517	Driver selected seat position
SAE J941	Eyellipse (ISO 4513/BS AU 176)
SAE J1052	Driver and Passenger head position contours
SAE J287	Hand controls reach envelopes (ISO 4040/BS AU 199)

The corresponding International standards (ISO) and British standards (BS) are given in parenthesis after the description. These and other associated standards are described in detail in Roe (1993) and can be found in the SAE Handbook (1996). They form the basis for occupant packaging in the automotive industry throughout the world. There are also standards relating to display and control design, fields of view, mirror design, vibration, and the thermal environment in a vehicle. Good coverage of these topics and associated standards can be found in Peacock and Karwowski (1993).

In areas where ergonomics research is less complete (e.g. control and display design, particularly where emerging technologies play a role), there exist a huge number of design recommendations and guidelines. They range from the very specific (e.g. it should be 13.7 mm above the ground level) to the very general (e.g. it should be comfortable) making use in the automotive design context sometimes difficult. Some useful sources are listed below:

Sanders and McCormick (1992)	Displays, controls, workstation layout
Peacock and Karwowski (1993)	Occupant packaging, displays and controls, the ageing driver
Defence Standard 00-25	Displays, controls, anthropometry, noise, vibration, workstation layout
Pheasant (1996)	Anthropometry
Campbell *et al.* (1998)	Design guidelines for in-vehicle information systems

The use of recommendations and guidelines should be approached judiciously. They can be extremely useful for making early design decisions but their appropriateness and success should always be verified using the other methods described below. There are a number of reasons for caution. They are often relevant to very specific groups of users and very specific tasks making their use, with populations and in contexts at variance with those in which the original data was derived, invalid. Equally, the situation in which the data were derived may mean that the guidelines are too general and may be of little relevance to the particular context of the driving task and vehicle environment. Very often the background detail is not included so the user of the guidelines is unable to make an informed choice concerning their applicability. Even where guidelines are apparently completely appropriate (e.g. sill heights and door apertures for ease of ingress/egress) there is little additional support given in situations where compromises to optimum ergonomics must be made. For example, in the design of a 'sports car', the levels of comfort and usability are compromised in order to meet the user populations' expectations of aesthetic and driving performance. A further problem with guidelines and recommendations is that where technologies are emerging, as they are in the area of in-car telematics, guidelines and recommendations are emerging much more slowly so it may simply not be possible to find an appropriate recommendation or guideline. Further information on guidelines for the use of emerging technologies in vehicles may be found in Section 9.8.

9.4.2 Anthropometry

Anthropometry can be defined as the measurement of human body dimensions. Static anthropometry is concerned with the measurement of human subjects in rigid, standardized positions (e.g. static arm length being equivalent to its anatomical length) and static anthropometric data are used in designing equipment for the workplace where body movement is not a major variable, e.g. seat breadth, depth and height. Dynamic anthropometry is concerned with the measurement of human subjects at work or in motion (e.g. functional arm reach is a factor of the length of the upper arm, lower arm and hand, as well as the range of movement at the shoulder, elbow, wrist and fingers). Dynamic anthropometric data can be used to establish control locations using reach envelopes for the hands and feet and locations of head restraints, seat belts and air bags using data concerning the arcs described by various parts of the body under crash conditions. Biomechanics is the measurement of the range, strength, endurance, speed and accuracy of human movements and such data are also used in the design of controls to establish satisfactory ranges of control movement and operative forces.

Anthropometric and biomechanical data are usually specified in terms of percentiles. The population is divided into 100 percentage categories, ranked from least to greatest, with respect to some specific type of body measurement. For example:

- 5th percentile stature is a value whereby 5% of the population are shorter and 95% are taller;
- 50th percentile stature is the median stature;
- 95th percentile stature is a value whereby 95% of the population are shorter and 5% are taller.

The reader is referred to Roebuck (1995) for a full description of the variety of methods used to collect anthropometric data. The strategies for using the data in design are described below:

(a) Find the relevant data for the intended occupants with respect to their race, occupation, age, sex, disability. The data should be task specific so that, for example, arm reach to a lever that will be pushed is quite different to arm reach for operating a push button.
(b) Make any necessary allowances for secular growth and clothing (e.g. 10 mm per decade for stature in USA and UK, ~45 mm for female shoes, ~25 mm for male shoes)
(c) Establish your design limits. Traditionally these have been stated as 5th percentile values for females and 95th percentile values for males. The authors consider that these limits are

somewhat out-of-date given the concern for quality of life, high productivity and safety and the authors recommend using 1st percentile female to 99th percentile male values wherever possible. This is particularly important when several dimensions are critical for accommodation. Multivariate accommodation is discussed in Section 9.4.6.

(d) Design for extreme individuals when appropriate. To establish the minimum clearance (e.g. the door should be a minimum of *x* cm high) use the upper percentile value (e.g. 99th or greater percentile male value for the relevant dimensions). To establish maximum reach or strength, use the lower percentile value (e.g. 1st percentile female).

(e) Design for adjustable range where minimum fatigue, optimum performance, comfort and safety is required (e.g. vehicle seats, steering wheel, seat belt mountings). Use 1st percentile female to 99th percentile male values wherever possible.

(f) Design for the 'average' person when adjustability is not feasible (e.g. height of exterior door handles) but never use 'averages' for clearances, reach or strength. The 'average value' should be used on the basis that it would cause less inconvenience and difficulty to the user population than one which was larger or smaller.

Anthropometric and biomechanical data are extremely useful to the designer at the early stages of design or when a novel design is being considered. However, it is very important that any new design or modification is studied using mock-ups and the evaluation of prototypes. The data can tell you where a person can reach but it does not tell you how the design, location and direction of travel of a variety of controls can affect driving performance, comfort and safety. Limitations to the use of percentiles have already been discussed under Fallacies 7 and 8 above.

Sources of civilian anthropometric data include Bodyspace (Pheasant 1996) which presents data for a variety of nationalities including the UK, US, Swedish, Dutch, French, Polish, Brazilian, Sri Lankan, Indian, Hong Kong Chinese and the Japanese. Adultdata (Peebles and Norris, 1998) and Childata (Norris and Wilson, 1995) can be obtained through the UK Department of Trade and Industry and present data for a variety of nationalities. Some specialist surveys have examined driver anthropometry in the UK (MIRA Survey, Haslegrave, 1980), France (Rebiffe *et al.*, 1984) and the US (Sanders, 1977). A source book on Indian anthropometry has recently been produced (Chakrabarti 1997). Another very useful source is PeopleSize (Open Ergonomics 1999), an interactive computer based package that can also provide information on multivariate accommodation. Figure 9.1 shows one of the People Size screens, from which the designer clicks on the desired dimension and the selected percentile values are displayed underneath. Table 9.1 presents an example of People Size data for large US male and small UK female values.

Anthropometric methods are currently changing with the introduction of body scanners. Such systems allow the collection of thousands of data points for the human body and this type of data is very appropriately starting to be used in human modelling systems (see Section 9.4.6 below).

9.4.3 2-dimensional manikins

2-dimensional (2D) plastic manikins are often found in design studios, whereas sets of anthropometric data are rarely available unless the company employ an ergonomist. These manikins are typically full size and are overlaid on engineering drawings in order to examine

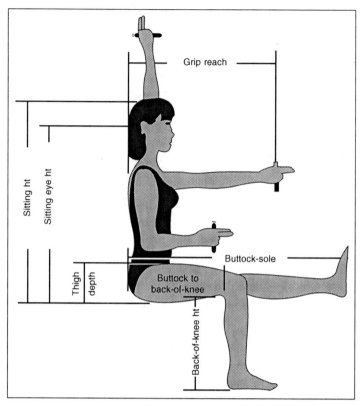

Figure 9.1 An example of a People Size interactive screen interface. Stature, their shoulder height may differ by as much as 122 mm. Furthermore, this torso proportion was found to have virtually no correlation with either stature or weight

occupant accommodation. In the authors' experience, such manikins are often used without a knowledge of what they represent in terms of age range, occupational group, nationality or posture (e.g. sitting upright or slumped). For example, the authors have found that British vehicle manufacturers who market their vehicles in the United States were unaware that the sitting height (vertical distance from the compressed seat surface to the top of the head) of their 95th percentile adult male manikin was 50 mm shorter than the 95th percentile erect sitting height recorded by the National Health Survey conducted in the United States some 30 years ago (Stoudt, Damon, McFarland and Roberts, 1965). The manikin was designed to have a slumped sitting height which, according to the US survey, was equivalent to only a 60th percentile erect sitting height. Furthermore, if the manikins are based on old data then they need to have allowances made for secular growth.

Another major problem with 2D manikins is that they can be used in a very simplistic way. For example, designers may have 50th and 95th percentile adult male manikins and a 5th percentile adult female manikin. The nature of these manikins gives support to the notion that people come either tall and long limbed, short and short limbed or somewhere in between. It has been repeatedly demonstrated that this is not true and that the inter-correlation between

Table 9.1 An example of anthropometric data from PeopleSize Pro software 1999, by permission of Open Ergonomics Ltd. (www.openerg.com). All dimensions are in mm

	British female 18–64 years 1st percentile	British female 18–64 years 5th percentile	US male 18–64 years 95th percentile	US male 18–64 years 99th percentile
Sitting height	741	763	984	1009
Eye height, sitting	625	648	867	893
Horizontal grip reach (from e.g. seat back/wall)	591	618	808	836
Thigh depth (maximum), sitting	115	124	214	238
Buttock to sole of foot (leg straight), sitting	927	962	1189	1239
Buttock to back of knee (popliteal tendon), sittting	417	439	589	619
Back of knee height (popliteal tendon), sitting	339	356	496	515

Notes:
(1) Measurements are for sitting erect and without shoes or heavy clothing.
(2) Multiple accommodation – if you specify 95th percentile for Sitting Height, Grip Reach, AND Back of Knee Height, only 89% of people are accommodated. In order to fit 95th percentile in all these three dimensions, you must specify 98th percentile for each dimension.

body dimensions is rather poor. For example, Haslegrave (1986) reported that seated shoulder height varies from 30.6 to 39.5 percent of stature, which means that among men of average stature, their shoulder height may differ by as much as 122 mm. Furthermore, this torso proportion was found to have virtually no correlation with either stature or weight. The manikin designer can resort to other techniques to ensure that the manikins are statistically correct, for example by calculating median values or using regression equations to describe component body dimensions for groups of men or women of a given stature and weight. Whichever method is chosen, to define a variety of statistically 'correct' manikins, there is still the problem of estimating the percentage of people accommodated by a particular design. A common mistake made by many designers is to use the 5th percentile female stature and 95th percentile male stature manikins to assess a workstation, assuming that if both of these manikins can be accommodated then so can 95% of the adult population (see Fallacy 8, above). This is an incorrect assumption as it implies that those people 'designed out' due to their sitting height, hip breadth or leg length, for example, are greater than 95th percentile male values are all the same people. Similarly, all those with sitting eye height, arm length or leg length smaller than 5th percentile female values are assumed to be the same individuals. As these dimensions are not strongly correlated then these assumptions are incorrect.

9.4.4 Package drawings

The 2D package drawings often provide the first visualizations of the proposed vehicle occupants. They are produced after the product planning stage, when the market specification has taken place and the basic parameters of the vehicle are known. Typically, they used to be hand drafted but most are now computer generated from packages such as CATIA. The drawings are produced in side elevation, plan and front/rear views and show outlines of the extremes of the chosen occupant population (normally 5th percentile female to 95th percentile male) 'packaged' with the major mechanical components (see Figure 9.2). The drawings, in most cases, are provided by the packaging department (an engineering department) to provide the 'hard points' that the stylists must not impinge upon in their design.

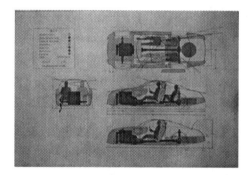

Figure 9.2 Package drawing showing plan, side and front views

Some legislative issues such as vision can be examined in the 2D package drawings and some of the SAE procedures referred to in Section 9.4.1 are carried out at this stage (e.g. establishing reach zones).

The 2D manikins of occupants suffer, of course, with all the complications described in the previous section and it is generally agreed that the dimensions of the SAE manikins used for legislative purposes are inappropriate for the derivation of occupant 'hard points' (e.g. roofline) as input to the styling process. Most automotive companies have produced their own manikins based on much more recent data from the sources referred to in the previous section. Initial occupant packaging remains, in general, the domain of packaging engineers, with some input from ergonomists in some companies, and the complexities of anthropometry are often not fully understood (see 9.4.2 for discussion of anthropometry). Therefore, in terms of occupant packaging the manikins used may still be inadequate. Other factors can further reduce their accuracy in 'real life' terms as further changes are made in the vehicle, which had not been anticipated and allowed for in the initial package. For example, the 95th percentile manikin is often used to define the height of the roof lining and the sunroof (which encroaches around 25 mm into this already limited headroom) is subsequently designed. This results in an even larger percentage of the potential market being inadvertently 'designed out'. It is possible that issues such as this are not discovered or realized (if indeed, they are) until too many other aspects of the design and manufacture are fixed, and the error is carried forward.

In smaller companies and consultancies, and in the student environment, package drawings are often used in a much more creative way to examine, using the three views concurrently, to evaluate some of the dynamic aspects (e.g. pedal travel and ingress/egress) of driving and using the vehicle. This is normally done with jointed movable manikins. The shortcomings of using this method, without validation by other methods offering much more realism, are clearly apparent. However, in many situations where time and resources are extremely limited it can provide some reasonable ergonomics data very quickly.

9.4.5 'Quick and dirty' mock ups

'Quick and dirty' ergonomics methods are essentially the formal methods described in 9.4.7, 9.4.8 and 9.4.9, but applied less formally and with fewer constraints. They are used, as the name implies, to provide rapid information at an acceptable level of accuracy for the current needs. The information may be used to inform, or resolve, the actual design or to inform the building of CAD or physical models to be used in formal methods. The methods and their practice are described in some detail in Jordan *et al.* (1996) and Jordan (1998). They are used in situations where information is required that does not exist in any of the known sources, and where the resources of time and/or money are extremely restricted. The 'quick and dirty' mock-up is a frequently used tool in both the industry and student automotive design projects. The mock-up is generated from available materials (wood, foam, steel), and whilst it may look nothing like the finished design, the critical aspects will be correct, as fixed at that point in time.

Figure 9.3 shows a 'quick and dirty' mock up used in a student project at Coventry University. The project had a six-month time scale from start of design research to the finished model of a sports car, so a complex and time consuming evaluation was not possible. The generic 'sports car' is highly compromised in ergonomics terms in order to achieve the desired aesthetic and performance requirements of the purchasing population. The student project was designed specifically for an existing marque and as a consequence had to incorporate manufacturer specific characteristics which further compromised the ergonomics. These characteristics typically reduce the ease of ingress and egress (e.g. wide sills for structural support and low seats to lower the vehicle's centre of gravity), cause additional blind spots in the driver's visual field (e.g. low seat, wide steep pillars and steeply raked windscreen for aerodynamics/aesthetics reasons; high rear end to accommodate mid-engine for optimum centre of gravity and handling performance) and compromise fit for large males (low roofline for aerodynamics/aesthetics reasons).

The design was low, with low, deep, high sided seats (because it's a sports car), a high wide sill (for chassis strength), an offset footwell (to accommodate wheels/tyres), scissor doors and an engine placed at the rear of the car in a high position. The mock-up was built to examine issues concerning the seat and steering wheel adjustment ranges, ingress/egress and rearwards visibility.

The mock-up was built of wood and foam and included an accurately dimensioned door, door aperture ('A' and 'B' pillars), sill, pedals, engine cover and rear window. The seat was an existing seat set at the correct height for the design. The door action and its range of movement were as it would be in the finished vehicle but it was not possible to give it an appropriate weight. The subjects were those which were available (students) who covered the extremes of the chosen user population (5th percentile UK female dimensions to 95th percentile UK male

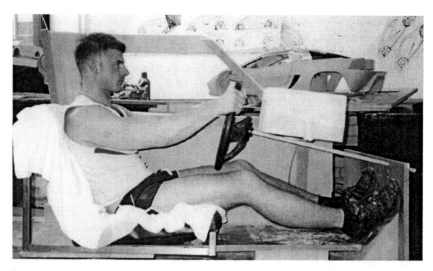

Figure 9.3 An example of a 'quick and dirty' mock up

dimensions). The sports car was to be an expensive one so the purchasing population, in reality, will be older than the student population and consequently less physically able. Results gained from the students were, therefore, likely to more favourable than if it had been possible to use an accurate representation of the population.

The evaluation of ingress/egress revealed that the combination of the scissor door, the low seat and the low, wide sill produced a very awkward movement that resulted in the subject's head making contact with either the roofline or the door, as they exited the vehicle. A design decision was to cut away some of the roofline (i.e. make it part of the door) to improve the access. There is a weight cost in making the door larger which in this study it was not possible to explore. The evaluation was also used to identify where to place the necessary handles on the door to enable closing and where to place grab handles on the vehicle to support ingress and egress. Examination of rearwards visibility over the engine cover led the student designer to increase the height of the roofline at the rear of the vehicle to accommodate a larger rear window for improved visibility. It is intended that the student will use the buck for more 'quick and dirty' experimentation on visibility of displays and reach to controls when interior design concepts are being explored.

In this study useful ergonomics information was derived very quickly and design decisions were able to be made in a much more informed manner than if the student had used only a set of 2D package drawings to evaluate the user issues. However, there are limitations to the accuracy of the data collected, as described above, and the information should always be used with caution and the necessary consideration for the shortcomings. Whenever time and resources do allow, the findings should always be validated when a more realistic physical representation is available.

9.4.6 Human modelling computer aided design systems

These systems are intended to be used as a predictive tool for the assessment of the capabilities of people when interacting with the designed physical environment. Being computer based, they allow rapid modelling of concept and development models of a vehicle either from engineering drawings, digitizers or by importing data files from other packages such as ALIAS (used by stylists) or CATIA (used by engineers). The basic functionality that is required of human modelling systems is listed below, in each case followed by a brief discussion of the relevant issues:

The 3D modelling of people of the selected sex, age, nationality and occupational groups. This is achieved using published anthropometric data, if indeed it exists for the population being examined. The current databases have several shortcomings, basically because they were established with little consideration for the needs of 3D human modelling systems. For example, surveys record external body dimensions whereas computer models need joint-to-joint dimensions in addition. The limited number of anthropometric dimensions recorded in surveys leave many gaps which make a fully defined 3D computer model difficult to produce. The relatively recent technique of body scanning, whereby thousands of data points can be recorded from the surface of the body, now makes it possible to model individual people with considerable accuracy.

A knowledge base of comfort angles for the major joints of the body. Human models come with various numbers of joints. Those with relatively few (e.g. less than 20) do not have detailed models of the hands or spine. With such details the number of joints can be well over 100. This large number of degrees of freedom in the human model's posture poses problems for the user who has to decide how to position the model realistically. The problem is made easier in some systems with the provision of automated reach tests, inverse kinematics and grasping behaviours such that the model's hand can reach, grasp and operate specified handles. This is done automatically ensuring that the various joint angles do not exceed maximum or comfortable ranges as specified in the published literature. The inter-relationships between joints, such as the knee and hip, are typically not considered when using comfort angle recommendations. For example, the range of comfortable backrest angles is affected by any constraints to the knee angle, such as is experienced in a low sports car seat.

The ability to model the proposed vehicle package in 3D together with the simulation of

ranges of adjustment to be incorporated into the design. This 'working' model of the product being developed is an essential part of a human modelling ergonomics design system as the human model needs to interact with the design in order to assess the physical characteristics of the interface. The requirements for an ergonomics model of a prototype design are, however, substantially different to the needs of other forms of CAD systems as the extremely detailed geometric information from an engineering CAD package is rarely required for ergonomics evaluations. Furthermore, human modelling systems should be used at the concept stages in design in order to help define the initial design specification, rather than just evaluate it at a later stage after the engineering criteria have been satisfied. Engineering CAD models rarely have the functionality of the various components under investigation embedded in their data structure (e.g. seat adjustment ranges, mirror rotation constraints) so these must be added to the ergonomics model. In many cases, it will be easier to create specific models for an ergonomics evaluation rather than simply transfer in detailed engineering models.

The ability to assess the kinematic interaction between the models of people and the workstation, specifically in terms of the issues of user fit (e.g. headroom and legroom), reach (e.g. to the steering wheel, gear selector, and pedals) and vision (e.g. of the road environment, both directly and in the mirrors, and the instrument binnacle). Figure 9.4 shows a SAMMIE model of the prototype Fiat Punto car where a large male driver is simulating reversing the car, simultaneously assessing reach to the clutch pedal with one foot, reach to the steering wheel with one hand and reach to the gear selector with the other, twisting in the seat and assessing vision around the head restraint, past the rear seat occupants and through the rear

Figure 9.4 SAMMIE simulation of a large male driver reversing in the concept model of the Fiat Punto

window to the road environment. The same analysis can be conducted with a small female driver with the seat, steering wheel and head restraint adjusted to suit her needs, within the ranges specified by the prototype design.

The ability to make iterative modifications to the design to achieve optimum compromises. Some systems provide information on static strength or calculate torque loads on certain joints providing information to help identify more efficient designs in this respect. Human modelling systems have most to offer at the concept stage of design when they can be used to explore possible options for a design. Design is all about working within constraints, and sometimes challenging these constraints, to achieve the best compromises.

The authors have always advocated that human modelling systems should not replace user trials with full size mock-ups, unless the design or the design modifications are so simple as to not warrant concern. In-depth user trials can reveal problems with so many more issues including long-term discomfort, effects of fatigue, negative transfer of training, error rate, performance and the acceptance of the product. Many designers, engineers and ergonomists are expectantly waiting for the all-singing, all-dancing human modelling system to appear. The likelihood of such a system being developed in the near future seems remote. However, the advantage of using human modelling systems is that it is possible to build full size mock-ups with the confidence that few, if any, modifications will be necessary to physically accommodate the users. The detailed evaluation of criteria such as those above can proceed without delay and without the extra costs of getting 'the basics' right.

Current systems used extensively in vehicle design include SAMMIE, JACK, SAFEWORK and RAMSIS (see Porter *et al.*, 1999 for further details). Comparisons between some of the older systems can be found in Dooley (1982), Rothwell (1985), Porter *et al.* (1993, 1995) and Das and Sengupta (1995) and it is not the intent of this section to present a detailed description of each human modelling system. It is important to appreciate that the quality of a product's ergonomics has more to do with the design team's judgement and ability to incorporate sound ergonomics principles in the design than to the use of any specific human modelling system (Das and Sengupta, 1995). Such systems do not automate the design process by creating ergonomics solutions to a set of specified inputs; rather, they should be regarded as tools to be used by the design team. The various systems cost from as little as a few hundred US dollars up to 60 000 dollars for a software licence. Some systems can run on a PC but many require a Sun or Silicon Graphics workstation or equivalent. Usability is a key requirement of such systems and a fast response time often requires a high specification computer.

Human-modelling CAD systems have the potential to offer considerably more validity as a simulation of people than the traditional 2D manikins which are overlaid on drawings. There is an important distinction between the evaluation of which percentile values are accommodated for a particular design dimension and the evaluation of what percentage of the population will be accommodated in all respects. 3D human modelling systems offer significant advantages in both respects and, in particular, the latter. Roebuck (1995) discusses two statistical methods by which some human modelling systems predict the percentage of the target population that will be accommodated by a particular workstation design. CAR and MDHMS use Monte Carlo methods to generate a large number of theoretical human models, each one of which represents a possible case that could occur in a population of people without violating any of the underlying anthropometric statistics for that population. Another statistical approach is Principal Component Analysis, a version of which is used in the SAFEWORK system.

SAMMIE is currently developing a dataset of body scans and anthropometric dimensions, including joint centres and joint mobility, from a carefully selected sample of people representative of the British and European population. Each person can then be modelled individually within SAMMIE and automatically positioned in a prototype workstation model according to a range of pre-defined criteria. Evaluations of fit, reach, vision and the required postures will then be conducted automatically resulting in the identification of those individuals who failed to successfully complete any of these tests.

9.4.7 Fitting trials

This method for determining optimum dimensions is very useful as anthropometric and biomechanical data do not take into account other factors such as ease of operation, comfort, visual requirements and safety aspects. The method is described below:

(a) Select a group of subjects, preferably from the user population or similar. Use between 20–30 subjects who have been chosen to represent the population as a whole (i.e. wide range of stature, weight or whatever physical dimensions are most critical to the evaluation of the design). Record the relevant anthropometric data for each individual so that the position of each subject within the estimated distribution for all users is known.

(b) Construct a mock-up of the workplace. Allow adjustment to be available for the major features of the design (e.g. seat height, steering wheel height, gear lever location). An idea of the useful range of adjustment can be provided from relevant anthropometric data.

(c) Describe the task(s) to be performed in detail. These should be as realistic as possible.

(d) Decide in which order the adjustments will need to be evaluated. Those parts of the design which are fundamental to the user's task should be examined first, e.g. a typical sequence might be: seat adjustment in relation to accelerator heel point and headroom, steering wheel adjustment, gear lever adjustment, lateral pedal adjustment, secondary controls, etc. Each component's position will be established one at a time, in the order defined, using the method detailed below.

(e) For each subject, move the adjustable component (such as the seat, steering wheel or gear lever) at discrete increments throughout its range of travel, commencing from one extreme to the other and back again. At each position, the subject is asked if that position is considered by himself to be satisfactory or unsatisfactory with respect to the task at hand. The starting position of the adjustable component should be balanced across all the subjects so that half of them commence their subjective judgements starting at one extreme, whilst the other half start at the other extreme. By adopting this method, the subjective information is recorded in the form of tolerance ranges.

(f) If one position is found to be satisfactory for all subjects, then this component can be fixed at this position. Otherwise, it is necessary to allow for a suitable range of adjustment for the component to position it within a range of positions that will be considered satisfactory by all subjects.

(g) Repeat stages (e) and (f) above for the other adjustable components of the design in the order established in stage (d), having made the necessary changes to the mock-up in the light of any previous findings.

(h) Make all the necessary modifications to the mock-up and evaluate as a complete package.

Porter and Gyi (1998) used a variant of the above method to identify preferred family car driving postures using a highly adjustable driving rig (see Figure 9.5). Two videos of the 2.5 hour (60 mile) test route used regularly by the Vehicle Ergonomics Group in their road trials were made for use with the rig, giving a driver's view of the road, with a voice-over of instructions about the route to guide the driver of when to slow down, change gear, etc.

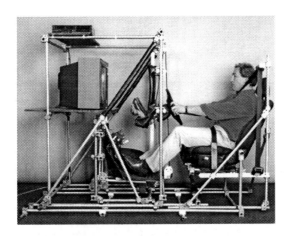

Figure 9.5 The experimental driving rig

The method of fitting trials was carried out to obtain the preferred driving posture by varying the height and horizontal location of the steering wheel and pedals with respect to the seat. The seat tilt angle was also adjustable. For each of these adjustments the component was moved by the experimenter at discrete increments throughout its range of travel from one extreme to the other and back again, in a balanced order. When a satisfactory position was reached, it was temporarily fixed. Following adjustment of all the controls the positions were fine tuned until satisfactory. A 10–15 minute driving simulation at the rig was then carried out to further confirm that this posture was optimum and then relevant measures regarding the positions of the controls from a fixed reference point were documented.

Each subject's driving posture was then measured whilst partially depressing the accelerator, placing the hands on the steering wheel and looking ahead as though they were driving on a road.

Actual observed postures were compared with recommendations from the literature as shown in Table 9.2. Knee angle and foot-calf angle were very similar to the theoretical recommendations of Rebiffe (1969) and Grandjean (1980). However, generally subjects preferred to sit with a smaller trunk-thigh angle than previously recommended. Neck inclination, arm flexion, and elbow angle were greater than the ranges of any previous recommendations.

Care should be taken when using such data for individual joints as a significant positive correlation was identified for the trunk-thigh and knee angles. This means that the preferred trunk-thigh angle is dependent upon the knee angle, and vice versa. For example, none of the authors' subjects who drove with a comparatively large knee angle (125–136 degrees) had a small trunk-thigh angle (89–96 degrees). It was also found that if the driving workstation of a

Table 9.2 Comparison of observed postural angles for comfort (in degrees) with the literature

	Rebiffe (1969)	Grandjean (1980)	Observed postures (n = 55)	95% confidence limits
Neck inclination	20–30	20–25	30–66	29–63
Trunk-thigh angle	95–120	100–120	90–115	89–112
Knee angle	95–135	110–130	99–138	103–136
Arm flexion	10–45	20–40	19–75	16–74
Elbow angle	80–120	–	86–164	80–161
Foot–calf angle	90–110	90–110	80–113	81–105
Wrist angle	170–190	–	–	–

car were designed around the middle range for preferred trunk-thigh angle (97–104 degrees) and knee angle (114–124 degrees), only 29% ($n = 16$) of the sample would be able to sit in their preferred driving posture.

The subjects' preferred positions of the controls were recorded from the driving rig and converted to standard SAE dimensions (see Figure 9.6 and Table 9.3). These values were directly compared with actual vehicle dimensions from a sample of 32 well-known cars. The observed maximum values with reference to the H-point calculated from the rig exceeded these vehicle dimensions, implying that at present no car on that list will fit all users comfortably. The driving rig data also identified a need for both extensive horizontal and vertical adjustment in the steering wheel in order for individuals to obtain their optimum postures. It is suggested that designers use the range of calculated SAE dimensions shown in Table 9.3 to enable a large variety of postures and body sizes to be accommodated.

Table 9.3 Driving rig values for selected SAE dimensions

Driving rig values	L11	L40	L53	H30	H17
Mean	438	16	738	301	628
Maximum	602	25	889	335	689
Minimum	322	5	577	283	580
Std deviation	48	4	67	11	24

9.4.8 User trials

User trials is a generic term that includes a wide variety of activities, including both the 'quick and dirty' assessments and the method of fitting trials discussed earlier. These two methods have been dealt with separately because their focus is upon providing initial data to inform design decisions regarding the dimensions and adjustment ranges for the driving package. Following this early preparatory work, a prototype design or series of designs are proposed and modelled as full size static bucks and, subsequently, as full size working prototypes (often referred to as 'ride and drive' prototypes). At these stages, a formal ergonomics evaluation of the proposals is required and, when this involves a representative sample of the user population

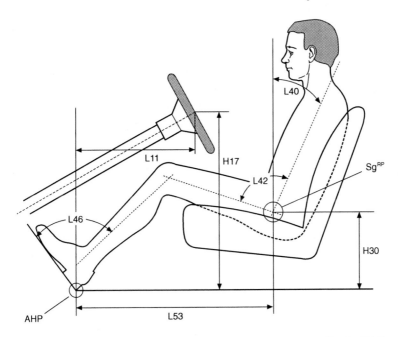

Figure 9.6 Vehicle seating configuration (based upon *SAE Handbook*, 1985)

performing or simulating appropriate tasks, these evaluations are referred to as user trials. If the results of such trials reveal unacceptable levels of occupant accommodation or poor levels of perceived quality, then modifications should be made before progressing further. The criteria which are used to define whether the design is acceptable or unacceptable should be set before the trials commence. Indeed, they should be set before design work commences.

Care must be taken to ensure a representative sample of users, otherwise the results of the trial cannot be assumed to be able to cover all potential users. For the purposes of assessing occupant accommodation, it is clearly very important to identify the appropriate nationalities and age range for the proposed vehicle. Once this has been established and relevant anthropometric data have been sourced or specifically collected, then the selection of subjects for the user trials can proceed to cover a wide range of percentile values for important body dimensions and a full range of ages. This selection method is described in more detail in the Fiat Punto case study (see 9.5.1). This is one of the essential aspects of performing a user trial as it provides quite a different input to that of primarily listening to the needs of senior management within the manufacturing company. These individuals may not actually form part of the intended user population for a specific vehicle (in terms of body size or age), in which case their personal views should be treated with some caution.

The tasks performed or simulated need also to be specified with care, whether dealing with a static buck in the design studio or a 'ride and drive' prototype being driven on the roads. In particular, they need to take account of the environmental issues. For example, ease of ingress/egress needs to consider realistic scenarios such as parking in a multi-storey car park where the close proximity of other cars prevents the doors being fully opened during these tasks. Similarly, parking at the kerbside and exiting or entering the vehicle to or from the pavement, particularly when there is a steep camber to the road, can have quite adverse consequences which need to be examined at this stage of design in order to be able to make useful design improvements. Clearly, driving tasks are difficult to simulate in a static buck unless a driving simulator is used. Such simulators can vary enormously in their functionality and realism as well as their cost.

User trials can be used to collect both subjective and objective data. Examples of subjective data include reports of comfort/discomfort, ease of reach and use, quality of visibility, preferences, perceptions of build quality, safety and performance. Objective data can include clearances, postural measurements, muscle activity, pressure distribution, time taken to complete a task and the number of errors. The integration of both types of data is to be encouraged but it is important to know which data are most relevant for any given question. For example, when considering comfort issues, subjective data will always provide the best information whereas objective data will provide the best data on actual performance. Having said this, much research has taken place on making a car feel 'sporty' in terms of throttle response and other characteristics, rather than just providing raw performance.

Road trials are one form of user trial which is discussed in more detail in Section 9.5.1.

9.4.9 Owner questionnaires

A questionnaire provides a structured way of obtaining information via a number of questions with pre-defined answer categories. These may vary from selecting appropriate multiple choice items to ranking their preferences for various features. An example of the former is shown in Figure 9.7 which shows a series of questions and answer categories related to the features of a seat. Rating scales are also commonly used for assessing the nature and magnitude of the respondents' opinions, for example, concerning their discomfort (see Section 9.5.1 for an example of a comfort/discomfort rating scale). Open-ended questions can also be included in a questionnaire, where the respondents answer in their own words, although this requires more in-depth analysis. Usually, such open questions are asked in pilot questionnaires which are distributed to a small sample of people to check that the questions set were clear and not ambiguous. The responses to any open-ended questions can then be divided into several categories which then form the answer categories in the subsequent large scale survey of owners and users. The questionnaire can also be used as the basis of a structured interview where the experimenter asks the questions and fills in the responses.

Questionnaires are often used when it is important to obtain the views of a large number of people, particularly when they are spread geographically and they would be difficult and expensive to interview individually. The Vehicle Ergonomics Group has conducted owner questionnaire surveys of many vehicles, often as their first involvement with a manufacturer. It is important to understand which features are popular with owners, and which require improvement. A survey of owners who purchased the vehicle within the last 12–18 months can be a very effective way of starting the ergonomics input to a face-lift model or a replacement model. Whilst a typical postal questionnaire may only be rewarded with a 10–15% response rate, the authors have found that preparing an independent questionnaire and sending it on behalf of the manufacturer to named individual purchasers, can often result in around a 60% response rate. The biggest weakness of an owner questionnaire is that it does not provide information from those people who did not choose to buy it (which must be considered to be crucial information

Seat Feature Assessment

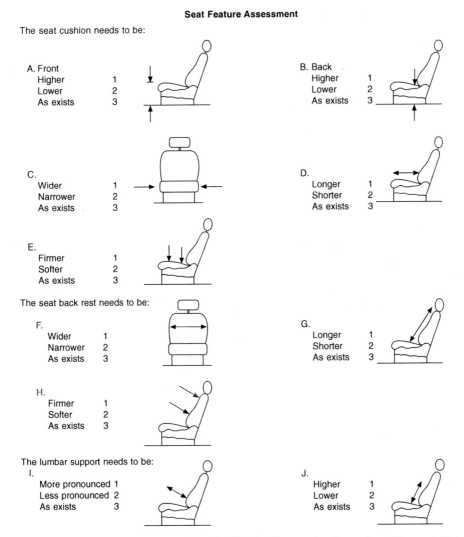

The seat cushion needs to be:

A. Front
   Higher 1
   Lower 2
   As exists 3

B. Back
   Higher 1
   Lower 2
   As exists 3

C.
   Wider 1
   Narrower 2
   As exists 3

D.
   Longer 1
   Shorter 2
   As exists 3

E.
   Firmer 1
   Softer 2
   As exists 3

The seat back rest needs to be:

F.
   Wider 1
   Narrower 2
   As exists 3

G.
   Longer 1
   Shorter 2
   As exists 3

H.
   Firmer 1
   Softer 2
   As exists 3

The lumbar support needs to be:

I.
   More pronounced 1
   Less pronounced 2
   As exists 3

J.
   Higher 1
   Lower 2
   As exists 3

Figure 9.7 Seat Feature Assessment used by the Vehicle Ergonomics Group, Loughborough University

for improving the market share of the vehicle, yet it is so rarely collected). Another problem is the low response rate as it is not always clear what information is missing from those who chose not to return the questionnaire. The authors have found evidence that postal questionnaires focusing upon occupant accommodation and comfort have a greater percentage of small and large respondents than would be expected from data on the general population. Presumably, these individuals experience more problems and are, therefore, more motivated to reply. One of the big advantages with owner questionnaires is that the respondents have intimate experience of their vehicle. In the authors' experience, most owners have been impressed that they have been contacted for their personal feedback. If the manufacturer is committed to constant improvement in the product, then these owners should have a pleasant experience when inspecting the new model in the showroom a few years later.

9.4.10 Strengths and weaknesses of the various methods

A variety of methods have been presented that provide information that can be used to evaluate and help improve standards of occupant accommodation. The method(s) chosen depend upon the time available, the stage in the development cycle, the importance of 'getting it right', the resources and expertise available. A comparison of the strengths and weaknesses of the methods discussed is provided in Table 9.4. A combination of methods will provide complementary information, particularly if the strengths of one method help to overcome the weaknesses of the other and vice versa.

9.5 Case studies

This section presents two recent projects where the quality of the occupant accommodation was particularly important to the design team. The first case study describes the authors' involvement with the Fiat Punto. The importance of 'designing-in' people was identified by the Chief Executive of Fiat and this car quickly became Europe's best selling car in its sector. This case study is followed by a description of the development of Coventry University's Lightweight Sports Car (LSC), a research project to examine how ergonomics and design methods can work together effectively.

9.5.1 The ergonomics development of the Fiat Punto

The Fiat Punto was voted Car of the Year 1995 by an international panel of 56 senior motoring writers from 21 European countries. Comfort was one of the main criteria for the judges and this section briefly describes the input that was provided by the Vehicle Ergonomics Group during the design and development of this car in order to ensure high levels of occupant accommodation and comfort.

Human modelling predictions
The authors' involvement commenced with the assessment of the proposed driving package using SAMMIE. Engineering drawings of the proposed interior packaging were provided by Fiat Auto and a computer model was constructed. Assessments of the accommodation offered were made by examining issues of fit, reach, vision and overall posture (see Figure 9.4) using a variety of man models depicting male and female drivers of several European nationalities. A large number of human models were constructed varying in dimensions from 1st percentile UK female to 99th percentile UK male. In addition, several 'worst case' models were used such as the evaluation of reach with a long legged driver (e.g. 99th percentile male leg length) with short arms (e.g. 60th percentile male arm length). The anthropometric survey of RAF aircrew (Simpson and Hartley, 1981) contains a large number of scatter-diagrams which present 2000 individual data points for various pairs of body dimensions and this information was used to determine the body proportions of the 'worst cases'. This computer based analysis concluded that the prototype Punto design performed well although recommendations to further improve the package were presented to the manufacturer. Examples of the use of SAMMIE in the design of driver's cabs for trams and trains can be found in Porter *et al.* (1996).

Table 9.4 Strengths and weaknesses of the various methods and tools

Method /tool	Strengths	Weaknesses
Design Recommendations and Checklists	Quick Easy to use	Relevance to specific users, tasks or vehicle type may be dubious May have little scientific validity No account taken of compromises Either too specific (i.e. should be 457.2 mm) or too general (e.g. should be comfortable).
Anthropometry and Biomechanics	Quick Good for novel designs Useful for assessing the influence of age, sex, race, etc. upon design	May be a lack of data relevant to user or task Data may be out of date Data often relate to standardized postures, not necessarily working postures Design may become too academic, mistakes being hard to identify
3D human modelling CAD Systems	User and task specific predictions, quick and accurate for geometric issues such as fit, reach and vision Enables effective communication at an early stage Compromises can be objectively explored	Expensive to set up (hardware, software, training), but very cost-effective thereafter Does not assess personal preferences, psychological space, fatigue, task performance
Mock-ups and Fitting trials	Control selection of users and their tasks Study comfort and performance over time Sound basis for identifying good and poor designs using both objective and subjective methods Essential for novel designs Compromises can be investigated Design problems are quickly identified	Can be time consuming and expensive Can be difficult to obtain representative subjects May not be a very realistic simulation of task or environment
Owner Questionnaires and Interviews	Valuable information direct from the user population Small details may be detected which the casual observer may have overlooked User involvement	User may take poor design for granted Opinions can be strongly biased Cannot be used for novel designs until after production Biased sample – does not include those people who chose not to use the existing equipment Biased sample – low response rate from postal questionnaires, who returns them? No detailed assessment of body size, performance or comfort
User trials and road trials	Control selection of users and their tasks Study comfort and performance over time Sound basis for identifying good and poor designs using both objective and subjective methods Allows comparative testing	Can be time consuming Require production and/or prototype vehicles to test Can be difficult to obtain representative subjects

Testing of prototype and competitor cars
The seating in the Punto was subsequently developed using extensive road trials in order to assess the subjective comfort of two camouflaged prototype Puntos (see Figure 9.8). One of these prototypes was fitted with the basic 'Functionale 1' seat and the other with the high specification 'Estetico 1' seat which included seat height adjustment. Two competitor cars in the same market sector were also assessed to provide comparative data. Discomfort data for the driver's seats were provided by 20 carefully selected members of the public who covered a wide range of body sizes and who drove each car on a specified test route for 60 miles, taking 2.5 hours to complete.

Figure 9.8 One of the camouflaged Fiat Punto prototypes used in the road trials

Selection of subjects
The selection of subjects (i.e. drivers) was primarily based upon age and body size for a number of relevant dimensions. 10 males and 10 female subjects were finally selected, from a sample of some 60 potential drivers, to systematically cover the range of percentile values for these important dimensions (see Figure 9.9). To make the sample as representative of the population as possible, care was taken to select one male and one female subject (i.e. 10% of the male and female samples) for every decile category (e.g. 10th percentile or less, or above 10th to 20th percentile) for stature. Thus 10% of the male drivers selected fell into a stature range that 10% of the adult male population would also fall into. The 20 subjects covered the age range 19 to 72 years, with a mean age of 40 years.

Subjective data
Engineers are typically reluctant to place much confidence upon subjective data. Certainly, people are poor at estimating of dimensions, weights, forces and so on and much greater accuracy can be gained using objective data from suitable sensors. However, the 'gold standard' for assessing occupant comfort must rely upon subjective reports from a wide variety of driver sizes and ages throughout an extended period of driving the vehicle in the environment in which it will be used. No laboratory test yet devised can provide such detailed and valid information (see Gyi and Porter, 1999, for a discussion of the problems in using interface pressure as a predictor of car seat discomfort).

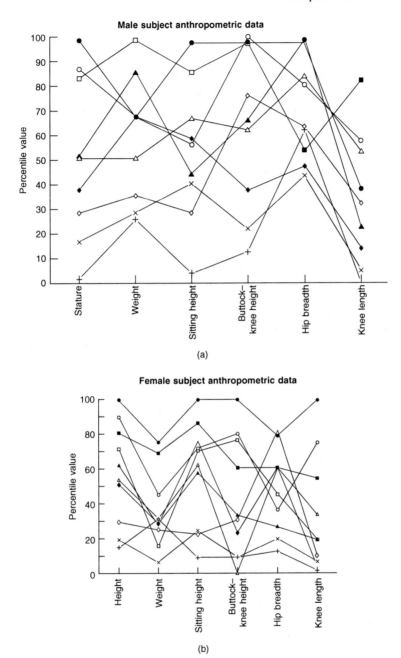

Figure 9.9 Percentile values of the subject sample for a variety of body dimensions

In order to collect high quality subjective data concerning feelings of comfort/discomfort, considerable care was taken to avoid other factors which may have influenced the findings. Each subject drove one car a week, whenever possible on the same day of the week and at the same time of day for all 4 cars. This reduced the likelihood of any reported discomfort being unduly influenced by changes in work or leisure activities during the study, such as sporting activities, or by traffic densities and so on. The subjects were always accompanied during the road trials with the same experimenter who gave directions on the route (identical for each journey) and administered the questionnaires at appropriate times. The road trial questionnaires included a seat feature checklist by which they could comment upon the suitability of various seat dimensions and comment whether they were satisfactory or would like them to be longer/ shorter, wider/narrower, higher/lower, or firmer/softer. A diagram of the seat was provided and subjects were asked to assess different parts of the seat in terms of their hardness/softness and the degree of support that each area offered. Other questionnaires covered the range and ease of making adjustments to the driving package, the seat belt design, the ease of reach and use of controls, the visibility and interpretation of displays, the mirror design, and ease of ingress/ egress.

Reported comfort/discomfort
The most important questionnaires contained the body part discomfort diagrams (see Figure 9.10). These bipolar comfort/discomfort rating scales were completed for 20 body areas at 15, 45, 75, 105 and 135 minutes of driving. The 7 point scale used was: 1 very comfortable; 2 comfortable; 3 fairly comfortable; 4 neutral; 5 slightly uncomfortable; 6 uncomfortable; 7 very uncomfortable. The data analysis examined the differences between the reports of comfort/ discomfort for each car in several ways. These included looking at the distribution of the scores on the 7-point scale for each body area at each of the 5 time periods; examining the percentage of drivers who reported discomfort (i.e. scores of 5, 6 or 7) at each time period; and calculating the number of minutes of reported discomfort over the 2.5 hours, again for each body area. This latter measure was calculated by assuming that each report of discomfort in a particular body area was experienced for a duration of 30 minutes (the rating scale being administered at the mid-point of each 30 minutes epoch). This measure has been found to be very useful in highlighting differences between seats and cars in over 100 such evaluations conducted by the Vehicle Ergonomics Group since 1981.

Road trial 1
The first set of road trials revealed some interesting findings. The 'basic' Functionale 1 seat in the Punto was found to be more comfortable than the 'stylish' Estetico 1 seat and the competitor car that had been previously considered by several of the Fiat management to be the 'best in group' (the Honda Civic) was shown to be clearly the most uncomfortable at the end of the trial. The other competitor car (Renault Clio) was judged to be the most comfortable to drive. The analysis then compared reported areas of discomfort in the 2 Punto prototypes with comments about the seat design and driving package so that design recommendations could be provided. The comparative data from 4 cars was extremely useful in identifying those features that caused problems and in suggesting the appropriate changes.

Design changes
Several areas for improvement were identified for the Punto seats and design changes were made to the foam dimensions and profile, as well as to the fabric and the stitching. These changes were expected to improve the support to the low back and to create a more even

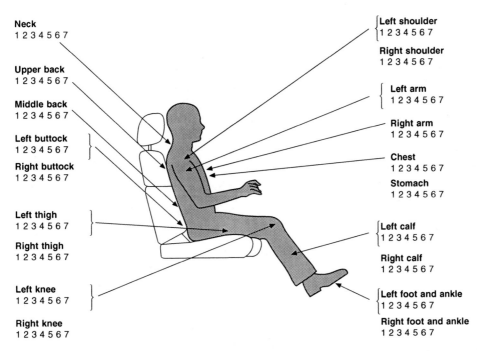

Neck
1 2 3 4 5 6 7

Upper back
1 2 3 4 5 6 7

Middle back
1 2 3 4 5 6 7

Left buttock
1 2 3 4 5 6 7

Right buttock
1 2 3 4 5 6 7

Left thigh
1 2 3 4 5 6 7

Right thigh
1 2 3 4 5 6 7

Left knee
1 2 3 4 5 6 7

Right knee
1 2 3 4 5 6 7

Left shoulder
1 2 3 4 5 6 7

Right shoulder
1 2 3 4 5 6 7

Left arm
1 2 3 4 5 6 7

Right arm
1 2 3 4 5 6 7

Chest
1 2 3 4 5 6 7

Stomach
1 2 3 4 5 6 7

Left calf
1 2 3 4 5 6 7

Right calf
1 2 3 4 5 6 7

Left foot and ankle
1 2 3 4 5 6 7

Right foot and ankle
1 2 3 4 5 6 7

You have now been sitting in the vehicle for approximately 15 minutes. Would you now describe your feelings of comfort in each body area, shown in the illustration below, using the following scale.

1   Very comfortable
2   Comfortable
3   Fairly comfortable
4   Neutral
5   Slightly uncomfortable
6   Uncomfortable
7   Very uncomfortable
Please circle the appropriate number for each area.

Time

Figure 9.10 Body part discomfort diagrams used in the road trials for comfort evaluation

pressure under the thighs and, additionally for the Estetico, to make the seat effectively wider.

Showroom assessment
The revised designs (Functionale 2 and Estetico 2) were then evaluated in the laboratory against the earlier prototypes and the competitor's seats using the method of paired comparisons (Guildford, 1954). This method provides only 2 seats at a time for assessment, the subject choosing their preferred one from each of all possible pairings. With 6 seats there were 15 possible pairings and the strength of this method is that the subject makes a large number of simple decisions, rather than one complex decision involving all 6 seats together. This static laboratory evaluation involved 30 subjects covering a wide range of sizes. The results showed that the overall preferred seat was the Estetico 2 with the Functionale 2 a close second. The very firm seat in the Honda was the least preferred again.

Road trial 2

Having successfully passed this 'showroom' test, another set of road trials were conducted using the same 4 cars as before. This final evaluation showed that the modifications made to the Punto seats had achieved their objectives as they now had the fewest reports of discomfort (see Figure 9.11). The Estetico 2 seat had comparatively low reports of discomfort in nearly all body areas and was regarded as the most comfortable seat. The Functionale 2 seat also performed well and was the second most comfortable seat, but comparatively high levels of neck discomfort were reported due to the poor seat belt fit (no height adjustment to the B pillar anchorage or to the seat height). All production models in the Punto range now offer height adjustable front seat belts as a consequence of this finding. Figure 9.11 also shows comparatively high reports of discomfort in the right knee for drivers of the Estetico 2. This was due to a lack of clearance with the centre console on the prototype car (which were all left hand drive models). The reports of right shoulder discomfort in both prototype Punto cars was due to the stiff action of the gear lever.

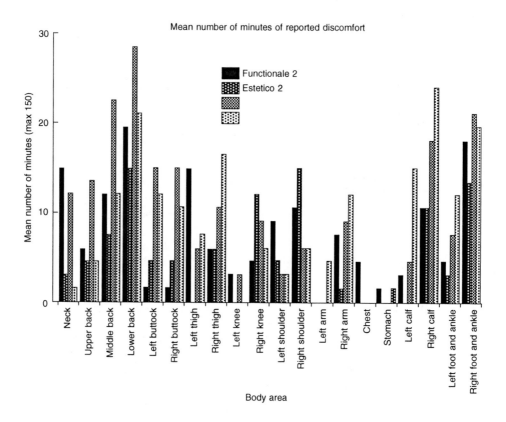

Figure 9.11 Mean number of minutes of reported discomfort during the 2.5 hour road trials. The cars evaluated included 2 competitor cars which remain anonymous

9.5.2 The Coventry School of Art and Design lightweight sports car

A lightweight sports car (LSC) has recently been developed at Coventry School of Art and Design. In the concept design stages the traditional design practice of designing a car from the 'outside-in' was ignored in favour of designing the car from the 'inside-out', using both old and new technologies and methods (Porter and Saunders, 1996). The brief specified a sports car weighing less than 500 kg and costing less than £10 000, to be designed and built within 20 months. The weight, price and time constraints meant that it was essential that design, engineering and ergonomics were carried out concurrently.

As a completely new vehicle there were none of the conventional vehicle design constraints (e.g. number and configuration of wheels, number and configuration of passengers or a fixed user population and design for them before having to consider any engineering hard points). However, it must be noted that in a general sense 'sports cars' are the type of car where occupants are most compromised because they are bought, in the main part, by enthusiasts for engineering performance and aesthetics reasons.

Identification of potential purchasing population
SAMMIE was used to define space envelopes including seat travel for the chosen user population. The necessary 'minimal' design of the sports car meant it would be competing with the likes of the Caterham 7 and the population was chosen to reflect that sports car buying population. The users were defined by the design team as ranging from a small Japanese male (5JM), using 5th percentile values where appropriate (which is equivalent, for example, in stature to a 21st percentile British female; Pheasant, 1996) to a large American male (99AM) using 99th percentile values where appropriate. Sports cars tend to be bought by males, hence the relatively large 'smallest' user and the extremely large 'largest' user.

Package drawings
The generated space envelopes were then used, in the traditional fashion on paper, in conjunction with basic engineering information (engine, drive train and wheels) in different configurations. These covered rear, front and mid-engined, three and four wheels, two and three occupants in order to assess the vehicle 'footprints' in plan view and to choose one theme on which to develop the design (Figure 9.12). The decision was made based on possible aesthetic, engineering feasibility, weight and potential cost. A conventional side-by-side, rear engined package was chosen.

Human modelling and design
Once the package had been defined, the designers were able to begin designing an aesthetic theme based upon the brief, the package and their initial concept sketches. It must be noted that as an 'open-top' sports car the occupants were always a part of the aesthetic and almost all the sketch development included the occupants. SAMMIE 'men' were imported into Alias in typical sports car driving postures. This enabled the concurrent 3D CAD modelling alongside the conventional 2D aesthetic theme sketching programme (Figure 9.13). Alias is a 3D design tool which allows the modelling of complex curved surfaces and allows easy manipulation of form, colour and finish. The car was designed in Alias to 'look good' and to fit these computer manikins. All posture evaluations at this stage were carried out visually with the detailed assessments taking place later on during the seat design stage.

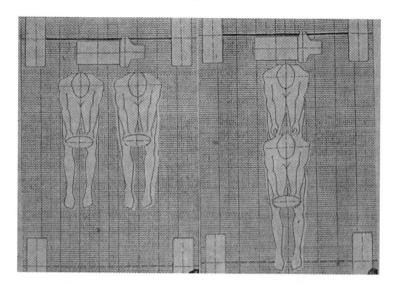

Figure 9.12 An example of vehicle footprints from which the basic vehicle layout was chosen

Quick and dirty ergonomics

A full size buck (Figure 9.14) was concurrently built with seats, pedals, chassis tubes and nose cone for early ergonomics fit and visibility evaluations as well as aesthetic judgements. The provision of ergonomics constraints early in the project allowed an aesthetic theme to be developed which the designers were happy with, whilst accommodating an appropriate population. The method of providing 10th scale space envelopes provided the perfect means of conveying the ergonomics data to the designers in a form they not only understood but on to which they could sketch directly the aesthetic themes. The seating buck was also used to establish the acceptability of the exterior and interior visibility (i.e. of the instruments) and an acceptable steering wheel position. An adjustable steering wheel was considered inappropriate for cost and practical reasons. It is also not anticipated that the car will be used for long, or frequent, journeys.

Seat design

The nature of the vehicle combined with weight and cost constraints of the vehicle meant that a glass fibre, fixed back angle, bucket style seat was the most sensible solution. The design was to be in-house and further work was carried out using SAMMIE to establish the required seat adjustment. The studies resulted in the proposal of a trunk-thigh angle of 110 degrees (this falls within the acceptable limits as recommended by Porter and Gyi, 1998) and a forward adjustment of 200 mm. The recommendation also included the raising of the seat combined with a tilting forward action to make the backrest more upright and to lower the seat front, as forward movement occurred. This has the effect of making forward sight lines for large and small members of the specified population approximately equivalent as well as improving pedal reach for the smaller driver. The seat adjustment was combined with the pedal size and between pedal distance specified by Henry Dreyfuss Associates (1993).

(a)

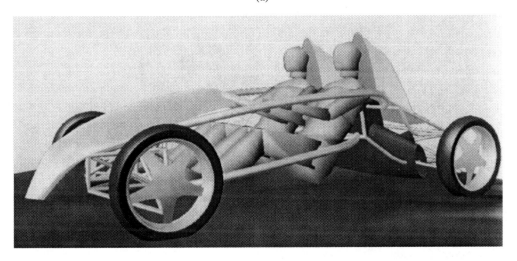

(b)

Figure 9.13 Design development through sketching and CAD

Some informed compromises were made as a consequence of engineering and aesthetic requirements. The seat travel was reduced to 175 mm and the pedals were moved inboard (to accommodate inboard suspension) by 130 mm from the ideal. Figure 9.15 shows the developed package. It should be noted that the pedal spacing is still particularly wide for a sports car and that the accelerator pedal remains to the right of the seat centre line. It was possible to confirm 'showroom' acceptability of this package using the seating buck and a test population covering

Figure 9.14 Design development through sketching and CAD

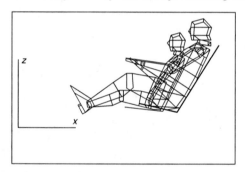

Figure 9.15 Side view of the concept seating package showing a Japanese and an American male driver with their appropriate 5th and 99th percentile body dimensions, respectively

both size extremes of the population. It is anticipated that further tests will take place at the running prototype stage.

The seat design was subsequently developed in clay based on basic anthropometric dimensions; minimum (e.g. seat length) and maximum (seat width and seat back height). The designers were reluctant to include a head restraint as a part of the design, claiming that it would spoil the 'look' of the vehicle. Interestingly, the development of the head restraint in clay in the chassis received immediate approval, the perceived aesthetic was one where height was given to balance a low and wide vehicle. The shape and aesthetic of the seat were developed with constant reference to both ends of the target population resulting in a seat which has 'showroom' acceptable comfort for both small and large users. Limited cushioning was also developed, its shape and positioning based upon the pressure distribution plots of Gyi and Porter (1999). Figure 9.16 shows the prototype vehicle.

Figure 9.16 The prototype vehicle exhibited at the Birmingham Motor Show in 1996

This project demonstrates the benefits of establishing a clear specification of the anthropometry of the potential owners for a vehicle at the earliest stages of design. It may not always be appropriate to consider the 'standard' 5th percentile female to 95th percentile male values. More importantly, it shows that when the ergonomics input occurs at the appropriate time in the design cycle then the design can perform at an engineering level and remain aesthetically uncompromised.

9.6 Future trends

The importance of considering ergonomics issues throughout the design and development process will become even greater in the future, for the following reasons:

- There is an increasing world-wide market. A greater range of nationalities give rise to increasing variation in body size and proportions. There are many other issues to consider, including stereotypes and cultural preferences;
- Driving exposure is likely to increase. For example, increased transnational mobility within the European Union will lead to greater annual mileage for many people, whilst worsening traffic congestion will result in longer journey times;
- Cars are getting smaller for environmental reasons. If many drivers are complaining today about a lack of legroom and headroom then it is clear that much more care must be taken with vehicle packaging in the future. Variations of body size, weight, and fragility must be understood as an airbag that is designed for a large, fit male will not work very successfully for a small, elderly female with osteoporosis;
- Vehicle development cycles will continue to shorten. Some ergonomics methods require time in order to provide high quality information. This is particularly true regarding the development of seating where the acid test is the extent to which the occupants report

discomfort during a representative journey length, often 2 hours or more. Such road trials often take something of the order of 3 months to complete with a carefully selected sample of drivers driving prototype and competitor vehicles. There is therefore a strong incentive to develop rapid predictive methods and many manufacturers are experimenting with seat pressure systems as an objective indicator of seat comfort. If objective predictive methods are going to be used increasingly before they have been properly validated then there is considerable concern that the quality of seat design may actually decline;

• Virtual reality (VR) will be used increasingly as a predictive tool. There may be the expectation that many more sophisticated ergonomics issues can be examined using immersive virtual environments. Too much confidence may be placed upon the accuracy of such evaluations based only on one person's double-guess of the issues. For example, the posture that a tall male VR operator may find personally acceptable when 'immersed' in the body of a small female is unlikely to be the same as that chosen by a real small female in a real vehicle;

• Legislation will continue to be introduced that is likely to disadvantage a substantial percentage of vehicle occupants. For example, roll-over protection in car rear seats may be assessed using a 50th percentile US male dummy. It would be interesting to hear the views of the 50% of US males whose heads would be crushed before this dummy's head is endangered;

• Fleet managers of company cars may increasingly adopt a single manufacturer sourcing for company cars in the UK in order to attract the greatest discounts. What will be their hidden cost related to days off work with back problems, and possibly the increased likelihood of accidents, in those cases where the car does not fit the driver? Until vehicles are designed to comfortably and safely accommodate all people, then the individual's freedom of choice of make and model will be essential;

• In-car telematics, which are designed to provide support to the driver (e.g. systems which sense the distance between a vehicle and the closest physical constraints when reversing a vehicle into a parking space) are removing responsibility from the driver. Arguably they make the car safer to drive but they may also de-skill the driver. Reversing a vehicle into a parking space is a relatively simple skill to lose, but having total vehicular control removed from the driver (e.g. a motorway with an electronically controlled convoy system to improve efficiency of movement of people and goods) is a more complex issue. How does the driver adjust from the controlled to the non-controlled situation?

9.7 Strategies for improving occupant accommodation and comfort

Few vehicle manufacturers have a reputation for providing excellent standards of occupant accommodation. Porter (1994) concluded that a corporate strategy to make improvements in this respect should include the activities listed below.

Concept design

• Make a clear statement of who the intended user population will be, in terms of gender, nationalities, age range, body sizes.

- Collect data concerning this population from published studies or, if not available, consider obtaining information directly from the population.
- Design for this population using relevant anthropometric data for body sizes at the earliest stages of design. Allow for secular growth if the data are old. Ensure there is sufficient adjustability in the driving package.
- Predict postural problems using human modelling systems.
- Improve communication within the design team so that effective solutions can be developed between those people responsible for engineering, styling and finance.
- Provide ergonomics expertise within the company by setting up training programmes or employing qualified ergonomists.
- Initiate research to improve databases and explore new concepts for reducing discomfort

Development

- Evaluate postural comfort using selected subjects in a prototype vehicle as soon as possible. Static prototypes are useful for assessing many aspects of occupant packaging but road trials are essential for examining and improving occupant comfort.
- Evaluate competitor vehicles to identify the strengths and weaknesses of the in-house design.
- Reiterate the design until an acceptable standard of postural comfort is obtained.

Product launch and after

- Inform the potential purchasers of the care that has been taken to provide high levels of postural comfort.
- Evaluate success of the vehicle packaging by distributing a detailed ergonomics questionnaire to a sample of the actual user population as well as a sample of people who are using competitor vehicles.
- Feed this information back to the design team for future use, such as face-lifts and new models.

9.8 Further reading

The focus of this chapter is occupant packaging and some associated issues have been discussed and referred to (e.g. control and displays and emerging technologies). Ergonomics is a broad discipline and there are many areas of ergonomics which impact upon automotive design. Further reading in areas related to occupant packaging are given below:

Controls, displays and workstation layout
See Section 9.4.1.

Human-machine interface for in-vehicle information systems
European Statement of Principles on Human Machine Interface for In-Vehicle Information and Communication Systems (1998)

Campbell, Carney and Kantowitz (1998)
DoT Driver Information Systems (1994)
Parkes and Franzen (1993)
Hancock and Parasraman (1992)

Fatigue and driving
Hartley (1995)

Comfort and performance
SAE (1996)

Vehicle crashworthiness and occupant safety
See Chapter 7 of this book

General ergonomics methods
Wilson and Corlett (1995)

9.9 Author details

The Vehicle Ergonomics Group is based in the Department of Design & Technology at Loughborough University. This Group was established by Professor Mark Porter in 1981 and it specialises in seating, interior packaging, control and display design, access and egress, external visibility, mirror design and occupant posture, safety and musculo-skeletal health. This Group has worked with numerous vehicle and component manufacturers worldwide at all stages of the design process, from concept designs using SAMMIE, expert evaluations of design bucks, road trials with camouflaged pre-production prototypes using members of the public, through to owner evaluations of their cars after purchase. Recent consultancy projects include the following:

EH101 helicopter	*accommodation of multi-national aircrew;*
Brussels Tram 2000	*design of the driver's cab and interior seating;*
European Fighter Aircraft	*control design for emergency use by multi-national aircrew;*
Lantau Line train (Hong Kong)	*design of the driver's cab, interior seating and emergency evacuation;*
Fiat Punto and Brava cars	*development of the driving package and seating;*
Amsterdam and Rotterdam Trams	*design of the driver's cab;*
Rolls-Royce Motor Cars	*evaluation of pre-production prototypes, interior packaging for new models, control and information management systems.*

Dr Samantha Porter is a Principal Lecturer in Ergonomics at Coventry School of Art and Design. She teaches ergonomics to Transport and Automotive Design students at undergraduate and postgraduate levels. Her research is focused on developing computer methods for supporting the automotive styling process, working closely with manufacturers including BMW, Ford and the Rover Group. Dr Porter also contributes to the work of the Vehicle Ergonomics Group.

9.10 References

Campbell, J.L., Carney, C. and Kantowitz, B.H. (1998). Human Factors Design Guidelines for Advanced Traveler Information Systems (ATIS) and Commercial Vehicle Operation (CVO). Publication No. FHWA-RD-98-057, Federal Highway Administration, US Department of Transportation.

Chakrabarti, D. (1997). Indian anthropometric dimensions for ergonomic design practice. National Institute of Design, India.

Damkot *et al.* (1984). The Relationship between Work History, Work Environment and Low Back Pain in Men. *Spine*, pp. 9, 4, 395–399.

Daniels, G.S. (1952). The 'average man?' Technical Note WCRD 53-7. WPAFB, Ohio: Wright Air Dev. Center, USAF (AD-10203).

Das, B. and Sengupta, A.K. (1995). Computer-aided human modelling programs for workstation design, Ergonomics, pp. 38, 9, 1958–1972.

Defence Standard 00-25. Human Factors for Designers of Equipment. Her Majesty's Stationery Office.

Dooley, M. (1982). Anthropometric modelling programmes – a survey. IEEE Computer Graphics and Applications, pp. 2, 17–25.

DoT, 1994, Driver Information Systems. Code of Practice and Design Guidelines.

European Statement of Principles on Human Machine Interface for In-Vehicle Information and Communication Systems, 1988, Version 1 13/2/98, HMI Experts Task Force, CONVERGE Project.

Frymoyer et al. (1983) Risk factors in low back pain. An Epidemiological Study. *American Journal of Bone and Joint Surgery*, pp. 65, 213.

Grandjean, E. (1980). Sitting posture of car drivers from the point of view of ergonomics. In: *Human Factors in Transport Research (Part 1)*, Oborne, D. and Levis, J.A. (eds), pp. 205–213, Taylor & Francis.

Gyi, D.E. and Porter, J.M. (1995) Musculoskeletal troubles and driving: a survey of the British Public. In: Robertson, S. (ed.), *Contemporary Ergonomics*, Taylor & Francis. pp. 304–309.

Gyi, D.E. and Porter, J.M. (1999). Interface pressure and the prediction of car seat comfort. *Applied Ergonomics*, pp. 30, 2, 99–108.

Hancock, P.A., and Parasraman, R. (1992). Human factors and safety in the design of intelligent vehicle-highway systems (IVHS), *Journal of Safety Research*, pp. **23**, 181–198.

Hartley, L., (ed.), (1995). *Fatigue and Driving*, Taylor & Francis.

Haslegrave, C.M. (1980). Anthropometric profile of the British car driver, *Ergonomics*, pp. 23, 436-67.

Haslegrave, C.M. (1986). Characterising the anthropometric extremes of the population. *Ergonomics*, pp. 29, 2, 281–301.

Henry Dreyfus Associates (1993). The Measure of Man and Woman. The Whitney Library of Design, New York.

Jordan, P.W., Thomas, B., Weerdmeester, B.A., Mclelland, I.L. (1996). Usability Evaluation in Industry. Taylor & Francis.

Jordan, P.W. (1998). *An Introduction to Usability*. Taylor & Francis.

Kelsey, J.L., Hardy, R.J. (1975). Driving of motor vehicles as a risk factor for acute herniated lumbar intervertebral disc. *American Journal of Epidemiology* (102), 1, pp. 63–73.

McConville J.T., 1978, Anthropometry in sizing and design, In: Anthropometric Source Book, Volume 1: Anthropometry for designers, NASA Reference Publication 1024, Washington, DC: Scientific and Technical Information Office.

Norris, B. and Wilson, J.R. (1995). CHILDATA The handbook of child measurements and capabilities – data for design safety. Government Consumer Safety Research. Department of Trade and Industry.

Open Ergonomics Ltd, (1999). PeopleSize Pro software, Loughborough, Leics, UK (www.openerg.com).

Parkes, A.M. and Franzen, S. (eds), (1993), *Driving Future Vehicles*, Taylor & Francis.

Peacock, B. and Karwowski, W. (1993). eds. *Automotive Ergonomics*, Taylor & Francis.

Peebles, L. and Norris, B. (1998). ADULTDATA The handbook of adult anthropometric and strength measurements – data for design safety. Government Consumer Safety Research. Department of Trade and Industry.

Petherick, N. and Porter, J.M. (1996). Driver discomfort and accommodation problems. Unpublished report, Vehicle Ergonomics Group, Loughborough University, Leics., UK.

Pheasant, S.T. (1996). *Bodyspace: Anthropometry, Ergonomics and Design* (2nd edition), Taylor & Francis.

Pietri, F., Leclerc, A., Boitel, L., Chastang, J-F., Morcet, J-F. and Blondet M. (1992) Low-back pain in commercial travellers. *Scand. J. Work Environ. Health*, pp. 18, 52–58.

Porter, J.M. and Gyi, D.E. (1998). Exploring the optimum posture for driver comfort. *International Journal of Vehicle Design*, pp. 19, 3, 255–266.

Porter, J.M. and Gyi, D. (1995). Low back trouble and driving. *Proceedings of the 2nd International Scientific Conference on Prevention of Work-related Musculoskeletal Disorders*, PREMUS'95, Montreal, Canada, September 1995, pp. 117–119.

Porter, J.M. and Porter, C.S. (1997). The interface between designers and ergonomists, Proceedings of the 13th Triennial Congress of the International Ergonomics Association, Tampere, Finland, vol. 2, pp. 240–242.

Porter, J.M. and Porter, C.S. (1998). Turning automotive design 'inside-out'. *International Journal of Vehicle Design*, pp. **19**, 4, 385–401.

Porter, J.M. (1994), Seating Design: Current problems and future strategies. *Automotive Interiors International, Autumn edition*, pp. 6–19, Turret Group plc.

Porter, J.M. (1995). The ergonomics development of the Fiat Punto–European Car of the Year 1995. Proceedings of the IEA World Conference 1995, Rio de Janeiro, eds. de Moraes, A. and Marino, S., Associacao Brasileira de Ergonomia, pp. 73–76.

Porter, J.M., Case, K. and Freer, M.T. (1996). SAMMIE: a 3D human modelling computer aided ergonomics design system. *Co-Design Journal*, s2 07, 68–75.

Porter, J.M., Case, K. and Freer, M.T. (1999). Computer aided design and human models. In: *Handbook of Occupational Ergonomics*, eds. Karwowski, W. and Marras, W., pp. 477–497, CRC Press, Florida.

Porter, J.M., Case, K., Freer, M.T. and Bonney, M.C. (1993) Computer-Aided Ergonomics Design of Automobiles. In: *Automotive Ergonomics*, eds. Peacock, B. and Karwowski, W.,Taylor & Francis, pp. 43–78.

Porter, J.M., Freer, M., Case, K. and Bonney, M.C. (1995). Computer Aided Ergonomics and Workspace Design. In: *Evaluation of Human Work: A Practical Ergonomics Methodology*, eds. Wilson, J.A. and Corlett, E.N., Taylor & Francis Ltd., pp. 574–620.

Porter, J.M., Porter, C.S. and Lee V.J.A. (1992). A Survey of Driver Discomfort. In: Lovesey, E.J. (ed.), *Contemporary Ergonomics* (1992). Taylor & Francis Ltd., pp. 262–267.

Porter, S. and Saunders, S. (1996). The Use of Computer Aided Styling Tools in the Automotive Design Process. *Co-Design Journal*, 07/08,1996, pp. 56–61.

Rebiffe, R. (1969). The driving seat: its adaption to functional and anthropometric requirements. In: *Sitting Posture*, Grandjean, E. (ed.), pp. 132–147, Taylor & Francis.

Rebiffe, R., Guillien, J. and Pasquet, P. (1984) Anthropometric survey of French drivers. Laboratoire de physiologie et de biomecanique de L'association Peugeot-Renault.

Roe, R.W. (1993). Occupant Packaging. In: *Automotive Ergonomics*. eds: Peacock, B. and Karwowski, Taylor and Francis, pp. 11–42.

Roebuck, J.A. (1995). Anthropometric Methods: Designing to Fit the Human Body, Human Factors and Ergonomics Society, USA.

Rothwell, P.L. (1985). Use of man-modelling CAD systems by the ergonomist. In: *People and Computers: Designing the Interface*, eds. P. Johnson and S. Cook, pp. 199–208. Cambridge University Press.

SAE Handbook (1996). Society of Automotive Engineers.

SAE, 1996, SP-1155, Automotive Design Advancements in Human Factors; Improving Driver's Comfort and Performance.

Sanders, M.S. (1977). Anthropometric survey of truck and bus drivers. Department of Transportation, Federal Highway Administration, Bureau of Motor Safety, Washington, D.C. 20590 (available through National Technical Information Service, Springfield, Virginia 22151).

Sanders, M.S. and McCormick, E.J., Human Factors in Engineering and Design (7th edn) McGraw-Hill, SAE (1990). 1990 SAE Handbook, On Highway Vehicles and Off Road Highway Machinery, Vol 4, Warrendale PA: Society of Automotive Engineers.

Simpson, R.E. and Hartley, E.V. (1981). Scatter diagrams based on the anthropometric survey of 2000 Royal Air Force Aircrew (1970/71), Royal Aircraft Establishment Technical Report 81017, Farnborough, Hampshire, England.

Thompson, D.D. (1995). An ergonomic process to assess the vehicle design to satisfy customer needs. *Int. J. of Vehicle Design*, 16,2/3, pp. 150–157.

Tovey, M.J. (1992). Intuitive and objective processes in automotive design. *Design Studies*, Vol. 13, 1, pp. 23–41.

Walsh, K., Varnes, N., Osmond, C., Styles, R. and Coggon, D. (1989). Occupational causes of low-back pain. *Scand. J. Work Health*, pp.15, 54–59.

Wilson, J.A. and Corlett, E.N. (1995). *Evaluation of Human Work: A Practical Ergonomics Methodology*, eds. Taylor & Francis.

10. Suspension systems and components

Brian Hall, PhD, BScEng, CEng, MIMechE

The aim of this chapter is to:

- Introduce the basic features of vehicle suspension systems;
- Indicate simple methods for the analysis of vehicle suspension systems and their components;
- Demonstrate the design requirements for vehicle suspension systems and how to achieve them;
- Suggest methods by which the designer can improve the occupant perception of ride by the control of suspension design.

10.1 Introduction

It is probably true to say that the average member of the vehicle owning public is unaware of the range of duties performed by a vehicle suspension. Certainly many would recognize the importance of the suspension for ride, but fewer would identify its importance in the handling of a vehicle. In reality a vehicle suspension is required to perform effectively under a range of operating conditions including high levels of braking and accelerating, cornering at speed and traversing rough terrain – manoeuvres which are required to be done in comfort and with safety. These requirements present the chassis engineer with some challenging problems and introduce some unavoidable design compromises. There are a number of excellent texts (Adler, 1996; Bastow *et al.*, 1993; Dixon, 1996; Gillespie, 1992; Heisler, 1989; Hillier, 1991; Milliken *et al.*, 1995; Reimpell *et al.*, 1998; Wong, 1993 which address a number of these issues. The aim here is to give a broad treatment of the main issues for the aspiring automotive designer.

The chapter begins by identifying the functions of suspensions and goes on to introduce the student to the mechanics of suspension systems. Suspension kinematics and kinetics are analysed and suspension components and characteristics are discussed. There is a section on ride which includes vehicle excitation induced by road surfaces, vehicle modelling and human perception of ride. The chapter concludes with a section outlining some of the developments in controllable suspensions.

10.2 The role of a vehicle suspension

The principle requirements are:

- To provide good ride and handling performance – this requires the suspension to have vertical compliance providing chassis isolation and ensuring that the wheels follow the road profile with very little tyre load fluctuation;

- To ensure that steering control is maintained during manoeuvring – this requires the wheels to be maintained in the proper positional attitude with respect to the road surface;
- To ensure that the vehicle responds favourably to control forces produced by the tyres as a result of longitudinal braking and accelerating forces, lateral cornering forces and braking and accelerating torques – this requires the suspension geometry to be designed to resist squat, dive and roll of the vehicle body;
- To provide isolation from high frequency vibration arising from tyre excitation – this requires appropriate isolation in the suspension joints to prevent the transmission of 'road noise' to the vehicle body.

It will be seen that these requirements are virtually impossible to achieve simultaneously, leading to design compromises with less than ideal performance.

10.3 Factors affecting design

Suspension design like other forms of vehicle design are affected by the reduced development times dictated by market forces. This means that for new vehicles, refined suspensions need to be designed quickly with a minimum of rig and vehicle testing prior to launch. Consequently considerable emphasis is placed on computer-aided design requiring the use of multi-body systems analysis software of which ADAMS (Ryan, 1990) is typical. This software enables many 'what–if' scenarios to be tested quickly without the need for a lot of development testing, but they do require sophisticated mathematical models to be developed for various components and sub-systems.

In addition to the functional constraints placed on a given design, suspensions are also required to meet certain performance targets which vary across the range of vehicles. There are also other limitations such as cost, weight, packaging space, requirements for robustness and reliability, together with manufacturing, assembly and maintenance constraints.

10.4 Definitions and terminology

There is a lot of terminology associated with suspension design[1] which may appear novel to the student meeting the subject for the first time. Most of this will be described as it arises, but it may prove useful to introduce the vehicle axis system and terminology associated with wheel position at this stage.

10.4.1 Vehicle axis system and terminology

In simple studies of whole vehicle braking, accelerating and turning analyses, it is appropriate to position the origin of the vehicle axes at the centre of gravity (CG) of the whole vehicle.

---

[1]See Gillespie (1992) for a comprehensive list of Society of Automotive Engineers (SAE) vehicle dynamics terminology.

However, in suspension design we are concerned with the movement of the vehicle body (the sprung mass) relative to the other moving parts of the suspension and wheels (the unsprung masses). So in this case it is usual to place the vehicle axis system (a right-handed set of axes) at the CG of the sprung mass as shown in Figure 10.1.

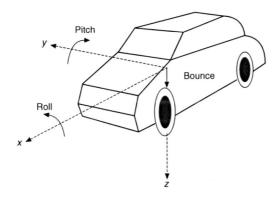

Figure 10.1 Sprung mass axes and displacements relevant to suspension analyses

Since the sprung mass is treated as a rigid body, it has six degrees of freedom (DOF) comprising three translations and three rotations. Only three of these are relevant for suspension studies, bounce, roll and pitch as shown in Figure 10.1.

10.4.2 Definitions for wheel orientation

Since one of the functions of a suspension system is to maintain the position of the wheels constant relative to the road[2] throughout the motion of the suspension it is important to identify how the wheel position is defined. Figure 10.2 provides these definitions.

Camber angle is the angle between the wheel plane and the vertical – taken to be positive when the wheel leans outwards from the vehicle.

Swivel pin (kingpin) inclination is the angle between the swivel pin axis and the vertical. The swivel pin inclination has the effect of causing the vehicle to rise when the wheels are turned and produces a noticeable self-centring effect for swivel pin inclinations greater than 15°.

Swivel pin (kingpin) offset is the distance between the centre of the tyre contact patch and the intersection of the swivel pin axis and the ground plane – taken to be positive when the intersection point is at the inner side of the wheel. The swivel-pin offset reduces steering effort because the wheel tends to roll during turning. With zero offset, the kingpin axis intersects the centre of the tyre contact patch. If the wheel is steered under these conditions there is significant tyre scrub at the front and rear of the contact patch leading to a significant steering effort. The disadvantage of offset is that longitudinal forces at the tyre contact patch due to braking, or striking a bump or pothole is transmitted through the steering mechanism to the steering wheel.

---

[2]Some beneficial and some detrimental effects can result from changes in wheel position for certain operating conditions.

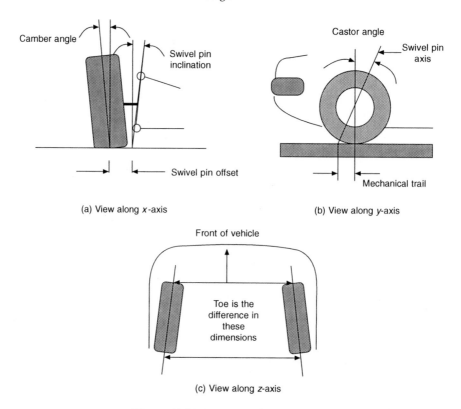

Figure 10.2 Wheel position definitions

Castor angle is the inclination of the swivel pin axis projected into the fore–aft plane through the wheel centre – positive in the direction shown. Castor angle provides a self-aligning torque for non-driven wheels.

Toe-in and *Toe-out* is the difference between the front and rear distances separating the centre plane of a pair of wheels, quoted at static ride height – toe-in is when the wheel centre planes converge towards the front of the vehicle as shown in Figure 10.2(c).

Suspension travel can result in changes in wheel orientation relative to the ground and consequently to steering effects unrelated to those initiated by the driver of the vehicle. When these arise from vertical travel of the unsprung mass they are referred to as *bump steer* effects. Roll of the sprung mass can induce roll steer and flexibility in the suspension mechanism can also give rise to compliance steer.

10.5 The mobility of suspension mechanisms

Suspension systems are in general three-dimensional mechanisms and as such are difficult to analyse fully without the aid of computer packages. Their analysis is complicated by the inclusion of many compliant bushes which effectively result in links having variable lengths. Notwithstanding these complications it is possible to gain an appreciation of the capabilities and limitations of various mechanisms used in suspension design by neglecting bush compliances and concentrating on the basic motion of suspension mechanisms.

A fundamental requirement of a suspension mechanism is the need to guide the motion of each wheel along a (unique) vertical path relative to the vehicle body without significant change in camber. This requirement has been addressed by employing various single degree of freedom (SDOF)[3] mechanisms which have straight line motion throughout the deflection of the suspension[4].

Despite the apparent complexity of some suspension systems, a basic understanding of their kinematics can be derived from a two-dimensional analysis, i.e. by considering the motion in a vertical transverse plane through wheel centre. Fundamental to this analysis is an understanding of how the number of degrees of freedom (*mobility* in mechanisms parlance) of a mechanism are related to the number of links and the types of kinematic constraint imposed on them. In general the aim is for a SDOF or a mobility of one. Mechanisms which have a mobility of zero are structures, i.e. not designed for motion, while those having two degrees of freedom require two prescribed inputs to position them uniquely. This is not desirable for suspensions.

Most of the kinematic connections between the members of a suspension mechanism can be reduced down to the *kinematic pairs* shown in Figure 10.3. Each has an associated number of degrees of freedom and can be classified as lower pairs (connections having a SDOF) or higher pairs (more than one DOF). It has been shown[5] that the mobility M, of a plane mechanism forming a closed kinematic chain, is related to the number of links n, the number of lower pairs j_l and the number of higher pairs j_h. According to the Kutzbach criterion:

$$M = 3(n - 1) - j_h - 2j_l \tag{10.1}$$

For spatial (three dimensional) mechanisms there is an equivalent equation [Suh *et al.* (1978)] The use of equation 10.1 can be illustrated with reference to the double wishbone and MacPherson strut suspensions[6] whose kinematics are represented two dimensionally in Figure 10.4. Both suspensions can be seen to represent a single closed kinematic chain.

In the case of the double wishbone suspension (Figure 10.4(a)), there are four links, AB, BC, CD and DA forming a *four-bar chain*, i.e. $n = 4$. Each of the four joints are of the revolute type (lower pairs) and hence $j_l = 4$. There are no higher pairs and therefore $j_h = 0$. Substituting into equation 10.1 gives $M = 3 \times (4 - 1) - 0 \times (2 \times 4) = 1$, i.e. a SDOF mechanism.

In the kinematically equivalent mechanism for the MacPherson strut (Figure 10.4(b)) the telescopic damper is replaced with an extension of the wheel attachment to pass through a trunnion (a 2DOF joint) at C. The mechanism thus has three links AB, BC and CA, i.e. $n = 3$. There are two lower pairs (one at A and one at B) and one upper pair (at C). This gives $j_l = 2$ and $j_h = 1$. Hence $M = 3 \times (3 - 1) - 1 - (2 \times 2) = 1$, i.e. a SDOF mechanism.

While mobility analysis is useful for checking for the appropriate number of degrees of freedom, it does not help in developing the geometry of a mechanism to provide the desired motion. For suspension mechanisms this process is called position synthesis and requires the use of specialized graphical and analytical techniques (Erdman *et al.*, 1984), aided by computer

---

[3]Degree of freedom has been defined in the Chapter 8.
[4]A number of these can be found in Dixon (1996).
[5]See Suh *et al.* (1978) for further discussion and examples.
[6]These suspensions are discussed in Section 10.6.

Name of pair	Geometric form	Degrees of freedom
Revolute R	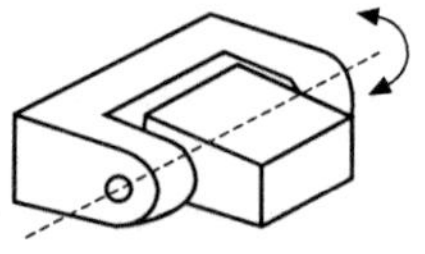	1
Prism P	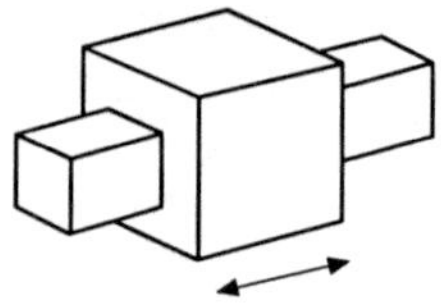	1
Cylinder P	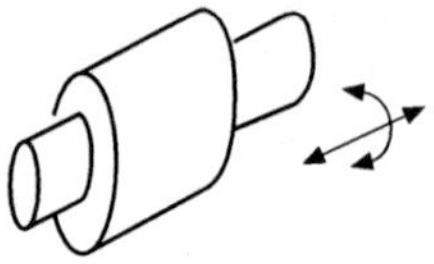	2
Sphere S	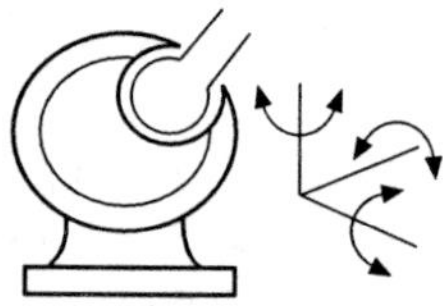	3

Figure 10.3 Kinematic pairs

software. This departure from the well established suspension types is only required when it is necessary to produce enhanced suspension characteristics, e.g. to produce changes in camber and toe under certain operating conditions to improve handling.

10.6 Suspension types

There is a range of generic suspensions which are commonly used. In this section we will describe some of them and discuss some of their important features. Some of the associated diagrams aim to convey only the essential kinematic principles.

Factors which primarily affect the choice of suspension type at the front or rear of a vehicle are engine location and whether the wheels are driven/undriven and/or steered/unsteered. In general, suspensions can be broadly classified as *dependent* or *independent* types.

With *dependent suspensions* the motion of a wheel on one side of the vehicle is dependent on the motion of its partner on the other side, that is when a wheel on one side of the vehicle

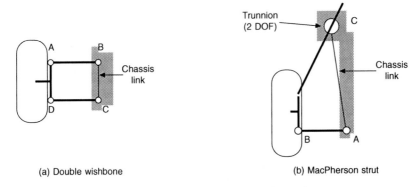

(a) Double wishbone   (b) MacPherson strut

Figure 10.4 Two-dimensional kinematics of some common suspension mechanisms

strikes a pot-hole the effect of it is transmitted directly to its partner on the other side. This has a detrimental effect on the ride and handling of the vehicle.

With *independent suspensions* the motion of wheel pairs is independent, so that a disturbance at one wheel is not directly transmitted to its partner. This leads to better ride and handling capabilities.

10.6.1 Dependent systems

As a result of the upward trend in vehicle refinement, these are not so common on passenger cars. However, they are still commonly used used on commercial and off-highway vehicles. They have the advantages of being relatively simple in construction and almost completely eliminate camber change thereby reducing tyre wear.

They are occasionally used in conjunction with non-driven axles (dead axles) at the front of some commercial vehicles, but are more common at the rear of front-wheel drive light commercial and off-road vehicles. This type of system is also used in conjunction with rear driven axles (live axles) on commercial and off-highway vehicles.

There are a number of ways of mounting a solid axle. The following two examples are very common.

(a) The Hotchkiss rear suspension (Figure 10.5)
In this case the axle is mounted on longitudinal leaf springs, which are compliant vertically and stiff horizontally. The springs are pin-connected to the chassis at one end and to a pivoted link at the other. This enables the change of length of the spring to be accommodated due to loading.

Earlier problems with inter-leaf friction (which affected ride performance) have been overcome by replacing each of the multi-leaf springs with a single tapered leaf. The requirements for good ride in passenger cars call for highly compliant leaf springs which lead to poor locational properties. Such flexible springs are also unable to control high braking and accelerating torques (leading to axle tramp). These latter problems can be overcome to some extent by using Panhard rods[7] to control lateral deflections and trailing arms to resist braking and accelerating

---

[7]Panhard rods to provide transverse stiffness without controlling suspension motion.

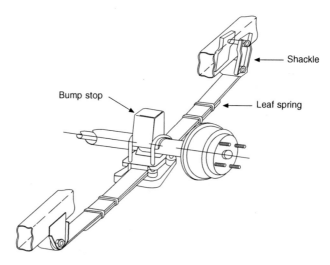

Figure 10.5 Hotchkiss rear suspension

torques. Despite these improvements this type of suspension is now rarely used on passenger cars.

However, for those vehicles where ride is not of primary importance they are still widely used. An example of this is the mid-range commercial van where load carrying capacity is important. In this case heavier or two-stage springing can be used to overcome the problems discussed above.

(b) Trailing arms (Figure 10.6)
Various configurations are possible, provided that they permit vertical and roll freedoms. Either coil or air springs can be used, the latter tending to give better ride performance. Lateral control can be provided by angling the upper links (as shown) or by using a Panhard rod. Compared to

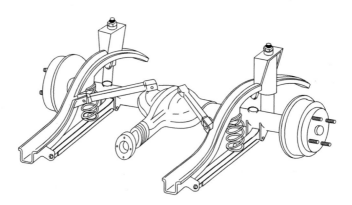

Figure 10.6 Trailing arm – rigid axle suspension

the Hotchkiss system, the four link design gives greater flexibility in the choice of roll-centre location (Section 10.8), anti-squat and anti-dive geometry (Section 10.10) and roll-steer.

10.6.2 Semi-dependent systems

In this form of suspension, the rigid connection between pairs of wheels is replaced by a compliant link. This usually takes the form of a beam which can bend and flex providing both positional control of the wheels as well as compliance. Such systems tend to be very simple in construction but lack scope for design flexibility.

An example of this form of suspension system is the trailing twist axle design shown in Figure 10.7.

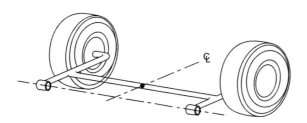

Figure 10.7 Trailing twist axle suspension

Additional compliance can be provided by rubber or hydro-elastic springs. Wheel camber is, in this case, the same as body roll.

10.6.3 Independent systems

This form of suspension has benefits in packaging and gives greater design freedom when compared to dependent systems. Some of the most common forms of front and rear designs will be considered. Both the MacPherson strut, double wishbone and multi-link systems are employed for front and rear wheel applications. The trailing arm, semi-trailing arm and swing axle systems tend to be used predominantly for rear wheel applications.

(a) MacPherson strut (Figure 10.8)
The vertical movement is constrained by the telescopic pivoted link (damper) and compliance is provided by a coil spring. Lateral constraint is provided by the lower transverse arm and longitudinal constraint is provided by the longitudinal link. Various options can be adopted for the constructional detail, these include replacing the lower transverse arm by an A-shaped member (A-arm) having the apex of the A connected to the knuckle and the base of the A connected to the chassis by two pin connections. This design obviates the need for the longitudinal member shown in the diagram since both longitudinal and transverse forces can now be reacted by the A-arm. Also the spring can be either co-axial or parallel with the damper.

Simplicity is the main benefit in this case. The disadvantages are:

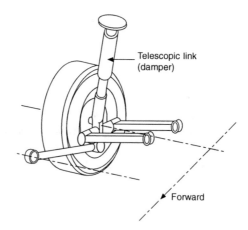

Figure 10.8 MacPherson strut suspension

- the installation height can be a problem when a low bonnet line is required by the vehicle stylist;
- the strut has to react against a moment imposed by wheel loading (Section 10.9), but this problem can be lessened by angling the suspension spring.

(b) Double wishbones (Figure 10.9)
This design produces a classic four-bar mechanism when viewed from the front of the vehicle. It has the knuckle located at the centre of the coupler link and is therefore capable of providing straight-line motion to the knuckle. However, because of packaging constraints it is normal to make the upper wishbone shorter than the bottom. The double wishbones provide the constructional strength to react transverse and longitudinal loads.

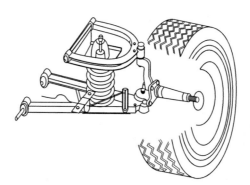

Figure 10.9 Double wishbone suspension

(c) Trailing arms (Figure 10.10)
This is used in rear axle applications with either front or rear wheel drive. It can be used with a variety of springs including torsion bars, coil springs, rubber springs or hydro-elastic types.

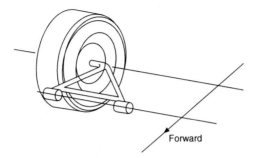

Forward

Figure 10.10 Trailing arm suspension

It is a relatively low cost form of suspension, but offers little flexibility in terms of kinematic design options, but there is scope for anti-lift control by adjusting the pivot position. Camber change is the same as body roll and caster change can be substantial. There is no toe change so roll steer is zero.

(d) Swing axles (Figure 10.11)
This is a very simple form of suspension used with driven axles. With short swing axles the camber changes and tyre scrub[8] can be considerable for the range of suspension travel. This form of suspension is particularly prone to suspension jacking[9].

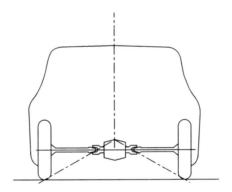

Figure 10.11 Swing axle suspension

(e) Semi-trailing arms (Figure 10.12)
This type of suspension is a cross between the swing axle and pure trailing arm designs. It allows a compromise between the control of camber and jacking. It is essential that the geometry is carefully selected to limit the amount of steer induced by the trail angle. However, this is a

---

[8]Tyre scrub is related to the horizontal transverse movement of the tyre relative to the road surface.
[9]Jacking is a form of instability induced into the suspension system during cornering (Gillespie, 1992).

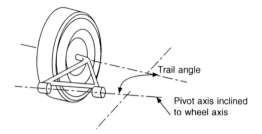

Figure 10.12 Semi-trailing arms

feature which can be exploited to provide small amounts of rear wheel steer to improve handling performance.

(f) Multi-link suspensions (Figure 10.13)
There is a wide variety of multi-link designs. Figure 10.13 shows a five link design which is used to control the separate functions required of a suspension. The ends of the links are mounted in flexible bushes, which are necessary because the mechanism is over-constrained kinematically.

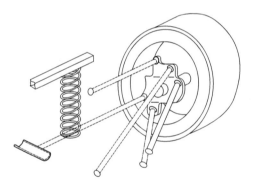

Figure 10.13 A multi-link suspension

10.7 Kinematic analysis

One of the first stages of suspension design (once the type of suspension has been selected) is to size the mechanism and ensure it is capable of fitting into the packaging envelopes. As part of this process it is necessary to check the geometry variations of the suspension over its operating range and ratios of spring and damper travel relative to wheel travel.

Comprehensive analysis of suspension motions requires the inclusion of joint compliances and the problem then becomes one of force-motion analysis. This requires the use of specialized computational software (e.g. Adams Ryan, (1990)) and associated modelling skills. A large number of multibody computer codes, many of which can be used for suspension analysis, are reviewed in Kortum *et al.* (1993). If joint compliances are neglected, the problem is simplified into a purely kinematic one and if, furthermore, the problem can be assumed to be 2-dimensional, an even more basic analysis can be carried out using graphical or computational analysis. The latter can be aided with general purpose analysis software (e.g. MathCAD, 1998). In this relatively brief treatment of suspension design fundamentals, graphical and computational analysis will be restricted to two-dimensional examples.

In graphical analysis of suspension motion it should be recognized that relationships between the relative motion of parts of the mechanism can be determined from sets of velocity diagrams. To cover the full range of suspension travel it is necessary to draw a number of diagrams corresponding to different positions of the mechanism. This has the advantage of providing a good 'feel' for what is happening, but the downside is the lack of accuracy and the tedium of drawing many diagrams. For the reader unfamiliar with velocity diagrams there are numerous texts in engineering dynamics which can be consulted, e.g. Meriam *et al.* (1993).

If a computational approach is adopted, there can be significant effort required to formulate the problem, but once this has been done mathematical software can be used to solve equations and present the results numerically or graphically. The significant benefits of this approach are improved accuracy and the ability to try out 'what–if' scenarios.

10.7.1 Graphical analysis

To illustrate the graphical approach consider the MacPherson strut in Figure 10.14(a). Assume that the aim is to determine (a) the suspension ratio R (the rate of change of vertical movement at D as a function of spring compression) and (b) the bump to scrub rate for the given position of the mechanism.

Begin by drawing the suspension mechanism to scale and assume the chassis is fixed. Let link AB have an arbitrary angular velocity $w_{BA} = 1$ rad/s in a clockwise direction. The velocity of B has a magnitude $V_B = \omega_{BA} r_{BA}$, i.e $V_B = 1 \times 331 = 331$ mm/s perpendicular to link AB and represented by the vector V_B in Figure 9.14(b). This vector is drawn to some scale from the pole of the velocity diagram O_V. Note a and c are also located at O_V since they have zero absolute velocity.

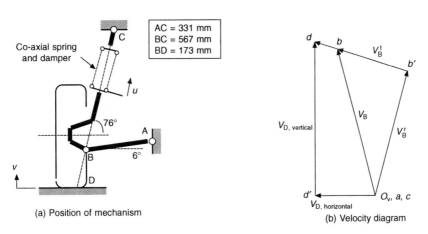

(a) Position of mechanism

(b) Velocity diagram

Figure 10.14 Example of graphical kinematic analysis

The velocity of B relative to C comprises a component parallel to BC (arising from the change in length of the equivalent link BC) and a component perpendicular to BC (the tangential component arising from the rotation of link BC about C). At this stage, magnitudes of neither of these components can be calculated, but by drawing a line from O_V parallel to BC and a line from b perpendicular to BC the two lines intersect at b'. Hence the magnitudes of the radial and tangential components of the velocity of B relative to C are established and can be scaled from the diagram. Their magnitudes are found to be $V^r_{BC} = 311$ mm/s and $V^t_{BC} = 113.2$ mm/s.

Since DB can be considered to be a rigid extension of link BC, the velocity of D relative to B, V_{DB} consists only of a tangential component V^t_{DB}. The magnitude of this can be determined by proportioning as follows:

$$\frac{db}{DB} = \frac{bb'}{BC} \text{ and introducing the data, } db = \frac{173}{567} \, 113.2 = 34.54 \text{ mm/s}$$

This establishes point d on the diagram. It is then possible to scale the vertical and horizontal components of d. These are found to be $V_{D,vertical} = 311$ mm/s and $V_{D,horizontal} = 147.6$ mm/s. Then

(a) $R = \dfrac{dv}{du} = \dfrac{dd'}{O_V b'} = \dfrac{311}{267} = 1.16$

(b)   Scrub to bump $= \dfrac{O_V d'}{dd'} = \dfrac{147.6}{311} = 0.47$

10.7.2 Computational (2-dimensional) analysis

The following example illustrates the computational approach using MathCAD software.

The mechanism shown in Figure 10.15 represents a double wishbone suspension and P is at the intersection of the tyre centre and the road surface. The primary (independent) variable is

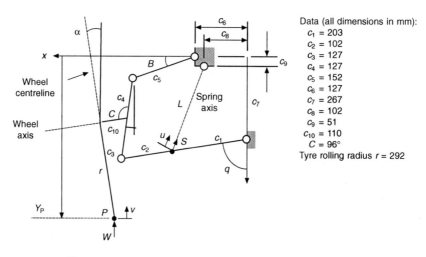

Data (all dimensions in mm):
$C_1 = 203$
$C_2 = 102$
$C_3 = 127$
$C_4 = 127$
$C_5 = 152$
$C_6 = 127$
$C_7 = 267$
$C_8 = 102$
$C_9 = 51$
$C_{10} = 110$
$C = 96°$
Tyre rolling radius $r = 292$

Figure 10.15 Example of computational kinematic analysis

q and the secondary (dependent) variables are A and B. (For details of this terminology and approach to kinematic analysis see Doughty (1988).)

The objectives are to determine how the camber angle α, and suspension ratio R (as defined in the previous example) vary for suspension movement described by q varying from $80°$ to $100°$, given that in the static laden position $q = 90°$.

(a) The solution begins by declaring the data and defining constants. Note: dimensions are not included in this solution, but MathCAD does allow this if required. Position equations (one in the x and the other in the y-direction) are written for the four-bar mechanism a, b, c, d and included in the *Given–Find* block of the program. This is used to iteratively solve the two non-linear simultaneous equations for the two secondary variables for each position of the primary variable (in steps of $1°$). Note the angles need to be expressed in radians and initial estimates are required to initiate the iteration procedure. The solutions for A and B at each angular position q are contained in the 21 two-element vectors making up the 2×21 matrix F. The vertical location Y, of the tyre contact point P, is expressed in terms of the primary and secondary variables, enabling the deflection v, to be determined from its mean position. This enables the graph of camber angle α, to be plotted as a function of v.

Data: $c_1 := 203$ $c_2 := 102$ $c_3 := 127$ $c_4 := 127$ $c_5 := 152$ $c_6 := 127$
   $c_7 := 267$ $c_8 := 102$ $c_9 := 51$ $c_{10} := 110$ $C := 96°$ $r := 292$

Constants $c_{12} := c_1 + c_2$ $c_{34} := c_3 + c_4$ $k_{dr} := \dfrac{\pi}{18}$

Solution estimates: $A := -10$ $B := 10$

Given

$$c_{12} \cdot \sin(q \cdot k_{dr}) - c_{34} \cdot \sin(A \cdot k_{dr}) - c_5 \cdot \cos(B \cdot k_{dr}) + c_6 = 0$$
$$c_{12} \cdot \cos(q \cdot k_{dr}) - c_{34} \cdot \cos(A \cdot k_{dr}) - c_5 \cdot \sin(B \cdot k_{dr}) + c_7 = 0$$
$$F(q) := \text{Find}(A, B)$$
$$q := 80..100 \quad i := 0..20$$
$$F(90)_0 := 6.03 \quad A_i := F(80 + i)_0 \quad B_i := F(80 + i)_1 \quad q_i := 80 + i$$

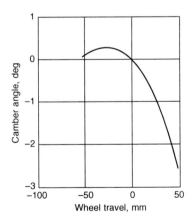

Camber angle (degrees) $\alpha_i := C - 90 - A_i$

Express angles in radians

$$q_{r_i} := q_i \cdot k_{dr} \quad A_{r_i} := A_i \cdot k_{dr} \quad B_{r_i} := B_i \cdot k_{dr} \quad \alpha_{r_i} := \alpha_i \cdot k_d$$

Vertical position of tyre contact point

$$Y_{P_i} := c_7 + c_{12} \cdot \cos(q_{r_i}) - c_3 \cdot \cos(A_{r_i}) + c_{10} \cdot \sin(\alpha_{r_i}) + r \cdot \cos(\alpha_{r_i})$$

Mean position of tyre contact point: $Y_{PO} := Y_{P_{10}} \quad Y_{PO} = 432.644$ mm

Deflection from mean position: $\qquad\qquad v_i := Y_{P_i} - Y_{PO}$

(b)   The second part of the solution begins by expressing the length of the suspension spring in terms of the primary variable and then proceeds to determine the velocity coefficients $K_{YP}(q) = \dfrac{dY_P}{dq}$ and $K_L(q) = \dfrac{dL}{dq}$. These allow the suspension ratio $R = \dfrac{K_{YP}}{K_L}$ to be determined.

Length of suspension spring

$$L_i := \sqrt{(c_1 \cdot \sin(q_{r_i}) - c_8)^2 + (c_7 + c_1 \cdot \cos(q_{r_i}) - c_9)}$$

Mean position of suspension spring $L_O := L_{10} \quad L_O = 238.447$ mm

Deflection from mean position $\qquad u_i := L_O - L_i$

Velocity coefficients $\qquad\qquad\qquad K_{A_i} := \dfrac{c_{12} \cdot \cos(q_{r_i} + B_{r_i})}{c_{34} \cdot \cos(A_{r_i} + B_{r_i})}$

$$K_{YP_i} := c_{12} \cdot \sin(q_{r_i}) + K_{A_i} \cdot (c_3 \cdot \sin(A_{r_i}) - c_{10} \cdot \cos(\alpha_{r_i}) + r \cdot \sin(\alpha_{r_i})$$

$$K_{L_i} := \dfrac{c_1 \cdot c_9 \cdot \sin(q_{r_i}) - c_1 \cdot c_7 \cdot \sin(q_{r_i}) - c_1 \cdot c_8 \cdot \cos(q_{r_i})}{\sqrt{\left[(c_1 \cdot \sin(q_{r_i}) - c_8)^2 + (c_7 + c_1 \cdot \cos(q_{r_i}) - c_9)^2\right]}}$$

Suspension ratio $\quad R_i := \dfrac{K_{YP_i}}{K_{L_i}} \qquad R_{10} = 1.607$ **at static ride height**

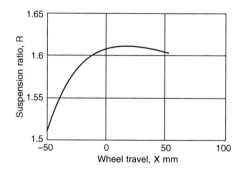

10.8 Roll centre analysis

Roll centre and roll axis concepts are important aids in studying vehicle handling, enabling simplifications to be made in load transfer calculations for cornering operations.

There are two definitions of roll centre, one based on forces and the other on kinematics. The first of these (the SAE definition) states that: *a point in the transverse plane through any pair of wheels at which a transverse force may be applied to the sprung mass without causing it to roll. The second states that: the roll centre is the point about which the body can roll without any lateral movement at either of the wheel contact areas.*

In general each roll centre lies on the line produced by the intersection of the longitudinal centre plane of the vehicle and the vertical transverse plane through a pair of wheel centres. The roll centre heights at the front and rear wheel planes tend to be different as shown in Figure 10.16. The line joining the centres is called the roll axis, with the implication that a transverse force applied to the sprung mass at any point on this axis will not cause body roll.

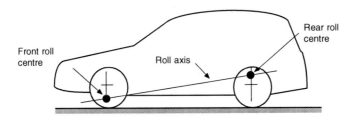

Figure 10.16 Roll axis location

As roll of the sprung mass takes place, the suspension geometry changes, symmetry of the suspension across the vehicle is lost and the definition of roll centre becomes invalid. Therefore, the limitations of roll centre analysis are:

* it relates to the non-rolled vehicle condition and can therefore only be used for approximations involving small angles of roll;
* it assumes no change in vehicle track as a result of small angles of roll.

For a given front or rear suspension the roll centre can be determined from the kinematic definition by using the Aronhold–Kennedy theorem of three centres[10] which states: *when three bodies move relative to one another they have three instantaneous centres all of which lie on the same straight line.*

To illustrate the determination of roll centre by this method consider the double wishbone suspension shown in Figure 10.17. Consider the three bodies capable of relative motion as being the sprung mass, the left hand wheel and the ground. The instantateous centre of the wheel relative to the sprung mass I_{wb}, lies at the intersection of the upper and lower wishbones,

---

[10]For examples illustrating the theorem see (Shigley *et al.* (1980)).

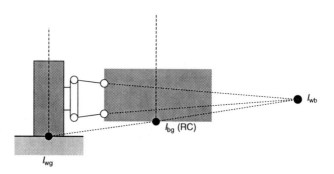

Figure 10.17 Roll centre determination for double wishbone suspension

while that of the wheel relative to the ground lies at I_{wg}. The instantaneous centre of the sprung mass relative to the ground (the roll centre) I_{bg}, must lie in the centre plane of the vehicle and on the line joining I_{wb} and I_{wg}, as shown in the diagram.

For a double wishbone suspension, I_{wb} can be varied by angling the upper and lower wishbones to different positions, thereby altering the load transfer between inner and outer wheels in a cornering manoeuvre. This gives the suspension designer some control over the handling capabilities of a vehicle. Figures 10.18 to 10.23 illustrate the locations of roll centres for a range of suspension types.

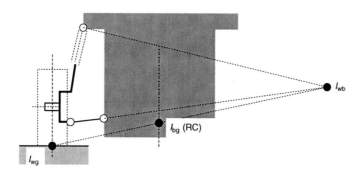

Figure 10.18 Roll centre location for MacPherson strut

In the case of the MacPherson strut suspension (Figure 10.18) the upper line defining I_{wb} is perpendicular to the strut axis. In the case of the trailing arm suspension (Figure 10.20) the trailing arm pivots about a transverse axis (forward of the wheel centre). In the front view (Figure 10.20(c)), the wheel is constrained to move in a vertical plane (with no transverse movement) and hence I_{wb} lies at infinity along the pivot axis (to the right). The roll centre therefore lies in the ground plane on centre-line of the vehicle. For the semi-trailing arm suspension (Figure 10.21) the pivot axis is inclined and intersects the vertical lateral plane through the wheel centre at I_{wb} a distance L from the centre plane of the wheel. The roll centre I_{bg} lies on the line connecting I_{wb} with the instantaneous centre of the wheel relative to the ground I_{wg}.

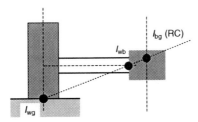

Figure 10.19 Roll centre location for swing axle suspension

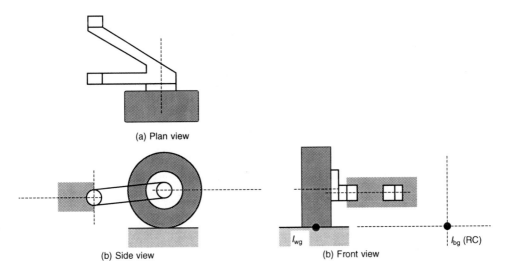

(a) Plan view

(b) Side view   (b) Front view

Figure 10.20 Roll centre location for trailing-arm suspension

Figure 10.22 shows the roll centre determination for a four link rigid axle suspension. In this case the wheels and axle can be assumed to move as a rigid body. The upper and lower control arms produce an instant centre at A and B respectively. Connecting these together produces a roll axis for the suspension. The intersection of this axis with the transverse wheel plane defines the roll centre.

Our final example illustrating roll centre location is the Hotchkiss rear suspension shown in Figure 10.23. The analysis in this case is somewhat different to the previous examples. Lateral forces are transmitted to the sprung mass at A and B. The roll centre height is at the intersection of the line joining these points and the vertical transverse plane through the wheel centres. The roll centre is of course at this height in the centre plane of the vehicle.

10.9 Force analysis

In this section simple methods of analysing the forces in suspension mechanisms resulting from vertical, lateral and longitudinal loading are introduced. The relationship between the vertical

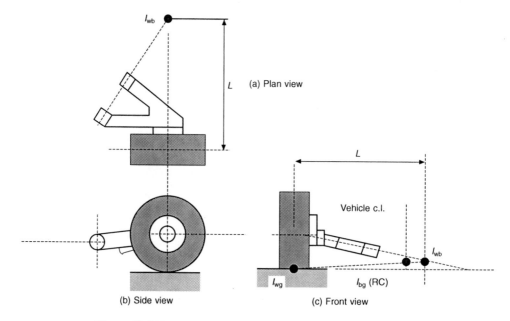

Figure 10.21 Roll centre location for semi-trailing arm suspension

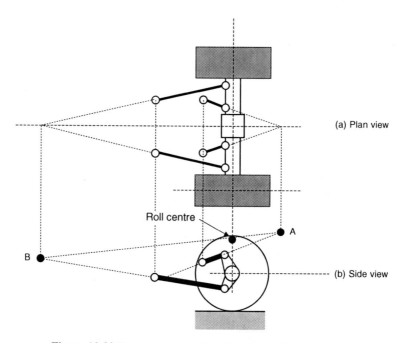

Figure 10.22 Roll centre for a four link rigid axle suspension

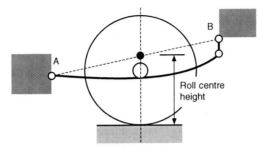

Figure 10.23 Roll centre location for a Hotchkiss suspension

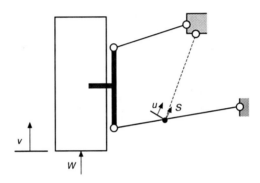

Figure 10.24 Notation for analysing spring and wheel rates in a double wishbone suspension

wheel loading and the spring forces is also discussed leading to the selection of suspension spring characteristics.

10.9.1 Relationship between spring and wheel rates

In general the relationship between spring deflections and wheel displacements in suspensions is non-linear, so that a desired wheel-rate (related to suspension natural frequency) has to be interpreted into a spring-rate. Consider the double wishbone suspension shown in Figure 10.24, where W and S are the wheel and spring forces respectively and v and u are the corresponding deflections.

Begin by defining the suspension ratio as: $R = \dfrac{S}{W}$ (10.2)

The spring stiffness is: $k_s = \dfrac{dS}{du} = d(RW) = R\dfrac{dW}{dv}\dfrac{dv}{du} + W\dfrac{dR}{dv}\dfrac{dv}{du}$ (10.3)

Using the principle of virtual work [Meriam *et al.* (1993)], $S\,du = W\,dv$ and hence equation 10.2 can be written

$$R = \frac{S}{W} = \frac{dv}{du}$$ (10.4)

Defining the wheel rate as:
$$k_w = \frac{dW}{dv}$$
(10.5)

Combining equations 10.3, 10.4 and 10.5 gives:

$$k_s = k_w R^2 + S\frac{dR}{dv}$$
(10.6)

Equations similar to 6 can be derived for other suspension geometries.

10.9.2 Wheel-rate for constant natural frequency with variable payload

The simplest model for representing vehicle ride is that of a single degree of freedom system (Section 10.13.4) in which the spring stiffness is that associated with wheel rate k_w and the mass m_s is a proportion of the total sprung mass. The undamped natural frequency is then:

$$\omega_n = \sqrt{\frac{k_w}{m_s}}$$
(10.7)

If k_w is maintained constant, the natural frequency decreases as the payload (and hence m_s) increases. It is possible to determine a variable wheel-rate which will ensure that the natural frequency remains constant as the sprung mass increases. Denoting the static displacement as:

$$\delta_s = \frac{m_s g}{k_w},$$
(10.8)

equation 10.7 can be written in terms of δ_s, i.e.

$$\omega_n = \sqrt{\frac{g}{\delta_s}}$$
(10.9)

From this it is seen that to maintain ω_n constant δ_s must be constant and hence the load/rate must be constant from equation 10.8, i.e.

$$\frac{W}{dW/dv} = \delta_s = \text{constant, or} \quad \frac{dW}{W} = \frac{dv}{\delta_s}$$

Integrating both sides gives:

$$\log_e W = \frac{v}{\delta_s} + c,$$
(10.10)

where c = constant. Assuming that the wheel load and suspension deflection at a nominal static load condition are $W = W_s$ and $v = v_s$ enables the c to be found, i.e.

$$c = \log_e W_s - \frac{v_s}{\delta_s}$$
(10.11)

Substituting into 10.10 and re-arranging gives:

$$W = W_s e^{\frac{v - v_s}{\delta_s}}$$
(10.12)

Equation 10.12 defines the required load-deflection relationship for tyre load as a function of tyre deflection v. The corresponding wheel-rate can be found by differentiating W with respect to v giving:

$$k_w = \frac{dW}{dv} = \frac{W_s}{\delta_s} e^{\frac{v-v_s}{\delta_s}} \tag{10.13}$$

Figure 10.25 shows typical graphs of wheel load and wheel rate as function wheel displacement for a natural frequency of 1.125 Hz. If the suspension ratio R and its derivative dR/dv are known as a function of wheel deflection, then the spring rate can be calculated, e.g. equation 10.6 can be employed for a double wishbone suspension and the numerical analysis of section 10.7 can be extended to provide R and dR/dv.

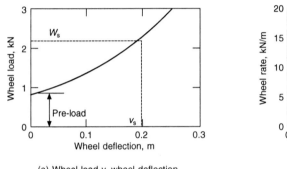
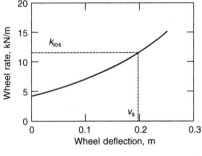

(a) Wheel load v. wheel deflection   (b) Wheel rate v. wheel deflection

Figure 10.25 Typical wheel load and wheel rate as functions of wheel displacement

10.9.3 Forces in suspension members

While computer packages are undoubtedly required for a comprehensive force analysis, some simple first estimates of loading of suspension members and chassis connection points can be carried out using graphical methods. In performing this analysis, it is assumed that the mass of the members is negligible compared to that of the applied loading. Friction and compliance at the joints are also assumed negligible and the spring or wheel rate needs to be known. Some basic principles of mechanics are employed in the analysis. In particular there is a need to be familiar with the use of freebody diagrams for determining internal forces in structures and the conditions for equilibrium of pin-jointed two- and three-force members. These conditions are summarized in Figures 10.26(a) and (b) respectively. In the case of three-force members, equilibrium requires the three forces to pass through a common point, i.e. be concurrent, and the vector sum of the forces must be zero. If one of the three forces is known the magnitudes of the other two can be found (graphically this involves drawing a triangle of forces).

(a) Vertical loading
As an example consider the double wishbone suspension shown in Figure 10.27. Assume F_W is the wheel load and F_S the force exerted by the spring on the suspension mechanism. Links

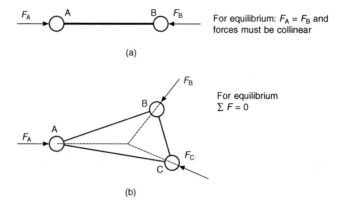

(a)

(b)

Figure 10.26 Equilibrium of two and three force members, (a) Requirements for equilibrium of two-force members (b) Requirements for equilibrium of a three-force member

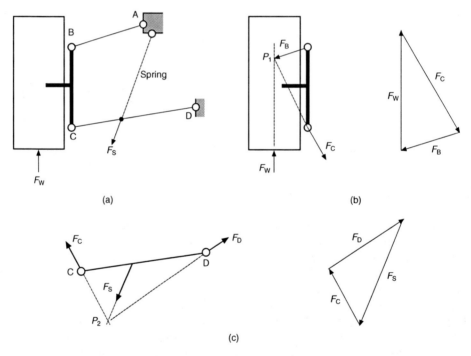

(a)

(b)

(c)

Figure 10.27 Force analysis of a double wishbone suspension (a) Diagram showing applied forces (b) FBD of wheel and triangle of forces (c) FBD of link CD and triangle of forces

AB and CD are respectively two-force and three force members. When the freebody diagram of the wheel and knuckle is considered (Figure 10.27(b)), the directions of F_W and F_B are known and together establish the point of concurrency P_1, for the three forces which act on the body. If the magnitude of F_W is known, the magnitudes of F_B and F_C can be determined from the triangle of forces. For the freebody diagram of link CD (Figure 10.27(c)), the point of concurrency is at P_2 and with F_C known, F_D and F_S can be found from the second triangle of forces. The corresponding chassis loadings comprise $F_A = F_B$, F_S and $F_D = F_C$. An analysis over the full suspension travel requires the graphical procedure to be repeated at suitable increments of suspension displacement. In order to define the applied loading at a given suspension position, it is necessary to know either the wheel or spring rate.

A similar analysis can be carried out for the MacPherson strut shown in Figure 10.28. In this case AB is a two force member and the point of concurrency of the forces F_W and F_B is at P. This means that the force F_C exerted on the strut at C acts through P. In analysing the forces exerted on the upper sliding part of the strut it is seen that the inclined force at C must be counteracted by a collinear spring force otherwise side forces and a bending moment act on the member. The solution is to set the axis of the spring coaxial with CP. This has the effect of reducing wear in the strut, but clearly bending effects are not completely eliminated for all suspension positions.

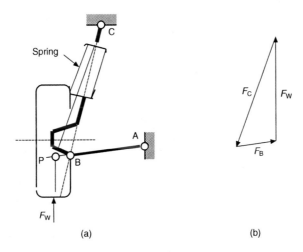

Figure 10.28 Force analysis of a MacPherson strut, (a) Wheel loading, (b) Forces acting on the strut

(b) Lateral and longitudinal loading
Lateral loading arises from cornering effects, while longitudinal loadings arise from braking, drag forces on the vehicle and shock loading due to the wheels striking bumps and pot-holes. The preceding principles can also be used to analyse suspensions for these loading conditions.

(c) Shock loading
Dynamic loading effects are very difficult to quantify, but experience has enabled a range of dynamic load factors to be established. These factors when multiplied be the static wheel loads give reasonable approximations for peak dynamic loads encountered by motor vehicles. Some typical values used by one manufacturer are given in Table 10.1.

By estimating the frequency of these occurrences over the life-time of a vehicle, it is possible to investigate possible modes of failure.

Table 10.1 Dynamic load factors

Load case	Load factor		
	Longitudinal	Transverse	Vertical
Front/rear pothole bump	3 g, at the wheel affected	0	4 g, at the wheel affected, 1 g at other wheels
Bump during cornering	0	0	3.5 g at wheel affected, 1 g at other wheels
Lateral kerb strike	0	4 g front and rear wheels on side affected	1 g at all wheels
Panic braking	2 g front wheels 0.4 g rear wheels	0	2 g front wheels, 0.8 g rear wheels

10.10 Anti-squat/anti-dive geometries

During braking and acceleration there is a load transfer between front and rear wheels and the attitude of the sprung mass tends to change. When viewed from the side during braking there is a tendency for the sprung mass to dive (nose down) and during acceleration the reverse occurs, with the nose lifting and the rear end squatting. Since the load transfers occur through the suspension, it is possible to design the suspension mechanism to counteract this behaviour.

The same general principles apply to squat and dive analysis of the various drive-shaft and braking combinations which can arise. In each case the analysis requires an understanding of the forces acting. D'Alembert's principle (Meriam *et al.*, 1993) can be used to convert the dynamics problem into a statics one, thereby simplifying the solution. Space limitations restrict our attention to the case of a four-wheel drive vehicle having outboard brakes.

10.10.1 Determination of anti-dive geometry – outboard brakes

Consider the freebody diagram of a vehicle during braking as shown in Figure 10.29. The

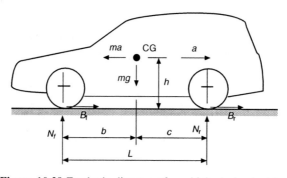

Figure 10.29 Freebody diagram of a vehicle during braking

D'Alembert force (a pseudo-force sometimes called the *inertia force*) ma, tends to oppose the deceleration. The forces at each pair of wheels comprise normal and braking components. Assume that there is a fixed braking ratio k, between front and rear braking forces:

$$k = \frac{B_f}{B_f + B_r} \tag{10.14}$$

Under braking conditions the vertical loads on the axles differ from the static values. Take moments about the rear tyre contact point giving:

$$N_f L - mah - mgc = 0$$

Re-arranging gives:

$$N_f = \frac{mgc}{L} + \frac{mah}{L} \tag{10.15}$$

The first term on the right hand side is the static load and the second term is the increase in load, i.e. load transfer, due to braking. The corresponding vertical force at the rear is

$$N_r = \frac{mgb}{L} - \frac{mah}{L} \tag{10.16}$$

The overall effect is an increase in load at the front and a decrease at the rear producing a tendency for dive. Consider now the front suspension with inclined links such that the wheel effectively pivots about O_f in the side view (Figure 10.30). The suspension spring force S_f, may be expressed as the static load S_f plus a perturbation dS_f, due to braking, i.e. $S_f = S_f + \delta S_f$ where $S_f = \frac{mgc}{L}$

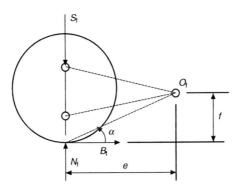

Figure 10.30 Front wheel forces and effective pivot location

Under static load conditions ($a = 0$), the spring load is $S_f = \frac{mgc}{L}$. Taking moments about O_f produces: $N_f e - S_f e - B_f f = 0$. Substituting for N_f and S_f and setting $dS_f = 0$ for zero dive gives:

$$\frac{mahe}{L} - B_f f = 0 \qquad (10.17)$$

But $B_f = mak$. Substituting this into equation 10.17 and re-arranging gives

$$\frac{f}{e} = \frac{h}{kL} = \tan \alpha \qquad (10.18)$$

If O_f lies anywhere on the line defined by equation 10.18, the condition for zero deflection at the front suspension is satisfied. If O_f lies below this line, i.e. on a line inclined at an angle α' to the horizontal, then the percentage anti-dive is defined as:

$$\left(\frac{\tan \alpha'}{\tan \alpha} \right) \times 100\% \qquad (10.19)$$

A similar analysis for a rear suspension having the geometry shown in Figure 10.31 leads to an additional equation:

$$\frac{f}{e} = \frac{h}{L(1 - k)} = \tan \beta \qquad (10.20)$$

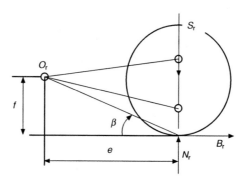

Figure 10.31 Rear wheel forces and effective pivot location

If O_r lies on the line defined by equation 10.20 there is no tendency for the rear of the sprung mass to lift during braking.

It follows that for 100% anti-dive, the effective pivot points for front and rear suspensions must lie on the locus defined by equations 10.18 and 10.20 (as shown in Figure 10.32). If the pivots lie below the locus less than 100% anti-dive will be obtained. In practice anti-dive rarely exceeds 50% for the following reasons:

- Subjectively zero pitch braking is undesirable;
- There needs to be a compromise between full anti-dive and anti-squat conditions;
- Full anti-dive can cause large castor angle changes (because all the braking torque is reacted through the suspension links) resulting in heavy steering during braking.

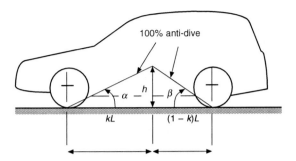

Figure 10.32 Suspension pivot locii for 100% anti-dive during braking (outboard brakes)

10.10.2 Determination of anti-pitch geometry – acceleration

The analysis for anti-squat suspensions is similar to that for anti-dive, except now the direction of the D'Alembert force is reversed. Furthermore the braking forces are replaced by tractive forces (opposite in direction) which may be applied to either front or rear wheels or both for the case of four-wheel drive. It should be noted that anti-pitch geometry can only be applied to the suspension at which the drive is applied.

Consider the case of a four-wheel drive vehicle with independent suspension. Assume the vehicle geometry shown in Figure 10.29. The forces at front and rear wheels are as shown in Figure 10.33. Note that the drive torque is reacted at the powertrain, producing a drive torque on the half-shafts and hence the FBDs of the wheels. Assume the tractive effort is split in the ratio

$$\lambda = \frac{T_f}{T_f + T_r} \tag{10.21}$$

Taking moments about O_f (Figure 10.33(a))

$$T_f f_f + N_f e_f - S_f e_f - M_f = 0 \tag{10.22}$$

where $T_f = \lambda ma$

$$N_f = \frac{mgc}{L} - \frac{mah}{L}$$

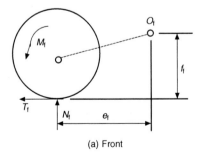

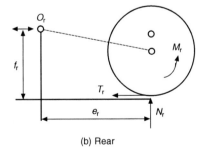

(a) Front   (b) Rear

Figure 10.33 Wheel forces and effective pivot locations

$$M_f = T_f r = \lambda mar$$

It follows that the change in the front spring force is:

$$\delta S_f = ma \left[\frac{\lambda(f_f - r)}{e_f} - \frac{h}{L} \right] = k_f \delta_f \tag{10.23}$$

where k_f = front suspension stiffness.

A similar analysis is used for the rear suspension (Figure 10.33(b)). Taking moments about O_r produces

$$T_r f_r - N_r e_r + S_r e_r - M_r = 0 \tag{10.24}$$

where $T_r = (1 - \lambda)\, ma$

$$N_r = \frac{mgb}{L} + \frac{mah}{L}$$

$$M_r = T_r r = (1 - \lambda)\, mar$$

The change in rear spring force is

$$\delta S_r = ma \left[-\frac{(1 - \lambda)(f_r - r)}{e_r} + \frac{h}{L} \right] = k_r \delta_r \tag{10.25}$$

where k_r = rear suspension stiffness.
The pitch angle is

$$\theta = \frac{\delta_r - \delta_f}{L} \tag{10.26}$$

positive clockwise. Substituting from equations 10.23 and 10.25 gives:

$$\theta = \frac{ma}{L} \left[-\frac{\lambda(f_f - r)}{e_f k_f} + \frac{h}{Lk_f} - \frac{(1 - \lambda)(f_r - r)}{e_r k_r} + \frac{h}{Lk_r} \right]$$

Zero pitch occurs when $\theta = 0$, i.e. when the term in square brackets is zero. This indicates that the anti-squat and anti-pitch performance depends on the following vehicle properties – suspension geometry, suspension stiffnesses (front and rear) and tractive force distribution.

For a solid axle the drive torque is reacted within the wheel assembly, i.e. it is an internal moment as far as the freebody is concerned. In this case $M = 0$ and equation 10.26 can be modified by setting $r = 0$ for the appropriate solid axle(s).

10.11 Lateral load transfer during cornering

During cornering, centrifugal (inertia) forces act horizontally on the sprung and unsprung masses. These forces act above the ground plane through the respective mass centres causing moments to be generated on the respective masses. These in turn lead to changes in vertical loads at the tyres which affect vehicle handling and stability. In general the vertical loads on the outer wheels increase while those on the inner wheels decrease.

The process of converting the transverse forces into vertical load changes is termed *lateral load transfer*. Figure 10.34 shows a two-axled vehicle undergoing steady-state cornering. The lateral acceleration is assumed to be perpendicular to the vehicle centre-line. This neglects any change in attitude angle the vehicle body has to its direction of travel due to steering, slip angles at the tyres, roll and compliance steer.

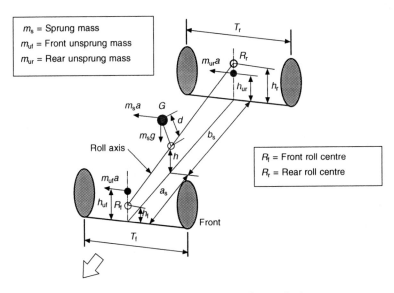

Figure 10.34 Steady-state cornering analysis

Notation and assumptions in the analysis are:

- G is the sprung mass centre of gravity;
- The transverse acceleration at G due to cornering is 'a';
- The sprung mass rolls through the angle ϕ about the roll axis;
- The centrifugal (inertia) force on the sprung mass $m_s a$ acts horizontally through G;
- The gravity force on the sprung mass $m_s g$ acts vertically downwards through G;
- The inertia forces $m_{uf} a$ and $m_{ur} a$ act directly on the unsprung masses at the front and rear axles. Each transfers load only between its own pair of wheels.

The analysis is split into four steps:

1. Load transfer due to the roll moment

Replace the two forces at G with the same force at A plus a moment (the roll moment) M_s about the roll axis, i.e.

$$M_s = m_s ad \cos \phi + m_s gd \sin \phi \approx m_s\, ad + m_s g\phi \tag{10.27}$$

where ϕ is treated as a small angle. M_s is reacted by a roll moment $M\phi$ (at the suspension springs and anti-roll bars) and distributed to the front and rear suspensions. The relationship between $M\phi$ and ϕ is assumed to be linear for small angles of roll, i.e.

$$M_\phi = k_s\phi \tag{10.28}$$

where k_s = total roll stiffness.
From equations 10.27 and 10.28 one obtains

$$\phi = \frac{m_s a d}{k_s - m_s g d} \tag{10.29}$$

M_ϕ can be split into components $M_{\phi f}$ and $M_{\phi r}$ at the front and rear axles such that

$$M_\phi = M_{\phi f} + M_{\phi r} = k_{sf}\phi + k_{sr}\phi \tag{10.30}$$

where k_{sf} and k_{sr} are the roll stiffness components at front and rear axles ($k_{sf} + k_{sr} = k_s$). The front load transfer due to the roll moment is then:

$$F_{fsM} = \frac{k_{sf}\phi}{T_f} = \frac{k_{sf} m_s a d}{T_f (k_{sf} + k_{sr} - m_s g d)} \tag{10.31}$$

Similarly, the rear load transfer due to roll moment is:

$$F_{rsM} = \frac{k_{sr}\phi}{T_r} = \frac{k_{sr} m_s a d}{T_r (k_{sf} + k_{sr} - m_s g d)} \tag{10.32}$$

T_f and T_r are the front and rear track widths of the vehicle.

2. Load transfer due to sprung mass inertia force

The sprung mass is distributed to the roll centres at front and rear axles (see the brief discussion on dynamically equivalent bodies in Chapter 8). The respective masses at front and rear are:

$$m_{sf} = \frac{m_s b_s}{L} \text{ and } m_{sr} = \frac{m_s a_s}{L} \tag{10.33}$$

The centrifugal force at A is distributed to the respective roll centres at the front and rear axles as follows:

$$F_{fs} = m_{sf}a \text{ and } F_{rs} = m_{sr} = m_{sr}a \tag{10.34}$$

and the corresponding load transfers are:

$$F_{fsF} = \frac{m_{sf} a h_f}{T_f} \text{ and } F_{rsF} = \frac{m_{sr} a h_r}{T_r} \tag{10.35}$$

3. Load transfer due to the unsprung mass inertia forces

The respective load transfers at the front and rear axles due to the unsprung mass inertia forces are:

$$F_{\text{fuF}} = \frac{m_{\text{uf}}\,ah_{\text{uf}}}{T_{\text{f}}} \text{ and } F_{\text{ruF}} = \frac{m_{\text{ur}}\,ah_{\text{ur}}}{T_{\text{r}}} \tag{10.36}$$

4. Determine the total load transfer

Combine the load transfers due to roll moment with those due to inertia forces on the sprung and unsprung masses using equations 10.31, 10.32, 10.35 and 10.36, i.e. the load transfers for front and rear wheels are:

$$F_{\text{f}} = F_{\text{fsM}} + F_{\text{fsF}} + F_{\text{fuF}} \tag{10.37}$$

for the front wheels and

$$F_{\text{r}} = F_{\text{rsM}} + F_{\text{rsF}} + F_{\text{ruF}} \tag{10.38}$$

for the rear wheels.

10.12 Suspension components

So far attention has been concentrated on suspension mechanisms, their kinematics and geometric requirements. In this section the other essential components of suspension systems will be discussed. These include springs and dampers, both of which have a profound effect on ride and handling performance.

In addition to the constraints imposed by suspension performance requirements, designers of these components have a range of other constraints to consider. These include cost, packaging, durability and maintenance. Because of the hostile environment in which suspension components operate and the high fluctuating loads (and hence stresses) involved, fatigue life is one of the designer's prime concerns.

10.12.1 Springs – types and characteristics

Suspension systems require a variety of compliances to ensure good ride, handling and NVH performance. The need for compliance between the unsprung and sprung masses to provide good vibration isolation has long been recognized. In essence, a suspension spring fitted between the wheel and body of a vehicle allows the wheel to move up and down with the road surface undulations without causing similar movements of the body. For good isolation of the body (and hence good ride), the springs should be as soft as possible consistent with providing uniform tyre loading to ensure satisfactory handling performance. The relatively soft springing required for ride requirements is normally inadequate for resisting body roll in cornering, therefore it is usual for a suspension system to also include additional roll stiffening in the form of anti-roll bars. Furthermore, there is the possibility of the suspension hitting its stops at the limits of its travel as a result of abnormal ground inputs (e.g. as a result of striking a pothole). It is then necessary to ensure that the minimum of shock loading is transmitted to the sprung mass. This requires the use of additional springs in the form of bump stops to decelerate the suspension at its limits of travel. Finally, there is also a requirement to prevent the transmission of high frequency vibration (>20 Hz) from the road surface, via the suspension to connection points on the chassis. This is achieved by using rubber bush connections between suspension members.

It follows therefore that the compliant elements required in suspension systems are: suspension springs, anti-roll bars, bump-stops and rubber bushes. In this section our attention is confined to the types and characteristics of suspension springs and the operation of anti-roll bars. Further details of suspension components can be found in Bastow *et al.* (1993), Dixon (1996), Gillespie (1992), Milliken *et al.* (1995), Heisler (1989), Hillier (1991) and Reimpell *et al.* (1998).

10.12.2 Suspension springs

The main types of suspension spring are:

- Steel springs (leaf springs, coil springs and torsion bars);
- Hydropneumatic springs.

(a) Steel springs

(i) Leaf springs
Sometimes called semi-elliptic springs, these have been used since the earliest developments in motor vehicles. They rely on beam bending principles to provide their compliance. They are a simple and robust form of suspension spring still widely used in heavier duty applications such as lorries and vans. In some suspensions (e.g. the Hotchkiss type) they are used to provide both vertical compliance and lateral constraint for the wheel travel. Size and weight are among their disadvantages.

Leaf springs can be of single or multi-leaf construction. In the latter case (Figure 10.35(b)) interleaf friction (which can affect their performance) can be reduced with the use of interleaf plastic inserts. Rebound clips are used to bind the leaves together during rebound motion. The change in length of the spring produced by bump loading is accommodated by the swinging shackle. The main leaf of the spring is formed at each end into an eye shape and attached to the sprung mass via rubber bushes. Suspension travel is limited by a rubber bump stop attached to the central rebound clip. Structurally, leaf springs are designed to produce constant stress along their length when loaded.

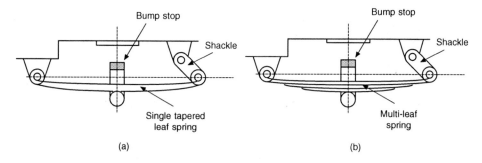

Figure 10.35 Examples of leaf spring designs, (a) single leaf type, (b) multi-leaf type

Spring loading can be determined by considering the forces acting on the spring and shackle as a result of wheel loading (Figure 10.36). The spring is a three force member with F_A, F_W and F_C at A, B and C respectively. The wheel load F_W, is vertical and the direction of F_C is parallel to the shackle (a two-force member). The direction of F_A must pass through the intersection of the forces F_W and F_C (point P) for the link to be in equilibrium. Knowing the magnitude of the wheel load enables the other two forces to be determined. The stiffness (rate) of the spring is determined by the number, length, width and thickness of the leaves. See Anon (1996) and Fenton (1996) for stiffness and stress formulae. Angling of the shackle link can be used to give a variable rate (Milliken *et al.* (1995)). When the angle $\theta < 90°$ (Figure 10.36), the spring rate will increase (i.e. have a rising rate) with bump loading.

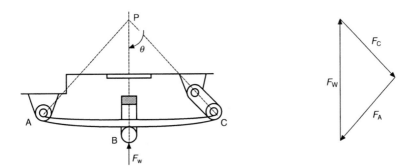

Figure 10.36 Leaf spring loadings, (a) wheel load, (b) force on the spring member

(ii) Coil springs

This type of spring provides a light and compact form of compliance which are important features in terms of weight and packaging constraints. It requires little maintenance and provides the opportunity for co-axial mounting with a damper. Its disadvantages are that because of low levels of structural damping, there is a possibility of surging (resonance along the length of coils) and the spring as a whole does not provide any lateral support for guiding the wheel motion.

Most suspension coil springs are of the open coil variety. Which means that the coil cross-sections are subjected to a combination of torsion, bending and shear loadings. Spring rate is related to the wire and coil diameters, the number of coils and the shear modulus of the spring material. Cylindrical springs with a uniform pitch produce a linear rate. Variable rate springs are produced either by varying the coil diameter and/or pitch of the coils along its length. In the case of variable pitch springs, the coils are designed to 'bottom-out' as the spring is loaded, thereby increasing stiffness with load.

Coil spring design is well covered in the literature. In addition to Wahl's classic text (Wahl, 1963), the reader is also recommended to consult Anon (1996) and chapters in Shigley *et al.* (1983), Mott (1985) and Milliken *et al.* (1995).

(iii) Torsion bars

This is a very simple form of spring and consequently very cheap to manufacture. It is both wear and maintenance free. Despite its simplicity it cannot easily be adopted for some of the more popular forms of suspension.

The principle of operation (Figure 10.37) is to convert the applied load F_W into a torque $F_W \times R$ producing twist in the bar. A circular cross-section bar gives the lowest spring weight for a given stiffness. In this case simple torsion of shafts theory can be used to determine the stiffness of the spring and the stresses in it. As the lever-arm rotates under load, the moment-arm changes somewhat, requiring twist angle corrections (for large rotations) in design calculations (Shigley *et al.*, 1983 and Milliken *et al.*, 1995). In general, stiffness is related to diameter and length of the torsion bar and the torsion modulus of the material. Some bending (a moment $F_W \times L$) is induced in the twist section of the member and supports should be included to minimize this.

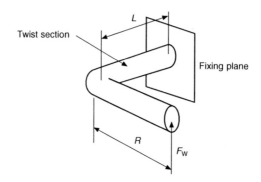

Figure 10.37 Principle of operation of a torsion bar spring

(b) Hydropneumatic springs
In this case the spring is produced by a constant mass of gas (typically nitrogen) in a variable volume enclosure. The principle of operation of a basic diaphragm accumulator spring is shown in Figure 10.38. As the wheel deflects in bump, the piston moves upwards transmitting the motion to the fluid and compressing the gas via the flexible diaphragm. The gas pressure increases as its volume decreases to produce a hardening spring characteristic.

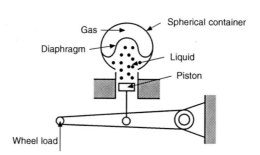

Figure 10.38 Principles of a hydropneumatic suspension spring

The principle was exploited in the Moulton–Dunlop hydrogas suspension where damping was incorporated in the hydropneumatic units. Front and rear units were connected to give pitch control. A detailed description of this system can be found in Heisler (1989).

A further development of the principle has been the Citroen system which incorporates a hydraulic pump to supply pressurized fluid to four hydropneumatic struts (one at each wheel-station). Height correction of the vehicle body is accomplished by regulator valves adjusted by roll-bar movement or manual adjustment by the driver. A detailed description of this system and how it operates can be found in Heisler (1989).

In general, hydropneumatic systems are complex (and expensive) and maintenance can also be a problem in the long term. Their cost can, however, be off-set by good performance. The two systems discussed in the preceding paragraphs are covered by patents, but there is still scope for development of alternative hydropneumatic systems. Some of these are incorporated into controllable suspensions discussed in Section 10.14.

10.12.3 Anti-roll bars (stabilizer)

These are used to reduce body roll and have an influence on a vehicle's cornering characteristics (in terms of understeer and oversteer). Figure 10.39(a) shows how a typical roll bar is connected to a pair of wheels. The ends of the U-shaped bar are connected to the wheel supports and the central length of the bar is attached to the body of the vehicle. Attachment points need to be selected to ensure that bar is subjected to torsional loading without bending. If one of the wheels is lifted relative to the other, half the total anti-roll stiffness acts downwards on the wheel and the reaction on the vehicle body tends to resist body roll. If both wheels lift by the same amount the bar is not twisted and there is no transfer of load to the vehicle body. If the displacements of the wheels are mutually opposed (one wheel up and the other down by the same amount), the full effect of the anti-roll stiffness is produced. The total roll stiffness k_{rs} is equal to the sum of the roll-stiffness produced by the suspension springs $k_{r,sus}$ and the roll stiffness of the anti-roll bars $k_{r,ar}$, i.e. these springs are effectively in parallel. Typical contributions to total roll stiffness as a function of body roll are shown in Figure 10.39(b).

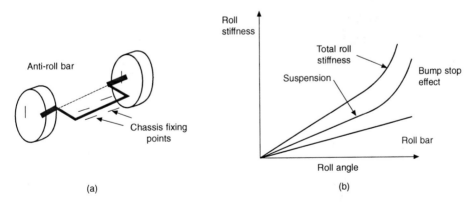

(a) (b)

Figure 10.39 Anti-roll bar geometry and the effect on roll stiffness, (a) anti-roll bar layout, (b) roll bar contribution to total roll stiffness

10.12.4 Dampers – types and characteristics

Frequently called shock absorbers, dampers are the main energy dissipators in a vehicle suspension. They are required to dampen vibration after a wheel strikes a pot-hole and to provide a good compromise between low sprung mass acceleration (related to ride) and adequate control of the unsprung mass to provide good road holding.

Suspension dampers are telescopic devices containing hydraulic fluid. They are connected between the sprung and unsprung masses and produce a damping force which is related to the relative velocity across their ends. The features of the two most common types of damper are shown in Figure 10.40. Figure 10.40(a) shows a dual tube damper in which the inner tube is the working cylinder and the outer cylinder is used as a fluid reservoir. The latter is necessary to store the surplus fluid which results from the difference in volumes on either side of the piston (as a result of the variable rod volume). In the monotube damper (Figure 10.40(b)) the surplus fluid is accommodated by a gas-pressurized free piston. An alternative form of monotube damper (not shown in Figure 10.40), uses a gas/liquid mixture as the working fluid to absorb the volume differences. Comparing the two types of damper shown in Figure 10.40, the dual tube design offers better protection against stones thrown up by the wheels and is also a shorter unit making it easier to package. On the other hand, the monotube strut dissipates heat more readily.

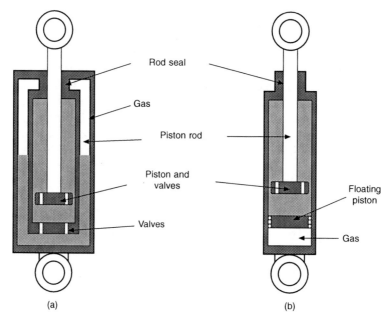

Figure 10.40 Damper types, (a) dual tube damper, (b) free-piston monotube damper

In dealing with road surface undulations in the bump direction (damper being compressed) relatively low levels of damping are required when compared with the rebound motion (damper being extended). This is because the damping force produced in bump tends to aid the acceleration of the sprung mass, while in rebound an increased level of damping is required to dissipate the energy stored in the suspension spring. These requirements lead to damper characteristics which are asymmetrical when plotted on force-velocity axes. The damping rate (coefficient) is the slope of the characteristic. Ratios of 3:1 for rebound to bump are quite common (Figure 10.41). However, one manufacturer has discarded this reasoning and uses linear dampers on one of its vehicles (Bosworth, 1996).

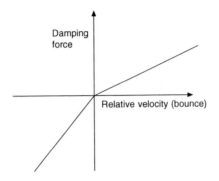

Figure 10.41 Non-linear damper characteristics

The characteristics called for in damper designs are achieved by a combination of orifice flow and flows through spring-loaded one-way valves. These provide a lot of scope for shaping and fine tuning of damper characteristics[11]. Figure 10.42 illustrates how the force-velocity characteristics can be shaped by the combined use of damper valves. At low relative velocities damping is by orifice control until the fluid pressure is sufficient to open the pre-loaded flow control valves. Hence the shape of the combined characteristic.

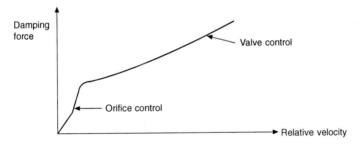

Figure 10.42 Shaping of damper characteristics

A driver operated adjustment mechanism can be used to obtain several damping characteristics from one unit. Typical curves for a three position adjustable damper are shown in Figure 10.43.

---

[11]For a more detailed account of damper operation and characteristics see Bastow *et al.* (1993).

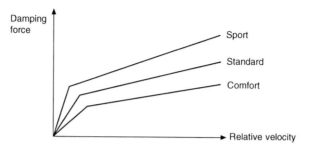

Figure 10.43 Operating modes for adjustable dampers

A continuous electronically-controlled adjustment forms the basis of one type of controllable suspension designed to improve both ride and handling.

10.13 Vehicle ride analysis

Ride comfort is one of the most important characteristics defining the quality of a vehicle. It is principally (although not exclusively) related to vehicle body vibration, the dominant source of which is due to road surface irregularities. In order to design vehicles which have good ride properties it is essential to be able to model ride performance in the early stages of vehicle development. This requires a basic understanding of road surface characteristics, human response to vibration and vehicle modelling principles. These are introduced briefly in this section.

10.13.1 Road surface roughness and vehicle excitation

In general, road surfaces have random profiles, which means that when they are traversed by a vehicle, the vertical height of the surface (above a reference plane) cannot be predicted in advance. For this reason they are described as non-deterministic. There are, however, certain properties of random functions which can be described statistically, e.g. the mean and mean square value can be determined by averaging and frequency content can be determined by methods based on the Fourier transform. Road surfaces can be considered as being made up from a large (theoretically infinite) number of sinusoidal profiles of different wavelength and amplitude. It can be shown (Wong, 1993) that the frequency characteristics are described by the power spectral density $S(n)$ of the height variations as a function of the spatial frequency n. The units of n are cycles/m and those of S are m³/cycle. From large amounts of measured road data it has been established (Dodds *et al.*, 1973) that S and n are related and can be approximated by:

$$S(n) = \kappa n^{-2.5} \tag{10.39}$$

where k = the roughness coefficient. For a vehicle traversing a road surface at a velocity V m/s, the spatial random profile is converted into a random time-varying input to the wheels of the vehicle. It may be shown (Wong, 1993) that its spectral density is given by

$$S(f) = \frac{S(n)}{V} \tag{10.40}$$

where $n = f / V$. Hence

$$S(f) = \kappa V^{1.5} f^{-2.5} \tag{10.41}$$

where the units for $S(f)$ are m²/Hz. Typical values of κ for a motorway, a principal road and a minor road are $0.25 \cdot (10)^{-6}$, $4 \cdot (10)^{-6}$ and $15 \cdot (10)^{-6}$ m²/(cycle/m)$^{1.5}$ respectively for $0.01 < n < 10$ cycle/m (Dodds *et al.*, 1973). The variation of $S(f)$ for a vehicle traversing a poor minor road at 20 m/s is shown in Figure 10.44.

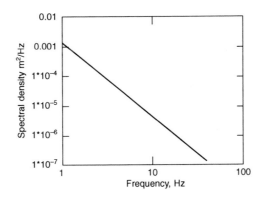

Figure 10.44 Spectral density of a road input as a function of vehicle speed (poor minor road)

10.13.2 Human perception of ride

Ride is related to the level of comfort perceived by a person travelling in a moving vehicle. This is influenced mainly by the levels and frequencies of vibration of the vehicle's passenger compartment in relation to human body sensitivity to vibration. In order to understand the human perception of ride it is therefore necessary to examine this sensitivity to vibration. One of the most important aspects of this is how the human body responds as a whole when subjected to vibration.

10.13.3 Human response to whole body vibration

The human body is a highly complex physical and biological system which differs somewhat from person to person. From a vibration point of view it can be considered to be a complex assemblage of linear and non-linear elements which result in a range of body resonances from approximately 1 to 900 Hz. From the ride point of view we are concerned with whole-body vibration of a seated person and in this case the most important of these resonances occurs in the frequency ranges 1–2 Hz (head–neck) and 4–8 Hz (thorax–abdomen).

It is generally agreed (e.g. Butkunas, 1967) that for the average passenger car, the perception of vibrational motions diminishes above 25 Hz and merges with the perception of audible sound. This dual perception (vibration and sound) which persists up to several hundred Hz is related to the term harshness. An additional factor in assessing human response to whole body vibration is that of motion sickness (kinetosis). The symptoms are well known and arise from low frequency vibration (<1Hz) of the type most commonly experienced on ships, but which can also be important in vehicle ride evaluations.

In general the tolerance to whole-body vibration decreases with time, however in the case of low frequency vibration related to motion sickness there is evidence (Reason, 1974) that tolerance increases with time as the spatial senses adapt to the conditions.

Guidance on exposure limits for whole-body vibration are given in the International Standard ISO 2631 (ISO, 1978) and the equivalent British Standard BS 6841 (BSI, 1987). These standards relate to whole-body vibration from a supporting surface to either the feet of a standing person or the buttocks of a seated person. The latter being of relevance in vehicle ride assessment. The standards specify criteria for health (exposure limit), working efficiency (fatigue-decreased proficiency boundary) and comfort (reduced-comfort boundary (RCB)), the latter being of particular relevance to ride assessment. The criteria are specified in terms of

- the direction of vibration input to the human torso;
- the acceleration magnitude;
- the frequency of excitation; and
- the exposure duration.

Anatomical sets of axes are defined to relate the direction of vibrational input to the body. For a seated person these are shown in Figure 10.45. The general form of the RCBs for different exposure times are shown in Figure 10.46(a) and (b). These indicate that the most sensitive frequency range for vertical vibration is from 4–8 Hz corresponding to the thorax–abdomen resonance, while the most sensitive range for transverse vibration is from 1 to 2 Hz corresponding to head–neck resonance. ISO 2631 also presents severe discomfort boundaries in the range from 0.1 to 0.63 Hz to account for motion sickness. In this case the most sensitive range is from 0.1 to 0.315 Hz.

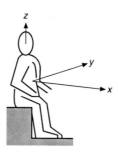

Figure 10.45 Anatomical axes for vibration imparted to humans

The standard is applicable for periodic, random and transient translational vibration. In the case of broad-band excitation the standards recommend that either

(a) the r.m.s. value of the acceleration in each third octave band is compared with the appropriate limit at the centre frequency of each band, or

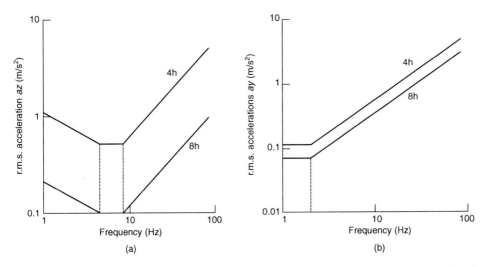

Figure 10.46 Whole-body RCB vibration criteria, (a) RCB for vertical (*z*-axis) vibration (b) RCB for lateral (*x* and *y* axis vibration)

(b) the overall acceleration signal for the 1 to 80 Hz range may be frequency-weighted according to the curves of Figure 10.46(a) and (b) to give weighted r.m.s. a_z, and a_x/a_y values respectively.

The overall weighted values are then compared with the permissible values in the 4 to 8 Hz range for a_z and in the 1 to 2 Hz range for a_x/a_y.

10.13.4 Analysis of vehicle response to road excitation

Ride performance is assessed at the design stage by simulation of vehicle response to road excitation. This requires the development of a vehicle model and analysis of its response. In assessing ride performance, the response is weighted in accordance with the ISO criteria of the previous section to account for human response to vibration.

Vehicle models
Models of varying complexity are used in analysing ride. For a passenger car, the most comprehensive of these has seven degrees of freedom (Figure 10.47). These comprise three degrees of freedom for the vehicle body (pitch, bounce and roll) and a further vertical degree of freedom at each of the four unsprung masses. This model allows the pitch, bounce and roll performance of the vehicle to be studied.

The suspension stiffness and damping rates are derived from the individual spring and damping units using the approach discussed in Section 10.9. Various tyre models have been proposed (Bohm *et al.*, 1987). The simplest of these uses a point-contact model to represent the elasticity and damping in the tyre with a simple spring and viscous damper. Since tyre damping is several orders of magnitude less than suspension damping, it has little impact on ride performance and is usually neglected.

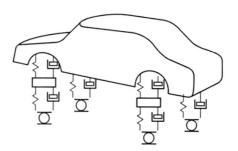

Figure 10.47 Full vehicle model

Much useful information can be derived from simpler vehicle models. The two most often used for passenger cars are the half-vehicle model (Figure 10.48(a)) and the quarter vehicle model (Figure 10.48(b)). These have four and two degrees of freedom respectively. Because of the reduced number of degrees of freedom certain information is unobtainable from these models. In the case of the half vehicle model, roll information is lost and for the quarter vehicle model pitch information is also lost.

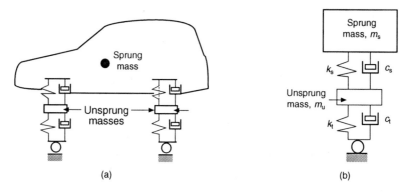

(a)   (b)

Figure 10.48 Half and quarter vehicle models, (a) half vehicle model, (b) quarter vehicle model

10.13.5 Response to road excitation

Pitch and bounce characteristics
These can be investigated by considering the free vibration of a simplified form of the half vehicle model (neglecting the unsprung masses). The effective stiffness at each wheel-station is obtained by replacing the suspension stiffness and tyre stiffness by an equivalent stiffness. Since these stiffnesses act in series so the equivalent stiffness k is determined from:

$$k = \frac{k_s k_t}{k_s + k_t} \qquad (10.42)$$

These simplifications do not have a serious affect on the vibration modes and their frequencies.

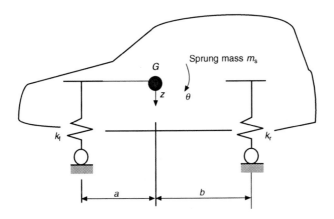

Figure 10.49 Notation for pitch–bounce analysis

Consider the simplified model shown in Figure 10.49 noting that the generalized coordinates are z and θ. Using the notation shown and drawing the free-body diagram, it can be shown that the equations of motion are:

$$m_s\ddot{z} + (k_r + k_f)z + (k_r b - k_f a)\theta = 0$$

$$\text{(10.43), (10.44)}$$

$$m_s r_y^2\ddot{\theta} + (k_r b - k_f a)z + (k_f a^2 + k_r b^2)\theta = 0$$

The stiffnesses k_f and k_r are determined from equation 10.42 for front and rear ends of the vehicle, while r_y is the radius of gyration of the sprung mass about the transverse axis through its centre of gravity G. Letting

$$A = \frac{k_r + k_f}{m_s}, B = \frac{k_r b - k_f a}{m_s} \quad \text{and} \quad C = \frac{1}{m_s r_y^2}(k_f a^2 + k_r b^2),$$

equations 10.43 and 10.44 become:

$$\ddot{z} + Az + B\theta = 0$$

$$\text{(10.45), (10.46)}$$

$$\ddot{\theta} + \frac{B}{r_y^2}z + C\theta = 0$$

Equations 10.45 and 10.46 are uncoupled if $B = 0$, i.e. if $k_r b = k_f a$. Then pitch and bounce motions are uncoupled. In this case the natural frequencies for pitch and bounce are: $\omega_{n,\text{bounce}} = \sqrt{A}$ and $\omega_{n,\text{pitch}} = \sqrt{C}$ respectively and when the wheels strike a bump on the road surface only pitching motion will tend to be excited resulting in a poor ride[12]. Generally it is desirable to have coupled bounce and pitch motions for vehicle applications.

---

[12]Pitching motion is less acceptable than bounce motion from the human response point of view.

In this case it is necessary to solve equations 10.45 and 10.46 simultaneously. Using the method outlined in Chapter 8, it can be shown (Wong, 1993) that two natural frequencies are given by:

$$\omega_{n1}, \omega_{n2} = \frac{1}{2}(A + C) \mp \sqrt{\frac{1}{4}(A - C)^2 + \frac{B^2}{r_y^2}} \tag{10.47}$$

and the oscillation centres O_1 and O_2 corresponding to the two frequencies are located at distances

$$l_1 = \frac{B}{\omega_{n1}^2 - A} \tag{10.48}$$

and

$$l_2 = \frac{B}{\omega_{n2}^2 - A} \tag{10.49}$$

respectively from G. The first of these values will be negative and the other positive. According to our sign convention for displacements, the one having the negative sign will be located to the right of G and the one having the positive sign will be to the left of G as shown in Figure 10.50. A road input at either the front or rear wheels will in this case excite both pitch and bounce.

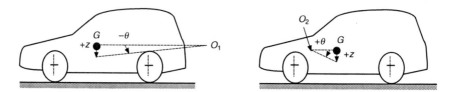

Figure 10.50 Location of oscillation centres

If the inertia coupling ratio $\dfrac{r_y^2}{ab} = \ell^{13}$, O_1 and O_2 then coincide with the rear and front suspension spring attachment points respectively. The 2-DOF model can then be represented by two concentrated masses m_f and m_r connected together by a massless link where

$$m_f = \frac{m_s b}{a + b} \tag{10.50}$$

and

$$m_r = \frac{m_s a}{a + b} \tag{10.51}$$

There is then no coupling of motions between the front and rear suspensions (desirable for good ride) and the system behaves as two separate SDOF systems. The corresponding natural frequencies of these are:

---

[13]The intertia coupling ratio ranges typically from 0.8 for sports cars to 1.2 for some front wheel drive cars.

$$\omega_{nf} = \sqrt{\frac{k_f (a + b)}{m_s b}} \tag{10.52}$$

and
$$\omega_{nr} = \sqrt{\frac{k_r (a + b)}{m_s a}} \tag{10.53}$$

In selecting front and rear-end natural frequencies it is normal to make the front end frequency slightly lower than that at the rear, i.e. the corresponding periodic times T_{nf} and T_{nr} are such that $T_{nf} > T_{nr}$ giving a 'flat' response to road inputs (Milliken *et al.*, 1995). This can be understood by appreciating that an input disturbance from the road to a moving vehicle affects the front wheels first and the rear wheels later (the delay time being dependent on the speed of the vehicle and the wheelbase). This tends to produce a pitching motion of the vehicle body which is undesirable. Making $T_{nf} > T_{nr}$ tends to produce in-phase motion of front and rear ends (bounce motion of the vehicle body) soon after a disturbance reaches the rear wheels.

Suspension performance analysis
Despite its simplicity, the quarter vehicle model is capable of explaining many of the design conflicts associated with the choice of spring and damping parameter values. It must be remembered that this model has two degrees of freedom and hence two natural frequencies. For a typical automobile, the lower of these frequencies lies in the range from 1 to 2 Hz; while the other frequency is around 10–11 Hz. The lower of the two frequencies is associated with resonance of the sprung mass (affecting ride performance) and the other frequency is associated with unsprung mass resonance, or wheel-hop (affecting tyre load fluctuation and hence handling performance).

The suspension designer usually has little influence over the magnitudes of both sprung and unsprung masses, tyre stiffness and suspension working space, leaving only the selection of characteristics and parameter values for suspension springs and dampers to achieve the desired suspension performance. Within the limitations of the quarter vehicle model, it is possible to investigate the effect these have on ride and handling performance. A simple assessment of the requirements can be obtained from the linearized quarter vehicle model shown in Figure 10.48. In developing this model it is necessary to use a linear approximation to the damper and suspension spring characteristics. Furthermore, if no road-tyre separation (tyre limiting) is assumed, it is possible to determine the frequency response of the model using the approach outlined in the Chapter 8. Ride performance can be assessed from the displacement transmissibility between the road and the sprung mass (a measure of vibration isolation) while handling performance can be assessed from the displacement transmissibility between the road and the unsprung mass (an approximate measure of tyre force fluctuation).

In order to investigate the effect of suspension spring stiffness on suspension performance, consider a linearized quarter vehicle model (neglecting tyre damping). Using the notation of Figure 10.48 together with the following fixed parameter values: $m_u = 40$ kg, $m_s = 260$ kg, $k_t = 130$ kN/m and $c_s = 1200$ Ns/m. Defining the suspension stiffness variation in terms of the ratio $r_s = k_t/k_s$ and taking values of $r_s = 5$, 8 and 12, the sprung and unsprung mass transmissibilities can be obtained as shown in Figure 10.51. In interpreting these results it is important to appreciate that a large value of r_s corresponds with a soft suspension and vice versa. Figure 10.51(a) indicates that the lowest transmissibility (best ride) is produced with the softest suspension (in this case $r_s = 12$), whereas in Figure 10.51(b) the lowest transmissibility at the wheel-hop frequency (good road holding) and in the mid-frequency range between the two resonances, requires a hard suspension (in this case $r_s = 5$). For this latter case the transmissibility for the body mode is relatively large.

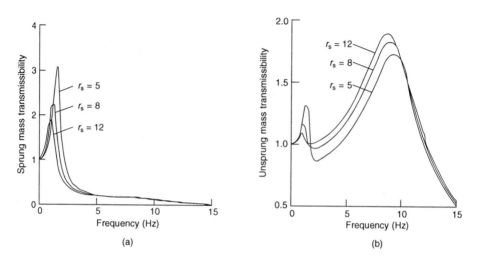

Figure 10.51 Effect of suspension stiffness on sprung and unsprung mass transmissibilities, (a) sprung mass transmissibility, (b) unsprung mass transmissibility

The effect of suspension damping on suspension performance can be investigated with a similar analysis, except that in this case the fixed parameters are taken to be: $m_u = 40$ kg, $m_s = 260$ kg, $k_t = 130$ kN/m and $k_s = 13$ kN/m. The variation in suspension damping can be described in terms of damping ratio defined as $\zeta = c_s \sqrt{2(m_s k_s)}$. Taking $\zeta = 0.1, 0.25, 0.5$ and 1 the sprung and unsprung mass transmissibilities shown in Figures 10.52(a) and (b) are obtained. Figure 10.52(a) indicates that control of the sprung mass resonance requires high levels of damping, but results in poor isolation in the mid-frequency range. On the other hand in Figure 10.52(b) it is seen that the wheel-hop resonance also requires high levels of damping for its control, but with the same penalties in the mid-frequency range. The solution is to select a damping ratio which provides reasonable control over the resonances and provides good isolation in the mid-frequency range. An average value around 0.3 is commonly adopted for passenger cars.

Refined analysis calls for the inclusion of suspension spring and damper non-linearities, tyre limiting, random road excitation, assessment of ride, tyre force fluctuation and clearance space limitations in the modelling. This analysis is very non-linear, calling for simulations in the time domain. In these simulations road inputs can be derived from a filtered random variable (Cebon *et al.* 1983) which accounts for road surface description and vehicle speed with corrections for possible tyre limiting effects. Assessment of ride is made by determining the r.m.s. ISO-weighted acceleration response of the sprung mass denoted by the Discomfort Parameter D. The weighting (a similar principle to the A-weighting of sound as described in Chapter 8) is

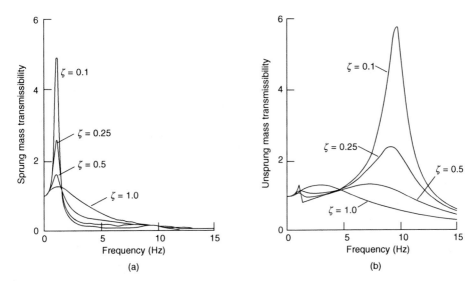

Figure 10.52 Effect of suspension damping on sprung and unsprung mass transmissibilities, (a) sprung mass transmissibility, (b) unsprung mass transmissibility

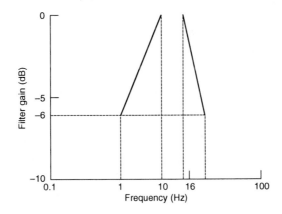

Figure 10.53 ISO weighting characteristic for vertical vehicle body acceleration

derived from the ISO whole-body vibration curves as discussed in Section 7.4.2. This leads to a filter having the characteristics shown in Figure 10.53. This type of analysis is a built-in feature of software such as VDAS [Horton (1992)].

Denoting the r.m.s. tyre force variation as L and the suspension clearance space as C, it is possible to draw a conflict diagram (Figure 10.54) which shows how these variables are influenced by the choice of parameter values to maintain a constant value of C^{14}. The ideal suspension

---

[14]Equal workspace comparisons of suspension performance are used because they indicate how well a given suspension space is being used. In practice this space is limited.

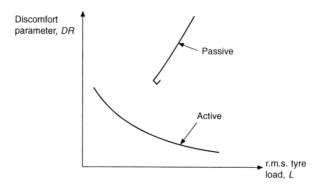

Figure 10.54 Conflict diagram for constant suspension workspace

(having good ride and handling) minimizes D and L. This corresponds to a point on the curve closest to the origin. Since each point in the curve is associated with a particular suspension stiffness and damping, these values can be used to select the appropriate parameters for a particular vehicle. Unfortunately, when the vehicle traverses a different type of road and/or at a different speed, the performance locus changes and the suspension settings are no longer 'optimal'. It is these limitations which have driven designers to consider the development of controllable suspensions, where parameter values can be continuously adjusted.

10.14 Controllable suspensions

With passive suspensions the 'control' force exerted simultaneously on the sprung and unsprung masses is:

$$u = c_s(\dot{z}_1 - \dot{z}_2) + k_s(z_1 - z_2) \tag{10.54}$$

while controllable systems aim to provide a control force u, which is able to provide a better performance. This usually requires the feedback of information from the dynamics of the vehicle together with a control law which provides a demand signal to control some form of actuator. An active system is able to reduce the sprung mass resonance more effectively than a passive system because the sprung mass acceleration is being continually monitored and the actuator delivers a force to minimize it. Control of the wheel-hop frequency is, however, not possible because the forces required would have to react against the spring mass and necessarily increase its acceleration. Overall, active suspensions are able to effect a significant improvement in ride performance. These can be represented on the conflict diagram as shown in Figure 10.54.

The potential benefits of controllable suspensions are not confined to improving just the individual performance at each wheelstation, but offers also the possibility of controlling ride height, roll, dive and squat, giving generally improving vehicle safety. At the present time only a limited number of proposals have been implemented on production cars and these have tended to be associated with luxury vehicles where the increased cost can be more easily absorbed.

Several types of controllable suspension have been proposed. In terms of the classification proposed by Crolla *et al.* (1989) the most common proposals fall into the category of fully active, slow active and semi-active. In terms of quarter vehicle models these can be represented diagrammatically as shown in Figures 10.55 to 10.57.

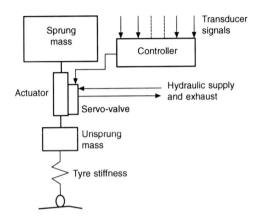

Figure 10.55 Fully active suspension

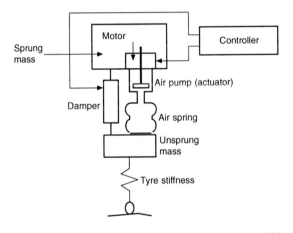

Figure 10.56 Slow active suspension (Sharp *et al.*, 1988)

The first of these diagrams shows a fully active system in which the functions of a conventional spring and damper have been replaced by a controllable actuator which, in most of the proposed schemes, is hydraulic. In terms of speed of response, it is classified as having a high bandwidth, i.e. up to 60 Hz. The actuator is driven by an on-board pump controlled by signals derived from transducers fitted to the sprung and unsprung masses. These signals are processed in a controller according to some control law to produce a controlled force at the actuator. With practical limitations taken into account, ride can be improved by 20–30% for the same wheel travel and dynamic tyre load when compared with a passive suspension (Crolla *et al.*, 1989). Despite these significant performance improvements, power consumption, hardware costs and system

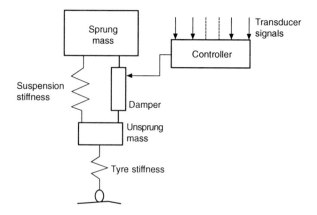

Figure 10.57 Semi-active suspension

complexity have made this form of suspension unviable. Since the ends of the actuator are connected to the sprung and unsprung masses they provide a direct link for high frequency vibration transmission above the operating bandwidth. Despite these limitations, fully active systems provide a benchmark for other forms of controllable suspensions.

The second alternative is classified as slow active and as the name implies has a low bandwidth (up to approximately 6 Hz). The aim of this form of suspension is to control the body mode to improve ride. Above its upper frequency limit it reverts to a conventional passive system which cannot be bettered for control of the wheel-hop mode. Such systems require much less power than the fully active system, with simpler forms of actuation. The potential performance gains are less than those for a fully active systems, but the viability is much improved (Sharp *et al.*, 1988) and there is not the problem of high frequency vibration transmission because the actuator is in series with a passive spring.

The third alternative is similar to a passive system except that the passive damper is replaced with a controllable one. This can be thought of as an actuator with limited capability. It is designed to produce a controlled force when called upon to dissipate energy and then switches to a notional zero damping state when called upon to supply energy. It has been shown (Crolla *et al.*, 1988) that the performance potential of this suspension closely approaches that of a fully active system under certain conditions, but the hardware and operational costs of this type of suspension are considerably less than that for the other two types. One of the disadvantages of the system is that its performance is impaired by changes in payload which alter the suspension working space. This problem can be overcome by combining the controllable damper with some form of self-levelling system. This has been achieved in a single prototype suspension strut linked to a gas spring. See Hall *et al.* (1991) for a description of the system.

An alternative to this continuous damper control is to be found in switchable dampers, which switch in discrete steps from one damping characteristic to another. The switching criteria being decided by some simple control law. This form of control requires relatively little sophistication in hardware or control laws, but performance gains are modest.

In all cases, the type and number of measurements (related to hardware cost) is dictated by the control laws used. This is one area where there is a lot of on-going research activity. A representative selection of this work is contained in the following papers: Wilson *et al.* (1986), Cebon *et al.* (1996), Pilbeam *et al.* (1993), Kim *et al.* (1993), Ting *et al.* (1995).

10.15 References

Adler, U. (ed.). (1996). *Automotive Handbook* (4th edn). Robert Bosch GmbH (Distributed by SAE).

Anon. (1996). *Spring Design Manual.* SAE.

Bastow, D. and Howard, G. (1993). *Car Suspension and Handling* (3rd edn), Pentech Press, London and SAE, Warrendale USA.

Bohm, F. and Wilhemeit, H-P (eds). (1997). 'Tyre Models for Vehicle Dynamic Analysis'. *Proc 2nd Int Colloquium on Tyre models for vehicle dynamic analysis,* Berlin 1997. Supplement to Vehicle System Dynamics, vol 27. Swets and Zeitlinger.

Bosworth, R. (1996). 'Rover's System Approach to Achieving First Class Ride Comfort for the New Rover 400'. *Automotive Refinement (Selected papers from Autotech95),* pp.113–122, IMechE.

British Standards Institution, B.S. 6841 *Guide to the measurement and evaluation of human exposure to whole-body mechanical vibration and repeated shock.* BSI 1987.

Butkunas A.A. (1967) 'Power spectral density and ride evaluation'. *Sound and Vibration,* 1 (12), pp. 25–30.

Cebon. D and Newlands, D.E. (1983). The artificial generation of road surface topography by the inverse FFT method'. *Proc IAVSD-IUATM 8th Symposium on the dynamics of vehicles on roads and tracks,* ed. J.K. Hedrick, pp. 29–42, Swets and Zeitlinger.

Cebon, D; Besinger, F.H. and Cole, D.J. 1996. Control strategies for semi-active lorry suspension. *Proc IMechE, Part D, J Automobile Eng;* 210, pp. 161–178.

Crolla, D.A. and Abdoul Nour, A.M.A. (1988). Theoretical comparisons of various active suspension systems in terms of performance and power requirements. *Paper No C420/88, Int Conference on Advanced Suspensions,* IMechE.

Crolla, D.A. and Sharp, R.S. (1989). Active suspension control, *Seminar Paper No. C399/3, Autotech 89,* NEC Birmingham, IMechE.

Dodds, C.J. and Robson, J.D. (1973). 'The description of road surface roughness *J Sound and Vib.* **31** (2), pp. 175–183.

Dixon, J.C. (1996) *Tyres Suspension and Handling* (2nd edn), Edward Arnold. Distributed by SAE.

Doughty, S. (1988). *Mechanics of Machines,* John Wiley.

Erdman, A.G. and Sandor, G.N. (1984). *Mechanism Design: Analysis and Synthesis,* (Vol. 1) and *Advanced Mechanism Design,* (Vol. 2). Prentice-Hall.

Fenton, J. (1996). *Handbook of Vehicle Design Analysis,* Mechanical Engineering Publications.

Gillespie, T.D. (1992). *Fundamentals of Vehicle Dynamics.* SAE.

Hall, B.B. and Tang, J.S. (1991). 'A combined self-energising self-leveling semi-active damper unit for use in vehicle suspensions'. *EAEC paper 91064, 3rd EAEC Conference on Vehicle Dynamics and Powertrain Engineering,* Strasbourg, June.

Heisler, H. (1989) *Advanced Vehicle Technology,* Edward Arnold.

Hillier, V.A.W. (1991) *Fundamental of Motor Vehicle Technology.* 4th edn, Thornes.

Horton, D.N L. (1992) *VDAS – Vehicle Dynamics Analysis Software,* Version 3.4, University of Leeds (Department of Mechanical Engineering).

International Standards Organisation. ISO 2631-1 (1978). *Guide for the evaluation of human exposure to whole-body vibration.* ISO.

Kim, H. and Yoon, Y-S. (1993). 'Neuro controlled active suspension with preview for ride comfort'. *SAE Technical Paper No 931969.* SAE.

Kortum, W. and Sharp, R.S. (1993). 'Multibody Computer Codes in Vehicle System Dynamics'. *Supplement to Vehicle System Dynamics, Volume 22,* Swets and Zeitlinger.

Mathsoft. (1998). *MathCAD 8.0*, MathSoft Inc Massachusetts, USA.

Meriam, J.L. and Kraige, L.G. (1993). *Engineering Mechanics, Vol 1 Statics* and *Vol 2 Dynamics*. 3rd edn, John Wiley.

Milliken, W.F. and Milliken, D.L. (1995). *Race Car Vehicle Dynamics*. SAE.

Mott, R.L. (1985). *Machine Elements in Design*. Charles E Merrill.

Pilbeam, C. and Sharp, R.S. (1993). 'On the preview control of limited bandwidth vehicle suspensions'. *J Automobile Eng*; Proc IMechE, Part D, **207**, pp. 185–194.

Reason, J. (1974). *Man in Motion: the Psychology of Travel*. Weidenfeld and Nicholson. London.

Ryan, R. (1990). 'Multibody Systems Analysis Software'. *Multibody Systems Handbook* (W. Schielen (ed.)), Springer Verlag, Berlin.

Sharp, R.S. and Hassan, J.H. (1988). 'Performance predictions for a pneumatic active car suspension'. Proc IMecE, 202, D4, pp. 243–250.

Shigley, J.E. and Uicker, J.J. (1980). *Theory of Machines and Mechanisms*. McGraw-Hill.

Shigley, J.E. and Mitchell, L.D. 1983. *Mechanical Engineering Design*. McGraw-Hill.

Suh, C-H. and Radcliffe, C.W. (1978). *Kinematics and Mechanisms Design*. John Wiley.

Ting, C-S; Li, T-H.S. and Kung, F-C. (1995). 'Design of fuzzy controller for active suspension system'. *Mechatronics* **5**(4), pp. 365–383.

Wahl, A.M. (1963). *Mechanical Springs*, 2nd edn, McGraw-Hill.

Wilson, D.A., Sharp, R.S. and Hassan, S.A. (1986). 'The application of linear optimal control theory to the design of active automotive suspensions'. *Vehicle System Dynamics*, **15**, pp. 105–118.

Wong, J.Y. (1993). *Theory of Ground Vehicles*. (2nd edn), SAE.

10.16 Further reading

Bastow, D. and Howard, G. (1993). *Car Suspension and Handling* (3rd edn), Pentech Press, London and SAE, Warrendale USA.

This book provides a comprehensive coverage of suspension systems and components. It is very good on descriptive detail without burdening the reader with a lot of analysis. The appendix contains a very useful section on suspension calculations with numerous worked examples.

Dixon, J.C. (1996). *Tyres Suspension and Handling*, (2nd edn), Edward Arnold. Distributed by SAE.

This book contains two relevant chapters. The one dealing with suspension components contains very little on detail design of components, but does provides the analytical foundation of suspension kinematics missing in the previous text. The other chapter addresses the whole range of suspension characteristics with supporting analysis. This tends to be somewhat sketchy in parts and the associated diagrams are not always explicit. Each chapter concludes with a set of exercises.

Gillespie, T.D. (1992). *Fundamentals of Vehicle Dynamics*. SAE.

This book contains chapters on ride and suspensions. The one on ride covers sources of excitation, vehicle response to vibration, the effects of suspension stiffness and damping, the benefits of active suspensions and the human perception of ride. It includes a number of simple worked examples.The chapter on suspensions is notable for its clear description of suspension types and properties, anti-pitch, anti-squat and roll centre analyses.

Heisler, H. (1989). *Advanced Vehicle Technology*, Edward Arnold, 1989.

This book has one chapter devoted to suspensions. It gives an excellent descriptive treatment (supported by numerous diagrams) of suspension geometries, roll centre analysis, body roll effects, pitch and dive behaviour and suspension components. The discussion extends to commercial vehicles unlike the other texts.

Milliken, W.F. and Milliken, D.L. (1995). *Race Car Vehicle Dynamics*. SAE.

Although aimed primarily at racing vehicles, this book contains a lot of material that is applicable to ordinary road-going vehicles. This includes excellent, detailed chapters on ride and roll-rate, suspension geometry, suspension springs and dampers, all of which are well supported by clear diagrams. Most of these chapters contain numerous worked examples.

Reimpell, J. and Stoll, H. (1998). *The Automotive Chassis: Engineering Principles*. Arnold.

This is a unique book addressing both practical and analytical aspects of chassis design. There is detailed discussion of suspension types, suspension components (spring and damper types and characteristics), tyres and tyre characteristics, axle kinematics and elastokinematics, A particularly attractive feature of the book is the detailed discussion of data for well-known vehicles in current production. This welcome new text is essential reading for anyone involved in suspension design.

Wong, J.Y. (1993). *Theory of Ground Vehicles*. (2nd edn). SAE.

This book has a chapter on vehicle ride characteristics which covers human response to ride, vehicle ride models, random vibration (including road surface description and vehicle response analysis) and active/semi-active suspensions. The section on random vibration, which is useful for students who are new to the subject, is concerned particularly with road profile descriptions and inputs to vehicle models. The chapter concludes with a set of problems for the student.

11. Control systems in automobiles

H. Morris, MSc, BSc, CEng, MIEE

The aim of this chapter is to:

- Introduce the concept of electronic vehicle control systems;
- Indicate the process through which the designer can create automatically controlled systems that aid drivers to control vehicles;
- Suggest areas in vehicle design where electronic control systems can improve vehicle performance;
- Demonstrate the integration of vehicle control systems.

11.1 Introduction

Legal and marketing pressures produced by demands for improvements in environmental operation, vehicle performance, safety of operation, ergonomic design and communication with external vehicular sources have lead to the ongoing development of control systems in a number of areas of automobile design. Generally, these application areas are in:

- Engine management systems to improve fuel economy and exhaust gas emissions;
- Steering and braking systems;
- Vehicle performance and drivability;
- Driver and passenger safety and comfort;
- Information provision.

This chapter is intended to introduce the basic principles of a number of different control systems used in automobiles. It is not intended to be exhaustive and the readers attention is drawn to the bibliography at the end of the chapter for more comprehensive and technology specific references. The aim is to show the burgeoning use of Electronic Control Units (ECUs), employing various forms of microcontroller to implement different forms of control strategy in examples drawn from the five areas referred to above. An explanation is made of the fundamentals of operation of each example and why the adoption of microcontroller based solutions has produced performance levels not previously possible with mechanical or early analogue electronic type solutions. Before discussing the principles of examples of current technology, there is a need to consider the basic structures and types of control system to be encountered. Control systems may be categorized by the strategy they employ into one, or a combination of, the following types:

(1)   Open-loop control;
(2)   Feedforward control;

(3)   Closed-loop or feedback control;
(4)   Sequential control.

11.1.1 Open-loop control

Figure 11.1 shows an open-loop control system. This arrangement is generally unsatisfactory because of the way disturbances affect the controlled variable. e.g. relating ignition timing (the manipulated variable) directly to engine speed (the input variable) would ignore the disturbing effect of load on the optimum ignition timing point for the engine. If some direct or indirect means of measuring the disturbance is used to counter the latter effect, the form of the control system changes.

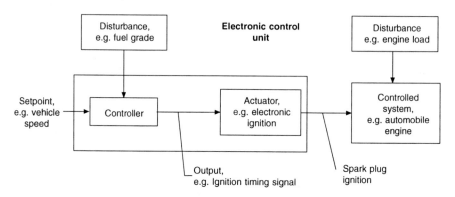

Figure 11.1 Open-loop control system

11.1.2 Feedforward control

Figure 11.2 shows a feedforward control system in which the input demand is modified by the output from a disturbance measurement system to correct for the effects of the disturbance. In the example of ignition timing, engine load is estimated indirectly by measurement of inlet manifold vacuum. The optimal relationship between ignition timing and load for a particular engine speed is not linear so one of the best microcomputer-based ways of modifying ignition timing with load for a defined speed range is to use a look-up table with input variables of engine speed and inlet manifold pressure. The values in this table are determined during test-bed trials and implanted into a Read Only Memory (ROM) module particular to an engine variant so that a standard Electronic Control Unit (ECU) development may be employed for an automobile model range, differing electronically only in the ROM used. Rather more sophisticated compensation algorithms are present within Electronic Spark Advance systems to counter the effects of engine warm-up, overheating, idling performance and knock control, but the underlying weakness remains of lack of feedback of engine performance measures to alter timing. Factors like engine wear and spark plug deterioration will detract from optimum ignition timing operation in what is still effectively an open-loop system, but long engine life and regular maintenance intervals together with increased spark plug life in modern transistorized ignition systems make these forms of control systems remain attractive.

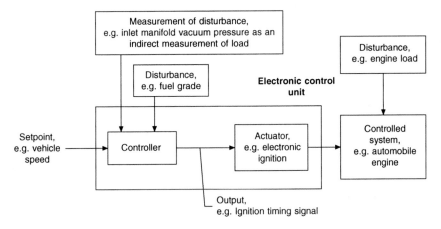

Figure 11.2 Feedforward control system showing compensation offset

11.1.3 Closed-loop control

This deficiency is overcome if we adopt a closed-loop control system structure such as that shown in Figure 11.3. This is sometimes referred to as an error-driven system since a measurement of the controlled variable is subtracted from a desired level, known as the set-point, and this difference or error is used by the controller to apply corrective action to the controlled system via what is generally termed an actuator. Closed-loop systems may be shown to be relatively insensitive to parameter changes in their forward path i.e. controller, actuator and controlled system, so the problems of engine wear or spark plug deterioration could be overcome by use of feedback or measurement of the controlled variable. Closed-loop systems are only as good as the feedback transducers or sensors employed and a great deal of development effort has gone into producing robust, stable, reliable and low cost sensors for automobiles. Some simplified examples of automotive closed-loop systems in current use are shown in Table 11.1, together with the associated sensor type.

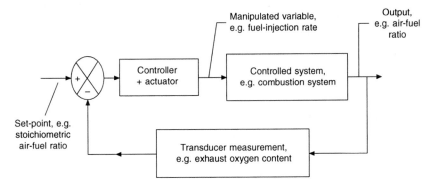

Figure 11.3 Closed-loop control system structure

Table 11.1 Examples of automotive closed-loop control systems

Control system	Indirectly controlled variable	Directly controlled variable	Manipulated variable	Sensor	Actuator
Fuel injection system	Air–Fuel ratio	Exhaust oxygen content	Quantity of injected fuel	Zirconia or Titania based electro-chemical	Fuel injector
Knock control	Knock	Knock sensor output	Ignition timing	Piezo-electric accelerometer	Ignition coil switch. transistor
Anti-lock braking system	Wheelslip limit	Wheelspeed	Brake line pressure	Magnetic reluctance	ABS solenoid valve

Various forms of controller exist to produce closed-loop control. Most readers will be familiar with the domestic central heating system which, typically, is a closed-loop system with a thermostat sensor producing a fully on/fully off cycle to the burner. If the room temperature profile is examined it is a cyclic variation of a degree or so, as the room heats and cools under the on/off cycle. The magnitude of this cyclic temperature variation is determined by what is termed the hysteresis of the thermostat, i.e. the thermostat contacts will close at one room temperature to switch on the burner but not open to switch off the burner until room temperature has risen a degree or so. This cycle is termed a limit cycle because of the limitation in room temperature variation produced by the thermostat action. Limit cycle control is present in a number of automobile control systems such as radiator cooling fan control, engine air–fuel ratio control, hydraulic pressure control and adaptive suspension height control. The basic choice between this form of control and what follows is largely determined by the type of the actuator or final control element, e.g. it is much easier to switch on and off a radiator cooling fan than to produce a continuously variable speed fan.

The alternative to using limit cycle action is to formulate a control algorithm within the closed-loop controller which gives a continuously variable output to an actuator. The controller endeavours to minimize the error between the set-point and the controlled variable as well as reject the effects of disturbance. Additionally, the algorithm must be constructed to give good dynamic response to the effects of change of set-point or disturbance. The traditional algorithm employed is that of three-term control or PID control, where PID is an acronym for Proportional, Integral, Derivative (Bissel, 1992). Before the digital age, this algorithm was produced using analogue circuits to give a controller output of

$$e_o = K_P e + K_D \frac{de}{dt} + K_I \int e \, . \, dt$$

Here the terms K_P, K_D and K_I are the proportional gain, derivative gain and integral gain terms respectively. These act independently on the error signal e to give three separate actions. Proportional action gives an immediate output if an error appears in the system whereas integral action takes a time dynamically to produce a reaction to such an error.

Figure 11.4 shows the output from a PI controller to a step input error function. The reason for inclusion of integral action is to remove steady state error from a system, as the integral term

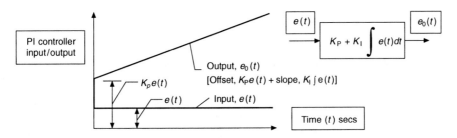

Figure 11.4 Input/output relationships for a PI controller

will continue to generate a changing output signal with time, provided the error remains finite, i.e. the output generated by the integral term will cause the system to respond and the controlled variable to change to a point where the steady state error is zero. Derivative action uses the rate of change of the error signal such that with an increasing error signal, when the rate is positive, the controller output is boosted so speeding up system response. Under conditions of reducing error signal as the system output approaches the set-point, the rate is negative, the controller output is reduced so reducing overshoot of system response. This has the effect of improving the system response to sudden changes in load or set-point change but also acting as a braking effect on the response when the system approaches its desired set-point level. Derivative action must be used with caution since unfiltered feedback signals or systems with noise pollution will upset the contribution made by this term to the controller output and give erratic behaviour to the system.

Moving to the digital age the PID algorithm may be implemented as

$$e_o = K_P e_k + K_D \left[\frac{e_k - e_{k-1}}{T} \right] + K_I [u_{k-1} T + e_k T]$$

where

$$u_{k-1} = \sum_{i=0}^{i=k-1} e_i$$ i.e. the sum of previous error signals, e_k and e_{k-1} are the present and previous error signals, and T = sampling interval present in the digital system.

Figure 11.5(a) shows a continuous error signal which, in the nature of digital systems, is sampled at a fixed repetition rate. Output signal e_o is generated once per sampling period using

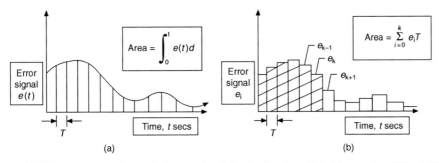

Figure 11.5 Continuous and sampled error signal levels showing digital integration (Bissel,1992)

present error signal magnitude e_k, previous error signal magnitude e_{k-1} and the integral sum of previous error signals u_{k-1}. The process of digital sampling and integration of the error signal e_i is illustrated in Figure 11.5(b).

Sampling time T must be chosen to be sufficiently short to prevent the introduction into the control loop of significant time delay which could lead to either unacceptable dynamic response or instability (closed loop oscillations of the controlled variable). This time is chosen in relation to the bandwidth of the system to be controlled. A rule of thumb guide is to select a sampling frequency of at least 50 times that of the system bandwidth if the loop is to be designed using classical control techniques. This figure may be reduced to as low as 5 times the system bandwidth if special digital control design procedures are adopted for selection of the terms in the controller algorithm, although the dynamic performance of the latter design is likely to be inferior to that of the faster sampling system (Bissell,1992). The selection of a fast sampling rate will have implications on processor speed and the form of programming language used in the microcontroller for control functions.

One further problem, common only to digital implementations of PID, is integral wind-up which occurred in early digital designs. Here an actuator which fails to open or close far enough to eliminate the difference between the set-point and the controlled variable would create an extremely large and cumulative integral contribution to the controller output. When the set-point is changed to a lower and more attainable value this large digital number takes a long time to 'unwind' and the system takes an uncharacteristically long period to reach a steady state (Chesmond, 1982). The answer to this problem is to jacket the contribution of the integral term to within bounds rather like the saturation limits of an analogue amplifier. An example of this problem might be in a cabin air-conditioning system where on either an extremely hot or cold day the system fails to achieve the cabin pre-set temperature. A rapid change of environmental temperature would lead to a poor system response in an algorithm prone to integral wind-up.

11.1.4 Sequential control

Sequential control is present in a number of control systems within automobiles. Sequences of events which may be open, feed-forward or closed-loop in structure are found in examples of manufacturers electronic spark advance and electronic fuel injection systems. For instance, during cranking of the engine on start-up, the electronic control unit (ECU) controlling fuel injection produces an injection duration using the formula

Injection period = basic cranking injection duration + temperature correction
+ battery voltage correction.

During this initial phase an appropriate cranking duration related to cranking speed is modified by information from the intake air and coolant temperature sensors so this system algorithm belongs to the feedforward category where the disturbance effect is measured and an allowance made for its effect. A correction may also be made for battery voltage variations since the injector speed of response is a function of battery voltage. Engine speed feedback sensing, after successful starting, enables the ECU to progressively reduce the injection duration; the ECU producing a large enrichment of the fuel supplied during the initial phase.

Following the start-up phase, the ECU moves into an engine warm-up enrichment phase which, as the coolant temperature rises, reduces the injector duration.

In the final stage of the sequence, what may be termed normal or steady-state operation is reached at the end of the warm-up phase and stoichiometric control or control of the air–fuel mixture to a ratio of 14.7:1 ($\lambda = 1$) is produced by use of closed-loop control application to the injection duration. This is formed by using information from an exhaust gas oxygen concentration sensor and modifying the injection duration by a correction coefficient constrained to a range of between 0.8 and 1.2. Figure 11.6 shows the output of a zirconia exhaust gas oxygen sensor as a function of air/fuel ratio. This device is an galvanic device producing a low d.c. output voltage which is a function of the difference in oxygen content between the oxygen levels present in the exhaust manifold to that present in free air. Both zirconia and titania sensors require operating temperatures above 280°C and it is necessary to either mount them in the exhaust downpipe or heat them with an electric heating element to bring them quickly up to a working temperature, enabling stoichiometric control action to be operational within minutes of starting a cold engine. Incorporating such a characteristic, which is similar in switching action to a room thermostat, in a closed loop system leads to limit cycle control action akin to that of the on/off action of a domestic central heating system. The result of such action, as with the heating system, is tight control of the air–fuel ratio.

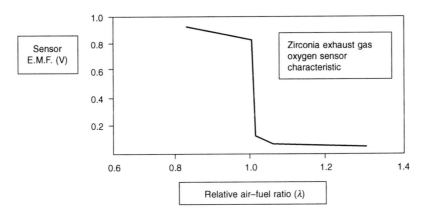

Figure 11.6 Output characteristic of a zirconia exhaust gas oxygen sensor (Chowanietz, 1995; Adler, 1988)

This chapter has so far introduced the concepts of conventional control techniques and used examples found in automobiles to demonstrate their application. They have all been what are termed single-input/single-output systems and the complexity of, for instance, engine management systems has only been hinted at in the way they employ a mixture of control strategies. Their implementation has been made possible, in part, by developments in sensor technology. What has also arisen in many applications are control systems which merge the input information from multiple sensors. This is termed sensor fusion where, with a degree of *artificial intelligence* (A.I.) incorporated into the controller, the collective sensor information is used in a more powerful way. Increased processing power offers not only the digital implementation of conventional control techniques as described earlier but also the more powerful alternative of artificial intelligence control. This last class of control systems includes neural networks, rule based systems and fuzzy logic controllers which have been developed to overcome the deficiencies of conventional control techniques. These systems will figure largely in future automobile electronic developments, especially the latter form of A.I., the basics of which are described in reference (Chuen, 1990).

11.1.5 Microcontrollers for automotive applications

A microcontroller is a single-chip microcomputer which incorporates microprocessor(s), memory (RAM and ROM), input/output circuits and timers into a single LSI device. There are many versions of this device used in automotive applications ranging from devices such as the Motorola MC 68705R3, an 8-bit unit with 112 bytes of RAM, 4 Kbytes of EPROM, four I/O ports, a timer and a four-channel analogue to digital converter (ADC) to the Motorola MC 68332, a 32-bit unit, 2 Kbytes of RAM and 2 timers but without ROM and I/O ports (Motorola, 1998). The strength of the latter 32-bit unit is its mathematical capabilities, speed and precision. Recall that amongst the attributes required of the ECU are an ability to implement different forms of control strategy at various stages via mathematically based algorithms, as well as merge the inputs of multiple sensor inputs into a knowledge base from which intelligent actions are formed. This is only possible with microprocessor based solutions.

Careful consideration has been made of the instruction set adopted to make the microcontroller relatively easy to program at an assembly language level and the system is designed to optimize compiler performance when programming in a high level language such as C++. This means that a minimum amount of op-code is generated following program writing and compilation, and the programmer, familiar with the high level language, is spared the complexities of the device at the *op-code* level. As an example of the instruction set for Motorola's M68332 Microcontroller the TABL instruction is a *Table Look-up and Interpolate* command which is useful in applications such as ignition timing or injector on-time. Both require a single output quantity to be derived from a measurement of at least two inputs. Where speed of calculation or non-linearity of the relationship between input and output is a difficulty, a look-up table (previously mentioned) is created in a reserved part of memory (ROM). This need only have a limited number of values in it, say 16×16, and interpolation between points may be achieved using the TABL instruction to achieve better resolution.

Figure 11.7 shows the general structure of a microcomputer-based electronic control unit where the inputs are a mixture of analogue and digital signal levels. The outputs may be required to be able to drive relatively large loads such as motors and solenoids, and power driver technology has been developed by companies like Motorola in which devices incorporate both current monitoring and temperature sensing for fault reporting to diagnostic routines run by the ECU or off-vehicle equipment.

11.2 Automotive application of sensors

What follows here is a tabular survey of the sensors found in the five application areas to be described in this chapter. It is not intended to be fully comprehensive or discuss the principles of operation, references will be given so that the reader may follow up an interest in device principles of operation. The objectives are to show the range of variables and the large number

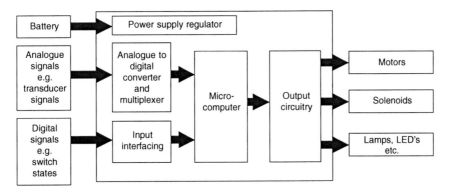

Figure 11.7 Microcomputer-based electronic control unit

of sensors involved with an application and how some applications might share common sensor information. It is important to emphasize that not all variables which it is desired to measure are directly available from the output of a transducer or sensor. In these cases indirect variable measurements are made and the variable of interest computed from it. There is often a dynamic or non-linear relationship between these and a computational task is required to relate them, hence the need for a microcomputer.

11.2.1 Engine management systems

Table 11.2 Application of engine management sensors (Chowanietz, 1995; Jurgen, 1995; Nwagboso, 1993).

Measured variable	Direct/indirect measurement	Sensor technology/ reference	Sensor mounting location
Intake manifold absolute pressure	Indirect measurement of engine load or mass air–flow intake	Wheatstone bridge arrangement of thick film resistors bonded onto a thin alumina diaphragm	Within intake manifold
Mass airflow	Direct and indirect measurement of fuel injector basic pulse width	Various forms including 'flap' type, 'hot-wire', Karman vortex and thick-film diaphragm	Within air intake
Temperature	Direct measurement at various locations	Thermistor or thermocouple depending on temperature range	Intake air, outside air, catalytic converter, engine coolant, hydraulic oil
Engine speed and crankshaft reference position	Direct measurement	Magnetic reluctance or Hall effect device	Flywheel on end of engine crankshaft

(Contd)

Measured variable	Direct/indirect measurement	Sensor technology/ reference	Sensor mounting location
Battery voltage	Direct measurement	Resistive attenuator	
Throttle position	Direct measurement	Potentiometer	Accelerator pedal
Knock (engine cylinder pressure oscillations during ignition)	Direct measurement	Piezoelectric accelerometer type.	Cylinder block or head
Oxygen concentration in exhaust gas (Lambda sensor)	Direct measurement	Zirconia or Titania based exhaust gas oxygen sensors	Exhaust manifold (normal operation above 300°C)

11.2.2 Chassis control systems

Table 11.3 Chassis control sensors and associated technology (Chowanietz, 1995; Jurgen, 1995; Nwagboso, 1993)

Measured variable and application	Direct/indirect measurement	Sensor technology/ reference	Sensor mounting location
Wheelspeed and engine speed, (ABS, TCS and electronic damping)	Direct measurement	Magnetic reluctance or Hall effect device	Brake assembly and crankshaft flywheel respectively
Steering wheel angle, (Electronic damping)	Direct measurement	Potentiometer or optical encoder	Steering shaft
Throttle position	Indirect measurement of vehicle accel.	Potentiometer	Accelerator pedal
Chassis and wheel acceleration, (electronic damping)	Direct	Piezo-electric accelerometer	Engine compart-ment and wheel assembly
Brake system pressure (electronic damping)	Indirect measurement of vehicle decelerat-ion	Flexing plate sensor with strain gauges mounted on plate	Brake master cylinder
Steering shaft torque (Electric power assisted steering)	Direct measurement	Optical device relying on steering shaft distortion under driver's twisting action	Steering shaft

11.2.3 Vehicle safety and on-board navigation systems

Table 11.4 Safety and on-board navigation sensors
(Chowanietz, 1995; Jurgen, 1995; Nwagboso, 1993)

Measured variable	Direct/indirect measurement	Sensor technology/ reference	Sensor mounting location
Vehicle deceleration (air-bag systems)	Direct measurement	'G' sensor (Piezo-electric accelerometer)	Single-point electronic sensing, location in dashboard or steering wheel
Wheelspeed and engine speed (Vehicle nav. systems)	Direct measurement	Magnetic reluctance or Hall effect device	Brake assembly.

Note the commonality of use of sensors for different applications such as ABS and on-board navigation. This points to the need to develop systems which can share common sensor information in an efficient manner (Rzevski, 1995).

Sensor research and development is being directed towards:

- Smaller, less intrusive, more reliable and cheaper sensors;
- Integration with single-chip microcomputer based products to improve sensor output and communication between transducers and associated control system.
- Employing *databus* or *multiplex* techniques where signals share conductors reduces the wiring loom size found in conventional systems;
- The use of multiple sensor data, combined with artificial intelligence, to provide sensor fusion;
- The use of multiple sensor data, combined with artificial intelligence, to provide sensor fusion.

11.3 Engine management systems

Efforts to improve fuel economy and reduce exhaust gas emissions have produced a revolution in the way engines are controlled and managed. Gone are the contact-breaker ignition systems, mechanical timing advance control and carburettors. These systems all have deficiencies in terms of fuel economy and exhaust gas emissions and have been replaced by all embracing engine management systems. Modern automobile engine management systems consist of a microprocessor-based electronic control unit (ECU) and a large number of electronic and electromechanical sensors and actuators. It is the function of this unit to: (i) Maintain accurate control of the air–fuel ratio via an electronic fuel injection system (EFI), (ii) Ensure accurate and precise ignition timing over a range of engine operating conditions, (iii) Monitor and control additional parameters such as idle speed, exhaust gas recirculation, air conditioner operation and fuel evaporative emissions.

11.3.1 Electronic fuel injection (EFI)

Figure 11.8 illustrates a basic electronic fuel injection system. EFI allows precise and fast control of fuel injected into each cylinder by control of the 'on-time' period of the solenoid operated injectors which consist of a spray nozzle and a solenoid-operated plunger connected to each cylinder. Fuel pressure in the delivery pipe to the injectors is maintained constant by a fuel pressure regulator in the fuel line in a circulatory system which has fuel constantly flowing around it when activated.

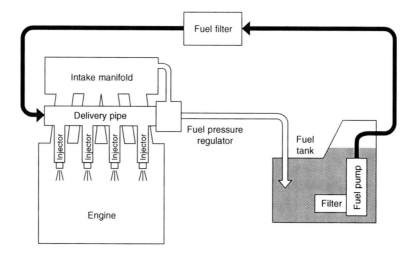

Figure 11. 8 Injection system fuel delivery (Chowanietz, 1995)

Solenoid operated fuel injectors have opening and closing times of between 0.5 and 1 ms. Considering an engine operating speed of 6000 rpm where the revolution period is 10 ms this gives an adequate control range of between 1 and 10 ms for the injector on-time.

Most fuel injection systems now use multi-point or sequential fuel injection, with one fuel injector near the intake valve (or valves) of each cylinder. At a device level, Motorola offer a fuel injector IC package which is interposed between the microcontroller and the injectors and provides the high solenoid drive current required whilst incorporating both over-voltage and short-circuit protection, a link being formed via an interrupt request line to fault reporting diagnostic routines. Figure 11.9 illustrates the arrangement.

Combining the ignition coil-on-plug with a direct fuel injector unit to give better dynamic response is the design objective of the next generation of EFI.

11.3.2 Types of EFI control system

The functions of fuel-air ratio control using sequential fuel injection to the engine cylinders and distributorless ignition timing control are integrated into engine management systems currently in use. Two types of EFI control system are common. The primary input signals to both systems

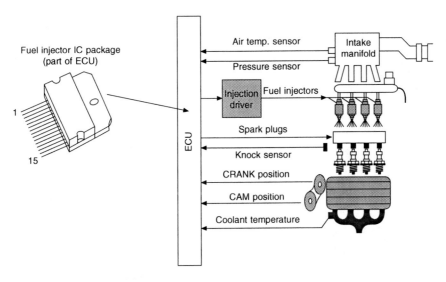

Figure 11.9 Power driver application (Motorola, 1998)

are engine speed and intake air mass, and it is the way the latter signal is derived that distinguishes the two types of EFI system; speed-density EFI and mass air-flow EFI (Chowanietz,1995; Jurgen, 1995; Adler, 1988).

Speed-density EFI
Mention has already been made of the inlet manifold absolute pressure (MAP) sensor's role in ignition timing related to engine load but in this system it has another function. The basic fuel injection opening period or pulse width is related directly to the mass of air flowing into the engine since the fuel-air ratio must be maintained constant in steady-state operation and the mass of air-flow is related to the manifold absolute pressure by the equation

$$m_a = \frac{V_d n_v p_i}{R T_i}$$

where V_d is the displacement of the cylinder, n_v is the volumetric efficiency or the fraction of V_d actually filled on each stroke, p_i is manifold absolute pressure, R is a constant and T_i is the intake air temperature. Since n_v is a non-linear function of engine speed and unique to a particular engine design, a look-up table using the two variables of engine speed and MAP sensor output combined with intake air temperature is normally used to generate a basic injector opening period.

Mass air-flow EFI
In this system a direct measurement of the quantity of air drawn into the engine on each intake stroke is made by an air-flow sensor (AFS). Various forms of air-flow sensor exist from simple flap-type, hot-wire and Karman vortex devices, all of which are placed directly in the inflow air-stream. Direct measurement in this manner is closer to the ideal of feedforward control than speed density EFI because factors like variation in volumetric efficiency and in engine displacement

due to speed and internal deposits are automatically compensated for. Both of these forms of EFI may be improved by the addition of an exhaust gas oxygen sensor to produce closed-loop control of the air–fuel ratio. Basically, either speed–density or mass air-flow may be used for EFI control but if the engine is to be controlled precisely around its stoichiometric or chemically perfect point the air–fuel ratio must be controlled to within 1%. This is only possible with closed-loop control, used in conjunction with speed–density or mass air-flow EFI.

Closed-loop control of air–fuel ratio
In a four-stroke spark-ignition (SI) engine it is the composition of the inducted air–fuel mixture and the timing of the ignition spark which control the combustion process, and in turn the performance and economy of the engine as well as the quantity of pollutants in its exhaust. The objective of low exhaust-gas emission levels dictates that the first function of the ECU must be to maintain the air–fuel ratio at 14.7:1 or stoichiometry (chemically perfect) as this condition is termed. Emission control strategies to control the emission of pollutants have led to the use of three-way catalytic converters. Figure 11.10 and Figure 11.11 show respectively, pollutant emissions and three-way catalytic converter conversion efficiency as a function of relative air–fuel ratio. In a closed-loop system, employing either speed–density or mass air-flow EFI, the fuel injection period computed by air intake measurement is modified or refined as a result of measurement of exhaust gas oxygen content. This takes the form of an injection period modification factor constrained to be within limits of between 0.8 and 1.2.

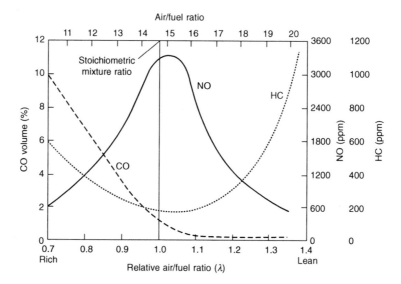

Figure 11.10 Pollutant emission as a function of relative air–fuel ratio, *l* (Chowanietz, 1995)

As may be observed, even with stoichiometric operation, oxidization is incomplete and instead of the ideal petrol burn in which the hydrocarbons of the fuel are combined with oxygen to give

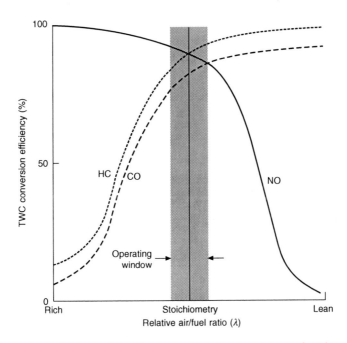

Figure 11.11 Conversion efficiency of the three-way catalytic converter as a function of relative air–fuel ratio (Chowanietz, 1995)

$$HC + O_2 + N_2 \rightarrow CO_2 + H_2O + N_2$$

the reaction is

$$n\,(HC) + O_2 + N_2 \rightarrow CO_2 + H_2O + N_2 + HC + NO_X + CO$$

The last three, HC, NO_X, and CO are what are termed *regulated pollutants* i.e. legislation is in place in various countries to control their levels. The unburned hydrocarbons (HC), oxides of nitrogen (NO_X) and carbon monoxide (CO) are all minimized through the combined action of the ECU to control air–fuel ratio, and the three-way catalytic converter to convert the three unwanted pollutants to carbon dioxide (CO_2), water (H_2O) and nitrogen (N_2).

The exhaust gas oxygen sensor (EGO sensor) provides an electrical feedback signal indicating whether or not air–fuel ratio is above or below the stoichiometric ($\lambda = 1.0$) level. ECU control of air–fuel ratio was mentioned in the section on control strategies and the limit cycle produced by the closed-loop control action has a frequency of between 0.5 Hz and 2 Hz, the period of the oscillation being due mainly to the time delay between a fuel injection change and being sensed by the EGO sensor. Since the ECU algorithm constrains the air–fuel correction coefficient to between 0.8 and 1.2 the perturbation of λ about the stoichiometric value of 1.0 is not great, and the average value of λ is maintained at this level.

11.3.3 Ignition timing control

Ignition systems have evolved from the electro-mechanical arrangements of points and distributor coil through a number of generations of development to either what is termed a distributorless ignition or the alternative, a coil-on-plug system. The first generation is shown in Figure 11.12 where basically the points current switching duty is replaced by a Darlington transistor power switch and the standard vacuum advance modifies ignition timing with load by moving the distributor body (Motorola, 1998).

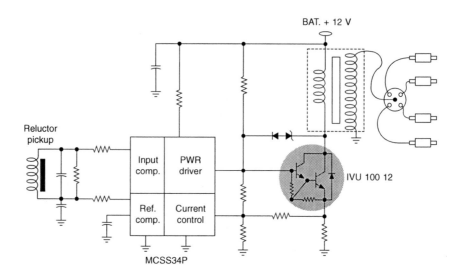

Figure 11.12 First generation electronic ignition (Motorola, 1998)

In the distributorless ignition system, shown with different forms of output power device in Figure 11.13, the distributor and rotor are replaced with a special ignition multi-coil. Two plugs are fired at once with this arrangement. However, one plug is in a cylinder under compression whilst the other is in the exhaust cycle so there is simply a wasted spark in this latter cylinder. To avoid the accelerated spark-plug erosion which accompanies this form of control a coil-on-plug design has been developed where a special ignition coil is provided per spark plug.

Figure 11.14 shows the arrangement where the ignition coil is powered from a +400 V d.c. source instead of the normal +12V battery supply, which allows the use of a physically smaller coil. The Insulated Gate Bipolar Transistor (IGBT) is gated with a multi-pulse signal, which generates a series of sparks during the combustion event. The voltage signatures of the ignition coil (reflecting the condition of the spark plug) are fed back to the engine controller for diagnostics. In both these modern generations of ignition system it is the engine management ECU which produces the ignition input to the ignition driver I.C.s.

The engine management system must repeatedly assess engine operating conditions of speed and load and compute the amount of ignition advance to achieve maximum brake torque (MBT) at full load. Accurate speed sensing and crankshaft top-dead-centre (TDC) position sensing are essential and either variable reluctance sensors or Hall effect devices are employed. The need to compute a different firing point for each speed of the engine arises primarily because the combustion process must be over at a fixed point in the crankshaft cycle, and if a

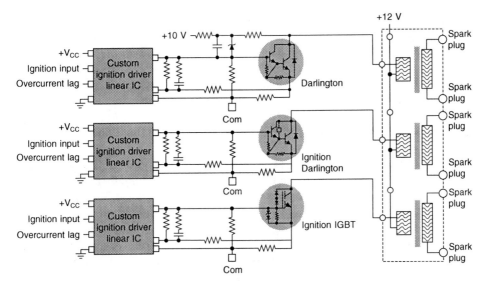

Figure 11.13 Distributorless ignition system (6-cylinder engine) (Motorola, 1998)

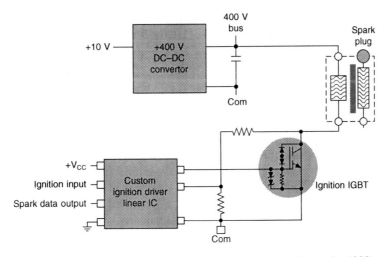

Figure 11.14 Coil-on-plug distributorless ignition system (Motorola, 1998)

fixed mixture burning period is considered then as engine speed increases the ignition or firing point must advance to ensure completion of burning by the time this fixed point is reached.

For example, a crankshaft rotating at 1000 rpm moves through 18° in 3 ms and at 3000 rpm moves through 54°, indicating a necessity to advance firing by 36°. The process is further complicated by load conditions. At lower than full-load a smaller mass of air–fuel mixture is needed per cycle and this results in a slower burning time. Timing must therefore be advanced to compensate for this. An indication of load is obtained from the inlet manifold absolute pressure (MAP) sensor and an ECU algorithm or look-up table relating engine speed and load then generates an appropriate ignition point in relation to TDC. Most MAP sensors rely on techniques developed for strain gauge applications. They consist of a Wheatstone bridge of thick film resistors deposited on a thin alumina substrate which, when subjected to the forces created by the inlet manifold pressure, produces a bridge output voltage proportional to the pressure. The amount of advance varies between different engine models and dynamometer test bed results are needed to establish the coefficients in the algorithm.

11.3.4 Knock sensing (Chowanietz, 1995 Jurgen, 1995; Adler, 1988; Sheingold, 1984)

Over-advance of ignition timing can lead to a phenomenon known as 'knock' (a metallic rattling sound) basically caused by a rising piston working against rapidly expanding cylinder gases which have been ignited prematurely. To safeguard against mechanical damage, but allow operation near to knock limits (a requirement in high performance engines), the ECU has a knock sensor so that ignition timing can be retarded to a point where knock disappears. Knock sensing may be by direct measurement of cylinder pressure, measurement of spark-plug ionization current or, the most common method, using a piezoelectric accelerometer mounted on a flexible part of the engine structure. After retardation of ignition timing and cessation of knocking, ignition timing is progressively re-advanced until it reaches its knock level and the cycle is repeated.

11.4 Electronic transmission control (Chowanietz, 1995; Jurgen, 1995)

Electronic transmission control designers faced a similar dilemma to those developing engine management systems. Instead of carburettors, contact-breaker ignition systems and mechanical timing advance mechanisms designers scrutinized the manual transmission system of clutch and gearbox to develop various solutions to reduce or eliminate the manual gear changing process. An additional dilemma was whether or not to adapt the mechanical components of currently developed automatic transmissions with their torque converter and epicyclical gear arrangements or go for a completely new technique such as the continuously variable transmission developed by DAF in Holland in 1955 and given the trade name 'Variomatic' (Chowanietz, 1995). The latter had limited torque capability but progressive development has seen it applied to small cars manufactured by Rover, Ford, Fiat, Subaru and Nissan, when incorporated with electronic control.

10.4.1 Electronic clutch control (Chowanietz, 1995)

This is a form of semi-automatic system which relieves the driver of the burden of depressing the clutch pedal when changing gear. The driver action of controlling engine speed via the accelerator pedal during gear change was retained in early versions but later developments use a form of electronic intervention which synchronizes engine speed with transmission speed during the gear change phase. In order to achieve this the normal throttle cable linking the accelerator pedal with the engine throttle disc is replaced by a closed-loop control system comprising an accelerator pedal position sensor and servomotor, both of which are connected to the ECU controlling the gear change process.

Figure 11.15 shows a block diagram of an Automatic Clutch and Throttle System (ACTS) for semi-automatic transmission. In this a further closed-loop servo system controls the movement of the clutch plate mechanism via an electro-hydraulic control valve and hydraulic power unit. In response to signals received from the engine and transmission speed sensors as well as the clutch release travel sensor and cylinder pressure the ECU provides a management system for the control of clutch engagement and disengagement to ensure safe and smooth gear changing under varying degrees of engine load and driver acceleration demands. Additional benefits accruing from the use of an ECU management system include improvements in safety of operation by prevention of engine starting when in gear and the selection of an inappropriate gear during gear change; both of which are signalled to the driver. Clutch wear may also be catered for by measuring the position of the clutch release lever at starting and using this as the reference for a fully engaged clutch.

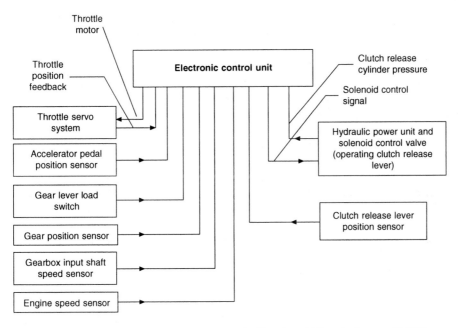

Figure 11.15 Block diagram of an Automatic Clutch and Throttle System (ACTS)

11.4.2 Electronically controlled automatic transmissions (Chowanietz,1995; Jurgen, 1995)

In this form of transmission system, gear selection as well as clutch control is made automatically by the ECU. The driver simply uses a selector to choose an operational mode from a standard selector mechanism, which in most cases replaces the gear lever found in manual transmission systems. In the case of a four-speed transmission system the Society of Automotive Engineers (SAE) recommends that the selector sequence be **P R N D 3 2 1,** as in Figure 11.16.

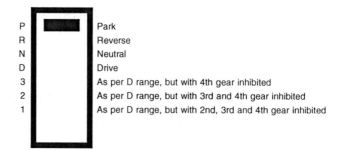

P	Park
R	Reverse
N	Neutral
D	Drive
3	As per D range, but with 4th gear inhibited
2	As per D range, but with 3rd and 4th gear inhibited
1	As per D range, but with 2nd, 3rd and 4th gear inhibited

Figure 11.16 Automatic transmission selector lever positions

There are a number of advantages of an automatic transmission over a manual transmission system and they include reduced driver fatigue, improved safety because both hands can remain on the steering wheel and engagement of the correct gear to suit prevailing driving conditions. The use of an ECU can improve automatic transmission performance by increased precision of gear changing, offers greater flexibility of shift pattern operation to suit the desire for different behaviour in circumstances demanding extra performance, economy or safety and simplifies the hydraulic control system compared to its mechanical predecessor. The disadvantages of an automatic transmission system are extra capital cost and reduced fuel economy, these being especially significant in Europe where the road conditions often require frequent gear changes and automobile engine size is smaller, thus highlighting the reduction in fuel efficiency.

The basic differences in component parts between a semi-automatic and an automatic transmission system relate to two major items. Designers have incorporated a unit known as a torque converter between the crankshaft and the gearbox input shaft. This unit serves to reduce or damp out the shocks caused by gear changing as well as multiply the torque available in a driveaway condition from standstill. There is a frictional loss in this item and during engine steady state or cruising conditions this would lead to poorer fuel economy than its manual transmission counterpart. However, to overcome this deficiency designers have refined the operation of the torque converter by using ECU based control techniques which recognize these conditions and effectively switch out the flexible coupling until conditions change. Part of the torque converter, known as a lockup clutch, is used to bypass the flexible coupling effect to produce full engagement between crankshaft and gearbox. In a further twist to this technique one system has been developed by Mitsubishi which modulates the degree to which the flexible coupling effect is removed in order to improve fuel efficiency but reduce gear change torque shock and vibration effects.

The second item of difference relates to the form of geartrain and clutch assemblies used in the transmission. The geartrain is normally of the compound epicyclical form and is combined with what may be termed friction elements. These are hydraulically operated multiplate brake-bands and multiplate clutches which engage to couple or lock the appropriate sets of planetary gear elements. The interface to these hydraulic elements is via solenoid valves electrically connected to the transmission ECU, so giving direct control of sequencing and timing of ratio changes to this unit.

The hydraulic pressure required to operate the various clutches and bands is produced by an oil pump which is driven by the engine. Oil pump output pressure (known as line pressure) is regulated by an ECU controlled solenoid valve and directed to the appropriate clutches and bands by shift solenoids. There are typically up to seven solenoid valves located in the valve body of the transmission. Three of the solenoids are termed solenoid shift valves and direct line pressure to enable shifting between 1–2, 2–3 and 3–4 gears respectively. A fourth solenoid controls the precise timing of clutch and brake-band engagement and a fifth modulates line pressure to the shift solenoids under the control of the ECU. This latter form of control is useful in reducing gearshift torque shock by temporarily reducing line pressure during gear changing, and also provides a means of increasing line pressure under increased load thereby producing greater brake-band and clutch engagement force. The remaining two solenoids are related to the control of the lockup clutch. One solenoid is simply an ON/OFF device directing line pressure to the torque converter lockup system and the second is modulated by a control signal from the ECU to give reduced line pressure to the lockup clutch in conditions requiring slip between the engine and gearbox. The latter is operational at low speed in crawling conditions and at high speed prior to full lockup where torque engagement shock is to be avoided.

11.5 Integration of engine management and transmission control systems (Chowanietz, 1995; Motorola, 1998; Jurgen, 1995)

Integration of these two systems produces an improvement of overall vehicle control. Digital data is interchanged between the ECUs to temporarily reduce engine torque during gear changes. This reduces clutch engagement shock and friction element wear, and increases efficiency and transmission fluid lifetime. Figure 11.17 shows a typical handshaking process between the two ECUs during either a gear change up or down.

During a change up the transmission ECU energizes line 1 to the engine management ECU which reacts by cutting off fuel injection and responding with a signal on line 2 to allow the gear change to proceed. During a change down the transmission ECU energizes line 3 to the engine management system which reacts by changing the ignition timing a few degrees to reduce engine torque, and responding again with an allow change signal on line 2. At the end of both gear change processes all lines are de-energized and both systems return to independent operation.

Figure 11.17 also indicates the degree of commonality of sensor information required by both systems, e.g. engine r.p.m. and throttle position sensing is required by both systems. The additional input signals to the transmission ECU help formulate the management of control functions. For instance, the multi-way switch reporting the position of the selector lever inhibitor switch is termed an inhibitor switch. If the lever is not in either Park or Neutral when starting, operation of the starter motor is inhibited and a warning buzzer sounded to draw the attention of the driver to the selector lever being in the wrong position. The hold switch is a latching push-button switch located on the selector lever to instruct the ECU to hold the transmission in a current gear ratio, a feature which is useful in descending a hill. The stoplight switch operates when the brakes are applied and if the transmission is in a lockup condition, the lockup clutch is then disengaged. The overdrive inhibit signal (O/D) comes from a separate cruise control unit and prevents the transmission from changing into overdrive (fourth gear) when cruise control is activated by the driver and the vehicle speed is more than a certain amount below the set cruising speed. The Automatic Transmission Fluid (ATF) thermosensor is a thermistor device

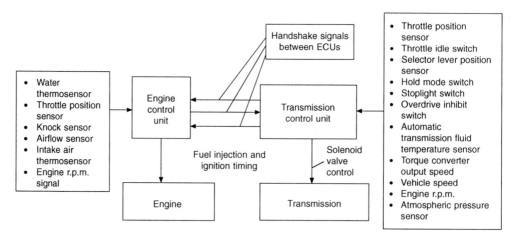

Figure 11.17 Integration of engine and transmission ECUs (Chowanietz, 1995)

and the ECU uses information from it at temperature extremes to modify the line pressure, thus accounting for changes in fluid viscosity. Finally, atmospheric pressure is sensed to detect when the vehicle is above 1500 m and because engines develop less power at high altitudes the automatic gear change points are modified to suit the change in performance.

Figure 11.18 shows a complete Powertrain Control Module (Motorola, 1998), integrating the functions described in the sections on engine management and transmission control as well as incorporating other control features. These include Exhaust Gas Recirculation (EGR) and Evaporative Emission Control (EEC) systems. The EGR system is designed to reduce NOX emissions by recirculating a small percentage of exhaust gas to the intake manifold to act as an inert dilutant in the intake mixture. The effect of this is to reduce the maximum combustion temperature and hence limit the formation of NO_X. The management of this function, i.e. the opening of the EGR valve and retardation of ignition timing to take account of the longer required combustion time, is controlled by the powertrain ECU and is dependent on engine temperature, load and speed. For satisfactory management a figure of 15% EGR is the maximum allowable that may be used.

Evaporative Emission Control is a system to collect, in a charcoal canister, fuel vapour evaporating from the fuel tank and to direct it to the intake manifold via a 'purge' valve which is under the management of the powertrain ECU. This prevents the escape of harmful fuel vapour emissions to atmosphere and improves the vehicle's fuel efficiency.

11.6 Chassis control systems

This section describes some of the systems which have been developed to improve control of the motion of a vehicle. These include anti-lock braking (ABS), traction control (TCS), power-assisted steering (PAS), electronic damping control (EDC) and power-assisted steering (PAS) systems. Additionally, a description of multiplex wiring is included to emphasize how the introduction of these and other systems of control might be interconnected in a vehicle, and

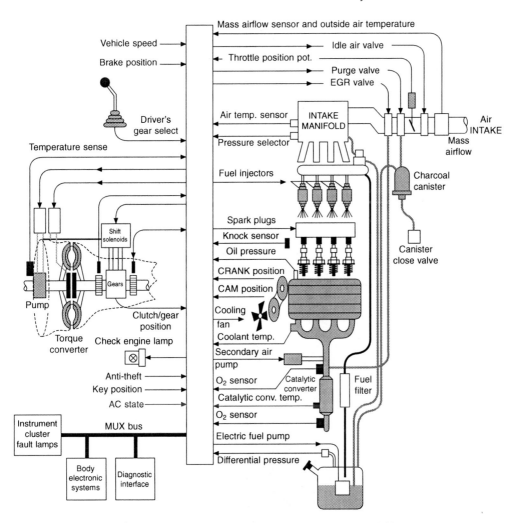

Figure 11.18 Powertrain control module (Motorola, 1998)

where considerations of reduction in size and weight of wiring harness, complexity of system failure diagnosis and system integration have led to the development of multiplex wiring systems.

11.6.1 Anti-lock braking systems (ABS) (Chowanietz, 1995; Jurgen, 1995)

Automobile braking systems without the benefit of ABS suffer from a number of problems which arise normally only during an emergency braking situation. These include:

• The vehicle skids, the wheels lock and driving stability is lost so the vehicle cannot be steered;

- If a trailer or caravan is being towed it may jack-knife;
- The braking distance increases due to skidding;
- The tyres may burst due to the excessive friction and forces being concentrated at the points where the locked wheels are in contact with the road surface;

With an ABS system when a driver applies the brakes in an emergency, the system regulates the brake pressure of the individual brake cylinders as a function of wheel acceleration or deceleration and avoids the problems described above.

Figure 11.19 shows the variation of the coefficient of friction (m) with slip ratio between a road surface and a vehicle tyre under different road conditions; slip being the ratio of wheel speed minus vehicle speed to vehicle speed. When braking, the slip ratio is deemed negative (i.e. wheel speed less than vehicle speed) and the coefficient of friction is a maximum, in a range of road conditions, between slip values of -0.15 to -0.3. Lateral grip, which is a measure of resistance to sideways forces on a tyre, is also shown in Figure 11.19 and optimum braking conditions for both longitudinal and lateral grip may be obtained if the slip ratio could be maintained around the value of -0.2. This is the function of ABS and the result of using the system is a reduction of stopping distance of between 15% in dry road conditions and 40% on a wet surface.

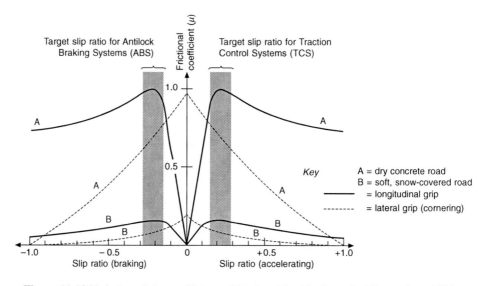

Figure 11.19 Variation of the coefficient of friction (μ) with slip ratio (Chowanietz, 1995)

ABS systems actually work by computing rate of change of velocity of the wheels of the vehicle to prevent them locking during braking and trying to maintain an average deceleration rate for the vehicle consistent with the optimum slip ratio. Figure 11.20 shows the component parts of a typical ABS.

Induction type wheel-speed sensors are located on the wheel assembly or differential. These sensors couple magnetically to a toothed wheel known as an impulse ring, the variation in

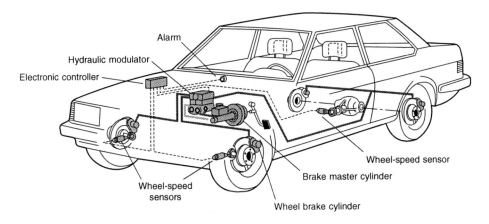

Alarm
Hydraulic modulator
Electronic controller
Wheel-speed sensor
Brake master cylinder
Wheel-speed sensors
Wheel brake cylinder

Figure 11.20 Antiskid braking system (ABS) (Bosch, 1984)

magnetic reluctance around a path linking the sensor and the impulse ring as the latter rotates producing a proportional variation in magnetic flux linking with the sensor coil. This results in a sensor induced a.c. voltage whose fundamental frequency is the product of the number of teeth on the impulse ring times the wheel speed in revs s^{-1}. The ABS ECU evaluates the signals produced by the wheel-speed sensors, calculates the wheel slip allowable for optimum braking and using solenoid valves regulates the braking pressure in the wheel brake cylinders to achieve the deceleration rate conducive with optimum braking. Bosch, who are market leaders in ABS systems, use a hydraulic modulator unit to achieve variation in braking pressure. This is a sealed unit consisting of solenoid valves interfaced to the ECU, hydraulic accumulators and a fluid return pump and is interposed between the brake master cylinder and the brake wheel cylinder hydraulic lines in each braking circuit. An ABS unit is described by the number of braking circuits that it regulates and two-, three- and four-channel systems are available. The hydraulic modulator unit has three operational positions controlled directly by the ECU. In the first of these the brake master cylinder is connected directly to the brake wheel cylinder and the ABS is inactive so that the driver has direct control of the braking effort. If the ECU detects excessive wheelslip the second operational position for the unit is activated which results in the pressure in the brake wheel cylinder being maintained at its current level. If the wheel slip continues at an excessive rate the ECU moves the hydraulic modulator unit to its third operational position in which the brake wheel cylinder is isolated from its direct connection to the brake master cylinder and connected to it via a return pump. This has the effect of reducing braking force and eliminating wheel lock. This whole process of the action of ABS and its effect on wheel-speed and braking pressure is shown in Figure 11.21.

In diagram 1 excessive wheelspin or deceleration is detected and the braking pressure is maintained at its current level. If the wheel continues to decelerate above a set level the pressure in the brake wheel cylinder is reduced to a new level, as in diagram 2, so that the wheel is not braked so much. If the wheel then accelerates again as a result of the reduced braking pressure and acceleration reaches a specific limit, brake wheel cylinder pressure is increased as in diagram 2. This control cycle of varying the hydraulic modulator operational position is executed approximately 4 to 10 times a second depending on the road conditions.

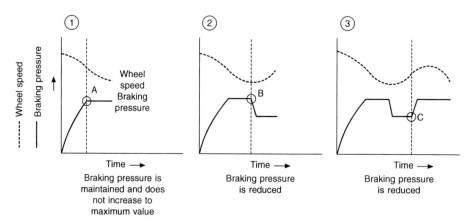

Figure 11.21 Wheel-speed and braking pressure during ABS-controlled braking (Bosch, 1984)

Testing of ABS systems (Jurgen, 1995)
Before a new ABS system can be launched, manufacturers perform stringent tests on specially prepared test tracks. These tests include:

- Straight line stopping;
- Braking in a turn;
- Braking on a surface with split friction coefficient (e.g. an oil patch hit by one side of the vehicle);
- Transitional road surface testing consisting of a checkerboard arrangement of low/high and high/low coefficient surfaces to check the dynamic response of the ABS;
- Lane change or obstacle avoidance manoeuvring while braking.

11.6.2 Traction control systems (Jurgen, 1995)

Traction control systems (TCS) are designed to prevent the drive wheels from wheelspinning during starting off or accelerating on a wet or icy surface. Vehicles with powerful engines are particularly susceptible to this phenomenon and results in reduction of either steering response on front-wheel-drive (FWD) or vehicle stability on rear-wheel-drive (RWD) vehicles. TCS operates to maximize adhesion to the road surface during acceleration whilst ABS has the same objective during braking. The two systems share common sensor information and may well be incorporated together in an ECU. The actuation part of the TCS varies for different systems and may be a permutation of fuel, ignition and driven wheel braking action to achieve reduction in driven wheel torque during wheelspin.

It is interesting to investigate TCS control strategies to achieve the apparently simple objective described above. Basically, the system should attempt to maintain the acceleration slip of the driven wheels at a value equal to the mean rotational velocity of the non-driven wheels plus a specified speed difference known as the slip threshold point. This means simply that if a vehicle's driven wheels are constantly at a faster speed than the non-driven wheels then the vehicle must be accelerating at a constant rate proportional to the difference in the two speeds.

This control objective must be qualified with a reference to road surface conditions or adhesion coefficient. For instance on dry road surfaces, maximum accelerative force is available at slip rates of 10 to 30%. On glare ice, maximum traction is achieved at levels between 2 and 5 percent so to cover these extremes TCS systems must respond to changes in adhesion coefficient and systems are designed for a slip rate range of between 2 and 20%. However, on loose sand or gravel and in deep snow the coefficient of adhesion increases continually with the slip rate as shown in Figure 11.22. For this reason, TCS systems incorporate slip-threshold switches to allow the driver to select a higher slip threshold or switch off the TCS.

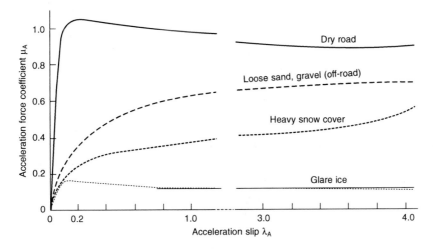

Figure 11.22 Adhesion force coefficient μ_A as a function of acceleration λ_A (Jurgen, 1995)

The control objectives of TCS are also modified by vehicle speed and curve recognition. Both of these variables can be derived from the speeds of the non-driven wheels. High vehicle speeds and low acceleration requirements on low coefficient of adhesion surfaces lead to a control strategy of progressively lower slip threshold setpoints as the vehicle speed increases, thus giving maximum lateral adhesion on the surface. The vehicle's acceleration rate and the engine torque provide the basis for reliable conclusions regarding the coefficient of adhesion or friction for the surface. The slip threshold is raised in response to higher friction coefficients to allow higher acceleration rates but there is a limit in TCS systems as discussed earlier.

Curve recognition or cornering detection also affects the control strategy for TCS. This strategy employs the difference in wheel speeds of the non-driven wheel speeds as a basis for reductions in the slip setpoint to enhance stability in curves.

11.6.3 Electronic damping control (Chowanietz, 1995; Jurgen, 1995)

The primary function of a shock absorber is to control vehicle movement against inertial forces, such as roll when the vehicle turns and pitch when the vehicle is accelerated or braked. The shock absorber has a secondary role of preventing vehicle vibration caused by a poor road surface. These are conflicting requirements of stability in the first case, requiring a hardness or stiffness, and comfort in the second case, requiring a softness of movement of the shock absorber.

Electronic damping control (EDC) is a means of attaining these twin objectives. Conventional shock absorbers use a spring and oil-filled damper arrangement and altering their characteristics on a continuously variable basis is now possible but remains difficult and expensive to achieve. Manufacturers therefore, have until recently opted for dampers which have at least three settings; 'soft', 'medium' and 'firm', one of these being selected at a time by the ECU of the EDC system. Other manufacturers, in developments analogous to the lateral thinking that went on in electronic transmission control where continuously variable transmission emerged as an alternative to an automatic gearbox, have produced electronically controlled suspension systems using air, nitrogen gas and hydraulic oil as a suspension agent.

As with other electronic control functions within vehicles it is the strategies controlling the damper system and the factors which modify them which make them interesting to study. Intelligent action by the ECU depends on a knowledge base created by sensor inputs and pre-programmed instructions. The selection of sensors used within an EDC system may be from a range measuring vehicle speed, engine r.p.m., brake system pressure, steering angle, chassis and wheel acceleration, throttle position, vehicle load and even road surface condition although the latter may be implied by processing signals from front and rear height sensors rather than direct measurement. For example, if the height sensor signal indicated a small high frequency but a large low frequency amplitude content this would imply a heaving or undulating road surface and would not require a softening of damper action. A large high frequency component would suggest a rough road surface and a requirement to soften damper action. The latter action is at odds with, for instance, damper action to prevent rolling during cornering. If the vehicle corners on a rough surface this conflict must be resolved by the ECU.

Longitudinal acceleration may be measured directly using an acceleration sensor, or can be inferred from brake system pressure and throttle opening angle. After processing this information allows cognitive action to be taken by the ECU on the vehicle damper system to prevent pitching during acceleration or braking.

Similarly, lateral forces on the vehicle may be inferred by the rate at which the steering wheel is being turned and the vehicle speed. This information is then used by the ECU to prevent rolling.

Figure 11.23 shows a typical schematic diagram of an electronically controlled damping system where the actuators are dampers each fitted with two ON-OFF fluid control solenoids that may be used to select one of four different damper settings (normal, soft, super-soft and firm). The ECU control modes input from the driver's console are either sport or smooth ride mode. Selection of the sport mode would then exclude selection of either soft or super-soft damper settings and give a harder but more stable ride.

11.6.4 Electronically controlled power-assisted steering (PAS) (Chowanietz, 1995; Jurgen, 1995)

These systems may be classified into hydraulic, hybrid or electric type structures. They have the common objective of reducing the effort required to turn the steering wheel to give a lighter and more direct feel to the steering. This reduction in effort may be quantified as a gain in control engineering terms and the steering process as a transfer function whose gain changes

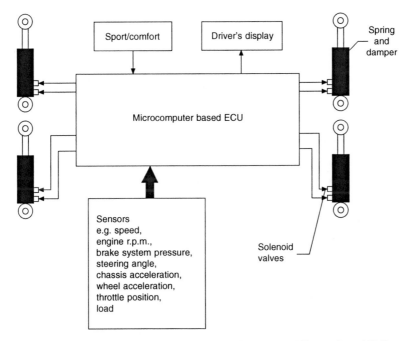

Figure 11.23 Electronically controlled damping system (Chowanietz, 1995)

with increasing vehicle speed, i.e. as the vehicle speed increases the effort required to turn the steering wheel decreases due to the lower adhesion to the road surface. This means that PAS systems must reduce the level of assistance as vehicle speed increases and this forms the control strategy for electronically based systems.

Electronically controlled hydraulic PAS
In the commonest form of this type of system, the ports of a solenoid valve are connected across the rack and pinion steering hydraulic power cylinder. Figure 11.24 shows the arrangement where with increasing vehicle speed the valve opening is extended by the ECU, thus reducing the hydraulic pressure in the power cylinder and increasing the steering effort.

The pump is driven from the engine and the bridge-like restrictions for control of the power cylinder are formed by the paths through the pump to port connections of a rotary valve. The latter is connected directly to the steering wheel and a small movement of this masks, or alternatively uncovers, port pathways for the high pressure hydraulic fluid to reach the power cylinder/solenoid valve parallel combination.

Hybrid PAS
These systems utilize a flow control method in which the hydraulic power steering pump is driven by an electric motor. The steering effort is controlled by controlling the rotating speed of the pump. This arrangement has the advantages of higher efficiency when driving at high speeds than the previous system using the bypass solenoid and since the pump is not driven directly from the vehicle engine there is a degree of freedom in the selection of its mounting location.

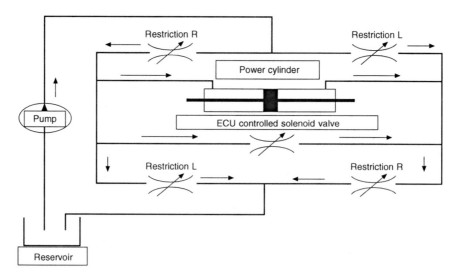

Figure 11.24 Hydraulic bridge circuit for electronically-controlled power steering showing flow paths (Jurgen, 1995)

Electric PAS

In this system the input to the rack and pinion steering system comes from a motor/reduction gearbox combination in which motor torque is applied directly to either the pinion gear shaft or to the rack shaft. The steering effort range is greater than with hydraulic systems, installations are cheaper and systems are more reliable. Power is only consumed when a movement to the steering wheel is made, unlike the hydraulic system where it is necessary to keep the pump operating continuously even when the steering wheel is not being turned. Figure 11.25 shows a schematic diagram of an EPAS in which a torque sensor is mounted on the column shaft connecting the steering wheel and worm wheel mechanism.

The electric motor in Figure 11.25 is coupled to the worm wheel mechanism through a reduction gearbox. The load torque T_L on the steering column is the load presented by the worm mechanism and the rack and pinion assembly to which it is attached. The amount of motor torque assistance given to the driver in turning the steering wheel is proportional to the motor current I_M. Since in a simple armature controlled d.c. motor the average current is given by

$$I_M = \frac{V_M - k \times N}{R},$$

where R is the armature resistance, N is the speed of the motor and V_M the motor voltage, it follows that the set point motor voltage must be a function of how much control effort is required from the d.c. motor. Imagine a driver turning a steering wheel at a constant rate, say in cornering. The d.c. motor, to give proper assistance, must turn at a speed proportional to this rate. Additionally, it must be remembered that at high vehicle speeds the assistance given to the driver must decrease in proportion to speed to improve steering stability, which means decreasing motor current or voltage as vehicle speed increases. These two control strategies are achieved by the control map shown in Figure 11.26, where an almost proportional relationship is illustrated

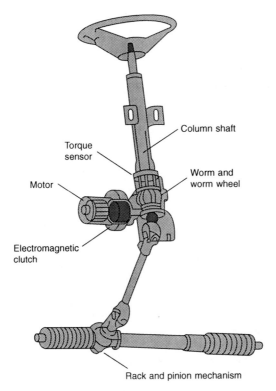

Column shaft

Torque
sensor

Worm and
worm wheel

Motor

Electromagnetic
clutch

Rack and pinion mechanism

Figure 11.25 Electric power assisted steering components (Jurgen, 1995)

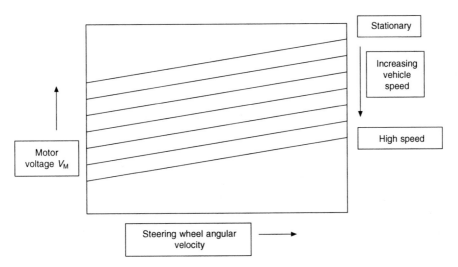

Stationary

Increasing
vehicle
speed

High speed

Motor
voltage V_M

Steering wheel angular
velocity

Figure 11.26 EPS control map showing relationship between steering wheel angular velocity, motor voltage and vehicle speed. (Jurgen, 1995)

linking the steering wheel angular velocity sensor output to the servo motor speed N, thus giving a motor voltage $k \times N$. The torque sensor on the column shaft provides an output T_M which is translated to a voltage $k_T \times T_M$ which is proportional to it but is also an inverse function of vehicle speed. These two voltages are then added to form the motor voltage V_M.

Further control strategies associated with improving the steering wheel responsiveness at high speed may also be employed to convert the motor into a generator in this condition to create a braking or reverse torque on the steering wheel. This has the effect of increasing the control effort or system damping required at high speed and improves the steering 'feel'.

11.7 Multiplex wiring systems (Chowanietz, 1995; Motorola, 1998)

These use one or more serial networks to achieve the following objectives:

- Reduction of the number of wires within the wiring looms of a vehicle;
- Improved system failure diagnosis;
- Distributed control centres in or off the vehicle which can talk and interact with one another;
- Improved manufacturing techniques and increased reliability due to a reduction in the number of wires and connectors;

Serial communication links are categorized by the Society of Automotive Engineers (SAE) according to the speed at which data may be transferred over them and each must adopt an agreed message exchange protocol. There are three classes or speed groups. Class A systems are 'low-speed buses', capable of transferring up to several thousand bits per second and suitable for use in the control of body electronic systems such as windows, doors and seats, etc. A Class B network can meet all of Class A requirements, but generally uses more complex error detection and checking methods. The data transfer rates are between 10 kbps and 125 kbps. A Class B system might be used for status and diagnostic data sharing between the engine control unit, the transmission control system, and the instrumentation console in a form of distributed control arrangement where the different ECUs 'talk' to each other and avoid duplicating sensor systems.

For high-speed real-time data communication, a Class C network would be used with a transfer rate between 125 kbps and 1 Mbps. Communication of data between systems such as engine control, traction control, active suspension and anti-lock braking systems requires high speed data links and low error rates. These demands make Class C systems, with their advanced data integrity features, the most costly of the three classes to implement.

Figure 11.27 shows three classes of multiplex network installed in a vehicle, connected to various control centres and linked to each other by 'gateways'.

Automotive manufacturers initially developed proprietary protocols for their serial data communications but standardization at national and international level has brought the following protocol standards:

- CAN: Controller Area Network, Class A, B and C;
- J1850: SAE Std., Class A and B;
- Various UART Based Protocols (Universal Asynchronous R_x/T_x).

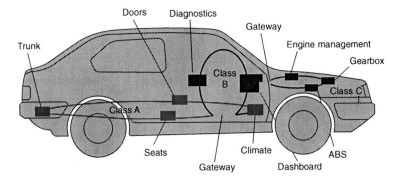

Figure 11.27 Multiplex networks installed in a vehicle (Motorola, 1995)

11.8 Vehicle safety and security systems

The systems to be described in this section are within a class referred to as body electronic systems since they are fitted within the passenger compartment of the vehicle. This is a burgeoning area of development and these systems are often seen by the purchaser of a vehicle as a product differentiator. They include instrumentation displays, heating, ventilation and air-conditioning (HVAC), central door locking (via remote keyless entry, RKE), anti-theft vehicle immobilization initiated by RKE, cruise control, on-board navigation or global positioning systems (GPS), interface connection points for easy plug-and-play of devices like pagers or cellphones and systems to supplement seat belt passenger protection in the form of air-bag and seat belt pre-tensioners. Since safety and security are critical features the electronic systems to implement the latter will be discussed in some detail and an overview given for some of the others.

11.8.1 Air-bag and seat belt pre-tensioner systems (Chowanietz, 1995; Motorola, 1998; Jurgen, 1995)

The component parts of these systems consist of crash detection sensors (typically a piezoelectric accelerometer formed by a silicon micromachined cantilever and signal conditioning amplifier all mounted on one substrate within a hermetically sealed case filled with silicon oil), a microcontroller which runs a crash algorithm to distinguish between crashes and normal vehicle dynamics, igniter (squib) triggering for the pyrotechnic inflator used for both air-bag deployment and seat belt tightening, and system monitoring. The allowable forward passenger travel with an air-bag system is 12.5 cm but with seat belt tensioning systems this figure drops to about 1 cm. Approximately 30 ms are required to inflate air-bags and the time required to tension a seat belt with a pyrotechnically activated seat belt retractor is approximately 10 ms. Thus, triggering commands must be given by the time maximally allowable forward displacement will be reached minus the activation time of the respective restraining device. Manufacturers have produced designs with varying numbers of both sensors and sensor mounting positions when employing electromechanical sensors to provide the required performance but current electronic based units are normally single-point systems such as the one illustrated in Figure 11.28.

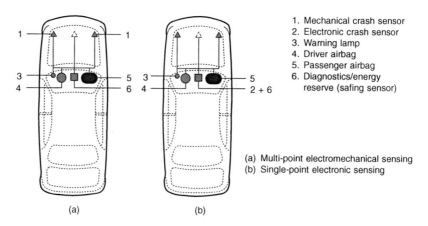

1. Mechanical crash sensor
2. Electronic crash sensor
3. Warning lamp
4. Driver airbag
5. Passenger airbag
6. Diagnostics/energy
   reserve (safing sensor)

(a) Multi-point electromechanical sensing
(b) Single-point electronic sensing

Figure 11.28 Crash sensing schemes (Jurgen, 1995)

When air-bag deployment is called for, the microcontroller or ECU turns on the firing current switches, allowing current to flow through the igniter, which initiates a gas generation reaction inside the inflation module. A central energy reserve capacitance maintains power for the system, allowing deployment even if the vehicle battery becomes disconnected during the crash. Figure 11.29 shows a block diagram of the microcontroller based system connected to a single-point accelerometer unit but this structure may be easily modified to add additional accelerometers for, say, side impact detection.

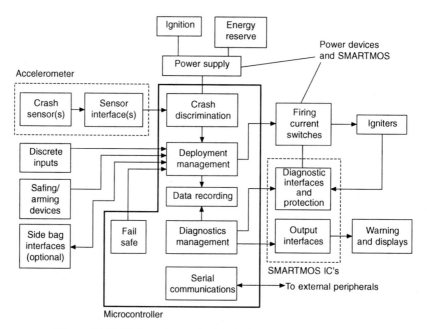

Figure 11.29 Air-bag electronics block diagram (Motorola, 1995)

A system readiness indicator is legally required for air-bag equipped vehicles and this means the periodic checking by diagnostic routines of the ignition loop and sensors. Reliability is the prime consideration in this system where it may lay dormant for years but be expected to be activated within milliseconds during crash conditions. Accidental triggering of the air-bag unit often occurred with DC-firing due to wiring short-circuit faults in the vehicle periphery (outside the ECU) and has led to the development of squib ignition by AC-firing, the firing pattern of pulses required being generated by the ECU.

11.8.2 Remote keyless entry and vehicle immobilization (Motorola, 1998)

This form of remote keyless entry security system combined with vehicle immobilization is becoming increasingly common, indeed insurance companies in countries such as Germany are insisting on immobilization as a prerequisite for granting insurance cover whilst companies in other countries are offering discounts to drivers whose vehicles are fitted with such devices. Figure 11.30 shows a block diagram of a typical system incorporating a 'smart' ignition key which allows several functions to be controlled. Besides its primary function of door lock control, each driver could have their own personalized key that would program several vehicle functions for that person such as seat position and mirror setting. Clever car thieves who captured the transmitted key codes of upmarket vehicles using data recording equipment and then used them to gain access have been thwarted by the development of rolling code generators that allow a unique code to be generated every time the RKE system is activated. Full immobilization includes shutting off the vehicle's fuel pump, fuel injectors, ignition, and starter

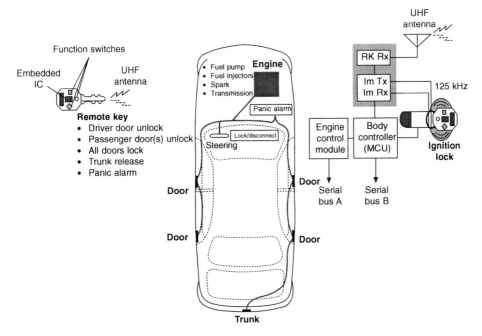

Figure 11.30 Remote keyless entry block diagram (Motorola, 1995)

while locking the transmission and steering, and causing the horn to sound and lights to flash. Additionally, the vehicle's cell phone could broadcast its position via a Global Position Satellite receiver to police authorities.

11.9 On-board navigation systems (Jurgen, 1995)

Modern developments in vehicle navigation systems use GPS information as a basis for positional information and is integrated with dead reckoning and map matching. GPS has the advantage of giving absolute position (albeit with an accuracy of up to 100 m) but is inadequate alone because of poor reception from satellite to vehicle caused by obstructions in the signal path. Dead reckoning is the process of calculating location by integrating measured increments of distance and direction of travel relative to a known location. The algorithm which gives a vehicle's coordinates (X_n, Y_n) relative to earlier coordinates (X_o, Y_o) is given by

$$X_n = X_o + \sum_1^n \Delta X_i = X_o + \sum_1^n \Delta l_i \sin \phi_i$$

$$Y_n = Y_o + \sum_1^n \Delta Y_i = Y_o + \sum_1^n \Delta l_i \sin \phi_i$$

where ϕ_i is the heading associated with Δl_i, the ith measured increment of travel. Figure 11.31 illustrates the concept and the integration process to calculate a vehicle's position.

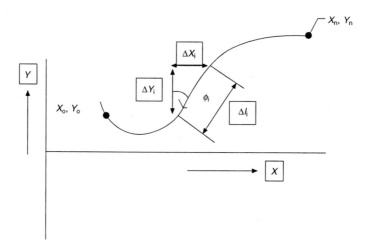

Figure 11.31 Path integration to produce dead reckoning information (Jurgen, 1995)

This technique is subject to a number of cumulative error sources such as tyre wear effects and tyre distortion due to either centrifugal force during cornering or change in tyre pressure in the course of use. (Error levels vary but could be as high as 3 to 4%.) Vehicle heading may be measured directly by a compass, a heading-change sensor based on gyroscopic principles or measured indirectly by differential odometer, the latter being formed from the signals picked up from the wheel speed sensors normally employed for ABS application. The majority of navigation systems use an on-board database of road map information and map matching, which is a form of artificial intelligence, is used to match the pattern of the apparent vehicle path with the road patterns of digital maps stored in the computer memory.

References and further reading

Adler, U. (ed.), (1988). *Automotive Electric/Electronic Systems*, Robert Bosch GmbH (Distributed by SAE). ISBN 3-18-419110-9.
Manufacturers detailed source material on their range of engine management systems.

Bissell, C.C. (1992). *Control Engineering*, Van Nostrand Reinhold (International). ISBN 0-278-00060-6.
A basic textbook on the concepts and modelling of control systems aimed at introducing students to a range of classical control techniques.

Bosch, Robert GmbH, (1984). *Automotive Electronics for Safety and Comfort* (Distributed by SAE).
A somewhat dated reference but containing, in layman's language, most of the fundamentals of the electronic based equipment found within a motor vehicle to ensure the safety and comfort of its occupants.

Chesmond, C.J. (1982). *Control System Technology*, Arnold. ISBN 0-7131-3508-5.
A text which bridges the gap between theoretical control engineering and the hardware of practical systems.

Chowanietz, E. (1995). *Automobile Electronics*, B-H. Newnes. ISBN 0-7506-1878-7.
A comprehensive introduction to automotive electronics which, besides providing a grounding in the subject, contains a number of case studies drawn from a wide variety of manufacturers' products. Highly recommended.

Chuen Chien Lee, (1990). *Fuzzy Logic in Control Systems, Parts 1 and 2, IEE Transactions on Systems, Man and Cybernetics*, Vol. **20**, No.2, March/April 1990, pp. 404–418 and 419–435.
Useful introductory reference to a branch of artificial intelligence which is finding applications in difficult control system problems.

Jurgen, R.K. (ed.) (1995). *Automotive Electronics Handbook*, McGraw-Hill. ISBN 0-07-033189-8.
A collection of articles by eminent authors from different parts of the auto industry. Extremely readable and containing good reference material.

Motorola (1998). *Automotive Design Solutions*, Internet address http://design-net.com/automotive.
Promotional material for Motorola products found on the Web, but which contains a good deal of insight into current microcontroller developments. Regularly updated and well worth the visit to the web site.

Nwagboso, C.O. (ed.), (1993). *Automotive Sensory Systems*, Chapman and Hall. ISBN 0-412-45880-2.
Another useful text containing contributions from various authors from the auto industry and academia but specifically on sensors.

Rzevski, G. (ed.), (1995). *Designing Intelligent Machines* (Vols.1 and 2), Butterworth-Heinemann ISBN 0-7506-2404-3 and ISBN 0-7506-2403-5.
A companion set of two books on the fundamentals of mechatronics, the fusion of mechanics and electronics in the design of intelligent machines. Contains the framework of a course in artificial intelligence.

Sheingold, D.H. (1981). *Transducer Interfacing Handbook*, Analogue Devices, Inc. ISBN 0-916550-05-2.
Contains the basics of transducer interfacing and the physical basis of transducers for measurement of temperature, force, pressure, flow and level.

12. The design of engine characteristics for vehicle use

Brian Agnew, PhD, CEng, FIMechE

The aim of this chapter is to:

- Introduce the types of power sources available to the designer;
- Indicate the important factors involved in the process of power source design;
- Demonstrate simple methods by which engine designs can be analysed.

12.1 Introduction

This chapter presents some aspects of the thermo-fluid mechanic processes that occur in a spark ignition engine. The purpose of this chapter is two-fold. Firstly the ideal Otto or constant volume air standard cycle is introduced as this represents the ideal thermodynamic cycle that closely resembles a spark ignition engine. The thermo-fluid dynamic processes within the cycle are identified which then allows the state variables of the working fluid to be determined at any point in the cycle. The cycle efficiency, specific work output and other characteristic features are identified. Real engine cycles are, however, significantly different from ideal cycles but the ideal cycle can form the basis of understanding that will lead onto the consideration of real cycle effects. The second purpose of this chapter then is to illustrate real cycle effects so that an ideal cycle can be modified to resemble an actual cycle. The differences between ideal and real cycles is illustrated by the application of fundamental thermodynamic and fluid dynamic equations that will give the astute reader an insight to the various processes that occur in a spark ignition engine. The chapter has been written with the intention of illustrating the application of fundamental processes to the analysis and design of an engine. To this end the use of computers in performing the analysis is avoided and the calculations illustrated can all be performed using a very basic calculator. The chapter ends with a description of the books recommended in a short reading list. All are well written and are very enjoyable to read furnishing the reader with some fascinating insights into the development of engine design from the 1920s to the present day.

12.2 The constant volume or Otto cycle

When beginning the analysis of a thermal power producing cycle the most important item to consider is the pressure–volume diagram associated with the cyclical processes. The area of the P–V diagram is an indication of the work produced per operating cycle which also produces an indication of the cycle power output. The different processes in the cycle can be illustrated and a measure of the cycle internal losses obtained. The pressure diagram in illustrating the relationship between pressure and crank angle is an indication of the main force acting on the structural components of an engine and is also a main contributor with inertial forces to the loading on the rotating elements. The bearings must be designed to withstand the cylinder pressure forces and the piston and con rod assembly must also withstand the net force due to gas pressure and inertia. A knowledge of the process paths within the cycle will also enable the thermodynamic state, notably the temperature of the working fluid to be determined. This is a major consideration when the thermal design of components such as the piston and the engine cooling system are considered.

A thermodynamic cycle of a spark ignition engine can be represented in Pressure–Volume coordinates as shown in Figure 12.1. This ideal cycle, known as the Otto Cycle after its inventor Gustav Otto, consists of an isentropic compression process 1–2, a constant volume

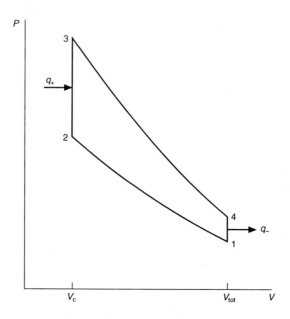

Figure 12.1 An ideal Otto cycle

heat addition process 2–3, an isentropic expansion 3–4, and a constant volume heat rejection process 4–1. The amount of heat received per unit mass of the working fluid per cycle is as shown in equation 12.1.

$$q_+ = c_v[T_3 - T_2] \tag{12.1}$$

Likewise the heat rejected from the cycle will be

$$q = c_v[T_4 - T_1] \tag{12.2}$$

The network produced by the cycle and the thermal efficiency of the cycle will then be:

$$W_{net} = Mc_v[T_3 - T_2 - T_4 + T_1] \tag{12.3}$$

$$\eta_{th} = 1 - \frac{|q_-|}{q_+} = 1 - \frac{(T_4 - T_1)}{(T_3 - T_2)} \tag{12.4}$$

Using the geometric relationships in the cycle the expression for the thermal efficiency then becomes as shown in equation 12.5.

$$\eta_{th} = 1 - \frac{1}{r^{n-1}} \tag{12.5}$$

The efficiency of the cycle is seen to depend upon the compression ratio r and the nature of the working fluid through the polytropic exponent n. The influence of compression ratio on the efficiency is shown in Figure 12.2 for different values of the exponent n. A value of n of 1.4 corresponds to that of an ideal gas with constant specific heats whilst the value of 1.3 is typical of that for products of combustion of petrol–air mixtures. For given values of r and n the efficiency is a constant quantity that is independent of the combustion process and therefore is also independent of the load.

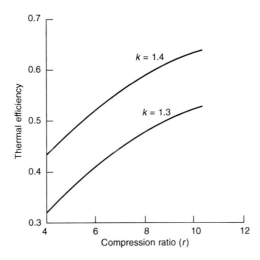

Figure 12.2 The influence of compression ratio r on the ideal cycle thermal efficiency for different values of the polytropic exponent

The mean effective pressure is often used as a measure of an engine's performance. It is obtained by dividing the work produced per cycle by the swept volume. The swept volume of each cylinder in the engine is

$$V_{sw} = V_1 - V_2 = \frac{\pi D^2 S}{4} \tag{12.6}$$

which produces for the mean effective pressure.

$$P_{MEP} = \frac{\int PdV}{V_{sw}} = \frac{P_1}{n-1} \frac{r^n}{r-1} \eta_{th} \left[\frac{P_3 - P_2}{P_2} \right] \tag{12.7}$$

The cycle mean effective pressure is seen to increase in proportion to the initial pressure of the cycle P_1. An increase in compression ratio r also increases the mean effective pressure.

An analysis of the expressions representing the cycle efficiency and mean effective pressure indicates that a high compression ratio is the most effective means of improving the characteristics of an engine. The maximum values of the compression ratio are however restricted by the necessity of ensuring normal combustion and restricting exhaust emissions below prescribed levels.

12.2.1 Performance at maximum power condition

In the case of an ideal cycle the state conditions at points 1 and 3 are fixed, state 1 by atmospheric considerations and state 3 by consideration of the maximum temperature or pressure that the engine structure can stand. It is therefore possible to change state 2 to lie anywhere between states 1 and 3. In the extreme if state 2 equals state 1 or 2 the work output of the cycle will be zero thus there will be a position of state 2 that will produce a maximum output for a given cycle. This next section explores this situation and establishes the condition for maximum power output.

If equation 12.3 is recast in terms of the cycle temperature ratio $T_3/T_1 = \gamma$ and the cycle compression temperature ratio $T_2/T_1 = \phi$ as shown below then it can be differentiated in terms of ϕ to produce a maximum.

$$\frac{W_{net}}{Mc_v T_1} = \theta - \phi - \frac{\theta}{\phi} - 1 \tag{12.8}$$

In deriving equation 12.8 use has been made of the second law of thermodynamics in establishing the relationship between T_2 and T_4 for this particular cycle. As the working fluid returns to its initial state after completing one cycle the entropy change around the cycle is zero.

$$\Delta S = \Delta S_{1-2} + \Delta S_{2-3} + \Delta S_{3-4} + \Delta S_{4-1} = 0 \tag{12.9}$$

$$0 = 0 + c_v \ln \left[\frac{T_3}{T_2} \right] + 0 + c_v \ln \left[\frac{T_1}{T_4} \right] \tag{12.10}$$

which leads to the following relationship

$$\frac{T_4}{T_1} = \frac{T_3}{T_2} = \frac{\theta}{\phi} \tag{12.11}$$

This equation could have been established equally well by examining the relationships between temperature and volume in the cycle but such an approach only applies to cycles with equal compression and expansion ratios and is therefore not generally applicable.

Differentiating equation 12.8 with respect to ϕ and equating the result to zero produces the following result.

$$\phi = \sqrt{\theta} \quad T_2 = \sqrt{T_1 T_3} \tag{12.12}$$

At this condition $T_4 = T_2$ the cycle efficiency will then be a function of the cycle temperature ratio θ only. This condition in some way explains the high noise output of sports car engines.

12.3 Deviations from the ideal cycles

The ideal cycle that has been examined so far is deficient in many areas when compared with a real cycle. The most noticeable omission of the ideal cycles is the gas exchange process in which the unburnt charge enters the engine and the products of the combustion process are rejected. In the four stroke engine the scavenge period occupies two piston strokes per cylinder, i.e. one revolution of the engine. The scavenging requires that the piston does work on the working fluid thus the power of the engine is more than halved because the cycle time has doubled and the network per cycle has decreased. If a directly proportional relationship existed between the ideal cycle and the actual cycle as implied by the statements above then the power output of the real cycle would be approximately as shown in Figure 12.3 being linear with the engine rotational speed. In fact the power output of a real cycle looks more like the second curve in Figure 12.3 exhibiting a maximum power point and then decreasing as the frictional effects and gas exchange processes detract from the cycle performance.

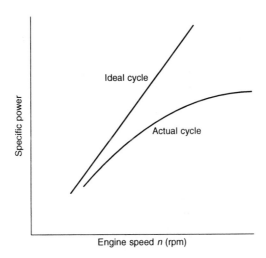

Figure 12.3 Specific power output for an ideal and an actual cycle

Experimental evidence indicates that the mean effective pressure of the practical cycle diagram is between 0.46 and 0.48 of the ideal cycle and the actual maximum cycle pressure P_3 is 85% of that for the ideal cycle. Clearly it is necessary to modify the ideal cycle analysis to bring it more into line with the real cycle. From a thermo-fluid dynamic approach there are three different phenomena that could account for the difference between the curves. These are the gas exchange process, progressive combustion and heat transfer. These will be examined in the following parts of this chapter.

12.3.1 The intake process

During the intake and exhaust periods of the cycle the piston moves at a finite speed not as in the ideal cycle infinitely slowly. Valves open and close to admit and discharge the working fluid at a finite rate which is controlled by pressure differences across the valves, the effective flow area and the piston speed. To accomplish the working cycle of an internal combustion engine it is necessary to intake fresh air and fuel into the working cylinder and expel the products of combustion formed during the previous cycle. The indicator diagram, Figure 12.4, shows schematically the process of gas exchange in a four stroke engine.

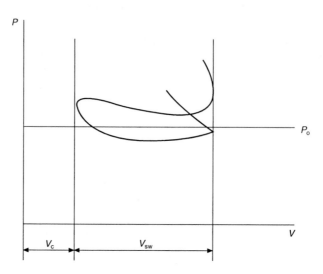

Figure 12.4 Indicator diagram of the gas exchange process

In an ideal cycle the intake and exhaust strokes occur at a constant cylinder pressure that is equal to the atmospheric pressure but in an actual cycle the intake stroke pressure will be below atmospheric pressure and the exhaust stroke pressure above atmospheric. The exact shape of the line representing the intake stroke depends on several factors including the engine speed, the hydraulic resistance of all the elements comprising the intake system, the cross sectional area through which the charge moves, the charge density, the presence of combustion residuals in the cylinder and heat transfer from the intake system walls that changes the charge density. When the inlet valve opens the pressure in the cylinder must drop below the ambient pressure by an amount equal to the intake system pressure loss before fresh charge will flow into the cylinder. The influence of each of the factors mentioned above can be determined if they are considered separately.

12.3.2 Cylinder pressure during the intake phase

The resistances in the intake system reduce the mass of fresh charge admitted into the engine cylinder by decreasing the charge density. The influence of hydraulic resistance can be judged if the pressure drop in the intake system or the pressure in the cylinder at the end of admission are known. The pressure in the cylinder during admission can be determined approximately if the entire process is considered as stable.

Using the Bernoulli equation between the inlet system and the engine cylinder

$$\frac{P_{in}}{\rho_{in}} + \frac{C_{in}^2}{2} + gz_{in} = \frac{P_{cy}}{\rho_{cy}} + (\beta^2 + c_f)\frac{C_{im}^2}{2} + gz_{cy} \qquad (12.13)$$

then assuming that $C_{in} = 0$ and $z_{cy} = z_{in}$ and neglecting the change in the density of the charge as it moves through the intake system produces

$$\frac{P_{in}}{\rho_{in}} = \frac{P_{cy}}{\rho_{cy}} + (\beta^2 + c_f)\frac{C_{im}^2}{2} \qquad (12.14)$$

then

$$\Delta P_{in} = P_{in} - P_{cy} = (\beta^2 + c_f)\frac{C_{im}^2}{2}\rho_{in} \qquad (12.15)$$

It follows from the above equation that the pressure drop in the intake system is proportional to the square of the velocity in the minimum cross section of the inlet system, and depends on the resistance coefficients of the system and the velocity drop. The continuity equation for the minimum section of the intake system, A_{im} and the cylinder will be

$$C_{im} A_{im} = C_{p.max} A_p \qquad (12.16)$$

The maximum piston velocity can be expressed as

$$C_{p.max} = \frac{R_c\, 2\pi n}{60}\sqrt{1 + \frac{R_c^2}{l_{rod}}} \qquad (12.17)$$

which leads to the following expression for C_{im}.

$$C_{im} = C_{p.max}\frac{A_p}{A_{im}} = \frac{2\pi R_c n}{60}\sqrt{1 + \frac{R_c^2}{l_{rod}}}\,\pi\frac{D^2}{4}\frac{1}{A_{im}} = C_1\frac{n}{A_{im}} \qquad (12.18)$$

Substitution of C_{im} shows that the pressure drop is proportional to the square of the engine speed and inversely proportional to the square of smallest area in the intake system, A_{im}.

$$\Delta P_{in} = (\beta^2 + c_f)\,\rho_{in}\,C_1^2\frac{n^2}{2}\frac{1}{A_{im}^2} = C_2\frac{n^2}{A_{im}^2} \qquad (12.19)$$

Experimental evidence shows that the pressure in the cylinder at the end of the intake stroke is approximately 80% to 90% of ambient pressure. In modern four stroke engines with overhead cams the possibility of increasing the area A_{in} is limited by the arrangement of the valves in the cylinder head. The total valve area may be increased by using four valves and inclined valves as in hemispherical heads.

12.3.3 Temperature at admission

Charge heating during the intake process
As the fresh charge is propelled through the intake system it comes into contact with the relatively hot surfaces of the inlet manifold and cylinder head. Heat is transferred into the fresh

charge increasing its temperature and reducing its density. This heat transfer process has a detrimental effect on charging of the cylinder. In the case of a carburetted engine where the mixture is formed external to the engine cylinder it is sufficient only to heat the fresh charge to vaporize the fuel. It is very difficult in practice to achieve only this degree of heating such that there will be an increase in the charge temperature. This increase can be expressed as follows:

$$\Delta T = \frac{q - \dfrac{h_{fg}}{1 + \alpha}}{c_p} \qquad (12.20)$$

in which q is the heat transfer to the mixture per unit mass of mixture and h_{fg} is the latent heat of the fuel, α is the air–fuel ratio and c_p is the specific heat of the mixture. In engines with multipoint fuel injection the fuel is usually injected directly into the inlet port. In this case the charge is not cooled in the inlet manifold by the latent heat of the fuel which is used instead to reduce the thermal loading of the inlet valve. It is difficult to give an exact value of the change in the charge temperature during the intake phase as it is dependent on many variables that are specific to individual engines. As a rule of thumb the temperature rise of a carburetted engine would be of the order of 0–20 °C and that of a fuel injected engine of 20– 40 °C

Residual gases
Not all the products of combustion leave the combustion chamber during the exhaust stroke. At TDC the clearance volume will be filled with exhaust gases that will constitute the residual fraction of the next cycle. During the intake stroke the residual gases expand, mix with the incoming charge and reduce the volume of fresh charge in the cylinder. The quantity of the residual gases usually expressed as a mol fraction of the fresh charge depends on how the cylinder is scavenged. In four stroke engines with a small degree of valve overlap (30–40°) the quantity of residual gas is:

$$M_{res} = \frac{P_{res} V_c}{R T_{res}} \qquad (12.21)$$

The value of P_{res} in the above equation is the pressure of the exhaust system and T_{res} depends upon the mixture composition, the expansion ratio and on the heat exchange during expansion and exhaust. At full load the value of the coefficient of residual gases γ_{res} lies between 0.06 to 0.1. When calculating the coefficient it may be assumed that $P_{res} = 1.1 - 1.25$ bar and the temperature $T_{res} = 900$–1000 K. The coefficient can be decreased in four stroke engines by increasing the valve overlap but this may adversely affect the low load running of the engine. Since the exhaust process in a four stroke engine can be assumed to have terminated at TDC then the residual gas coefficient can be expressed as:

$$\gamma_{res} = \frac{1}{r - 1} \frac{P_{res}}{P_{in}} \frac{T_{in}}{T_{res}} \frac{1}{\eta_v} \qquad (12.22)$$

If the valve overlap is large an additional scavenge effect may be produced that reduces the amount of the residual charge. This is described by using the coefficient of purging C_{pur}. Without scavenging $C_{pur} = 1$ and with absolute purging $C_{pur} = 0$. The residual gas coefficient then takes the form

$$\gamma_{res} = \frac{C_{pur}}{(r-1)} \frac{P_{res}}{P_{in}} \frac{T_{in}}{T_{res}} \frac{1}{\eta_v} \tag{12.23}$$

Substituting for the volumetric efficiency η_v (see later) produces an equation for the residual gas coefficient in terms of the temperature, pressure and experimental coefficients

$$\gamma_{res} = C_{pur} \frac{(T_{in} + \Delta T)}{T_{res}} \frac{P_{res} C_{pres}}{C_{ch} r P_a - c_{pin} C_{pur} P_{res}} \tag{12.24}$$

in which C_{ch} is the additional charge coefficient which is defined later. Without scavenging and additional charging of the engine then the residual gas coefficient takes the form shown below.

$$\gamma_{res} = \frac{T_{in} + \Delta T}{T_{res}} \frac{P_{res}}{r P_a - P_{res}} \tag{12.25}$$

Temperature at the end of admission
The fresh charge upon entering the engine cylinder is mixed at constant pressure with the residual gases. If the specific heat of the resulting mixture is assumed to be the same as the inlet charge then an energy balance based on the conditions before and after the in-cylinder mixing process allows the temperature of the mixture T_a to be determined.

$$c_p M_{in}(T_o - \Delta T) + c_{pres} M_{res} T_{res} = c_{pin}(M_1 + M_{res}) T_a \tag{12.26}$$

whence

$$T_a = \frac{c_{pin} M_1 (T_o + \Delta T) + c_{pres} M_{res} T_{res}}{(M_1 + M_{res}) c_{pin}} \tag{12.27}$$

Since

$$M_1 + M_{res} = M_1(1 + \gamma_{res}) \tag{12.28}$$

then

$$T_a = \frac{T_o + \Delta T + \dfrac{c_{pres} \gamma_{res} T_{res}}{c_{pin}}}{1 + \gamma_{res}} \tag{12.29}$$

12.3.4 Volumetric efficiency

The ability of the engine to draw in fresh charge is characterized by a factor called the volumetric efficiency. This is defined as the ratio of the volume of fresh charge present in a cylinder when actual compression begins, i.e. when the valves are closed, to the engine swept volume. According to this definition

$$\eta_v = \frac{M_{in}}{V_{cy} \rho_o} = \frac{V_{in}}{V_{cy}} \tag{12.30}$$

in which M_{in} is the quantity of fresh charge in the cylinder at the beginning of compression and V_{cy} is the clearance volume plus the swept cylinder volume. In four stroke engines in which the cylinder is scavenged during overlapping of the valves some of the admitted fresh charge is used for scavenging and does not participate in the compression and combustion processes. The amount of charge used for scavenging is appraised by the scavenging coefficient

$$C_{sc} = \frac{M_{tot}}{M_{in}} \tag{12.31}$$

where M_{tot} is the total amount of fresh charge admitted into the cylinder per cycle. For four stroke engines in which the valve overlap does not exceed 40–50°, the scavenge coefficient can be taken as 1.

Derivation of the volumetric efficiency
When the piston is at BDC at the end of the intake stroke the cylinder contains M_1 mols of fresh charge and M_{res} mols of residual gases. The Perfect Gas Law relates the conditions at this point as follows:

$$P_1 V_1 = R(M_1 + M_{res})T_1 \tag{12.32}$$

The total amount of fresh charge supplied refereed to the intake conditions can be expressed as

$$P_{in} V_{in} = R\, M_1 T_{in} = P_{in} \eta_v V_{cy} \tag{12.33}$$

In the general case M_1 may be greater than shown since additional charging may be possible from BDC to the moment the intake is closed. This is accounted for by the coefficient of additional charging

$$C_{ch} = \frac{M_{1a} + M_{res}}{M_1 + M_{res}} \tag{12.34}$$

and

$$M_1 + M_{res} = \frac{(M_{1a} + M_{res})}{C_{ch}} \tag{12.35}$$

Substituting produces

$$M_{1a} + M_{res} = \frac{C_{ch} P_1 V_1}{RT_1} \tag{12.36}$$

Hence

$$\frac{M_{1a} + M_{res}}{M_1} = \frac{C_{ch} P_1 V_1 T_{in}}{P_{in} V_{sw} T_a \eta_v} \tag{12.37}$$

but

$$\frac{V_1}{V_{sw}} = \frac{r}{r-1} \frac{P_1}{P_{in}} \frac{T_{in}}{T_1 (1 + \gamma_{res})} \tag{12.38}$$

Substituting from equation 12.31 into 12.30 and rearranging then produces the required result

$$\eta_v = C_{ch} \frac{r}{r-1} \frac{P_1}{P_{in}} \frac{T_{in}}{T_{in} + \Delta T + \dfrac{c_{pres}}{c_{pin}} \gamma_{res} T_{res}} \tag{12.39}$$

Referring to the previous section that defined the residual gas coefficient, the above equation can also be expressed as follows

$$\eta_v = \frac{T_{in}}{T_{in} + \Delta T} \frac{1}{r - 1}\left(C_{ch} r \frac{P_1}{P_{in}} - \frac{C_{pres}}{C_{pin}} C_{pur} \frac{P_{res}}{P_{in}}\right) \qquad (12.40)$$

Without scavenging and additional charging then this approximates to the following

$$\eta_v = \frac{T_{in}}{T_{in} + \Delta T} \frac{1}{r - 1}\left(r \frac{P_1}{P_{in}} - \frac{P_{res}}{P_{in}}\right) \qquad (12.41)$$

Factors affecting the volumetric efficiency
The previous work has shown that the volumetric efficiency is a function of the pressure P_1, the temperature T_1 at the end of admission, the charge temperature change due to heat transfer, the residual gas coefficient and the temperature and pressure of the residual gases, the compression ratio and the purging and additional charging coefficients. These variables in turn depend on a number of other variables and are also interrelated. It is then only possible to consider the influence of one of these variables if the influence of the effect this has on other variables is also considered. If the other parameters remain unchanged, the coefficient of admission will increase with increased compression ratio. In practice, an increase in the compression ratio also changes the residual gas fraction and the residual gas temperature. If the cylinder is fully scavenged it can be shown that an increase in compression ratio decreases the volumetric efficiency. The net effect on η_v will depend upon which change has the greatest influence and in practice it is found that the compression ratio has negligible effects on η_v.

The pressure P_1 has the greatest influence on the cylinder filling as P_1 depends on the resistance in the intake system and is proportional to the square of the charge mean velocity C_{im} in the narrowest cross section of the intake system. The value of P_1 is also affected by the design of the intake line (the arrangement of the valves, presence of turns, local resistances, etc.), the inner surface finish of the walls in the intake system, the position of the throttle and the engine speed. Figure 12.5 illustrates how the value of η_v depends on the mean velocity C_{im}

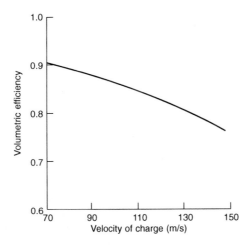

Figure 12.5 The influence of charge velocity at the inlet upon volumetric efficiency

of the fresh charge in the section of completely open inlet valves in a four stroke engine operating at the rated duty. A growth in the velocity reduces the volumetric efficiency, which must be taken into account when an increased charge rate is required.

Charging an engine at constant speed and variable load
When the load on a carburetted engine is decreased and the throttle correspondingly closed, the hydraulic resistance of the intake system increases and changes the gas exchange process. The influence of throttle position on the admission pressure P_a and other parameters is shown in Figure 12.6. As the throttle is closed the admission pressure reduces and the proportion of the expansion stroke occupied by the expanding residual gases increases thus delaying the start of the admission process. The residual gas coefficient also increases with throttle closure. Heating of the charge in the inlet system is reduced as the surface temperatures are lower with the reduced engine load. The net effect is a reduction in the volumetric efficiency.

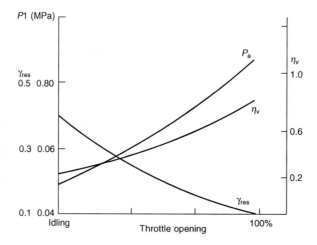

Figure 12.6 The influence of throttle position on various intake parameters

Influence of speed variation on charging four stroke engines
When the speed is changed in an engine operating under full load, the quality of charging is affected by the resistance in the intake system, heating of the charge and the presence of residual gases, as well as by the valve timing. Figure 12.7 illustrates the change in separate factors influencing η_v against a change of engine speed. The resistance of the system grows in proportion to the square of the speed, the result being an increase in ΔP and a reduction in P_1. Although the mean temperature of the heat transfer surfaces increases, the charge heating temperature ΔT drops because of the smaller time available for heat transfer. The residual gas coefficient slightly increases, which reduces the volumetric efficiency if the influence of valve timing and leakage of the charge past the rings is disregarded. When the speed increases η_v is first observed to grow and after reaching a maximum value decrease. At low engine speeds the increase in η_v with increased speed is due to the improved phase relationships of the intake and exhaust opening periods. As the speed further increases beyond the value at which η_v is a

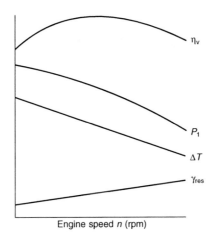

Figure 12.7 The influence of engine speed on several intake parameters

maximum the hydraulic resistance in the inlet system is beginning to dominate the process and reduces η_v.

12.4 The compression process

During the compression process the temperature and pressure of the mixture in the cylinder increase. The value of these properties at the end of compression depend upon the values at the beginning of the compression process, the compression ratio and the nature of the heat exchange between the fresh charge and the cylinder walls. Compression of the fresh charge produces conditions that are favourable for the process of ignition and combustion. During the beginning of the compression process the temperature of the mixture is below that of the cylinder surfaces. The temperature of the charge thus increases due to the addition of energy from the surfaces as well from the compression process. At a certain moment, the mean temperature of the charge and the walls are equalized, and heat is then rejected from the charge to the walls as the compression continues to completion. The nature of this compression process is shown in Figure 12.8. If the compression were to be adiabatic with an index of compression of 1.4 the process would follow line *a–b* in the figure. If the mean temperature of the chamber surfaces is T_w during the initial stages of compression the process path would follow a polytropic line with a variable index $n > k$ as shown in line *a–r*. From point *r*, where $T = T_w$ the compression takes place with a variable exponent $n < k$. The heat exchange during the second period is affected by the temperature difference $(T - T_w)$ which constantly increases and by the reduction in the surface area available for heat transfer. As a result the temperature and pressure of the charge at the end of compression shown as point *c* in the figure will be different from those obtained if the process had been adiabatic.

It is difficult to determine analytically the parameters at the end of the compression process taking into account the variable polytropic exponent. The temperature and pressure at the end of compression are usually computed using an average polytropic exponent n_1, which is taken to be constant during the entire process. Assuming that compression begins at BDC then

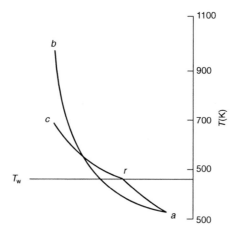

Figure 12.8 Indicator diagram illustrating the compression process

$$P_2 = P_1 r^{n_1}, \quad T_2 = T_1 r^{(n_1-1)} \tag{12.42}$$

Figure 12.9 shows values of P_2 and T_2 calculated using the above formulas for three values of n_1 when $P_1 = 0.9$ MPa and $T_1 = 323$ K. It can be seen that a change in the exponent n_1 within the indicated limits appreciably changes the values of P_2 and T_2. For this reason, the values of n_1 should be selected from available experimental data for engines similar in size and operating conditions to the one being designed.

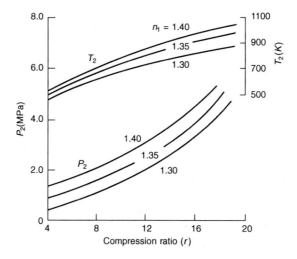

Figure 12.9 The influence of compression ratio on the compression pressure and temperature

The nature of the heat exchange and the little time available during which it occurs considerably reduce the total amount of heat exchange in high speed engines, which amounts to approximately

1.0 to 1.5 % of the heat introduced with the fuel. Therefore, in the absence of reliable data on the value of n_1, it can be determined from the mean adiabatic exponent using the formula

$$n_1 = 1 + \frac{\log_{10}\left[\dfrac{T_2}{T_1}\right]}{\log_{10} r} \tag{12.43}$$

12.4.1 Influence of various factors on the compression process

In the previous section it was mentioned that heat transfer from the engine structure to the fresh charge as it is compressed has an influence upon the compression process. It is the objective of this section to examine the factors that effect the heat transfer and also discuss other factors that have an influence on the compression process in real engines.

The condition of heat exchange during compression is determined by several factors. The temperature difference between the charge and the engine surfaces is clearly important but other factors to consider are:

1   The area of the heat transfer surfaces;
2   The amount of charge in the cylinder;
3   The time available for the heat transfer to take place;
4   The coefficient of convective heat transfer between the charge and the surfaces;
5   The amount of fuel evaporated and the latent heat of the fuel.

The final state of the fresh charge also depends upon the initial state and the magnitude (if any) of the blow-by of charge past the piston rings. In starting a cold engine, when the crankshaft rotational speed is low, the rings do not completely seal the combustion chamber so that as the compression takes a comparatively long time, appreciable leakage can occur through the piston rings. Here the mean polytropic exponent will be low which in turn reduces P_2 and T_2. The exponent is also affected by the cooling system. With air cooling, the engine temperature is relatively high so that relatively less heat will be rejected by the charge thus raising the value of n_1. If the engine is liquid cooled the surface temperatures will be relatively low resulting in a low value of n_1. The use of aluminium components such as pistons and cylinder heads will tend to reduce n_1 as this material has a relatively high coefficient of thermal conductivity.

12.5 Progressive combustion

Combustion in internal combustion engines does not occur instantaneously as implied by the ideal cycle model. Constant volume combustion can only occur if the process is instantaneous or if the piston stopped whilst the combustion takes place. In the first instance the stress on the mechanical parts produced by the very high rate of change of pressure would soon damage the engine and in the second a mechanism for controlling the piston motion would be very impractical. Combustion is usually initiated before the piston reaches the end of its travel (TDC) and finishes when the piston has moved beyond TDC on the return stroke. The time required for combustion depends on many factors including the combustion chamber geometry, the spark plug location, fuel type and in terms of crank angle degrees may occupy up to 80° depending on the engine speed and combustion chamber design.

In order to calculate the net work done by the engine during the combustion process it is necessary to define the point of ignition and the combustion duration and the relationship between the cylinder pressure and the volume during this phase of the cycle. The essential idea behind this aspect of the cycle analysis is that the pressure change at any instant is made up of two components, that due to the motion of the piston ΔP_p and that due to the combustion process at that instant ΔP_{comb} as shown below

$$\Delta P = \Delta P_p + \Delta P_{comb} \tag{12.44}$$

The first term in the above expression can be determined by consideration of the compression process to be reversible and adiabatic as characterized by the isentropic relationship

$$PV^\gamma = C \quad \text{or} \quad \log(P) = \log C - \gamma \log(V) \tag{12.45}$$

differentiating the logarithmic relationship then produces the change in pressure for a given change in cylinder volume with V and P referring to the condition at the initiation of the volume change.

$$\Delta P_p = -P\gamma \frac{\Delta V}{V} \tag{12.46}$$

The second term in equation 12.38 requires a little more analysis. Consider the events that occur in the combustion process in a closed cylindrical vessel shown in Figure 12.10. The vessel is initially filled with a combustible mixture of fuel and air of mass M at a pressure P_i and a temperature T_i. A spark occurs at the left hand end of the vessel and propagates with a characteristic velocity towards the right hand end. The speed at which the flame front progresses along the vessel will be due in part to the normal flame speed that occurs when a flame propagates into an unburnt mixture and it will also be due to the expansion of the products of combustion behind the flame front. It is assumed at any time during the combustion process the pressure in the vessel is uniform but the same cannot be said about the temperature. Imagine

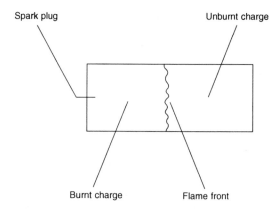

Figure 12.10 Flame propagation in a constant volume chamber

that the charge is sub-divided into n elements of equal mass. A temperature distribution will exist in the vessel because each of these elements will be at a uniquely different state just prior to combustion. For instance the first element to burn will initially be at the temperature T_i. As it burns the heat of combustion will be released raising the temperature to $T_i + \Delta T_c$. As the element is small it can be considered to burn at constant pressure. After combustion this element will then be compressed as the combustion of the remainder of the charge raises the chamber pressure to P_e the pressure at the end of combustion. The temperature of the first element to burn at the end of combustion will then be

$$T_1 = (T_i + \Delta T_{comb})\left[\frac{P_e}{P_i}\right]^{\frac{r-1}{r}} \tag{12.47}$$

The final element to burn on the other hand will be compressed by the rise in the chamber pressure due to the combustion of the previous elements and will then be heated by its own combustion process. In this case the final temperature will be

$$T_n = T_i\left[\frac{P_e}{P_i}\right]^{\frac{r-1}{r}} + \Delta T_{comb} \tag{12.48}$$

The final temperature of the first element to burn will be higher than that of the last element to burn and a temperature distribution will exist in the products of combustion as shown in Figure 12.11. If we arbitrarily choose any time after ignition, the flame front will be in a position shown with the reactants of mass m_r to the right and the mass m_p of the products of combustion to the left. Let T_p be the mean temperature of the products and T_r the temperature of the reactants. If the combustion process is adiabatic then an energy balance on the chamber will be as follows:

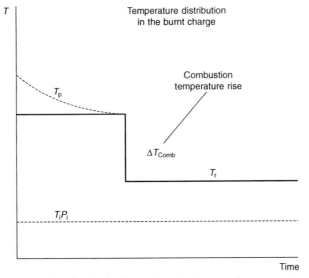

Figure 12.11 Temperature distribution in the combustion process in a constant volume chamber

$$Mu_r(T_i) = m_r u_r(T_r) + m_p u_p(T_p) \qquad (12.49)$$

Choosing the initial condition as a datum point from which to measure energy changes then produces

$$u_r(T_i) = 0 = u_r(T_e) \qquad (12.50)$$

and the energy equation then reduces to:

$$m_r C\upsilon_r(T_r - T_i) + m_p C\upsilon_p (T_p - T_e) = 0 \qquad (12.51)$$

Let

$$a = \frac{C\upsilon_r}{C\upsilon_p} \quad b = \frac{N_r}{N_p} \quad x = \frac{m_p}{M} \qquad (12.52)$$

Then equation 12.51 reduces to

$$a(1 - n)(T_r - T_i) + n(T_p - T_a) = 0 \qquad (12.53)$$

In addition the volume of the combustion chamber is constant so that the following applies

$$\frac{M R_i T_i}{P_i} = \frac{m_r R_r T_r}{P} + \frac{m_p R_p T_p}{P} \qquad (12.54)$$

which can be rearranged to produce

$$b\left(\frac{P}{P_i}\right)T_i = x(1 - x)T_r + xT_p \qquad (12.55)$$

Remembering that the reactants are compressed isentropically and eliminating T_p and T_r produces a relationship between the mass fraction burned, x, and characteristic pressures of the combustion process as shown.

$$x = \frac{(a - b)\left(\dfrac{P}{P_i}\right)^{\frac{r-1}{r}} + b\dfrac{P}{P_i} - a}{(a - b)\left(\dfrac{P}{P_i}\right)^{\frac{r-1}{r}} + b\dfrac{P_e}{P_i} - a} \qquad (12.56)$$

Typically in this type of combustion process a and b have values of 0.7 and 0.9 respectively thus the above equation can be simplified to the following with little loss of accuracy.

$$x = \frac{bP - aP_i}{b P_e - aP_i} \qquad (12.57)$$

If there were no change in the molar mass and the specific heats between the reactants and products an even simpler form of the equation would be produced

$$x = \frac{P - P_i}{P_e - P_i} \qquad (12.58)$$

The change in the pressure in the combustion chamber due to the combustion process is then

$$\Delta P_{\text{comb}} = (P_e - P_i)\Delta x \qquad (12.59)$$

In the above equation P_i and P_e are the pressures at the initiation and the end of a constant volume combustion process. Referring to the ideal cycle in which combustion occurs at TDC then $P_i = P_2$ and $P_e = P_3$. Consequently when combustion takes place at some volume V or when a fraction Δx of the reactants burns at volume V, the relation between ΔP_{comb} and Δx is simply written as

$$\Delta P_{\text{comb}} = (P_3 - P_2)\frac{V_3}{V}\Delta x \qquad (12.60)$$

in which P_2 and P_3 are the pressures that would occur if the cycle was an ideal cycle. Combining the previous results then leads to the required relationship linking the cylinder pressure change during the combustion process to that produced by the piston motion and that due to the combustion

$$\Delta P = -P\gamma\frac{\Delta V}{V} + (P_3 - P_2)\frac{V_{\text{TDC}}}{V}\Delta x \qquad (12.61)$$

In this equation the cylinder volume is assumed to change by an increment ΔV over a time period or crank angle period $\Delta\theta$ and an elemental mass of the reactants Δx burns during this time period.

To proceed further and produce a cylinder diagram it is necessary to integrate the above equation. The two relations that are required to do this are those that relate the combustion chamber volume to the crank angle and the mass fraction burned to crank angle. The geometric relationships between the crank, conrod and cylinder bore will provide the first of these relationships. Experimental evidence has shown that an equation of the form shown below is sufficiently accurate to describe the relationship between crank angle and mass fraction burned is of the form shown below.

$$x = \frac{1}{2}\left[1 - \cos\left[\left(\frac{\theta - \theta_i}{\Delta\theta_c}\right)\pi\right]\right] \qquad (12.62)$$

which produces the following relationship for Δx

$$\Delta x = \frac{\pi}{2}n\sin\left(\frac{\theta - \theta_i}{\Delta\theta_c}\pi\right) \qquad (12.63)$$

A simple and straightforward way of dealing with the combustion volume is to divide the combustion period into N steps and if the geometric relationship between the combustion chamber volume and the crank angle is known then the volume change over a crank angle interval is then

$$\Delta V = V_{\theta+\Delta\theta} - V_\theta \qquad (12.64)$$

A detailed description of the process can be found in the text by Heywood that is discussed later in this chapter.

12.6 The chemistry of the combustion process

The combustion of fuel in an engine combustion chamber is a complex chemical process that goes through several distinct phases. Initially the reactant species are broken down by the advancing flame front. A great deal of work has been performed in determining the products of combustion of hydrocarbon fuels. Much of this was stimulated by rocket research. At high pressures in the region of 50 bar and at high temperatures dissociation will occur and a significant amount of NO is present in the products. At high temperatures the species concentration is determined by chemical kinetics. For most species the chemistry is fast enough that local equilibrium may be assumed but this does not apply in the case of Oxides of Nitrogen. In this case the chemistry is not fast enough to assume that it is in equilibrium concentrations and once formed in the high temperature reactions its chemistry freezes during the engine expansion stroke so that there exist oxides of nitrogen in low temperature exhaust gases. This is important because these constituents together with unburnt hydrocarbons in the exhaust gases are responsible for the photochemical smog that blights several major cities world wide.

12.6.1 Chemical reactions in complete combustion

In complete combustion it is assumed that the reactants are completely oxidized so that the products resulting from the combustion of a hydrocarbon fuel in air consist only of carbon dioxide, water vapour and the nitrogen contained in the air. In this case the combustion equation for the complete combustion of iso-octane will be

$$C_8 H_{18} + 12.5 \, [O_2 + 3.76 \, N_2] \rightarrow 8 \, CO_2 + 9 \, H_2O + 47 \, N_2 \qquad (12.65)$$

This equation states that the complete combustion of iso-octane requires 12.5 mols of air which contains one mol of oxygen and 3.76 mols of nitrogen to produce 8 mols of carbon dioxide, nine mols of water and 47 mols of nitrogen. More generally the complete combustion of one mol of carbon requires the presence one mol of oxygen and 0.25 mols of oxygen for the combustion of one mol of hydrogen. The combustion equation for the reaction of a hydrocarbon fuel of the type $C_m H_n$ can then be written on the basis

$$C_m H_m + \left(m + \frac{n}{4} \right) [O_2 + 3.76 N_2] \rightarrow mCO_2 + \frac{m}{2} H_2O + \left(m + \frac{n}{4} \right) 3.76 \, N_2 \qquad (12.66)$$

and the theoretical air required for the combustion of one mol of the fuel is found from the expression

$$A_{th} = \frac{1}{0.21} \left(n + \frac{m}{4} \right) \qquad (12.67)$$

12.6.2 Excess air coefficient

The amount of air actually used in an engine can vary considerably from the stoichiometric depending on the type of combustion, the ignition process and the operating conditions. The ratio of the amount of air participating in the combustion to that theoretical required is termed the 'excess air coefficient' and is denoted by the symbol α which is defined as follows:

$$\alpha = \frac{A}{A_{th}} \tag{12.68}$$

Clearly when the mixture is stoichiometric $A = A_{th}$ and $\alpha = 1$. If insufficient air is available for complete combustion then $\alpha < 1$ and the mixture is said to be rich. In this case there will be insufficient oxygen for the complete combustion of the fuel so that the products of combustion will contain some unburnt fuel, carbon monoxide, CO, and the hydrogen radical H as well as small amounts of intermediate species that could include small amounts of intermediate species that could include O, OH, NO. Modem engines are constrained by the installation of catalytic converters for emissions control to operate with chemically correct (stoichiometric) mixtures but there is an increasing interest in development of engines that can operate with *n*-fixtures with excess air (lean). This has the advantage of reducing the formation of CO, NO_x and unburnt hydrocarbons until the onset of combustion instabilities. At the present state of the art homogeneous charge lean burn engines cannot meet the requirements of the emissions regulations but several manufacturers are conducting research into stratified charge direct injected engines.

12.6.3 Combustion modelling of the in-cylinder processes

With the concern about the effects of atmospheric pollution particularly NO_x, unburnt hydrocarbons and CO on the environment there has been developed a wide range of programs to study the formation of the products of combustion in the engine cylinder. Studies have shown that the combustion process can be adequately represented by considering the formation of 10 product species. These are CO_2, H_2O, N_2, O_2, CO, H_2, H, O, OH, and NO which can be described in the combustion equation as shown below:

$$C_\alpha\,H_\beta + x_{11}[O_2 + 3.76\,N_2] \rightarrow$$

$$x_1 CO_2 + x_2\,H_2O + x_3 N_2 + x_4 O_2 + x_5 CO + x_6 H_2 + x_7 H + x_8 O + x_9 OH + x_{10}\,NO \tag{12.69}$$

In order to fully describe the combustion process it will be necessary to simultaneously solve eleven equations for the eleven unknowns in the combustion equation, the mole fractions of the product species $x_1 - x_{10}$ and the number of mols of air x_{11} that react with one mole of fuel. Conservation of the atomic species C, H, N, O will furnish four equations, a fifth will stem from an energy (enthalpy) balance as the process is considered to be adiabatic with the enthalpy of the products being equal to that of the reactants. A sixth equation will stem from the mole fractions constraint i.e.

$$\sum_{i=1}^{10} x_i = 1 \tag{12.70}$$

The remaining equations come from the following partial pressure equilibrium equations describing hypothetical reactions that are deemed to occur in the overall process.

$$\tfrac{1}{2}H_2 - H, \quad \tfrac{1}{2}O_2 - O \tag{12.71}$$

$$\tfrac{1}{2}H_2 + \tfrac{1}{2}O_2 - OH, \quad \tfrac{1}{2}O_2 + \tfrac{1}{2}N_2 - NO \tag{12.72}$$

$$H_2 + \frac{1}{2}O_2 - H_2O, \quad CO + \frac{1}{2}O_2 - CO_2 \tag{12.73}$$

This and the method of solution of the resulting nonlinear equations is demonstrated below with an example concerned with the partial oxidation of methane CH_4 with oxygen.

12.6.4 Setting up the initial equations

The principal reactions in the production of synthesis gas by partial oxidation of methane are:

$$CH_4 + \frac{1}{2}O_2 \rightarrow CO + @H_2 \tag{12.74}$$

$$CH_4 + H_2O - CO + 3H_2 \tag{12.75}$$

$$H_2 + CO_2 - CO + H_2O \tag{12.76}$$

Assuming that the gases behave ideally so that the component activities are identical with component partial pressures, the equilibrium constants equations are respectively:

$$K_1 = \frac{P_{CO}P^2_{H_2}}{P_{CH4}P^{\frac{1}{2}}_{O_2}} \tag{12.77}$$

$$K_2 = \frac{P_{CO}P^2_{H_2}}{P_{CH_4}P_{H_2O}} \tag{12.78}$$

$$K_3 = \frac{P_{CO}P_{H_2O}}{P_{CO_2}P_{H_2}} \tag{12.79}$$

where the P_{xx} represent the partial pressures of CO, CO_2, H_2O, H_2, O_2 and CH_4.

The enthalpies of the various components for a specified temperature can be found in for instance the *JANAF Thermochemical Tables*. A fourth reaction may also occur at high temperatures:

$$C + CO_2 - 2CO \tag{12.80}$$

At relatively low temperatures (1500 K) any carbon formed would be deposited as a solid; the equilibrium constant equation for this reaction is given by

$$K_4 = \frac{P^2_{CO}}{a_c P_{CO_2}} \tag{12.81}$$

where a_c is the activity of carbon in the solid state. It is normal procedure not to include equation in the equilibrium analysis but to determine the equilibrium composition considering only the homogeneous gaseous reactions and then to determine the thermodynamic likelihood that solid carbon would appear as a result of reaction.

12.6.5 Method of solution

Let the following nomenclature be used

x_1 mole fraction of CO in the product mixture
x_2 mole fraction of CO_2 in the product mixture

x_3 mole fraction of H_2O in the product mixture
x_4 mole fraction of H_2 in the product mixture
x_5 mole fraction of CH_4 in the product mixture
x_6 number of mols of O_2 per mole of CH_4 in the reacting gases
x_7 number of mols of product gases per mole of CH_4 in the reactant gases

The combustion equation can then be stated as follows:

$$CH_4 + x_6O_2 = x_7(x_1CO + x_2CO_2 + x_3H_2O + x_4H_2 + x_5CH_4) \qquad (12.82)$$

Atom conservation balances
The number of atoms of each element in the reactants equals the number of each element in the products.

$$\text{Oxygen}: x_6 = \left(\frac{1}{2}x_1 + x_2 + \frac{1}{2}x_3\right)x_7 \qquad (12.83)$$

$$\text{Hydrogen}: 4 = (2x_3 + 2x_4 + 4x_5)x_7 \qquad (12.84)$$

$$\text{Carbon}: 1 = (x_1 + x_2 + x_5)x_7 \qquad (12.85)$$

Energy balance
Since the reaction is to be conducted adiabatically, that is no energy is added to or removed from the reacting gases, the enthalpy (H) of the reactants must equal the enthalpy of the products

$$[H_{CH_4} + x_6H_{O_2}]_{T_i} = x_7[x_1H_{CO} + x_2H_{CO_2} + x_3H_{H_2O} + x_4H_{H_2} + x_5H_{CH_4}]_{T_o} \qquad (12.86)$$

Mol fraction constraints
The formation of the combustion equation leads to the following constraints on the product species mole fractions:

$$x_1 + x_2 + x_3 + x_4 + x_5 = 1 \qquad (12.87)$$

Equilibrium equations
Two equilibrium equations follow directly from equations with the constituent partial pressure being replaced by their mole fractions

$$K_2 = \frac{P^2 x_1 X_4^3}{x_3 x_5} \qquad (12.88)$$

$$K_3 = \frac{x_1 x_3}{x_2 x_4} \qquad (12.89)$$

In addition there are five side conditions,

$$x_1 \geq 0, i = 1, 2 \ldots, 5 \qquad (12.90)$$

These conditions ensure that all mole fractions in the product mixture are non-negative, that is, any solution of equations that contain negative mole fractions is physically meaningless.

From physical chemical principles, there is one and only one solution of the equations that satisfies the above condition.

The seven equations may be written in the form

$$f_i(x) = 0, \quad i = 1, 2 \ldots, 7 \tag{12.91}$$

where

$$x = [x_1 x_2 x_3 x_4 x_5 x_6 x_7]: \tag{12.92}$$

as follows

$$f_1(x) = \frac{1}{2}x_1 + x_2 + \frac{1}{2}x_3 - \frac{x_6}{x_7} = 0 \tag{12.93}$$

$$f_2(x) = x_3 + x_4 + 2x_5 - \frac{2}{x_7} = 0 \tag{12.94}$$

$$f_3(x) = x_1 + x_2 + x_3 - \frac{1}{x_7} = 0 \tag{12.95}$$

$$f_4(x) = -28837x_1 - 139009x_2$$
$$- 7821x_3 + 18927x_4 + 8427x_5 + \frac{13492}{x_7} - 10690\frac{x_6}{x_7} = 0 \tag{12.96}$$

$$f_5(x) = x_1 + x_2 + x_3 + x_4 + x_5 - 1 = 0 \tag{12.97}$$

$$f_6(x) = P^2 x_1 x_4^3 - 1.7837 x_3 x_5 \, 10^5 = 0 \tag{12.98}$$

$$f_7 = x_1 x_3 - 2.6058 x_2 x_4 = 0 \tag{12.99}$$

The above system of simultaneous nonlinear equations has the form that is suitable for solution using the Newton–Raphson method which can be summarized as follows:

(1) Choose a starting vector $x_k = x_o =$ which is hopefully near a solution
(2) Solve the system of linear equations

$$\phi(x_k)\delta_k = -f(x_k) \tag{12.100}$$

where

$$\phi_{ij}(x_k) = \frac{\partial f_i}{\partial x_j} \quad i = 1, 2 \ldots 7 \quad j = 1, 2 \ldots 7 \tag{12.101}$$

and

$$f(x_k) = [f_1(x_k), f_2(x_k) \ldots, f_7(x_k)] \tag{12.102}$$

(3) Update the approximation to the root for the next iteration

$$x_{k+1} = x_k + \delta_k \tag{12.103}$$

(4) Check for possible convergence to a root. One test might be

$$|\delta_{ik}| < \varepsilon_2 \quad i = 1, 2 \ldots, 7 \tag{12.104}$$

If the above condition is true for all i then x_{k+1} is taken to be the root. If the test fails for any i then the process is repeated beginning with step 2. The iterative process is repeated until the above condition is met for some value of k or until k exceeds a specified limit.

The partial derivatives may be found for the case illustrated by partial differentiation of the seven functions $f_i(x)$ with respect to each of the seven variables. For example,

$$\frac{\partial f_1}{\partial x_1} = \frac{1}{2}, \quad \frac{\partial f_1}{\partial x_4} = 0, \quad \frac{\partial f_1}{\partial x_7} = \frac{x_6}{x_7^2}$$

$$\frac{\partial f_1}{\partial x_2} = 1, \quad \frac{\partial f_1}{\partial x_5} = 0$$

$$\frac{\partial f_1}{\partial f_3} = \frac{1}{2}, \quad \frac{\partial f_1}{\partial x_6} = -\frac{1}{x_7}$$

The method of solution outlined above is applicable to any number of equations or product species providing the input is properly defined. Computer libraries now contain general codes for solving any system of non-linear simultaneous equations that are suitable for this application.

12.7 Expansion and exhaust

During the process of expansion, also called the power or working stroke, the heat energy evolved by the combustion process is converted into mechanical work. In an actual cycle the working stroke begins at point c with the piston at TDC and ends during the exhaust, when the products of combustion are discharged. The combustion process begins with ignition at a point somewhat before TDC and continues as the piston movement goes beyond TDC into the expansion stroke to end at point z. It has been established that for maximum power the combustion process should be divided equally before and after TDC. Consequently if the combustion duration is $\Delta\theta_c$ crank angle degrees then the amount of ignition advance should be $\Delta\theta_c/2$ or half the combustion duration. In calculating the characteristics of the exhaust process point z in Figure 12.12 which represents the point of maximum cylinder pressure and temperature for an ideal cycle is taken as the point of reference. At the beginning of expansion combustion is taking place and the polytropic exponent has a negative value reaching 0 at the point when the cylinder pressure is a maximum. The exponent continues to increase until at the point of maximum cylinder pressure it has reached a value of 1. Thereafter the exponent continues to rise to reach a value between 1.0–1.5. It is very difficult to determine the different values of n thus it is usual to choose a mean value that can be applied to the entire expansion process. The conditions at the end of expansion can then be determined from the polytropic relationships using the condition at the end of combustion as the reference as shown below.

$$P_b = P_z \left(\frac{V_z}{V_b}\right)^n \tag{12.105}$$

$$T_b = T_z \left(\frac{V_z}{V_b}\right)^{n-1} \tag{12.106}$$

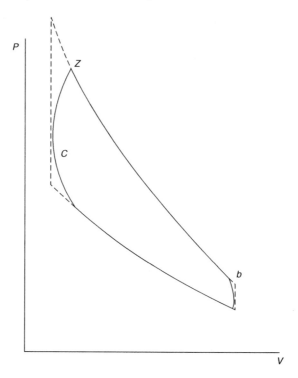

Figure 12.12 Comparison of actual and theoretical combustion cycles

In examining the process of progressive combustion it was established that a temperature gradient exists in a combustion chamber with the different elements of charge having different time temperature histories depending upon their point of ignition in the combustion process. As the concentration of product species is very temperature dependent it follows that a species distribution also occurs in the combustion process. To perform a detailed analysis of the emission formation it will be necessary to consider the history of several elements of mass as they are subjected to the state changes during the combustion process. The analysis will be constrained by the conservation equations of mass and energy as well as requirement that the volume of the combustion chamber be equal to the volume of the burnt and unburnt charge and the chamber pressure be uniform in space. The total emission concentration of any species will then be made up of the mass mean of the sum of the concentration of the different elements. The relationship between the mass fraction burned × the cylinder pressure and the crank angle established when the progressive combustion was considered could be used to establish the mass of an element of charge or a more complicated model could include a description of the mass burning rate in terms of the flame area and the flame velocity.

12.7.1 Exhaust

In a four stroke engine the exhaust gases flow past the exhaust valve at sonic velocity of 600 to 700 m/s until the pressure ratio across the valve has fallen below the critical value. This outflow is accompanied by a sharp noise. Approximately 60–70% of the exhaust gases are removed from the engine cylinder during this stage terminating near TDC. During the following stroke of the piston the products of combustion are exhausted at a much smaller velocity of the order of 200 to 250 m/s. The work done to expel the exhaust gases and the degree of cylinder scavenging depend on the timing of the exhaust process. The indicator diagram shown in Figure 12.13 illustrates three possible valve timings. When the valve opens early (point e') the work lost from the expansion process (area $e'bb'$) is large and cannot be compensated by the reduction in the pumping work during the following piston stroke. Opening of the valve late at point e'' reduces the lost expansion work but increases the pumping work (area $b''r''$) and the cylinder scavenging will be poor. The correct moment of opening of the exhaust valve is selected experimentally. The change of pressure during the exhaust depends on the gas exchange process. In the case of a naturally aspirated engine the line characterizing the exhaust process is above the admission line. The establishment of the cylinder pressure in the exhaust stroke is more difficult than the intake stroke because the cylinder pressure is the driving force for the removal of the burnt charge and it is significantly effected by the piston motion and the valve flow area. An analysis similar to that described for the intake stroke would therefore be subject to errors. An alternative method of analysis is thus presented below which could also be used for the intake process.

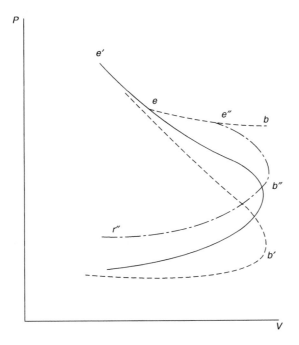

Figure 12.13 Indicator diagram showing three possible valve timings

Consider a control volume as shown in Figure 12.14 enclosing an engine cylinder and piston with the piston at the extreme of the expansion stroke. The exhaust valve is open and the piston retraces its path in the cylinder expelling a portion of mass dM from the cylinder. In the same

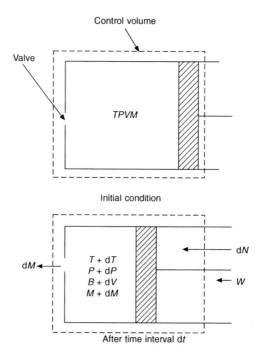

Figure 12.14 Control volume enclosing on engine cylinder and piston

time interval mass dN enters the control volume under the piston. The initial state of the substance in the control volume can be described in terms of the state variables T, P, V and M and after a time interval $d\theta$ these have changed to $T + dT$, $P + dP$, $V + dV$, $M + dM$. Some of the incremental changes will be negative. Work will be done by the piston during this process. If isentropic behaviour can be assumed than a statement of conservation of energy leads to the following equation

$$\frac{dP}{d\theta} = \gamma P \left[\frac{1}{M} \frac{dM}{d\theta} - \frac{1}{V} \frac{dV}{d\theta} \right] \qquad (12.107)$$

The equivalent equation for the intake process would be as shown in equation 12.108.

$$\frac{dP}{d\theta} = g \left[\frac{RT}{V} \frac{dM}{d\theta} - \frac{P}{V} \frac{dV}{d\theta} \right] \qquad (12.108)$$

These equations can be integrated once the relationship between V and θ and M and θ has been established. The geometric design of the crank shaft assembly and the cylinder bore, the slider crank formula, will produce the relationship between V and θ. The flow of charge through the valves is a function of the valve lift to head diameter ratio l/d and the pressure ratio across the valve. Experimental data illustrating the behaviour of flow through poppet valves is usually plotted in the form of a non-dimensional mass flow function against pressure ratio for different l/d ratios. Such data is shown in Figure 12.15. Clearly a knowledge of the valve cam profile is

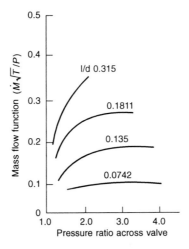

Figure 12.15 Mass flow function against pressure ratio for flow through an exhaust valve

required for this operation. It is possible using this data to undertake a stepwise analysis of the gas exchange process with a time increment $d\theta$ being chosen over which the flow is assumed to be steady. The total out flow can be determined from the data and the conditions in the cylinder readjusted to start the next time step.

12.8 Recommended reading

The following list consists of books that I have found to be well written, very informative and full of technical insights that puts them above the standard text book. All have been written by practising engineers of some standing in their National community who have been able to recount some of their considerable knowledge and experience in a concise and absorbing manner. The majority of them were produced before the computer had any significant influence on the design process and so are full of well founded fundamental engineering. From a chronological aspect the first is *W.O. Bentley* by Donald Bastow. It is possible in this text to understand the design process of the inter war period, a time when designs were laid out by single individuals and when personal characteristics could feature in the design. The preferred type of registration, drive or coupling and the influence this has on the subsequent design is well illustrated. This book is well appointed with technical drawings and design calculations which will be of interest to the student of engineering design. The second in the chronological list is that of Sir Harry Ricardo and J.G. Hempson. This is the last in a series of books of the same title produced by Sir Harry Ricardo and a co-author since the 1920s. The numerous reprints and new editions have been extensively rewritten to account for the development of the study of internal combustion engines. these books are very easy to read, they contain lavish engineering drawings of engines and associated equipment and prove an insight into a very interesting period of engine research and development. The book by Julius Mackerle is a rare source of data about a class of engines that is becoming less common as noise and emission regulations have largely restricted the development of air cooled engines. It offers a fascinating insight into design methods, both mechanical and thermal at a time of technical and political divide between the east and west. The engineering approach and solutions discussed and illustrated are fascinating. The remaining two books also include details of the fundamental aspects of engine design but they also include and can act as an introduction to computational methods in engine design. The text of Heyward is almost a compulsory text for the student of internal combustion engines containing a vast amount of data on both spark ignition and compression ignition engines. The process of combustion and emission formation is very well covered. The text by Ferguson is similar but is smaller and does not cover the material to the same extent of Heyward. It is useful, however, as a first step guide to the use of proprietary computer programs or the use of standard library sub routines in engine analysis.

Mackerle, Julius (1972). *Air Cooled Automotive Engines*. Charles Griffin and Co. Ltd.

Ferguson, Colin R. (1985). *Internal Combustion Engines*. John Wiley & Sons.

Heywood, John B. (1988). *Internal Combustion Engine Fundamentals*. McGraw-Hill.

Ricardo Sir Harry and Hempson, J.G. (1972). *The High Speed Internal Combustion Engine*. Blackie and Sons.

Bentley, W.O. (1978). *Engineer*, Donald Bastow, Haynes.

Notation

The following notation is used in this chapter:

A	area
c_f	resistance coefficient in the inlet system refereed to the narrowest cross section
c_p	specific heat at constant pressure
c_v	specific heat at constant volume
C	velocity
C_{ch}	coefficient of additional charging
C_{pur}	purge coefficient
$C_{p.max}$	maximum piston speed
C_{sc}	scavenge coefficient
d	valve head diameter
D	cylinder bore
g	gravitational constant
h	specific enthalpy
h_{fg}	latent heat of evaporation
k	polytropic exponent
l	valve lift
l_{rod}	length of connecting rod
M	mass
M	mass flow rate
M_{la}	mass with additional charging
n	engine rotational speed (rpm)
n_1	average polytropic exponent for a stated process
P	pressure
q	heat transfer
r	compression ratio
R	specific gas constant
R_c	crank radius
S	crank shaft stroke
T	temperature
u	specific internal energy
V	volume

| x | mass fraction burned |
| z | height above a datum |

Greek symbols

α	air–fuel ratio, excess air coefficient
β	head loss coefficient inlet tract to cylinder
γ_{res}	residual gas coefficient
$\Delta\theta_c$	combustion period
θ	crank angle, cycle temperature ratio t_3/T_1
η	efficiency
ρ	density
ϕ	cycle compression temperature ratio T_2/T_1

Subscripts

a	admission
c	clearance
ch	charge
cy	cylinder
comb	combustion
evap	due to evaporation
hex	due to heat exchange
i	ignition
im	at minimum cross section
in	inlet
is	inlet system
MEP	mean effective
o	ambient conditions
p	piston, products
pip	exhaust pipe
r	reactants
res	residual
sc	scavenge
sw	swept
th	thermal
tot	total
v	volumetric
w	wall
1,2,3,4	points in the ideal cycle
+	added to the cycle
−	rejected from the cycle

13. Transmissions and driveline

Nick Vaughan, BSc(Eng), PhD, CEng, FIMechE
Dave Simner, BSc, MSc

The aims of this chapter are:

- Demonstrate the need for transmission design and matching;
- Give examples of common gearboxes and transmissions available for vehicle design;
- Indicate the terminology and methods for transmission design;
- Aid the designer to understand the elements of the analysis of transmission systems.

13.1 Introduction

This section introduces the transmission systems that can be found in today's passenger car. Of course, many car derived components and systems can also be found in small commercial vehicles. Also larger derivatives, which have much in common, can be found in heavy goods and public service vehicles. We have endeavoured to introduce the main transmission types and some areas of technology that can be found within the units. In this chapter, however, we can only hope to introduce the subject of transmissions to you. In order to make up for this brevity we include references to other material so you are able to follow up any particular subject in greater detail.

It is probably worth stating that, in practice, the choice of transmission units for a particular vehicle is heavily influenced by what is in production and available. The cost of developing, and more importantly installing the equipment to manufacture a new gearbox would be prohibitive for a small specialist vehicle manufacturer. Equally, producing a special transmission to support a specific model would also be difficult to justify even for a large vehicle manufacturer.

Current developments are extremely interesting as technology, particularly electronic control, is very much blurring the distinction between the conventional classes of transmission. For example, automatic transmissions are often found now with a manual override function to allow the car to be driven using the gears selected by the driver. Conversely, manual gearboxes are having automation added to operate the clutch or shift the gears. These developments not only make the transmission interesting from an engineering perspective but also create marketing features from an area of the vehicle often hidden from view and largely ignored by the buyer until it causes a problem.

13.1.1 Definitions

Transmission – This term can be used to describe one unit within the driveline of a vehicle, often the main gearbox, or as a general term for a number of units.
Driveline – This includes all of the assembly(s) between the output of the engine and the road wheel hubs.

Powertrain – Essentially the driveline and engine together, but may also be taken to include other related parts of the vehicle such as the exhaust or fuel system.

Automatic transmission – Automatic transmissions come in various forms but have the common ability to change the ratio at which they are operating with no intervention from the driver.

Manual transmission – As the name suggests, drivers have to change the gear ratio setting rather than the transmission doing the job for them.

Continuously Variable Transmission (CVT) – CVTs are able to vary the ratio between input and output in a stepless manner rather than having a number of discrete ratios.

Infinitely Variable Transmission (IVT) – Essentially a CVT which has the additional ability to operate with zero output speed, hence negating the need for a separate starting device.

This chapter is going to look at the transmission systems used in cars. The rest of the driveline will not be considered in any detail so there will be no detail on such things as axles or 4 × 4 transfer gearboxes.

13.2 What the vehicle requires from the transmission

According to some engine colleagues, the transmission is a large, expensive bracket to stop the engine dragging on the road. However, we will hopefully demonstrate that transmissions are much more interesting than the other, less significant, part of the power train!

Essentially, the transmission or driveline takes the power from the engine to the wheels, in doing so actually makes the vehicle usable. The functions that enable this include:

- Allow the vehicle to start from rest, with the engine running continuously.
- Let the vehicle stop by disconnecting the drive when appropriate.
- Enable the vehicle to start at varied rates, under a controlled manner.
- Vary the speed ratio between the engine and wheels.
- Allow this ratio to change when required.
- Transmit the drive torque to the required wheels.

The transmission needs to perform all of the above functions and others in a refined manner. The structural aspects of the transmission, predominantly the casing, often contribute significantly to the structure of the powertrain and the vehicle as a whole. This is important when it comes to engineering for the lowest noise and vibration. The stiffness of the powertrain assembly itself is important in determining the magnitude and the frequency of the vibrations at the source (the engine). This stiffness (and indeed the strength) can also be important to the integrity of the vehicle in a crash. Particularly with front wheel drive vehicles, the way in which the body collapses on impact has to be engineered very carefully, and the presence of a large rigid lump such as the powertrain has a critical influence on the way this occurs. The size, shape and orientation of the unit also affect the intrusion into the passenger space after an impact.

13.2.1 The layout of the vehicle

The position of the powertrain components within the vehicle has implications both for the engineering of the vehicle and the driveline components including the transmission itself. Effects include:

- the space available for the powertrain and how it is packaged within the vehicle including the location of ancillary components;
- the weight distribution, since the powertrain components are relatively heavy;
- the structure to support the powertrain and react against the driving torques;
- vehicle handling and ride both from weight distribution and the location of the driven wheel set;
- safety structure and passenger protection.

The choice of vehicle layout is determined principally by the target market sector and brand image that the vehicle is required to project. Possible alternatives include saloons, ranging from large luxury saloons to micro or town cars, sports coupés or convertibles, estate cars or off-highway vehicles. In many cases the same vehicle platform will be used for several of these variants. The vehicle layout must also be sufficiently flexible to accommodate different engine and transmission options that are offered with many vehicles.

The main vehicle configurations in use are shown in Figure 13.1. The most widely used currently is the 'standard' front wheel drive layout shown in Figure 13.1(a). This has an engine mounted transversely to the vehicle axis with the transmission also transverse and in-line with the engine. The differential can be incorporated into the transmission casing. Another possibility is shown in Figure 13.1(b) with a longitudinal engine transmission assembly, again including a differential and the drive being taken to the front wheels. This configuration is used for larger front wheel drive vehicles where the size (i.e. length) of the engine gearbox assembly makes installation across the vehicle impossible. It also allows front, rear and four wheel drive vehicles to be developed easily from the same vehicle platform as the engine installation and front structure of the vehicle can remain the same in each. The main alternative, however, is the classic front engine rear wheel drive layout as in Figure 13.1(c). The engine and transmission are still in-line but mounted longitudinally with a connecting shaft to a separate rear mounted final drive and differential that are a part of the rear axle. A common variant amongst two seater sports vehicles is shown in Figure 13.1(d) with the engine and transmission transversely mounted

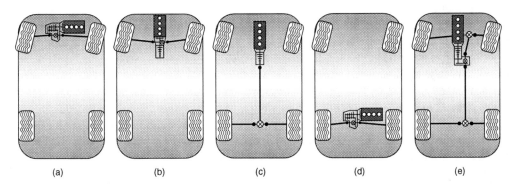

(a)   (b)   (c)   (d)   (e)

Figure 13.1 Some typical vehicle/powertrain configurations

to the rear of the vehicle and driving the rear wheels. If the engine is in front of the rear axle then this is usually referred to as a mid-engine layout. The final example shown in Figure 13.1(e) is a four-wheel (or all-wheel) drive power train frequently used in off-highway vehicles. The greater height of these vehicles allows the engine to be mounted above the front axle line with the front differential alongside. Variants of this also take the drive to the rear axle directly in-line from the gearbox. These are normally differentiated by virtue of the transfer gearbox design. 'Double offset' being the one illustrated and 'single offset' where the drive to the rear axle is in-line with the gearbox output shaft. It is also possible to derive four wheel drive configurations from the two wheel drive layouts. For example adding a longitudinal propeller shaft from the front differential of the standard transverse layout (Figure 13.1(a)) to an additional rear differential. There are many more front and rear wheel drive variants than those included here, but these few account for the vast majority of vehicles on the road.

The vehicle layout adopted has consequences for the transmission itself and the necessary controls and interconnections. These include the opportunity for the differential to be included in the same casing as the transmission and eliminate the need for an additional housing. However, the transmission and differential gears must then share the same lubricating fluid. For manual gearboxes the routing of the gear-change linkage can be more complicated for the mid-engined (and other rear-engined) layouts. These also have greater complication in ancillary and cooling system layouts that are discussed in more detail in the environmental considerations in Section 13.6. There is also a particular fuel economy advantage for transverse layouts that do not have to turn the drive direction through a right angle. This eliminates a bevel gear set that is less efficient than parallel shaft transfer gears.

13.2.2 Starting from rest

As the internal combustion engine cannot provide torque at zero speed, a device is required in the transmission that will enable the vehicle to start from rest and, when propulsion is not required, to disengage the drive between the engine and road wheels. Several devices are used in automotive transmissions to achieve this:

* The single plate dry friction clutch – used commonly with car manual gearboxes.
* The multiplate, wet (oil immersed) clutch – frequently used in motorcycles, variable transmissions, and some large, heavy-duty automatic transmissions.
* The fluid flywheel – rarely used today.
* The torque converter – used in the majority of automatic transmissions.
* Electromagnetic clutches – again used in some variable transmissions.

These devices are fitted between the engine output and transmission input. The design and application of the dry clutch and the torque converter are discussed in the sections on manual and automatic transmissions respectively. It should be pointed out that a smaller multiplate clutch is often used in automatic transmissions to disconnect or connect particular gears and hence allow the gear change required, these applications do not have the capacity of starting the vehicle from rest.

13.2.3 The vehicle requirement – what the powertrain has to deliver

If we consider the torque requirements (on the engine and driveline) there are a number of forces acting on the body of the vehicle that have to be overcome:

- The rolling resistance of the tyres.
- The aerodynamic drag of the vehicle body.
- Any resistance due to the climbing of an incline.
- Overcoming the inertia of the vehicle (as a whole) and the rotating parts, while the vehicle is accelerating.

This last point indicates that the engine also has to accelerate its own inertia, the effect of this is particularly significant in the lower gears.

Consider the first three of these that occur during steady state conditions:

Total running resistance force = F_{tot}

$$F_{tot} = F_{Ro} + F_{Ae} = F_{Cl}$$

where F_{Ro} = rolling resistance = fmg
$\quad\quad m$ = vehicle mass, kg
$\quad\quad f$ = coefficient of rolling resistance – typically approx 0.013 to 0.015 for
$\quad\quad\quad\quad$ normal road – however, it does increase with speed.
$\quad\quad g$ = 9.81 m/s^2 – gravitational acceleration
$\quad\quad F_{Ae}$ = aerodynamic resistance = $0.5\, \rho c_d A\, (v + v_h)^2$
$\quad\quad \rho$ = air density – typically 1.2 to 1.3 kg/m^3 (the latter is at 'standard
$\quad\quad\quad\quad$ temperature and pressure')
$\quad\quad c_d$ = drag coefficient, often around 0.3 to 0.4 for many cars.
$\quad\quad A$ = frontal area of a vehicle in m^2
$\quad\quad v$ = vehicle speed, v_h = headwind speed, in m/s
$\quad\quad F_{Cl}$ = climbing resistance = $mg \sin \beta$
$\quad\quad \beta$ = the gradient of the hill being climbed (degrees)

In addition to these, the engine also has to overcome any resistive forces from 'work' the vehicle may be doing, for example towing a trailer. While operating off road, a vehicle will have to also overcome the resistance provided by the soft ground. This can vary greatly and depends on the type of soil, how wet it is and other factors such as how disturbed or compacted the ground is. These additional forces acting on a vehicle can, in the extreme, be so large as to prevent the vehicle from moving, severely restrict the speed it is able to attain or exceed the available traction from the tyres.

Examples of how the rolling resistance and aerodynamic forces add up with increasing road speed, for a range of vehicles are illustrated on Figure 13.2. This assumes zero wind speed on level road. For example, if the vehicle were climbing an incline the lines would move up by a constant amount. A few interesting things can be seen on this graph:

- Firstly, just compare the overall resistance of the different cars. It can be seen that both the overall magnitude and the difference between vehicles increase significantly with speed.
- Compare the difference between the older design of the Mini and the more recent Lupo, this

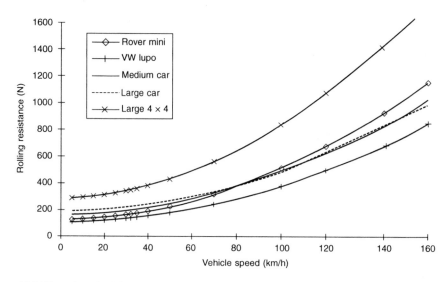

Figure 13.2 Plot of total rolling resistance vs road speed (level road) – data shown for various vehicles

becomes exaggerated at speed. The drag coefficients having a more pronounced effect as the speed increases.

- The very large load produced by the high weight combined with the large frontal area of the 4 × 4 vehicle.
- The difference between the medium and large can be seen to cross over as the speed increases. The heavier large car being the highest resistance load at low speeds, but then gaining an advantage at higher speeds because of the better aerodynamics – this is almost certainly helped by the longer length of the body and the body style.

The load on the transmission
The total rolling resistance that has to be overcome is the load acting on the vehicle. This is seen as a torque requirement at the driving wheels(s), which can be calculated if the dynamic rolling radius of the tyre(s) is known:

$$\text{Torque at the wheel (Nm)} = F_{tot} \times \text{rolling radius (m)}$$

Care should be taken as to how many wheels share the drive; hence the torque seen by any one part of the driveline may not be the whole figure. The main gearbox, however, invariably sees the whole of the engine torque so the number of drive wheels can be ignored when considering this.

This torque value can then be calculated back up the driveline, taking account of the transmission ratio(s) and efficiency to give the torque required at the engine (at the clutch or end of the crank). By considering the rolling radius of the road wheel and the ratios in the transmission, the engine speed can also be calculated.

There are a variety of conditions, at which the vehicle has to operate, that determine the gear ratios to be chosen. These are likely to be modified by certain practical considerations within the transmission, but for this purpose we can consider the initial requirements in order to determine the ratio set for the transmission.

13.2.4 Changing ratios – matching of the transmission to the vehicle

It is important to appreciate that the choice of gear ratios in a transmission is often dictated in practice by what is available or what is already in production. This situation occurs because of the large expense involved in engineering new gearsets, and installing or modifying the manufacturing plant to make the new parts. There are some cases that do necessitate a change, however. These may include a change in the engine, for example from petrol to diesel, or a significant change to the weight of the vehicle in which the gearbox is to be installed. Obviously, the finances available within the vehicle manufacturer and the volumes involved will have a very large influence in this decision. Where changes can be accommodated, they may be limited to one or two gear ratios leaving the intermediate ratios as is, hence not necessarily optimized. Finally, before looking at how the 'ideal' ratio may be chosen, the other limitation on ratio choice is the gear design itself. An example, on a first gear pair where there could be a limit on how small the drive pinion might be in order to withstand the shock loading which can occur in the gearbox.

There are a number of decisions that need to be made when deciding what gear ratios should be fitted in a particular transmission unit. A similar process has to be done for manual transmissions, automatics and CVTs. There is more flexibility in an automatic or a CVT because of the effect of the torque converter and/or the shift map. These, in addition to the gear ratios, influence the effective, overall ratio at any point in the operating regime. The factors, which have to be taken into account, are:

- The performance requirements of the vehicle.
- The weight, rolling resistance and other parameters of the vehicle.
- The restrictions that exist on the design of the transmission.
- Packaging restrictions in the vehicle and on the engine ancillaries, if the casing has to be altered.
- Availability – as discussed above.

The performance of a vehicle is very rarely, simply a matter of top speed and acceleration!
   Selection of the lowest ratio – 1st gear
This governs the starting performance of the vehicle and will depend on:

- Gradient of hill required to be climbed – worst case.
- Gross (fully laden) weight of the vehicle.
- Weight of any trailer required.
- Characteristics of the engine at low engine revs – i.e. minimum engine speed for effective air inlet 'boost' on pressure charged engines.

Selection of top gear ratio – typically 5th in passenger cars

- Engine characteristics.
- Economy requirements at cruise.
- In-gear performance – is the driver expected to change gear on overtaking?
- Top speed to be achieved in top or next gear (usually 4th) – is top gear an 'overdrive'?

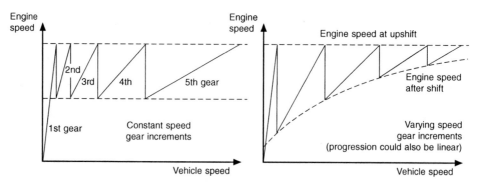

Figure 13.3 Engine speed vs vehicle speed for differing gear ratio progressions

The intermediate gears are usually spaced to provide an even, comfortable spread between these extremes. In theory the ratios are often chosen to give constant speed or varying speed increments between the gears. By using constant speed increments the engine would reduce by a consistent speed change each time the driver changed up. For example if a driver changed up while accelerating every time they reached, say 3000 rev/min, the engine speed would be the same after each gear shift. With variable speed increments this would not be the case, usually meaning that the change in engine speed with each gearshift would get progressively smaller as higher gears were engaged. The following figures illustrate this. The 'upshift' points are shown as constant for illustration although this is obviously not necessarily so in practice.

The particular vehicle requirements or limitations of the transmission selected can modify this spacing, for example due to:

- Complexity requirements – existing ratio sets may limit choice on new vehicles, especially for lower volume vehicles.
- In gear acceleration requirements – provision of particular characteristics at certain vehicle speeds, for example, achievement of 0–60 mph/100 kph without too many gear changes.
- Casing limitations on gear sizes.
- Emission and fuel economy requirements, i.e. engine conditions during the legislated drive cycle.
- Refinement issues at particular engine or driveline speeds.

All of these factors will influence the selection of the gear ratios in practice and possibly cause a compromise between the calculated, 'ideal' ratio set for a given car and what can be used on an existing vehicle.

Example of the considerations in matching a transmission to a vehicle
For this example we will look at some of the factors which would need to be considered when designing a gearbox for a road car, in this case a large 4 × 4.

Consider the rolling resistance of the 4 × 4 vehicle in Figure 13.2. Taking a rolling radius of 0.375 m for the tyre, the torque required at the wheel for any road speed within the range can be calculated. Consider Figure 13.4 – this is a fuel consumption chart for a large petrol engine. (A

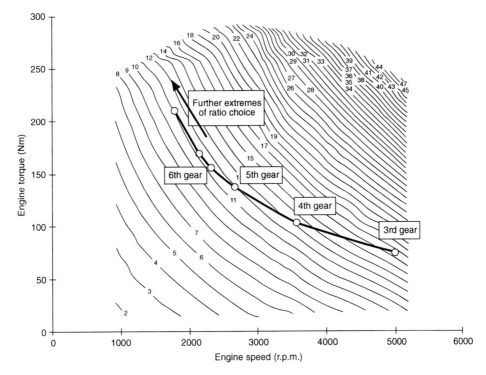

Figure 13.4 Chart of fuel (mass) flow for a large petrol engine – also shows engine conditions for a range of gear ratios.

line can be drawn through the top of the lines of constant fuel flow to indicate the max torque line.)

Taking some of the vehicle transmission details as:
Final drive ratio 4.2
Fifth gear 0.75
Fourth 1.0
Third 1.4
Plotting the engine conditions for 120 km/h (motorway), and assuming a loss of 5% in the transmission system gives the engine conditions shown on the graph for the different gears. The required tractive force at this speed is 1100 N; this equates to a torque of 413 Nm (total) required at the wheels. In theory, this would be a nominal 103 Nm at each wheel in the case of our 4 × 4 example.

We can also calculate how fast the wheels, transmission and engine would be rotating at this speed. The rolling radius (above) means the vehicle travels 1 km every 424 revolutions of the wheel. (Sometimes it is easier to consider the speed of the wheel for a given road speed, in this case 7.1 rev/min per 1 km/h.) This means our wheels will be rotating at 852 rev/min at 120 km/h.

From these figures the engine torque and speed at 120 km/h for the various gears quoted can be calculated. The operating points for the engine in the various gears show that as the vehicle changes up to 4th and 5th gear, the engine speed drops, the torque increases and the indicated fuel mass flow reduces. As we might expect, the vehicle uses less fuel in top than the lower gears.

What happens if we add an 'overdrive' sixth speed with a ratio of 0.6, or even 0.5? The line on the graph also indicates how the trend would continue if an overdrive ratio were to be added to the gearbox. The result indicates that if taken too far the fuel used would not necessarily continue to reduce. The engine conditions as the speed is reduced and the torque required increases are such that could find the engine to be unresponsive, requiring large throttle openings and even higher emissions due to the high engine load.

If we now consider how the tractive force ('effort') provided by the powertrain varies in each gear (by using the maximum torque values for the engine considered above). By taking account of the various gear ratios the force provided at the road can be compared with the road load (rolling resistance). In Figure 13.5, the original line from Figure 13.2 has been added (again considering the 4 × 4 vehicle). An allowance has also been made for the force required to climb hills of various gradients, so additional rolling resistance lines have been added for the different gradients.

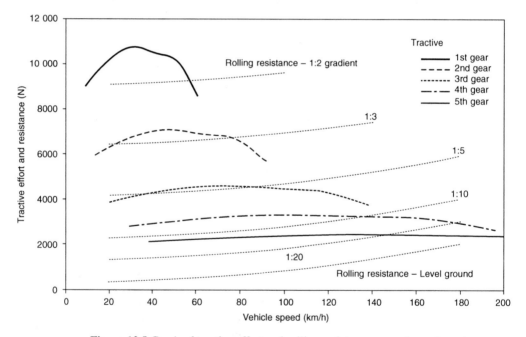

Figure 13.5 Graph of tractive effort and rolling resistance vs road speed

These graphs can be plotted easily for any vehicle/transmission/engine combination providing the basic information referred to above is known or can be estimated. The information provided is varied and useful; such as:

• The maximum speed attainable for different conditions and gears can be seen. In this example we may expect the vehicle to go faster in 4th than 5th as the tractive effort line for level ground crosses the available force line for 5th at a lower speed than in 4th (and before we run out of available engine speed in the lower gear.

- The maximum gradient that the vehicle could be expected to climb in any one gear can be estimated. Here it could be assumed that a 1 in 5 hill could just be climbed in 3rd gear – and at a maximum speed of about 80 km/h.
- Where the available force line is just above the required force, the close proximity of the two lines indicates that there is little if any available torque from the engine. So if the vehicle were on a 1 in 10 gradient at say 40 km/h in 4th gear, we might expect to be able to accelerate to nearer 120 km/h by looking at the graph. The two lines are quite close to each other, however, indicating that there is little additional torque available to accelerate the vehicle mass or accelerate the engine itself. At the very least we might expect the vehicle to be quite unresponsive.

It should be noted that in the lower gears these graphs can indicate that very steep hills can be climbed. In practice it may not be possible to actually start from rest on anything like these gradients because of the capacity of the clutch and the difficulty of achieving just the correct engine conditions. On two wheel drive vehicles the available grip from the tyres can also be a limiting factor. Even on the 4 × 4 example we are considering, in practice the low ratio in the transfer gearbox would be required at gradients much steeper than, say 1 in 3.

13.3 The manual gearbox

Most people who drive will be able to describe some aspects of one of these. As the name suggests, the driver has to change between one gear ratio and another, as the vehicle requires, when using this type of gearbox. The different gears have different ratios that allow different relative speeds between the engine and road wheels. There are several distinct types of these transmissions; including 'transverse' or 'transaxle' front wheel drive gearboxes and 'inline' gearboxes used in rear and four wheel drive vehicles. Four wheel drive vehicles will have an additional transmission unit on the rear of the gearbox to enable the drive of both front and rear axles.

Uses
Inline gearboxes are used in a wide range of vehicles from small passenger cars up to large trucks, while the vast majority of transverse gearboxes are used in passenger cars and small vans. It should be noted that manual gearboxes are nowhere near as common in the US and Japanese passenger car markets as they are in Europe. This is particularly the case with small to medium cars. In the past, the majority of larger passenger and commercial vehicles in Europe used manual gearboxes of one type or another. This particular area of the market is changing and becoming dominated by automatic transmissions.

Advantages
- Usually have high mechanical efficiency.
- Arguably the most fuel efficient type of transmission, although this depends on the driver selecting the most appropriate gear.
- Relatively cheap to produce – possibly only half of the equivalent automatic.

- Light weight – typically 50 to 70% of the equivalent automatic weight.
- Smaller and hence usually easier to package in the vehicle.

Disadvantages
- Some driver skill required – ask anyone who only drives autos!
- Emissions and fuel consumption can be heavily influenced by the driver's gear selection.
- Clutch operation and changing gears can be tiring, especially when in heavy traffic.
- Not suitable for all drivers, controls on larger vehicles can be heavy and most require some dexterity during operation.

13.3.1 The front wheel drive passenger car gearbox

Figure 13.6 is a cross section of a 'typical' front wheel drive transmission. The features it contains are typical of those found in many, if not all, gearboxes. The essential elements are the three shafts that take the drive from the engine (on the centre line of the crankshaft) to the output of the gearbox. From here, the driveshafts connect the gearbox to the (driven) wheel hubs. As mentioned in Section 13.2.1 this configuration of gearbox can also be used in rear wheel drive, 'mid engined' vehicles, the installations being very similar. Two of the obvious features of this

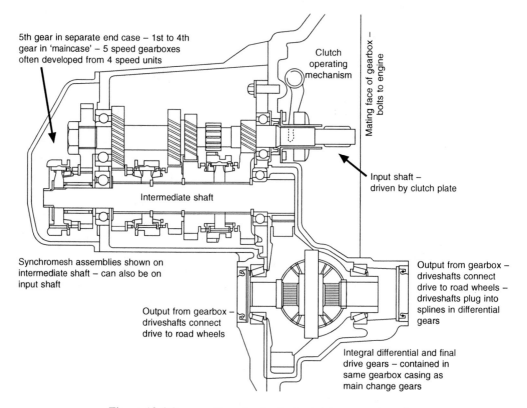

5th gear in separate end case – 1st to 4th gear in 'maincase' – 5 speed gearboxes often developed from 4 speed units

Clutch operating mechanism

Mating face of gearbox – bolts to engine

Input shaft – driven by clutch plate

Intermediate shaft

Synchromesh assemblies shown on intermediate shaft – can also be on input shaft

Output from gearbox – driveshafts connect drive to road wheels – driveshafts plug into splines in differential gears

Output from gearbox – driveshafts connect drive to road wheels

Integral differential and final drive gears – contained in same gearbox casing as main change gears

Figure 13.6 Cross section of a front wheel drive manual gearbox

gearbox are the integral differential and the final drive gear pair. The function of the differential is described in Section 13.6. Because of the final drive, the overall ratio of the front wheel drive gearbox has typical reduction ratios of around 12:1 in 1st gear and around 3:1 in top gear. In comparison, the typical rear wheel drive gearbox will have a reduction of 3 or 4:1 in 1st and an overdrive ratio of around 0.8 in top. Figure 13.7 is a schematic of the front wheel drive gearbox to show the drive path in the gearbox.

Drive passes from the engine, via the clutch to the input shaft. The various gear pairs then transmit torque to the intermediate shaft. The final drive pinion is on or part of this shaft and in

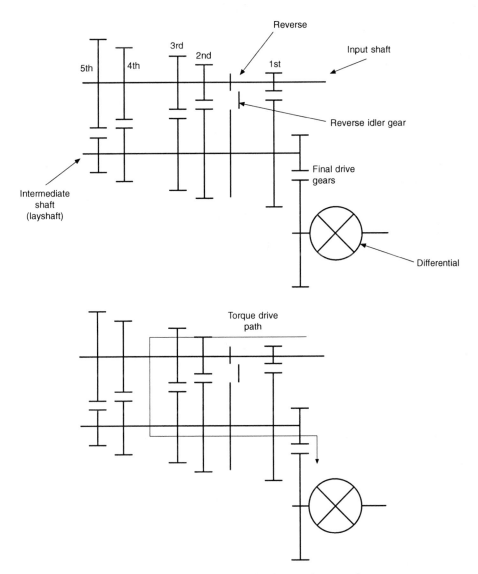

Figure 13.7 Schematic of a front wheel drive gearbox gear layout

turn drives the final drive wheel. The final drive wheel is part of the differential assembly. Each of the driveshafts is connected by a spline to the side gears in the differential. This allows the two driveshafts (and hence the wheels) to rotate at differing speeds to each other although the average speed will always be the same as the final drive wheel. It should be noted that the three shafts rarely lie in a single plane, and viewed from the end the shafts would lie in a 'V' shape. In this way the centre distance and relative position between the input and output shafts can be designed to suit the installation.

Drive is only engaged between the input and intermediate shafts using one set of gears at any one time. For example, when 1st gear is engaged, only that particular gear pair carries the drive between the two shafts. This is achieved using the synchromesh assemblies, which can be seen in the section illustration above. In this example the synchromesh assemblies are on the intermediate shaft, although they can be positioned on the input shaft.

The gearbox illustrated in this example is the simplest form of front wheel drive gearbox. Some gearboxes have two intermediate shafts, which although more complex (and hence tend to be more expensive), does shorten the design which can sometimes allow easier packaging in the vehicle.

13.3.2 The rear wheel drive car and commercial gearbox

Two key features distinguish the rear wheel drive gearbox from the front wheel drive gearbox discussed above; the lack of a differential, and the absence of the final drive reduction gear pair. Both of these will be found within a different unit in the vehicle and this can be seen in the diagrams in the section on vehicle/power train layout. As stated earlier, the ratios are typically, from 4:1 to about 0.8:1. As discussed below 4th gear is often a direct drive between the input and output and so gives a ratio of 1:1. In some gearboxes, however, this direct drive is provided in 5th gear to give a more efficient drive in that gear (there are no friction losses in the gear mesh points). This means that the final drive needs to be adjusted accordingly to provide the same overall gearing and the ratios in the lower gears will be higher to give the same ratio spread – a ratio of around 5:1 rather than 4:1 in first gear perhaps. Figure 13.8 is a cross section of a typical rear wheel drive gearbox, while Figure 13.9 shows a schematic of the gear layout.

It can be seen from the diagrams that the gearboxes have several separate sets of gears with various ratios. With the gearbox in a particular gear the power follows one of the possible paths through the gearbox. When the driver changes gear the power will flow along an alternative path. The schematic diagram below shows a simple 4-speed gearbox to illustrate the concept.

In these gearboxes the input shaft is driven, via the clutch, by the engine. The constant gears are continually in mesh and the input shaft always drives the layshaft. With no gear selected, i.e. neutral, all the gears on the mainshaft are free to rotate on the shaft and no drive can pass to the mainshaft. The synchromesh mechanism (discussed later) allows either the gears on the mainshaft to be connected to the shaft or the input and mainshaft to be connected together – shown below in Figure 13.10. This connection by the synchromesh allows the various gears to provide the different ratios of speed between the input and output shafts.

13.3.3 Gearchanging and the synchromesh

The mechanism usually used in modern manual gearboxes to allow the gears to be changed (when the clutch disengages the drive from the engine) is the synchromesh. This essentially

Typical design features of a rear wheel drive car gearbox:

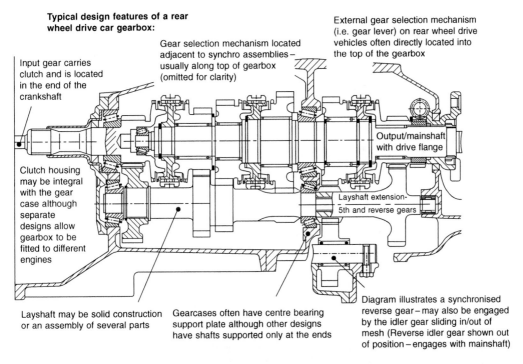

Input gear carries clutch and is located in the end of the crankshaft

Gear selection mechanism located adjacent to synchro assemblies – usually along top of gearbox (omitted for clarity)

External gear selection mechanism (i.e. gear lever) on rear wheel drive vehicles often directly located into the top of the gearbox

Output/mainshaft with drive flange

Clutch housing may be integral with the gear case although separate designs allow gearbox to be fitted to different engines

Layshaft extension-5th and reverse gears

Layshaft may be solid construction or an assembly of several parts

Gearcases often have centre bearing support plate although other designs have shafts supported only at the ends

Diagram illustrates a synchronised reverse gear – may also be engaged by the idler gear sliding in/out of mesh (Reverse idler gear shown out of position – engages with mainshaft)

Figure 13.8 Cross section of a rear wheel drive manual gearbox

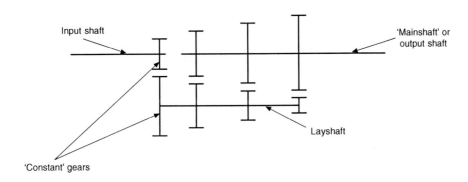

Input shaft

'Mainshaft' or output shaft

Layshaft

'Constant' gears

Figure 13.9 Schematic of a gear train layout in a rear wheel drive gearbox

allows the speed of the components to be matched before they are connected together as the new gear is engaged. If you consider the earlier diagram of a transverse gearbox (Figure 13.6), it can be seen that the synchromesh assemblies are on the intermediate shaft. This is not a requirement,

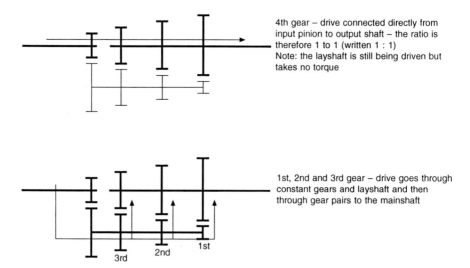

4th gear – drive connected directly from input pinion to output shaft – the ratio is therefore 1 to 1 (written 1 : 1) Note: the layshaft is still being driven but takes no torque

1st, 2nd and 3rd gear – drive goes through constant gears and layshaft and then through gear pairs to the mainshaft

3rd   2nd   1st

Figure 13.10 Diagrams to illustrate different gears selected in a simplified gearbox

and they could easily be on the input shaft or split between the two. Similarly, the synchromesh mechanisms can be on the layshaft or mainshaft in a 3 shaft, rear wheel drive gearbox.

A good starting point for synchromesh design is the paper written by Socin and Walters (1968). This not only explains the function of the synchromesh used in many gearboxes but also follows the main calculations that may be required during the design of single cone synchromesh.

In considering the function of a synchromesh, we need to refer back to the section on manual gearboxes above:

You may remember that in this schematic the drive goes via the layshaft, then the mainshaft gears and onto the mainshaft. Remember, the mainshaft gears are not permanently attached to the mainshaft – you can see this in the diagram of a manual gearbox above. Looking at the 1st and 2nd mainshaft gears for a moment you can work out that when you are in 1st gear the 1st mainshaft gear is coupled to and driving the mainshaft. When in second, the 2nd mainshaft gear is doing the same job. If you think about it, you can work out that the mainshaft, because it is connected to the wheels, carries on rotating at more or less the same speed both before and after any gearshift. Therefore if the driver changes gear, say from 1st to 2nd the layshaft, input shaft and clutch have to slow down so that the 2nd mainshaft gear can go at the same speed as the mainshaft. (See Figure 13.11.)

It is this speeding up and slowing down of the input side of the gearbox that the synchromesh does. The work done by the synchromesh assembly is to change the speed of the inertia on the layshaft and input shaft, which includes the clutch driven plate. The large majority of the inertia is found in the clutch plate.

The diagram below (Figure 13.12) is a cross section of single and triple cone synchromesh. Synchromesh are effectively a cone brake device and the action of the driver in applying force to the sleeve allows the two side of the assembly to tend towards the same speed. In the first part of the synchronising process, the sleeve applies a load onto the baulk ring, which is rotating at a different speed to the cone onto which it is pushed. The friction that is created by this causes the speed of the two parts to become the same. At this point the sleeve is able to move past the

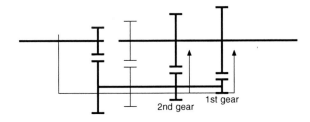

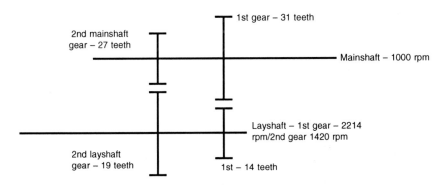

Figure 13.11 Diagram to show the gears involved and ratio implications of a gearchanging between 1st and 2nd gear

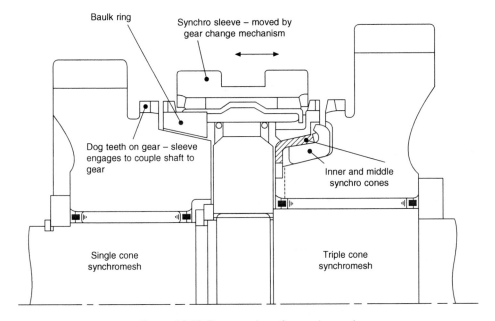

Figure 13.12 Cross section of a synchromesh

baulk ring and engage with the gear and the gearshift is complete. The detail of this process is covered in the paper referred to above.

13.3.4 Gear ratios – how they are achieved

Section 13.2 discussed why different ratios were needed and how they were selected. Given the overall design package of a transmission, the next task is to design the gear pairs within the casing to achieve the required ratios between the input and output shaft.

 As an illustration of how the ratios in a gearbox are achieved, included below are the gear tooth numbers used in a version of the Land Rover LT77 manual gearbox. This was used in the Land Rover, Range Rover Classic and a number of other vehicles in both 2 and 4 wheel drive versions.

The constant gears:
- 22 teeth on the input shaft/33 on the layshaft 'constant' gear (driven gear)
- Ratio – 0.666, i.e. layshaft rotates at 0.666 of the speed of the input shaft (slower)

3rd gear:
- 29 teeth on layshaft/27 teeth on the mainshaft gear
- Ratio – 1.074, i.e. mainshaft rotates at 1.074 of the speed of the layshaft (quicker)
- Combining the ratios gives 0.666 × 1.074 = 0.715, i.e. mainshaft/output shaft rotates at 0.715 of the input speed (slower)
- The inverse of this is normally quoted to cause a bit of confusion! So you would see 3rd gear quoted as being 1/0.715, which is 1.397

2nd gear:
- 19 teeth on layshaft/27 teeth on the mainshaft gear
- Ratio – 0.704, (slower)
- Combining the ratios gives 0.666 × 0.704 = 0.469, i.e. mainshaft/output shaft rotates at 0.469 of the input speed (slower)
- As with 3rd, by convention this is quoted as 2.132 – (1/0.469)

1st gear:
- 14 teeth on layshaft/31 teeth on the mainshaft gear
- Ratio – 0.452, i.e. mainshaft rotates at 0.452 of the speed of the layshaft (slower)
- Combining the ratios gives 0.666 × 0.452 = 0.301, i.e. mainshaft/output shaft rotates at 0.301 of the input speed (slower)
- As with 2nd and 3rd by convention this is quoted as 3.322 – (1/0.301)

5th gear:
Fifth gear on the LT77 is an 'overdrive' gear. This means that the output shaft of the gearbox rotates faster than the input. The numbers work out as:

- 37 teeth on layshaft/19 teeth on the mainshaft gear
- Ratio – 1.947, i.e. mainshaft rotates at 1.947 times quicker than the layshaft
- Combining the ratios gives 0.666 × 1.947 = 1.297, i.e. mainshaft/output shaft rotates at 1.297 times the input speed

- The inverse of this is again normally quoted, so you would see 5th gear quoted as being 1/1.297, which is 0.771

13.3.5 The clutch

On vehicles there is a requirement for a device to provide a coupling from the engine crankshaft to the transmission. This allows the engine to be started and run without the vehicle moving and the vehicle to be started form rest under control at various rates of acceleration. On manual gearboxes, the drive from the engine also has to be disconnected during gear changes.

Clutches are associated with manual gearboxes and are normally operated by the driver. Recent developments in both the passenger car and commercial truck market have meant an increasing number of vehicles are automating the function of the clutch. The vast majority, however, remain controlled mechanically by the driver.

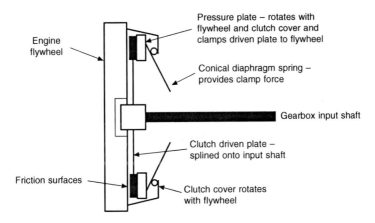

Figure 13.13 Diagram of a conventional single plate clutch

The spring pressure clamps the pressure plate onto the driven plate and the flywheel, with the assembly like this the drive is passed from the engine to transmission (see Figure 13.13). When the driver depresses the clutch pedal the movement is passed to the release bearing by either hydraulics or cable and the release bearing then pushes or pulls the diaphragm spring (depending on whether the clutch is a 'push' or 'pull' design), see Figure 13.14. The outer part of the release bearing is held (by the release lever) so it does not rotate, and the inner race of the bearing rotates with the diaphragm spring and clutch cover. The force applied and travel of the bearing is determined by the lever ratio built into the clutch pedal, hydraulics (or cable), and release lever. There is often significant development in this area as the loads and movement of the pedal have to be as low as possible for the driver while ensuring the clutch will operate correctly throughout the life of the clutch plate. As the clutch plate wears, the position of the various components and the load applied by the spring vary. Obviously, the load has to always be sufficient to clamp the driven plate and not allow any slip, and yet fully clear to allow the plate to rotate freely when the clutch pedal is depressed.

Figure 13.15 shows the whole clutch system. By depressing the diaphragm spring the pressure

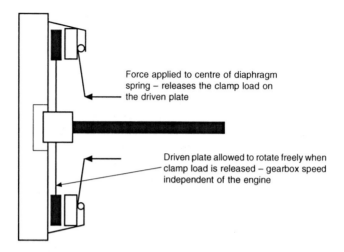

Figure 13.14 Diagram of clutch when released

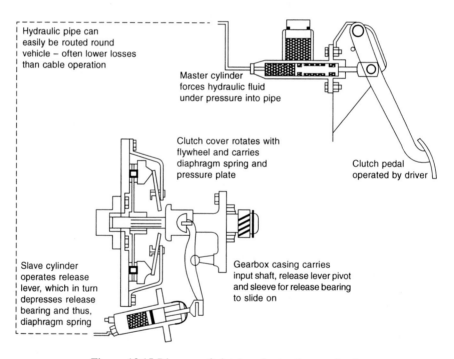

Figure 13.15 Diagram of clutch and actuation mechanism

on the driven plate is released. This actuation of the spring is achieved by hydraulics, cable or on older vehicles may have been mechanical linkage. When the load on the cover plate is released the driven plate is allowed to rotate freely inside the assembly and the drive to the transmission is disconnected.

13.3.6 Automated manual transmission

With the introduction of a number of vehicles recently, automation of synchromesh, 'manual' transmissions is becoming more popular. The reason for the development of these transmissions is twofold; firstly then can show an economy benefit over both manual and automatic transmissions. This is because they are more efficient than automatics and can be programmed to change gear more effectively than most drivers would. Secondly, automated manual transmissions are gaining in popularity in the performance car market, probably because of the links to Formula 1 racing and as a result of clever marketing! Examples include: BMW M3, MMC Smart, VW Lupo, Alfa 156.

These developments started some time ago with the introduction of automated clutches on several vehicles including the Renault Twingo, Saab 900 Sensonic and Ferrari. These cars retained the normal gear lever but automated the clutch so that no pedal was required. At start up they operate as an automatic with the control system actuating the clutch to achieve a start from rest when the accelerator pedal is depressed. During gear changes the clutch is operated in response to movement of the gear lever.

Consideration of the mechanics of the automated manual systems suggests that it may be difficult for these systems to replace the conventional automatic. The fundamental point is that the automated manual systems need to disconnect the drive from the engine to the transmission in order to achieve a gear change. With conventional automatics only a small reduction in the engine power is required to achieve a smooth transition form one gear to another because of the action of the torque converter. There are, however, twin clutch designs of transmission, which overcome this limitation by providing two parallel torque paths through the transmission where a gearchange simply switches from one path to another and engages one clutch rather than the other. This can be done without reducing the engine output (a 'hot shift'). This has been used in the past by large automotive gearboxes, but could be extended to the car market.

In the commercial market there are a number of manufacturers now producing automated manual transmissions for trucks. Whereas these developments have needed the driver to indicate the gear selection in the past, the latest developments have the intelligence to completely automate the gearchange. On heavy commercial vehicles this may need to include missing some gears, especially when unladen so the control software required is not trivial.

13.4 The automatic transmission

The concept of an automatic transmission offers considerable advantages to vehicle drivers since they can be relieved of the burden of selecting the right gear ratio. This burden, both mental and physical has become more significant with increasing traffic congestion. Any reduction in driver fatigue and increased opportunity for the driver to concentrate on other aspects of vehicle control must contribute to increased safety and a reduction in road traffic accidents. There are also benefits in terms of economy and emissions if an automated system can make a better selection of ratio than a non-expert driver does. There are several alternative solutions to achieve this automation including automated layshaft transmissions (described above), continuously variable transmissions (described in the next section) and the 'conventional' automatic transmission described here.

The term 'automatic transmission' (AT) is used to refer to a combination of torque converter with a ratio change section that is based on epicyclic gearsets. The use of these components can be traced back to the early days of automotive developments, and in a recognizable combination to the middle of the last century. Yet it is an area that is still seeing extremely rapid development today. The success of this combination lies in the simplicity of the torque converter as a device that inherently has ideal characteristics to start a vehicle from rest, and the opportunity that epicyclic gear sets provide to give relatively easy and controllable changes between ratios.

The controllability of these devices has allowed automatics to be developed with the good shift quality necessary to satisfy the driver's expectations for a gear change. Somehow, drivers of conventional manual shift vehicles are always more critical in judging the gear change of another driver rather than their own where a misjudged shift can be more easily forgiven. In just the same way they are more discerning in judging the quality of an automated gear change and thus high standards are required. In the past these have been virtually impossible to achieve from automated manual gearboxes. This situation is, however, changing with the greater use and sophistication of electronic controls.

The downside of an AT in comparison with a manual gearox alternative is greater cost, greater weight, larger size and lower efficiency. It has thus been used most in larger cars where these penalties are less significant and the driveability advantages most appreciated. This may well account for the large proportion of automatic transmissions used in the USA (approaching 90%) in comparison with Europe (around 20%). However, all these disadvantages have acted to maintain the pressure for development of the AT leading to modern designs that achieve a greater number of gear ratios within the same or even a reduced space envelope.

13.4.1 The Jatco JF506E – A state of the art transmission

An example of today's typical 5-speed automatic suitable for a medium to large size front wheel drive vehicle, where packaging and space constraints are severe, is shown as a sectioned view in Figure 13.16. The input shaft drives first into a 3-element torque converter (top right) directly to a pair of epicyclic gears that give a four-speed change section. A further fifth speed change is obtained on the secondary shaft with a final drive ratio providing the connection to the differential. The output drive shafts can then be taken either side of the differential shown at the bottom of the section. This layout can also be easily extended to give a four-wheel drive output with an additional gear section. The main components that make up this transmission are described in more detail below and the combined operation in Section 13.4.4.

This transmission is electronically controlled by a control unit interfaced with other vehicle systems, including engine management, via a CAN link (See Control Systems Chapter 11). The electronic control signals are passed to solenoid valves that apply hydraulic pressure to control clutches that select the required ratio. The programmed control strategy takes inputs from speed and temperature sensors to respond to the driver's demand. This demand comes principally from the accelerator pedal position but is modified by brake application and both the gear selector (D, 4, 3, 2) and a pattern selector (drive, sport and snow).

13.4.2 The hydrokinetic torque converter

Hydrokinetic drives involve the transfer of power through the 'kinetic energy' or velocity head

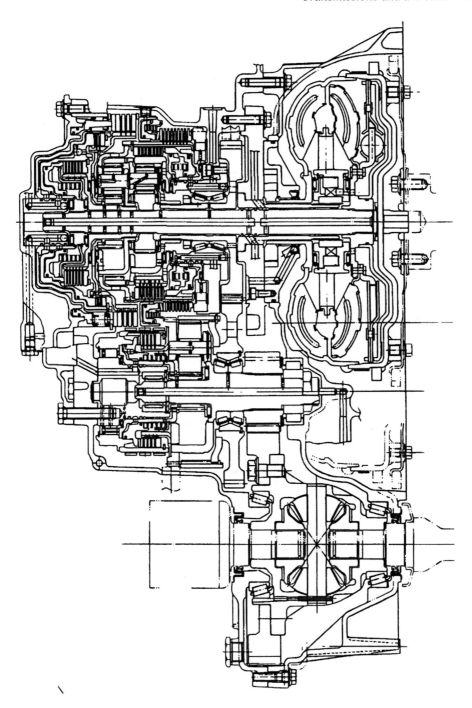

Fig. 13.16 Sectioned view of Jatco JF506E (courtesy of Jatco)

of a fluid. In such devices an impeller element creates the flow kinetic energy and a turbine element recovers the energy producing a torque output. There are two main types of hydrokinetic devices: fluid coupling and fluid converter. Both these families provide an automatic adjustment of ratio (input speed for a given output speed and load) and an infinite ratio capability that makes them highly appropriate as a 'starting device'. Their features include: stepless variation in torque and speed without external control, vibration isolation, shock load absorption, low maintenance and virtually wear free operation. Disadvantages include efficiency, design limitations, and great difficulty to control precisely.

The term torque converter is used here to discribe the converter-coupling as most frequently used in automotive applications. This is also known as a Trilok converter. It is so called because in a part of its operating range it gives a torque multiplication (behaving as a converter) and in the remainder it behaves as a coupling with a 1:1 torque ratio.

The basic equation defining the fluid torque acting on impeller or turbine is:

$$T = C\omega^2 D^5$$

where T = torque transferred
C = capacity constant
ω = rotational speed
D = diameter

The capacity factor C, is independent on the detailed geometry (blade angles etc.), fluid density and viscosity, and most importantly it varies with speed ratio.

13.4.2a Fluid coupling

Fluid couplings contain only two rotating, elements – impeller and turbine – within a toroidal casing as shown in Figure 13.17. Both these elements have radial vanes and the cavity is filled with hydraulic fluid. The impeller and casing are driven by the input and fluid trapped between the rotating vanes must also rotate and this in turn causes flow outwards to the largest diameter as a result of centrifugal action. This outward radial fluid flow is directed by the curvature of the impeller shroud back to the turbine section where the rotational component of velocity gives a torque reaction on the turbine blades as the fluid flow direction is changed. The fluid returns towards the centre-line of the assembly and re-enters the impeller at a smaller diameter.

Since there are only two elements there must always be an equal and opposite torque reaction thus input torque T_i, must balance output torque T_o:

$$T_o = T_i$$

Vanes in simple fluid couplings are radial and hence it can be an almost symmetrical device where the impeller and turbine functions can be reversed and torque transmitted in the reverse direction. However, it is also possible to use curved vanes that give asymmetry and a higher torque capacity in one sense. The transmitted torque will depend on the relative speed of the impeller and turbine. It will reduce to zero if they are rotating at the same speed and will reverse if the turbine rotates faster than the impeller. The relative speed may be expressed either as a speed ratio (ω_o/ω_i) or by a relative slip s, defined by:

$$s = \frac{(\omega_i - \omega_o)}{\omega_i} = 1 - \frac{\omega_o}{\omega_i}$$

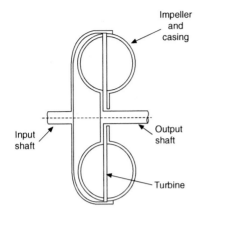
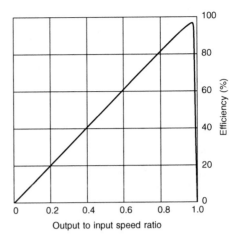

Figure 13.17 Fluid coupling and characteristics

The power transmission efficiency η, is also related to speed ratio as follows:

$$\eta = \frac{\text{power out}}{\text{power in}} = \frac{T_o \omega_o}{T_i \omega_i} = \frac{\omega_o}{\omega_i}$$

The efficiency characteristics are thus a linear function of speed ratio as shown. However, as the speed ratio approaches unity the torque transfer capability will reduce and the flow losses mean that the torque transfer falls rapidly to zero. This occurs in the region where slip is 2–5% (speed ratio 0.95–0.98), depending on the internal clearances within the coupling.

13.4.2b Fluid converter

The converter is like the coupling in having a turbine and impeller but in addition uses a third vane element called a reactor or stator that does not rotate. To prevent it from rotating it is connected via a tube concentric with the turbine output shaft to an internal part of the gearbox casing such as a bearing housing. The stator vanes re-direct the flow as in Figure 13.18 and *add* to the torque provided by the engine input to give a multiplying effect on the output torque (despite the apparent sequence implied by the flow path). The torque balance then becomes:

$$T_o = T_i + T_s$$

The efficiency is:

$$\eta = \frac{\text{power out}}{\text{power in}} = \frac{T_o \omega_o}{T_i \omega_i}$$

The blade angles of all three elements are curved to give the easiest flow path at a so-called 'design point'. This point usually represents a peak efficiency with respect to speed ratio as shown in Figure 13.19. At other conditions additional losses occur as flow meets the vanes at 'awkward' angles giving rise to shock losses. The blade curvature means that the converter is not symmetrical and will not transmit torque effectively in the reverse sense (negligible engine

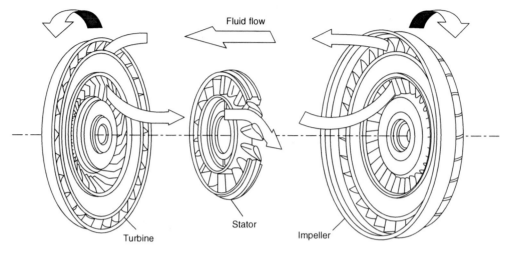

Figure 13.18 Torque converter 3D with flow path

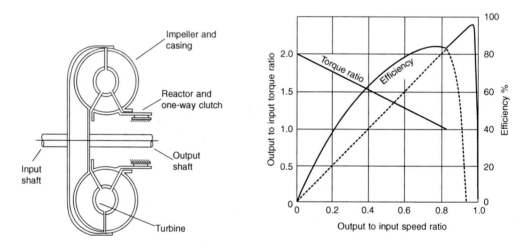

Figure 13.19 Torque converter and characteristics

braking). There is a compromise in design between achieving a high torque ratio at stall (zero output speed) but at the expense of efficiency. It is possible to achieve torque ratios of 5:1 but these days fuel efficiency has become increasingly important and automotive converters tend to operate around 2:1.

Figure 13.19 also shows that beyond the point of maximum efficiency the torque ratio tends below unity. This region is not attractive from an automotive viewpoint with reducing efficiency and hence the developments of the converter-coupling operation described in the next section.

13.4.2c Torque converter

The term torque converter is sometimes applied to the basic fluid converter described above but it is used here to describe the device that combines both converter and coupling operation. This combination is typically used in automotive applications and sometimes called a Trilok converter. The design is based closely on that of the fluid converter but with the addition of an overrun clutch (Heisler, 1989) connecting the reactor (stator) to its fixed reference frame. This prevents the reactor from rotating in one direction but will allow it to rotate freely in the other. Operation can be visualized during an acceleration sequence with an increasing vehicle speed when the operation initially follows that of the fluid converter. The reactor will be locked until a speed ratio is reached where the input and output torques are equal, and consequently the reactor torque has reduced to zero. In converter operation any further increase in speed ratio above this would give a reduced torque ratio that can only occur by a reversal of torque on the reactor (since the three component torques must still be in balance). In a converter coupling this cannot be reacted by the overrun clutch and the reactor will free wheel. Above this speed ratio the assembly is behaving as a two-element device and operates as a fluid coupling. This gives the combined characteristic shown by the full line in Figure 13.19 with increasing efficiency until the operating limits are reached as with a fluid coupling.

Further improvements in efficiency can be obtained if a lock-up clutch is used to mechanically lock the impeller and casing to the turbine and hence directly connect input and output shafts. This should only take place when the speed ratio is near unity, and needs to be controlled gradually in order to prevent any driveline shock that might be felt by the driver. This action can be actuated hydraulically as required by the transmission controller.

13.4.3 The epicyclic gear set – the key component in the automatic transmission

The epicyclic or planetary gear contains three sets of concentric gears meshing at two diameters as shown in Figure 13.20. These are connected to three external shafts distributing the transmitted torque between them. These three comprise the sun gear shaft, the annulus gear shaft and the carrier. The planet gears rotate about pinions mounted on the carrier, and this part of the assembly can be considered to behave as a single component. In automatic transmissions epicyclic gears are usually operated with one of these three shafts locked to the gearbox casing or frame leaving the remaining two to act as input and output shafts. Another alternative is for any two of the components to be locked together and the whole unit rotates as one, and although apparently trivial it is a convenient and common option.

An example of how the device works can be envisaged with the carrier shaft locked, leaving the sun and annulus free to rotate connected via the rotation of the planets about their now fixed centres. The peripheral speed of any of the planet gears must be the same at both the contact radius of the sun and the radius of the annulus. Thus the number of gear teeth on each must determine the relative speed of the sun and annulus, and since the tooth pitch or module must be the same, their relative diameters. They will of course rotate in opposite directions. This ratio is referred to as the fundamental ratio i for the epicyclic gear and:

$$i = -\frac{\omega_s}{\omega_a} = \frac{t_a}{t_s} = \frac{D_a}{D_s}$$

where: t = number of teeth
and subscripts a and s refer to annulus and sun respectively

Consideration of the torques acting on the gear teeth at the two meshing diameters indicates that these will be given by the inverse of the speed relationship. Also the tooth forces on the planet

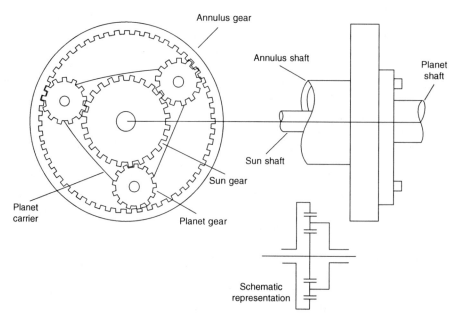

Figure 13.20 An epicyclic gear and stick equivalent

will be the same at both meshing interfaces but the highest tooth stresses will occur where there are fewer teeth to share this load. This must occur at the planet sun mesh and will occur on whichever is the smaller diameter of the two. Thus in sizing the gears for a required torque capacity this mesh region will indicate the limiting case.

Examination of the shaft speeds with other components locked allows the derivation of the overall kinematic relation for an epicyclic set as:

$$\omega_s = (1 + i)\omega_c - i\omega_a$$

There are limits in ratio that can be sensibly achieved with an epicyclic gear arising from a combination of physical packaging, tooth and pinion strength and the need to have a whole number of teeth on all components. This gives values for the fundamental ratio generally in the range:

$$2 \le i \le 4$$

The ease with which ratio changes can be implemented by a simple braking or clutching action can be illustrated by looking in more detail at a single epicyclic train. Figure 13.21 shows an epicyclic gear with the carrier and annulus shafts used as the input and output. In Figure 13.21(a) the sun gear is locked to the carrier and the whole assembly rotates as a single unit with a speed ratio of one. In 13.21(b) the sun gear is held stationery by the clutch and for a fundamental ratio $i = 2$, the annulus shaft will rotate at:

$$\omega_a = \left(\frac{1 + i}{i}\right)\omega_c = 1.5\,\omega_c$$

The change in ratio from 1 to 1.5 is a typical 'step' between gears. The relative rates at which the release and engagement actions take place on clutches and brakes can be controlled to give very

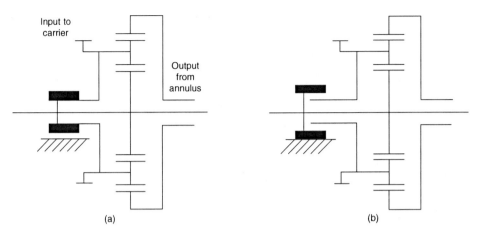

Figure 13.21 Single epicyclic sun locked to carrier and sun locked to frame. (a) Sun shaft locked to carrier shaft, (b) sun shaft held stationary

smooth transitions. The constructional details of typical clutches can be found in Heisler (1989). The ways that sets of epicyclic gears can be combined to give different overall ratios are described in the following sections.

13.4.4 JF 506E AT operation

The combined operation of these components can be seen by a more detailed examination of the JF 506E transmission introduced above. Figure 13.22 shows a schematic representation of the transmission with different gears selected. For clarity the main components are only represented on one side of the shaft centreline, effectively half of the transmission. The main torque and hence power flow path is shown by the heavy lines. There are two epicyclic gear sets on the primary shaft, labelled A and B, and a further reduction epicyclic on the secondary shaft.

In all gears the drive passes through the torque converter to the primary shaft with the stator free wheel becoming active at higher speed ratios. The drive from the parimary shaft comes from a path that is connected to both the annulus of epicyclic A and carrier of epicyclic B. this passes to the secondary shaft via transfer gearing onto the annulus of the reduction epicyclic set. The output then passes from the carrier of this epicyclic to the final drive and hence to the output shaft. In the first four gears and reverse, the sun gear of the reduction epicyclic is locked by the reduction brake. This gives a carrier rotation in the same sense as the annulus but with a reduction ratio of 0.8.

First gear – drive passes through epicyclic B with the annulus locked to frame through a hydraulic clutch and a free wheel clutch. The carrier is driven by the sun at about a third of the primary shaft speed and in the same direction.

Second gear – in this gear both of the primary shaft epicyclic play a part to give a combined ratio on the output to the transfer gear. The sun of epicyclic A is now locked and the torque reaction on the annulus of epicyclic B allows it to rotate with the carrier of epicyclic A in the free wheel sense of the overrun clutch. The combined motion increases the output to about a half of the primary speed.

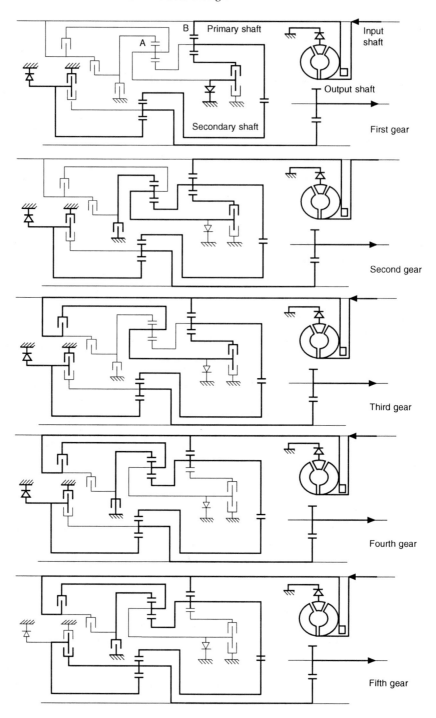

Figure 13.22 Power flow paths for different gears in JF506E (courtesy of Jatco)

Third gear – the annulus and the sun of epicyclic B are now locked together (via the carrier of A) and all of the components rotate at the same speed.

Fourth gear – drive comes entirely from epicyclic A with the sun gear locked. The annulus of this gear is driven by the carrier and rotates in the same sense at about 1.5 times the primary speed.

Fifth gear – the path from input to the secondary is the same as fourth gear but now the reduction brake is dis-engaged and the carrier and sun locked together. Again this means that the whole reduction epicyclic set will rotate as a single unit.

Reverse gear (not shown) – drive is obtained through the sun of epicyclic A with the carrier locked. This rotates the annulus in the opposite direction to the sun at a value just below half the primary speed. This is still subject to the reduction epicyclic ratio.

A set of manufacturer's ratios through the main section of the transmission (excluding final drive ratio) is given in Table 13.1.

Table 13.1 Typical ratio set

Gear	Ratio	Gear	Ratio
1	3.474	4	0.854
2	1.948	5	0.685
3	1.247	Reverse	2.714

13.4.5 Shift strategy

The basic ratio selection depends on a pre-determined shift strategy but there are also many subtleties in the way that changes are executed. The strategy is fundamentally a function of vehicle speed and the driver's accelerator demand and a typical example is shown in Figure 13.23. This shows shift-up and shift-down settings for each gear. There is obviously a need for some hysteresis in the change points between any two gears to prevent hunting phenomena developing. However, it is also obvious that for a given vehicle speed in many sections of the map a heavy-footed driver can easily invoke up and down shifts by moving the accelerator pedal.

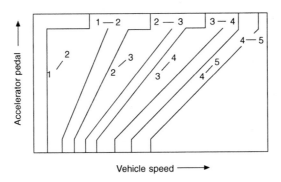

Figure 13.23 Shift strategy (courtesy of Jatco)

This basic shift strategy is modified by the pattern selection for the driver. In 'sport' mode higher engine speeds will be used, whereas in 'snow' mode lower engine speeds will be used. There are also driver selected gear hold positions (D, 4, 3, 2) that will limit the highest gear to be selected. In addition there are failure modes of operation that may either limit the gears that can be selected or hold a particular gear in order to provide safe operation. These are detected and appropriate actions invoked by the Automatic Transmission Control Unit (ATCU).

13.4.6 ATCU the controller

Automatic transmissions have been operated for many years with a hydraulic system providing both the logic and control actuating functions. The development of digital controllers and the necessary parallel development of low cost sensors, and electrical actuators have allowed the control logic to be implemented digitally. This allows considerably increased functionality together with greater flexibility and adaptability of the controller. Overall the classic combination of electronic 'brain' with hydraulic 'muscle'.

A block diagram showing the basic controller inputs and outputs for the JF506E is shown in Figure 13.24. There are also interconnections through the CAN bus, most importantly to the engine management but also brakes and the instrument pack. The flexibility of the bus system means that the path is open for integration with other vehicle systems (e.g. traction control) as a further development. In addition to driver control demands, the main inputs come from speed sensors. There are three of these, and they have been placed to give speed information on both sides of all epicyclic gears during a shift. These speeds are monitored and used by the controller to fine tune shift operations to give best shift quality. There is a control memory feature associated with this that learns about the hardware to adapt the control actions to give a consistently good shift performance. The main outputs from the controller operate solenoid valves, which activate clutches for each gear or control hydraulic system pressure.

The solenoids for gear change have an on/off action to directly open, or close, a hydraulic valve. When the valve is closed, pilot pressure is raised at one end of a sliding spool valve to move it against a spring return. Such a spool will usually be multi-functional and open or close a number of connections simultaneously. These may supply flow to raise pressure on a clutch or drain flow to derease clutch pressure, which is of course what is required in many gear change operations. They can also be used to provide other flow connections that may help in a sequence of operations or to inhibit actions like connecting the reverse gear when moving forward. The multi-function feature can allow a reduction in the number of components used and hence lower the cost.

The solenoids for pressure control also open and close a single directly acted valve. However, they are operated more rapidly to give alternate connection between a supply and a drain line. The proportion of time spent open relative to that closed gives a mean pressure level established between a minimum drain pressure and a maximum supply pressure. The open/close timing is obtained directly from a pulse width modulated (PWM) output from the ATCU. This transmission uses five on/off solenoids and four modulated solenoids that are used to control the hydraulic system line pressure and some individual clutch pressures.

The hydraulic system is supplied from a fixed capacity positive displacement pump. The pump outlet pressure in all automatic transmissions can have a significant effect on system efficiency, since the power absorbed by the pump is directly proportional to pressure, as well as

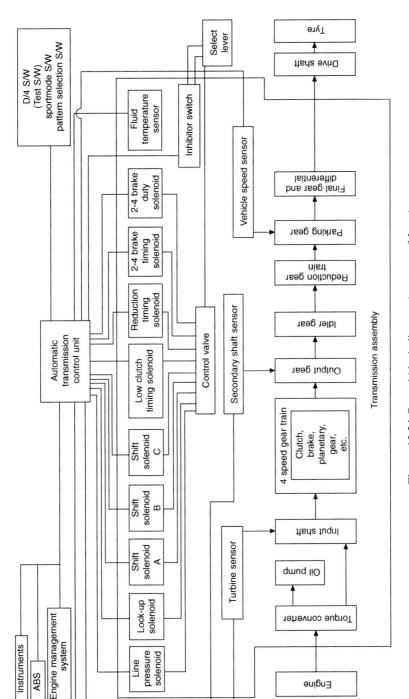

Figure 13.24 Control block diagram (courtesy of Jatco)

speed. It is thus kept as low as possible, but sufficient for the clutches to grip with a suitable margin for safety. Figure 13.25 shows the type of scheduling of system pressure with accelerator pedal position. This control can be considered typical but the mean level is changed with a number of conditions, such as starting from rest, or for acceleration then a higher mean level is set. If steady cruise conditons are detected by low accelerator pedal movements, then a cruise schedule will be adopted that reduces these levels in a process called 'cut-back' as shown. Reduced levels are also adopted during a gearshift operation to allow some controlled slip to take place across the clutches during the transition, and hence give a good shift quality. Also during gearshifts the ATCU requests the engine management system for a short period of reduced engine torque via the CAN link. There are also occasions when higher pressures can be selected when the driver indicates a need for engine braking by selecting a downshift at relatively high speeds. It should also be remembered that the hydraulic system provides a cooling function by removing heat from high temperature regions, as well as lubrication, and the actuation functions described here.

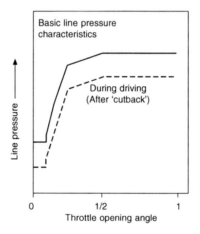

Figure 13.25 Line pressure control strategy (courtesy of Jatco)

Torque converter lock-up is also controlled by the ATCU. This locks the torque converter impeller to the turbine in regions of high-speed ratio. Again there is some hysteresis on this action so that once engaged it does not drop out too quickly. The lock-up clutch is activated by a pressure modulating type of solenoid to give a smooth and steady transition between states. The lock-up feature is normally only available in 4th and 5th gears but can be extended to both 2nd and 3rd gears if a high fluid temperature is detected, since this will eliminate any heat input via the torque converter losses.

The ATCU intelligence can be used in a similar way for the provision of this additional lock-up action. Monitoring of system responses allows detection of other failure states, and this can lead to action to protect the transmission, but still allow the vehicle to be driven, even with a reduced performance as described in Section 13.4.5. There are also low temperature conditions that can be detected from the oil temperature sensor and actions taken to try to encourage a quicker warm-up, both for the engine and transmission. For example, 5th gear may be inhibited at very low temperatures and the normal torque converter lock-up inhibited even at moderate temperatures. These features are all designed to enhance the overall system performance as well as refining the contribution mode by the transmission itself.

13.5 Continuously variable transmissions

The overwhelming majority of transmissions in road going vehicles are either manual or conventional automatic in design. These transmissions use meshing gears that give discrete ratio steps between engine and the vehicle speed. However, alternative designs exist that can transmit power and simultaneously give a stepless change of ratio; in other words a continuously variable transmission (CVT). Strictly speaking a CVT is a transmission that will allow an input to output ratio to change, continuously without any steps, in a range between two finite limits. An extension of this idea is a transmission that also allows a zero output speed to be included within this operating range. This can be considered as an infinitely variable transmission* (IVT). Since all vehicles need to come to rest then any transmission that inherently has an IVT characteristic is at an advantage. The use of the term CVT is used here in the more generic sense to include both types.

13.5.1 The rationale for the CVT

Transmissions that vary ratio continuously can be controlled automatically, and hence have the same advantages in terms of driving comfort as the more conventional automatic transmissions that were described above. In addition, they have the ability to vary the engine speed independently of vehicle speed. This brings the opportunity to optimize the engine operating point for any required output conditions and can offer either best economy or best power for acceleration. Figure 13.26 shows a typical petrol engine torque/speed map with contours of brake specific fuel consumption, and lines of constant power superimposed. Selection of the minimum fuel consumption for successively increasing output power gives the 'economy line' that is also shown. This indicates the most efficient point of engine operation for any output conditions and a CVT is required in order to enable these engine conditions to be matched to the vehicle output speed. A study of power levels used in typical vehicle operation indicates that the majority of time only low powers are required and that operation will take place predominantly along the low speed section of the economy line. This also indicates the need for a wide ratio spread in order to provide a good overdrive ratio that allows the low engine speeds to be used.

There are, in addition, performance benefits that can be obtained using a CVT since for ultimate acceleration the engine can be operated at its maximum power when required. In a vehicle with a discrete ratio transmission, the engine can only develop maximum power once for any ratio and there will be an interruption in power transmitted when the ratios are changed. Porsche showed (Kraxner *et al.*, 1999) that a CVT vehicle would out accelerate the same vehicle with a conventional manual transmission. It was further emphasized that it would only be the skilled drivers who could achieve the best from a manual transmission anyway.

---

*The idea of 'infinitely' can be easily accommodated when it is appreciated that transmission ratios are usually expressed in the form of input to output as a SPEED ratio, and that an infinite speed ratio is obtained when the output speed is zero.

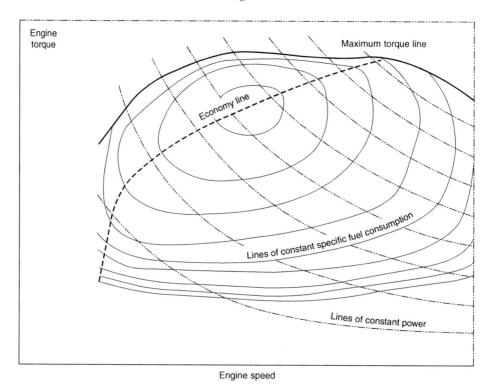

Figure 13.26 A typical petrol engine torque speed characteristic

Such transmissions have been under development since the early days of automobiles but have only recently begun to establish a place in the commercial market. Why is this, if there are potential benefits in terms of driving comfort, economy and performance? Different factors have contributed to this unpopularity at various times, and have included reliability, weight and cost. However, the main downside is the lower efficiency of all CVTs relative to geared transmissions. Losses in geared systems occur as a frictional torque required to rotate the transmission components. If an ideal speed ratio is defined for a gear pair then the output torque will be less than the ideal torque as a result of these losses, but the output speed will be the same as the ideal speed. However, all CVTs also incur some speed 'slip' between input and output as well as torque losses, and both output torque and speed will be less than the ideal. Since gears are very efficient it only takes a small additional loss for the effect to become significant.

The other factor is the way that the CVT powertrain is controlled and the resulting driver's feel or perception of the vehicle behaviour in response to accelerator pedal demands. This characteristic is called driveability, and is very much conditioned by the individual driver's previous experience and expectations. People more used to vehicles with manual transmissions are not as comfortable when driving a vehicle with any automatic transmission, because they feel they are less in control. There is also a significant compromise required between vehicle driveability and the possibility of achieving the desired economy line operation. A study of the engine map in Figure 13.26 shows that in most regions there is very little extra torque available between the economy line and the maximum torque curve. When a driver demands acceleration in this area then the system has to use this excess torque to first accelerate the engine before a significantly higher power can be transferred to the driving wheels. This gives a delay between a driver pressing the accelerator and the vehicle responding, this is also accompanied by the sound of the engine speeding up but nothing apparently happening. Many drivers find this disconcerting; particularly those used to manual transmissions. Road test reports have made comments about 'the rubber band effect' and '...the car's initial sluggish response...'.

Most vehicles fitted with CVTs have been with smaller engine sizes, generally below 1.5 litre, and not those noted as sporty cars even in standard form. In this category the performance margins are smaller and most manufacturers have not controlled the engine to rigidly follow the economy line and have compromised this to allow a larger torque margin for acceleration. The effect of transmission efficiency is also more significant in this category with a greater parasitic power lost. The result has been that most early production CVT vehicles have neither given better economy nor performance relative to vehicles with manual transmissions. However, a more valid comparison is with a conventional AT and here the CVT is the winner for both economy and performance (Liebrand, 1996). More recently larger engined vehicles like the Audi A6 have shown that CVT is not just a viable alternative to ATs but can deliver the improvements in both economy and performance even relative to the manual version (Audi's own figures). What is more, even road test reports in motor magazines have begun to report the virtues of CVTs as being realized '... a transmission that for once actually enhances the car's performance.' (*Autocar*, 2000).

13.5.2 Hydraulic transmissions

There are three main categories of transmission mechanism that can be used to provide the power and speed range for vehicle use: hydraulic systems, variable radius pulleys and traction drives. The first of these, described in this section, can be further sub-divided into categories of hydrostatic transmissions. Both of these are capable of giving a zero output speed and hence give IVT operation. The hydrokinetic drives have been described above in Section 13.4, and, as discussed above, torque converters are widely used in road going vehicles of all types.

Hydrostatic drives also rely on fluid flow to transmit power but it is the pressure level in the fluid that is significant rather than the flow velocity. A hydrostatic transmission comprises a pump unit supplying a motor unit, and both these are of the, so-called, positive displacement type. One at least must be variable capacity, usually this is the pump, and this is used to control the overall ratio. Hydrostatic drives are very widely used in agricultural and other off-road vehicles but have never been commercially used in automotive applications. It is generally the efficiency at low powers and potentially high noise levels that are the weak points of hydrostatic transmissions. However, they are still under development and, even recently, proposals have been made for buses and delivery vehicles with energy storage and hence a hybrid capability (Ifield website).

12.5.3 Variable pulley variator designs

The idea of a variable pulley system is a logical extension of a conventional V-belt fixed ratio drive. Figure 13.27 shows the principle, where there are two pairs of conical pulley sheaves and

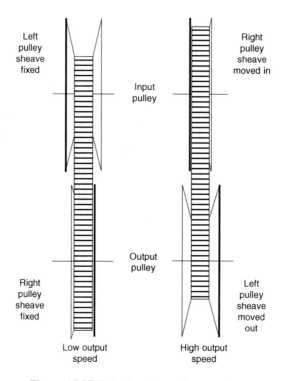

Figure 13.27 Variable pulley drive concept

a fixed length V-belt. One half of each pulley pair is moveable and their movement is synchronized. This allows the belt-rolling radius to be changed such that a relatively low output speed is obtained when its radius at the output is large. The output speed can be increased as this radius is made smaller and the rolling radius on the input is increased. As in a conventional V-belt, there is a frictional force between the pulley sheaves and the angled belt face that provides the transfer of tractive effort. There must be sufficient normal force between them to prevent gross slipping and the clamping forces that hold the pulley sheaves together also produce a tension load in the belt. The relative magnitude of these clamping forces can then also be used to control the belt position and overall ratio.

Early proposals, in the 1920s (Gott, 1991) relied on mechanical linkages to provide the movement of the pulley sheaves and maintain their relative position and pre-loading. The first commercially produced vehicle transmission (1958) was the DAF Variomatic and here the load was provided by pre-loaded springs in the pulleys, and the ratio change by a centrifugal effect working against these springs. The belt was fabric reinforced rubber, and two belts were used in parallel each driving a different wheel. This design was limited in its torque capacity to only smaller vehicles, and both wear and efficiency limitations restricted its broader application. Simple rubber belt designs have continued to evolve and are still widely used in two-wheel and some small city vehicle applications (Takahsashi *et al.*, 1991). Most modern variable pulley systems use steel belts to provide the power capability needed for automotive applications in a suitably compact package. The design and manufacture of the metal belts provides the heart of modern systems. The belt must have sufficient strength and rigidity to transmit the driving loads and yet also be flexible to keep the minimum rolling radius as small as possible. This has been achieved with two designs that also have significantly higher efficiency than the earlier rubber versions.

The first metal belt system was introduced into the market place in 1987 as the Transmatic, by Van Doorne's Transmissie (VDT) of the Netherlands. This transmission is the successor to the Variomatic but shares little more than the variable conical pulley concept with its predecessor. This is still the only automotive CVT in quantity production despite the recent launch of vehicles with both an alternative belt design (Audi) and a traction drive (Nissan). Over 3 million VDT units have been supplied for car use and current production of the belt is 500 000 units per year. Ford, Fiat, Honda, Mitsubishi, Nissan, Rover, Subaru, Toyota and Volvo have used it. Recent vehicles to use the this system have engine capacities in the 2-L range, representing a significant advance on the early small and super-mini (sub-compact) use. The capability for operating in extremes was shown by the successful demonstration of a version fitted in a Williams–Renault Formula 1 car before this technology was banned by the FIA in 1994.

The construction of the belt used in the VDT system is shown in Figure 13.28. The belt is assembled from two packs of flexible steel bands with a set of individual segment blocks retained by these bands. A typical design for an 85 kW, 165 Nm capability has 12 bands in each pack with about 430 segments in a belt that has a free length of 680 mm. The angled surface of the segments, below the shoulder, provides the contact surface with the pulley sheaves. Since the bands are free to slide relative to the segments and it is not possible for the segments to transfer load in tension, the only remaining mechanism is for the segments to transfer load in compression. Thus this design is often called a 'push belt'. Since the surface area between segments is relatively large the working levels of compression stress are low. When the bands have been placed in tension by the pulley clamping forces they will stretch and gaps open up between the segments

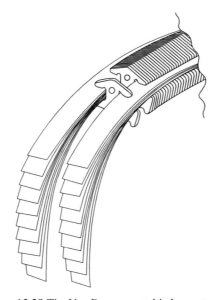

Figure 13.28 The Van Doorne metal belt construction

on the non-compression side of the belt. Many changes in detail design of the belt, particularly the segments, have been made since its introduction, and given improved performance and reduced manufacturing cost.

Another variable pulley system is based around the LuK-PIV chain as the flexible belt element. This is constructed in a way that is more reminiscent of conventional roller chain as used in motor and pedal cycles and is shown in Figure 13.29. Adjacent pairs of rocker pins are connected through a number of link plates, each set offset relative to the next set of link plates. The link lengths are not all identical and these give a staggered pitch for the pins that reduces acoustic noise. Some of the links near the outer edges of the chain are thicker to increase stiffness and reduce overall distortion under high load. The extended ends of the rocker pins act as the contact faces to the pulley sheaves. The ends of the pins have a crowned face rather than being flat, and this reduces the sensitivity to any misalignment but also reduces the active contact area.

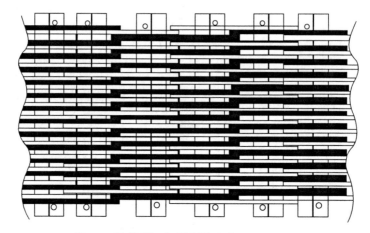

Figure 13.29 The LuK-PIV chain construction

13.5.4 Variable pulley transmissions

The variable pulley variators form the basis of a complete transmission system that also includes a hydraulic system, additional gearing and a starting device. A typical transmission based on the VDT push belt is shown in Figure 13.30, and is similar to that used by many major manufacturers. Until recently, all production variants used a clutch as a starting device, most frequently a wet plate clutch as in the Figure, although some low power designs have used an electromagnetic powder clutch. Audi have continued with this in their A6 2.4 Litre Multitronic, but the current trend is towards the use of a torque converter in place of the clutch. This gives a wider useable ratio range and has the start from rest and acceleration feel of a torque converter, and hence gives a better driveability feel for the transmission. However, it is less efficient and increases the torque in the pulley section, which will also increase the losses. The transmission shown also includes transfer and final drive gearing, including the differential appropriate for a front wheel drive vehicle. In the region of the clutch pack there is also an epicyclic gear that is used to give a forward and reverse shift by engaging the appropriate clutch.

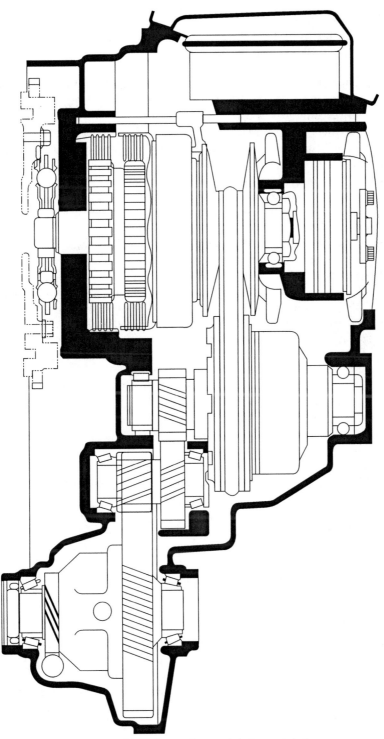

Figure 13.30 A complete push belt transmission

The hydraulic system provides similar functions to the more conventional ATs including lubrication, cooling and control. Both chain and belt systems use hydraulic pressure applied to pistons that are a part of the moving pulley sheaves. The supply is obtained from a pump, located towards the right of the transmission in Figure 13.30, and permanently driven from the engine via a quill shaft. This preferentially supplies the control system that will also determine the maximum operating pressure. Since it is better to keep lower operating pressures to improve efficiency, this will be controlled to the minimum necessary to provide the pulley clamping forces to prevent belt slip. Some designs use a single piston that provides both the base clamping force and ratio control by modulating the relative pressures on input and output pulleys. Others use a double piston design that separates the ratio and clamping functions. Although introducing mechanical complication, this provides the opportunity to reduce clamping pressures particularly in low throttle overdrive situations. This must be accompanied by a device that can raise the pressure rapidly, if the torque suddenly changes (as might occur if one wheel loses grip momentarily on a patch of ice).

The first versions of belt CVTs used hydraulic logic to control the transmission and schedule the ratio and system pressure with operating conditions. This has now changed, and all of the designs now being launched include electrohydraulic control for the same reasons of versatility and flexibility that were described above for ATs. A comparison carried out by LuK (Faust and Linnenbruegger, 1998) showed that a modern 5 speed AT required approximately 6 solenoids with 20 valves in comparison with their CVT design that needed only 3 solenoids and 9 valves.

The only control strategy that has been adopted until recently has been that of a full automatic with some driver selected inputs to hold low ratios to allow an increase in engine braking. As with conventional Ats, input signals come from driver inputs of accelerator and brake, as well as engine and output speeds. In all cases the transmission is scheduled to keep engine speeds as low as possible within the constraints and compromises of driveability and emissions. This gives the best fuel economy. However, several new systems (e.g.) Rover/MG, Audi and Nissan) have offered an additional option of a manual transmission emulation. This gives the driver the option to select between six pre-defined ratio settings to give the feel of a manual gearbox. The resulting ratio change is rapid and smooth and of course since there is no clutch there is no interruption in drive to the wheels. If accompanied by electric switches on the steering wheel it also follows the pattern established for Formula 1 and rally cars!

13.5.5 Traction drive designs

The other main mechanical alternative to pulley belt systems are those based on the idea of changing the axis of a rotating element between input and output discs. This concept is shown in Figure 13.31 by a single rotating roller that is held in friction contact between the two discs that have the same horizontal axis of rotation. In Figure 13.31(a) the axis of rotation of the roller is angled to the left and contacts the input disc at a smaller radius than the output disc, hence giving a reduced output speed relative to input. Figure 13.31(b) shows the roller axis rotated to the right giving the output disc an increase in speed. Drive is transferred via the contact patch between the roller and the two discs through a thin elastohydrodynamic liquid film, typically under 0.5 μm. This requires high loads holding the components together to give high contact patch pressures. The resulting high pressure in turn gives a significant increase in viscosity of the fluid in this region.

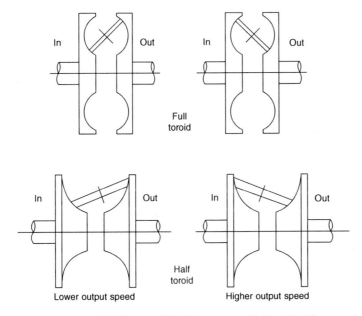

In Out In Out

Full
toroid

In Out In Out

Half
toroid

Lower output speed   Higher output speed

Figure 13.31 The toroidal drive concept (full and half)

The cavity between the discs is toroidal in shape and gives this name to variators based on this principle. There are in addition several other practical designs that include the Kopp variator for industrial use and the Milner ball drive (Akehurst *et al.*, 2001) used in lower power applications. The Figure shows a full and a half-toroidal cavity and both these form the basis for transmission designs. The power roller in the half toroid design is pushed outwards by the application of the clamping forces between the discs necessary to provide the surface contact pressures. The full toroid retains the power rollers within the cavity without generating these side forces but experiences a spin loss in the contact surface. This spin loss occurs because the surface of the power roller and the corresponding disc surface cannot be travelling at the same speeds throughout the contact patch.

In these traction drives both the materials used in the contact surfaces and the lubricating or traction fluid are important in giving good reliability and efficiency, ultimately the effectiveness of the transmission. High-grade bearing steels with very low impurity levels have been shown by NSK to give the best life in their half-toroidal variator. Special fluids have been developed by a number of companies including Monsanto, and more recently Shell. These have given good 'traction' coefficients in the order of 0.1 at high pressure whilst retaining good lubrication and low friction under more normal bearing conditions. It has also been demonstrated that cleanliness of the fluid is important in reducing the wear of the rolling components and hence better filtration than normal is required.

13.5.6 Toroidal transmissions

The first production toroidal transmission was lunched in the autumn of 1999, fitted in the Nissan Cedric/Gloria but only available in the Japanese market. This vehicle is fitted with a 3

Litre engine and the transmission is rated to an input torque capacity of 370 Nm. The variator section has dual toroidal cavities with two power rollers in each, as shown in Figure 13.32. The necessary contact surface loading is obtained through a combination of a preloaded spring establishing a minimum value and a loading cam that gives an additional component that increases with transmitted torque. This transmission also has a torque converter as the starting device and includes forward and reverse gearing via an epicyclic section. The torque converter has a maximum conversion ratio of 1.98 and the variator section has a ratio range of 4.4. The transmission is reputed to give a 10% improvement in fuel consumption relative to a conventional AT.

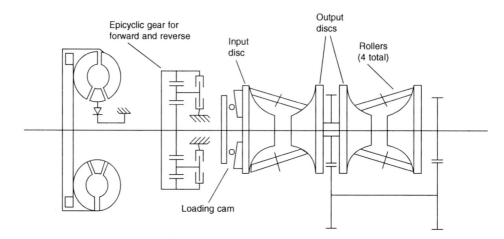

Figure 13.32 A dual cavity half toroidal transmission

The implementation of a full toroidal device in a transmission has been proposed by Torotrak and is currently under development for demonstration in sports utility vehicles. This is also a dual cavity with three rollers in each, a front rear layout and rated for 5 Litre engines producing around 450 Nm torque. In prototype form it has also been fitted to the 2 Litre Ford Mondeo in a front wheel drive configuration. This transmission is notable for the use of the split path principle to provide an IVT characteristic from a CVT variator.

The layout for the transmission is shown schematically in Figure 13.33 and shows two potential drive paths connecting the engine to the output, referred to as low and high regime. One path from the engine is via the variator and the sun of the epicyclic gear, and the second transferred from the engine directly to the planet carrier of the epicyclic gear via the low regime clutch. The output is taken from the annulus of the epicyclic gear and is a summation of the two inputs. A study of the epicyclic equation in Section 13.4.6 will show that it if the sun and carrier are driven at the right relative speeds, i.e. the sun is driven at about 3 times the carrier speed, then the resulting annulus speed will be zero. This will result in a zero output speed whilst the engine and internal components are still rotating and is called a geared neutral. If the sun speed is reduced relative to the carrier then the annulus and output will increase in speed in the same sense as the carrier. If, however the sun speed is increased relative to the carrier then the annulus will be driven in a negative sense relative to the carrier. Hence this configuration not only gives

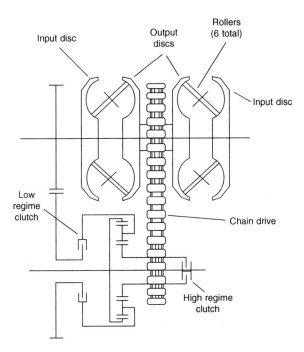

Figure 13.33 Torotrak transmission schematic layout

the transmission an effective starting device but also a reverse gear capability without any additional components.

Unfortunately it cannot then also meet the high overdrive ratio that is necessary for good economy and the range needs to be extended through what is called a mode change. In high regime the variator output is locked directly to the annulus and transmission output shafts by a multi-plate clutch. At the same time the second clutch disconnects the transfer drive input to the carrier of the epicyclic. The change between high and low regimes has been designed to take place when all three epicyclic components are rotating at the same speed, and corresponds to one end of the variator range. The full range of variator ratio is thus available in both high and low regimes, giving an overdrive capability of 100 km/h per 1000 rev/min.

A hydraulic system is again used for control purposes in both toroidal transmissions. The roller mechanisms transmit torque only when there is a steer effect trying to get them to move to new position and hence change ratio. Applying a load to the roller's mounting produces this steer effect, and the consequent reactions at the roller circumference produce a torque between the two discs. In both implementations the rollers are moved and loaded hydraulically but their control strategies are different. In the half-toroid the rollers are positioned in a closed loop hydromechanical system that sets a transmission ratio. In the full toroid the load is used directly to control the transmission torque rather than its ratio. It is this feature that makes control of the transmission around the geared neutral effective since a zero demanded output torque will give this condition directly. In both cases there is an electronic management system integrated with other vehicle systems and with electrohydraulic control inputs to the transmissions. Scheduling the control can again allow for a variety of different strategies including the discrete speed emulation, described above.

13.6 Application issues for transmissions

This section picks up on some of the important issues for transmissions that are not covered in the above sections. These include a number of design considerations required when a transmission is chosen for a particular vehicle.

13.6.1 The operating environment

The environment that any of the transmission units operate in can be a very major consideration when installing in the vehicle. It should be pointed out that any vehicle manufacturer will have their own standards to which they would expect a transmission to conform. For this reason, the material here can only ever be an approximate guide.

Firstly, considering the temperature constraints. Many transmission units are installed under the bonnet (hood), alongside, and attached to, the engine. The influence of the engine, cooling pack and exhaust system, together with any air ducting, can lead to extremely high air temperatures around the transmission. This air temperature can certainly reach similar temperatures to the required temperature of the transmission oil. Unless the transmission has cooling oil piped to a remote cooler or heat exchanger, the unit must rely on convection to remove any excess heat. This is obviously only possible if there is air of a lower temperature than the oil passing over the outside of the casing.

Ideally, the transmission oil temperature will warm up quickly on start up of the vehicle, particularly if the ambient temperatures are near or below freezing. This will enable the transmission to operate efficiently for as much of the time as possible. The transmission oil will obviously start near to the ambient air temperature and at low temperatures can be extremely viscous. This can be a severe problem with automatic and CVT transmissions particularly as the lubricant also acts as a hydraulic control fluid so fluidity is important.

At the other operating extreme we will want to prevent the oil getting too hot and compromising the durability of the oil itself or the components in the unit. Maximum operating temperatures are often around 140 or 150 °C, but it should be noted that these temperatures require a modern specification lubricant. Many older formulations will start to degrade much above 110 °C. It can sometimes be very difficult to control the lubricant to a reasonable upper operating limit when the vehicle is operating in high ambient temperatures and working hard. For example when towing a trailer with ambients around 50 °C in the Gulf States.

From the above discussion, it can be seen that the control of the lubricant temperature is not a trivial problem and many companies are working towards management of the transmission in much the same way as the engine. That is warming the unit up quickly with waste heat from the engine but then controlling the upper temperature, perhaps to a temperature below that of the engine coolant. Simply linking the cooling systems of the engine and gearbox is not necessarily the answer. Part of this development also entails careful consideration of the lubricant itself.

There is not the space here to discuss the choice of lubricant in any depth. It is perhaps, sufficient to highlight some of the main considerations.

- Cold temperature fluidity – Does the lubricant have to act as a hydraulic fluid? If so, how cold is the transmission expected to operate, and how critical is immediate pressure or oil flow from the pump?

- High temperature capability – What is the highest temperature the transmission can see in service? Remember it is very difficult to specify 'excursions' to high temperatures, as it is almost impossible to predict the duration and frequency over the life of a vehicle
- Viscosity grade and load carrying additives – What grade of oil is required to provide adequate gear and bearing life? In certain cases this is as dependent on the action of the additives as on the base oil.
- Air release and foaming – This tendency of the oil can be vital in certain cases. If the oil behaves badly, it can lead to oil being ejected from the breather system.
- Corrosion inhibition – Components within the transmission do get exposed to moisture, mostly from condensation. Because of this, the oil needs to prevent the surfaces from corroding, particularly while the transmission is stood for a long duration.
- Friction modification – The level of friction, which occurs at the rubbing surface of some parts, needs to be controlled within certain limits. An example of this is in automatic gearbox clutch packs, where too high a friction could cause problems as significant as too low a value.
- Viscosity modification – The viscosity grade of the base oil used may not be quite what is required. Additives can be used to improve the rate at which viscosity changes with temperature, providing that the oil is thicker at high temperatures to protect the running parts, while retaining the required fluidity at low temperatures.

Finally, other aspects of the operating environment also need to be considered. The corrosion of the outside of the transmission case is rarely too much of a concern, but if ferrous materials are used then they may need to be protected by painting. Similarly, immersion in water is unlikely to cause a significant problem, but for off road vehicles, this needs to be considered along with possible damage from rough terrain or sand etc. Tests on the transmission will be used to validate the design and could include salt spray, wading, and various types of off road driving. The environment within the engine bay may also require the outer transmission parts to resist degradation from fuels, antifreeze, brake fluid, etc.

13.6.2 Efficiency

Before continuing with this section, an important point to appreciate is that the efficiency (or losses) of a transmission unit can be considered in three ways:

- The load (torque) related efficiency, this occurs largely as a result of the friction losses at the gear mesh and can be considered in the form of a percentage efficiency or loss figure, i.e. a loss of 3% would be an efficiency of 97%, whichever is the most convenient.
- The parasitic losses can be considered to be independent of the applied torque, these losses can be considered to be 'drag torques' and take the form of a resistance within the transmission.
- The slip losses that may occur in transmission elements which do not involve a fixed gear ratio. Where the drive is transmitted by gear pairs, the input/output speed ratio is obviously fixed by the tooth numbers on the gears. Where the drive is transmitted by another means, the output speed is not necessarily a fixed ratio to the input.

The overall efficiency of a transmission unit needs to take all three aspects into account to arrive at an overall efficiency figure.

Although a number of authors in the literature consider all the losses in a transmission together, others have considered either the load related or parasitic losses in isolation. In common with this approach we will consider the losses separately. Merritt (1971) uses the definitions of oil churning losses and 'tooth friction' when discussing gear losses. He and other authors have chosen to elaborate on the friction losses at the gear mesh and assume the parasitic losses are small by comparison. Test work completed by the authors has demonstrated that this is not the case with the dip lubricated transmission units typical of those used in the automotive industry. The parasitic losses can be significant and need to be considered in a wide variety of transmission units. This is particularly the case when the gearbox is operated from a cold start where the lubricant is at or near the ambient air temperature. In winter, this could obviously be well below 0°C.

In automatic units, the losses associated with the oil pump are often the largest cause of parasitic loss. When the gearbox requires a high oil pressure the torque required to drive the pump can be significant proportion of the torque being transmitted. An example of this is a belt CVT operating at its low ratio at low vehicle speeds. The belt system requires high pressure but the transmitted torque is low due to the low road load so the pump load can be very significant.

Most transmission units use rolling element bearings rather than plain bearings. As with gears, bearings will have load related friction losses and parasitic losses due to the oil movement and windage of the rollers and cage. For prediction and comparison purposes, the bearing loss can be treated as part of the gear system losses. It can be assumed these are small, and either constant, or behave in a similar way as the oil churning or gear mesh losses relative to the 'control' variables of speed, torque and viscosity. Many authors consider the bearing losses to be an order of magnitude smaller than the corresponding gear losses.

This split of load related, parasitic and slip losses is important for other related areas of work. The load related, parasitic, and slip losses need to be treated separately in performance prediction and simulation work. Also an overall indication of the losses can often be derived for the transmission from simple parasitic loss testing at zero output torque (i.e. with the output disconnected so no absorbing dynamometer is required).

The efficiency of a transmission unit is particularly important during two operating conditions of the vehicle:

- Cold start, 'gentle' drive cycles, urban driving, test cycles, etc. The parasitic losses have an impact on fuel economy, as they are significant compared to the drive torque required by the vehicle.
- Arduous use, high speed, towing, etc. The friction (load related) losses are roughly proportional to the torque transmitted and can cause very high heat output from the unit. This, in turn, can lead to high oil temperatures and even oil breakdown or component failure due to insufficient operating oil film.

In summary, these loss mechanisms can be described in terms of the three categories discussed above:

Load related losses:
- Friction losses at the gear tooth mesh point
- Load related bearing friction losses

Parasitic losses:
- Oil churning where gears and shafts dip in the oil bath or foamed oil
- Oil displacement at the point where the gear teeth enter the mesh point
- Windage losses where gears operate in air or oil mist
- Oil seal drag
- Oil pump drag
- Parasitic losses in bearings due to oil displacement (and windage) within the bearings
- Drag in clutch packs in autos and CVT's (those not engaged)

Slip losses:
- Slip in the contact zone where drive is transmitted by friction (i.e. belt–pulley contact in a CVT)
- Slip that occurs in a fluid drive such as a torque converter.

In most non-pumped automotive transmissions the large proportion of the load related and parasitic losses come from the gear friction losses and the oil churning losses respectively. The pump losses must always be considered if the transmission has a high-pressure hydraulic circuit. In belt and toroidal CVT/IVT transmissions the slip losses can also be significant. As discussed in Section 13.4 the speed ratio (indicating slip) is always considered in torque converters.

13.6.3 Other transmission components

No review of transmissions would be complete without at least a brief mention of some of the other components that can be found within the driveline system of the vehicle:

- differentials
- breather systems
- the gearchange mechanism.

Differentials are used in order to allow the left and right hand wheels on any one axle to rotate at different speeds. This is essential on road going vehicles as the two wheels take different paths when the vehicle goes round a corner, hence travel different distances and the inner wheel will rotate slower than the outer. This effect can also be seen between the front and rear wheels of a four wheel drive vehicle so a centre differential is required between the front and rear axles.

As a result of the above requirement, differentials are part of the gearbox assembly on front wheel drive vehicles as the drive is taken directly to the wheels by two separate, output driveshafts. On rear and four wheel drive vehicles the differential is invariably part of a separate axle assembly. It should be noted that axle assemblies also contain part of the gearing required to reduce the speed of the rotating parts from engine speed to wheel speed. This is often referred to as the final drive and typically on a passenger car would be a reduction of around 4:1. Within an axle this would be achieved using a bevel gear pair which also turns the drive through 90°. The differential assembly then rotates with the output gear of this pair and allows the wheels to rotate as described above. It can be seen that the assembly often referred to as the 'differential' on a vehicle actually serves other purposes as well.

Breather systems are one part of the transmission system often ignored by many vehicle engineers, but can prove to be difficult to design. As the transmission warms up and cools down the air inside needs to be able to expand and contract. The breather allows this to happen without the air inside becoming pressurized, as this could push oil out past the oil seals, joints, etc. The problem with developing these is finding a position on the transmission and a design of breather, which allows the air to move in and out of the transmission without allowing water in or oil out. Considering that:

- oil inside the transmission can be extremely aerated
- there is often no part of the gearcase that does not have oil splashed on the inside
- the gearbox is often partly or wholly submerged during wading through flooded roads. It can be readily seen that the breather not only has an important task, but can prove difficult to engineer.

Sealing of the gearcase on the transmission is required to prevent oil leaking on to the ground and the subsequent loss of oil. The breather system has been considered above and is one part of the whole system. The other areas are where the shafts enter and exit the unit, the joints in the casing and the integrity of the casing itself. Proprietary 'rubbing' lip oil seals are used on most shafts (also called dynamic shaft seals). The temperature requirement of the application means that most are made from a synthetic material which has a thin lip contacting the rotating part. Important design aspects of these parts include the surface hardness, surface roughness and adequate oil supply to the lip; the contact area needs to be lubricated to avoid excessive wear. The manufacturers of these seals provide good literature to allow them to be applied correctly. To complete the sealing of the casing, the joints between the housings need to be sealed carefully, usually using silicone or RTV sealant applied at the point of assembly. Gaskets can also be used. The casings themselves can be a source of leaks if the castings used are allowed to be porous. This can occur in the casting process and needs to be avoided by both good design and manufacturing practice.

The gearchange on both manual and automatic transmissions require a mechanical connection between the driver's controls and the transmission unit. For manual gearboxes this is obviously used to shift the gearchange mechanism within the gearbox. On automatics it can sometimes do little more than engage the parking mechanism within the 'box'. With older automatic transmissions though, the shift mechanism can be mechanically linked to the internal hydraulics and used to directly influence the gearchange. The requirement to mechanically link the two is reducing because of the impact of electronic control. In practice, the difficulty of linking the controls (gearlever) with the transmission unit is much to do with the relative position of the two and the interface at the gearbox end. It goes without saying that rear or mid-engined installations can be difficult, as the connections on the power unit often face the rear of the vehicle. For manual transmissions, the use of solid rod connections is often favoured so the driver gets a more direct feel of the gearchange, although cable systems can provide a very good solution. Either solution will have two parts to it; one to transmit the axial movement (i.e. in and out of gear), and the other to transmit the 'cross gate' movement. The latter is where the gearlever is moved from the 1st/2nd position of the 'gate' over towards the 3rd/4th (usually the centre) and the 5th gear positions. Any solution to the gearchange will have to allow both movements to be transmitted to the mechanism in the gearbox.

References

Akehurst, S. Brace, C.J., Vaughan, N.D. and Milner, P.J. (2001). 'Performance investigation of a novel rolling traction CVT' SAE Int. Congress and Expo, Detroit, USA, March, 2001.

Autocar (2000). 6th September.

Faust, H. and Linnenbruegger, A. (1998). CVT Development at LuK, LuK 6th Symposium 1998, pp. 157–179.

Gott, P.G. (1991). *Changing Gears: The Development of the Automotive Transmission*. SAE ISBN 1 56091 099 2.

Heisler, H. (1989). *Advanced Vehicle Technology*. Butterworth-Heineman ISBN 0 7131 3660.

Ifield website – www.ifieldshep.com

Kraxner D., Baur, P., Petersmann, J. and Seidel, W. (1999). CVTip in Sports Cars, CVT '99, International Congress on CV Power Transmissions, Eindhoven, Netherlands, Sept. 1999, pp. 21–26.

Liebrand, N. (1999). Future Potential for CVT Technology, CVT '96, International Conference on CV Power Transmission, Yokohama, Japan, Sept. 1996.

Merritt, H.E. (1971). *Gear Engineering*, Pitman, ISBN 0 273 42977 9 – out of print but you should be able to find it in good technical/college libraries.

Socin and Walters (1968). *Manual Transmission Synchronizers*. SAE 680008.

Takahsashi, M., Kido, R., Nonaka, K., Takayama, M. and Fujii, T. (1999) Design and Development of a Dry Hybrid Belt for CVT Vehicles, CVT '99, International Congress on CV Power Transmissions, Eindhoven, Netherlands, Sept. 1999, pp. 254–259.

Further reading

Heisler, H. (1989) *Advanced Vehicle Technology*. Butterworth-Heineman ISBN 0-7131-3660
An excellent base text that describes operation and gives details of many of the components used in vehicles. There is a significant section on transmissions and the book is very well illustrated with many line drawings. Included in the transmission material is design and function detail of many of the components and sub systems found within the transmission system.

Vaughan, N.D. and Simner, D. (2001). *Automotive Transmissions and Drivelines*. Butterworth-Heinemann.
A text book aimed at degree and postgraduate levels that covers the material in this chapter in much more detail and also expands to include other driveline components. Many aspects of basic analysis are included with worked examples and a number of detailed case studies of typical transmission designs.

Gott, P.G. (1991). *Changing Gears: The Development of the Automotive Transmission*. SAE 1991 ISBN 1-56091-099-2
A fascinating historical view of transmission developments from the earliest days of automobile engineering. There is more emphasis placed on American manufacturers and the automatic transmission but the coverage is much wider, definitively comprehensive and embraces developments globally. Highly recommended.

Bibliography – other worthwhile references

Bosch Automotive Handbook. Good ready reference on many automotive subjects, regularly updated with new editions published.

Wong, J.Y.(1993). *Theory of Ground Vehicles*. Wiley, ISBN 0–471–52496–4

Bearing manufacturers – Most of the large manufacturers produce catalogues with extensive and useful reference material as an introduction. Particularly recommended are SKF, Timken, and NSK/RHP.

Further detail design guidance and information is often available from the component suppliers. The oil companies and clutch suppliers are good examples of this.

14. Braking systems

P.C. Brooks BSc, PhD, CEng, MIMechE
D.C. Barton BSc, MSc, PhD, CEng, MIMechE

The aim of this chapter is to:

- Aid the designer to understand the legal requirements of braking systems;
- Understand the basic requirements for braking systems to be successful;
- Understand the design process for achieving an efficient braking system;
- Appreciate the material requirements for efficient braking systems;
- Understand current developments in braking control systems.

14.1 Introduction

The safe and reliable use of a road vehicle necessitates the continual adjustment of its speed and distance in response to change in traffic conditions. This requirement is met in part by the braking system, the design of which plays a key role in ensuring a particular vehicle is suitable for a given application. This is achieved through the design of a system that makes as efficient use as possible of the finite amount of traction available between the tyre and the road over the entire range of operating conditions that are likely to be encountered by the vehicle during normal operation.

The purpose of this chapter is to introduce the reader to the basic mechanics associated with the deceleration behaviour of a road vehicle and provide insight to the many issues that must be addressed when selecting the brake rotor and friction materials. A complete coverage is not feasible within the confines of a single chapter and so a set of references and additional reading is provided at its end that points the interested reader to further sources of information.

The chapter commences with a review of the function of a brake system together with an outline of the principal components and their possible configurations. The subject of legislation is reviewed and its importance as a tool to aid the designer of a brake system is highlighted. Straight forward kinematic and kinetic analyses are used to address the fundamentals of the braking problem as a precursor to the analysis of brake proportioning, adhesion utilization and other related issues. A case study is built into this section of the chapter that illustrates the application of the theory and so reinforces understanding. The selection of appropriate materials from which to manufacture the friction pair is reviewed and problems linked to thermo-mechanical behaviour highlighted. The chapter concludes with a brief summary of more advanced topics, often linked to modern chassis control, that integrate the braking system with other chassis systems.

14.1.1 The functions and conditions of use of a brake system

In order to understand the behaviour of a braking system it is useful to define three separate functions that must be fulfilled at all times:

(a) The braking system must decelerate a vehicle in a controlled and repeatable fashion and when appropriate cause the vehicle to stop.
(b) The braking system should permit the vehicle to maintain a constant speed when travelling downhill.
(c) The braking system must hold the vehicle stationary when on a flat or on a gradient.

When simply stated, as above, the importance of the role played by the brakes/braking system in controlling the vehicle motion is grossly understated. Consideration of the diverse conditions under which the brakes must operate leads to a better appreciation of their role. These include, but are not limited to, the following:

- Slippery wet and dry roads.
- Rough or smooth road;
- Split friction surfaces;
- Straight line braking or when braking on a curve;
- Wet or dry brakes;
- New or worn linings;
- Laden or unladen vehicle;
- Vehicle pulling a trailer or caravan;
- Frequent or infrequent applications of short or lengthy duration;
- High or low rates of deceleration;
- Skilled or unskilled drivers.

Clearly the brakes, together with the steering components and tyres represent the most important accident avoidance systems present on a motor vehicle which must reliably operate under various conditions. The effectiveness of any braking system is, however, limited by the amount of traction available at the tyre–road interface.

14.1.2 System design methodology

The primary functions of a brake system, listed above, must be fulfilled at all times. In the event of a system failure, the same functions must also be performed albeit with a reduced efficiency. Consequently the braking system of a typical passenger car comprises a service brake for normal braking, a secondary/emergency brake used in the event of a service brake failure and a parking brake. Current practice permits service brake components to be used in the secondary/parking brake systems.

Irrespective of the detail design considerations all brake systems divide into the following subsystems:

(1) *Energy source*
    This includes all those components which generate, store or release energy required by the braking system. In standard passenger cars muscular pedal effort, applied by the driver, in combination with a vacuum boost system comprise the energy source. In the event of a boost failure, the driver can still apply the brakes by muscular effort alone. Alternative sources of energy include power braking systems, surge brakes, drop weight brakes, electric and spring brakes.

(2) *Modulation system*

This embraces those elements of the brake system which are used to control the level of braking effort applied to each brake. Included in this system are the driver, pressure limiting/ modulating values and, if fitted, antilock braking systems (ABS).

(3) *Transmission system*

The components through which energy travels to the wheel brakes comprise the transmission system. Brake lines (rigid tubes) and brake hoses (flexible tubes) are used in hydraulic and air brake systems. Mechanical brakes make use of rods, levers, cams and cables to transmit energy. The parking brake of a car quite often makes use of a mechanical transmission system.

(4) *Foundation brakes*

These assemblies generate the forces that oppose the motion of the vehicle and in doing so convert the kinetic energy associated with the longitudinal motion of the vehicle into heat.

There are four main stages involved in the design of a brake system. The first, and perhaps most fundamental stage, is the choice of brake force distribution between the axles of the vehicle. This is primarily a function of the vehicle dimensions and its weight distribution. Next is the design of the transmission system and this activity embraces the sizing of the master cylinder together with the front and rear wheel cylinders. Additional components, such as special valves that modulate the hydraulic pressure applied to each wheel are physically accounted for at this stage. The foundation brakes form the focus of the third stage of the process. As well as being able to react the applied loads and torques, the foundation brakes must be endowed with adequate thermal performance, wear and noise characteristics. The last phase in the process results in the incorporation of the pedal assembly and vacuum boost system into the brake system. To accomplish this design task, the engineer requires access to several fundamental vehicle parameters. These include:

- Laden and unladen vehicle mass;
- Static weight distribution when laden and unladen;
- Wheelbase;
- Height of centre of gravity when laden and unladen;
- Maximum vehicle speed;
- Tyre and rim size;
- Vehicle function;
- Braking standards.

It is essential to recognize that each of the preceding stages are closely linked and that the final design will take many iterations to realize. Thus any formal methodology must be designed so as not to compromise the overall system quality that could result from design changes at the component level. By way of example, a reduction in package space could lead to smaller diameter wheel brakes having to be fitted to the vehicle. This will change the brake force distribution unless checked, by say resizing the wheel cylinders, and in the worst case this could lead to premature wheel lock and a violation of the governing legislation.

14.1.3 Brake system components and configurations

The principal components put together comprise a conventional braking system are outlined below together with possible brake system layouts. The discussion of the components begins with the pedal assembly and moves through the brake system finishing with the foundation or wheel brakes.

Pedal assembly
A brake pedal consists of an arm, pad and pivot attachments. The majority of passenger cars make use of hanging pedals. A linkage is connected to the pedal and this transmits both force and movement to the master cylinder.

Brake booster
The brake booster serves to amplify the foot pressure generated when the brake pedal is depressed. This has the effect of reducing the manual effort required for actuation. Boosters are invariably combined with the master cylinder assembly. A vacuum booster employs the negative pressure generated in the intake manifold of a spark ignition engine whereas a hydraulic booster relies upon the existence of a hydraulic energy source and typically finds application in vehicles powered by diesel engines that generate only a minimal amount of intake vacuum.

Master cylinder
The master cylinder essentially initiates and controls the process of braking. The governing regulations demand that passenger vehicles be equipped with two separate braking circuits and this satisfied by the so-called tandem master cylinder. A tandem master cylinder has two pistons housed within a single bore. Each section of the unit acts as a single cylinder and the piston closest to the brake pedal is called the primary piston whilst the other is called the secondary piston. Thus if a leak develops within the primary circuit, the primary piston moves forward until it bottoms against the secondary piston. The push rod force is transmitted directly to the secondary piston through piston to piston contact thus allowing the secondary piston to pressurize the secondary circuit. Conversely, if the secondary circuit develops a leak then the secondary piston moves forward until it stops against the end of the master cylinder bore. This then allows trapped fluid between the two pistons to become pressurized and so the primary circuit remains operative.

Regulating valves
The dynamics of the braking process gives rise to need for some means of reducing the reducing the magnitude of the brake force generated at the rear of a vehicle under the action of increasing rates of deceleration. This need arises form the load transfer that takes place from rear to front during any braking event. This function is realized through the incorporation of some form of brake pressure regulating valve into the rear brake circuit. The exact nature of the valve depends upon the detail design but they fall into three generic types:

- Load sensitive pressure regulating valve: Valves of this type are fitted to vehicles that experience large in-service changes in axle load. The valve is anchored to the vehicle body and is also connected to the rear suspension through a mechanical linkage. This permits the valve to sense the relative displacement between the body and suspension and adjust the valve performance to effect control over the rear line pressure and so enable the rear brakes to compensate for the change in axle load;

• Pressure sensitive pressure regulating valve: Otherwise known as a pressure limiter, this type of valve isolates the rear brake circuit when the line pressure exceeds a predetermined value. They find application on vehicles that are characterized by a low centre of gravity and a limited cargo volume;

• Deceleration sensitive pressure regulating valve: This class of valve finds wide application. The actuation point is determined by the rate of deceleration of the vehicle and this is typically of the order of 0.3 g. A benefit of this type of valve is that it does provide for a degree of load sensitive operation as the overall deceleration of the vehicle is function of the vehicle weight and the line pressure. They are also sensitive to braking on a slope. Mathematical models of this class of valve are developed later in the text and their influence on the performance of a brake system is demonstrated.

Foundation brakes
Foundation, or wheel brakes, divide into two distinct classes, namely disc (axial) and drum (radial) brakes. Modern vehicles are invariably fitted with disc units on the front axle and there is a growing tendency to fit similar units to the rear axle. If drum brakes are fitted to the rear axle then these are typically of the Simplex type which employs a leading and trailing shoe configuration to generate the required brake torque. The torque output of this type of drum brake is not sensitive to change in vehicle direction. On vehicles fitted entirely with disc brakes, then a small drum unit is often employed to act as a parking brake on the rear axle of the vehicle. Issues surrounding the selection of the materials used to manufacture both discs and drums together with their friction material partners are discussed in more detail later in the text.

Brake system layouts
Legislative requirements demand a dual circuit transmission system to be installed on all road vehicles. Of the five possible configurations, two have become standard and these are known as the II and X variants shown in Figure 14.1. The II design is characterized by separate circuits

| 1 | Brake circuit 1 |
| 2 | Brake circuit 2 |

II variant   X variant

Figure 14.1 Common brake system layouts

for both the front and rear axles whilst in the X configuration, each circuit actuates one wheel at the front and the diagonally opposed rear wheel. The II design is often found on vehicles that are rear heavy and the X layout has application on vehicles that are front heavy.

$$S_4 = \frac{1}{2a_f}\left[U^2 + \frac{a_f^2(t_3 - t_2)^2}{4} - Ua_f(t_3 - t_2) \right]$$ (14.7)

14.2 Legislation

Without exception, motorized road vehicles, whether cars, buses or lorries, represent a potentially lethal hazard to other road users and pedestrians. Also the rise in 'green thinking' during the past decade has led to serious consideration of the impact of road vehicles on the environment in which we all must live. It is one of the many responsibilities of a government to ensure that all road vehicles are as safe as possible and that any adverse effects of the vehicle on the environment are minimized. This task is achieved through legislation which, in so far as the brakes are concerned, primarily sets the minimum standards for the performance of the systems and their components that combine to arrest the motion of a vehicle in a controlled manner. A design engineer has to take into account many factors associated with the mechanics of braking when designing a new brake system. In addition to these elements, conformity to the legislative requirements of the country or countries in which the vehicle is to operate is absolutely essential. Thus a working knowledge of the content and scope of such documents forms a very important part of the brake engineer's database of information.

There are many arguments both for and against legislation. However, despite some inevitable drawbacks standards and legislation form a necessary and desirable part of today's society and they are here to stay. Poorly written legislative documents may smother initiative and restrict technical progress by enforcing unrealistic standards and by failing to recognize the advance of technology. The phrasing of the documents is somewhat complex which can lead to difficulties in understanding their content. This is, to a large extent, unavoidable because they are legal documents and they must attempt to cover all eventualities, prevent ambiguity and close any loopholes. By default, proof of compliance with a national or international standard generates an overhead which is transferred to the consumer as an added cost. It is essential that this process does not inhibit either new or small manufacturers from entering the marketplace. Finally, national legislation can be used as an economic weapon (termed a technical barrier to trade), particularly by those countries operating an approval system, to protect their industry from world-wide intrusion into the local market. The job of the importer is made very difficult through the use of standards and test procedures that favour the home industry or by withholding interpretations, by allowing the test authority the use of subjective judgement and by showing a lack of co-operation in test scheduling and in the issue of approval documents.

Technical standards and legal requirements need to be kept under review in order to force quality upwards. They must set realistic standards for new and in-service vehicles that result in real improvements in safety and environmental protection. Bearing in mind the necessary legal constraints, they should be straight forward to understand and interpret as well as be universally acceptable to encourage free trade and so prevent the production of trade barriers. Standards and legislation must also be applicable to all types of vehicle and should not preclude innovative design by being so inflexible as to limit technical advance; ideally they must actively encourage the use of new technology. Given time, companies incorporate the formal test procedures in their design programmes and develop close working relationships with the national approval bodies. In principle this leads to improved export performance since approval obtained in the country of manufacture is automatically valid for all others. Also the clearly defined ground rules quite often act as an aid to product development.

With regard to the braking system, legislation first appeared in the form of the Motor Cars Order of 1904. Since this time, the range and complexity of the vehicles that populate the road network has markedly increased. Inevitably this has been accompanied by a similar increase in the size and complexity of the regulations pertaining to braking. The first major change away from the self-certification process came when the UK commenced along the road of Type Approval, favoured by continental Europe, by the incorporation of the Economic Commission for Europe (ECE) Regulations. These voluntary regulations attempt to harmonize vehicle legislation and they provide for the reciprocal acceptance and notification of vehicle systems and component approval. The relevant braking legislation is contained in Regulation 13. The next major change occurred when the UK became a member of the Common Market in which the acceptance of EEC Directives is binding on all member states. Member states are not allowed to impose more stringent standards than those contained in the Directives. They are, however, free to demand additional standards with regard to matters not covered by EEC legislation. EEC Directives differ from the corresponding ECE Regulations in one major aspect: Approvals issued by one member must be accepted by all others. The objective of either is, however, the same, namely the harmonization of differing technical requirements. In 1978 it became mandatory in the UK for all new cars to be type approved to the EEC Braking Directives. A similar exemption was granted in 1980 for vehicles approved to ECE Regulation 13. The current EEC Directive on braking is Directive 71/320/EEC as last amended by Directive 91/422/EEC.

Of the many requirements laid down in the EEC regulations, perhaps the single most important aspect that must be satisfied by a road vehicle relates to its use of available tyre-ground adhesion. Manufacturers have to submit adhesion utilization curves that demonstrate compliance with the limits defined in Figures IA and IB of EEC Directive 71/320 Annex II. The adhesion

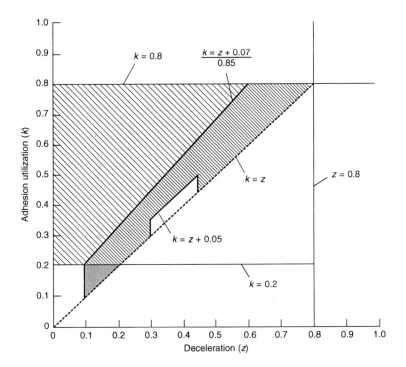

Figure 14.2 Adhesion utilization diagram for a category M_1 vehicle

utilization diagram that refers to a category M_1 vehicle (passenger vehicle with seating capacity up to eight including the driver) is shown in Figure 14.2, and is derived from that contained within EEC Council Directive 71/320/EEC as last amended by 91/422/EEC.

The Directive uses the letter k to indicate adhesion utilization and z for deceleration and it states that for all categories of vehicle for values of adhesion utilization between 0.2 and 0.8,

$$z \geq 0.1 + 0.85(k - 0.2)$$

For category M_1 vehicles, the adhesion utilization curve of the front axle must be greater than that of the rear for all load cases and values of deceleration between 0.15 g and 0.8 g. Between deceleration levels of 0.3 g and 0.45 g, an inversion of the adhesion utilization curves is allowed provided the rear axle adhesion curve does not exceed the line defined by $k = z$ by more than 0.05. The above provisions are applicable within the area defined by the lines $k = 0.8$ and $z = 0.8$.

Compliance of the braking system to the constraints defined in Figure 14.2 ensures that the rear wheels do not lock in preference to the front wheels and that the proportion of braking effort exerted at the front of the vehicle is limited so that the braking system does not become too inefficient. A detailed interpretation of this requirement is outlined later in the text.

14.3 The fundamentals of braking

14.3.1 Kinematics of a braking vehicle

Kinematic analysis of a braking vehicle
The distance travelled by a vehicle when braking, either during a stop or snub (when the final velocity is non-zero) is a basic measure of the effectiveness of a brake system. Before addressing issues coupled to the forces which act on a vehicle during a braking manoeuvre it is worthwhile to first consider the kinematic behaviour of the vehicle. A straight forward kinematic analysis assuming straight line (one dimensional) motion and constant deceleration provides a ready indication of stopping distance. Predictions of stopping distance made by this analysis find use in accident reconstruction.

With reference to Figure 14.3 the total distance travelled is made up of two parts. In part 1, the total distance travelled by the vehicle moving with constant velocity U is

$$S_1 = Ut_1 \tag{14.1}$$

In part 2 the vehicle is decelerated at a constant rate until such time as the vehicle comes to rest. The distance travelled is

$$S_2 = \frac{Ut_2}{2} = \frac{U^2}{2a} \tag{14.2}$$

Thus the total stopping distance is simply

$$S_i = S_1 + S_2 = Ut_1 + \frac{U^2}{2a} \tag{14.3}$$

The preceding analysis assumes the vehicle deceleration is achieved instantaneously and is sustained for the duration of the stop. No account is taken of driver reaction time, initial system

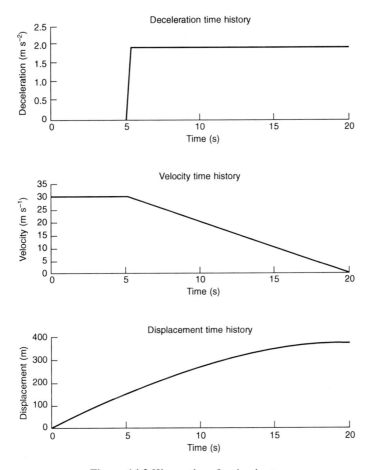

Figure 14.3 Kinematics of a simple stop

response time, deceleration rise time, change in deceleration during the period of actual braking and, if applicable, release time. These factors are now defined with reference to Figure 14.4, which illustrates the characteristics of a typical measured deceleration time history.

Driver reaction time
The time taken by the driver to respond to the danger, formulate an avoidance strategy and physically move his/her right foot from the accelerator to the brake, $(t_1 - t_0)$.

Initial system response time
The time taken from the point at which the brake pedal begins to move to the time at which a braking force is generated at the tyre road interface, $(t_2 - t_1)$.

Deceleration rise time
The elapsed time for the deceleration to reach the value, determined by the driver, for the stop in question, $(t_3 - t_2)$.

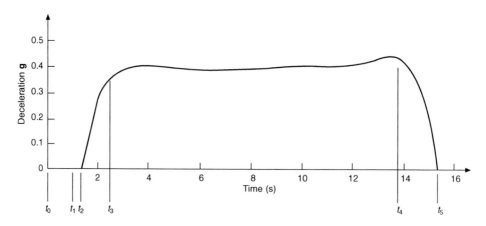

Figure 14.4 Typical measured deceleration time history

Braking time
The time taken from the point at which fully developed braking is reached to the time at which the vehicle stops or brake release begins, $(t_4 - t_3)$.

Release time
The elapse time between the point at which brake release starts to the time at which brake force generation ceases, $(t_5 - t_4)$.

The stopping time and distance are measured from the time t_0 to the time t_4 or t_5, if appropriate, whilst the braking time and distance is measured from time t_1. Inclusion of these factors modifies the simple kinematic analysis with the result that the predicted stopping times and distances are increased.

Additional deceleration forces, such as those arising from engine drag, aerodynamic drag, rolling resistance and gravity are not taken into account in the following analysis, taken from (Limpert, 1993), which is again built around one dimensional particle kinematics. With reference to Figure 14.5, the following expressions for the distance travelled during each stage of the stop can be derived. Note that as a stop is the subject of the analysis, the effect of release time $(t_5 - t_4)$ is ignored.

Distance travelled during reaction time S_1

$$S_1 = U(t_1 - t_0) \tag{14.4}$$

Distance travelled during the initial system response time S_2

$$S_2 = U(t_2 - t_1) \tag{14.5}$$

The distance travelled during the deceleration rise time S_3, assuming this to be linear, can be shown to be

$$S_3 = U(t_3 - t_2) - \frac{a_f (t_3 - t_2)^2}{6} \tag{14.6}$$

where a_f is the value of the mean fully developed deceleration. The distance travelled during

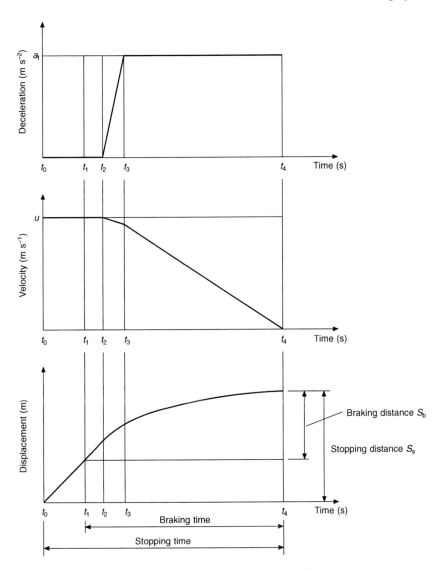

Figure 14.5 Four stage stop simulation

the course of braking, S_4, under the deceleration of a_f, which is assumed to remain constant, can be expressed as

$$S_4 = \frac{1}{2a_f}\left[U^2 + \frac{a_f^2(t_3 - t_2)^2}{4} - Ua_f(t_3 - t_2)\right] \tag{14.7}$$

Thus the total stopping distance, S_s, is simply

$$S_s = \sum_{i=1}^{4} S_i \tag{14.8}$$

and the braking distance, S_b, is

$$S_b = \sum_{i=2}^{4} S_i \qquad (14.9)$$

14.3.2 Kinetics of a braking vehicle

A general equation for braking performance can be easily derived through application of Newton's Second Law to a simplified free-body diagram of a vehicle in the direction of its travel, Figure 14.6. Assuming x is positive in the direction of travel, then

$$\Sigma F_x = M\ddot{x} \qquad (14.10)$$

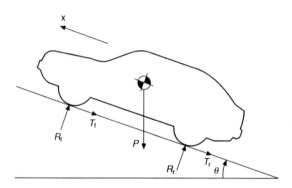

Figure 14.6 Free body diagram of a braking vehicle

and so

$$-T_f - T_r - D - P \sin \theta = M\ddot{x} \qquad (14.11)$$

where
M = Vehicle mass T_r = Rear axle braking force
P = Vehicle weight R_f = Front axle load
g = Acceleration due to gravity R_r = Rear axle load
T_f = Front axle braking force θ = Angle of incline

Note that the front and rear braking force terms, T_f and T_r, represent the sum of all the effects that combine to generate the forces which act between the front and rear axles and ground. These include the torque generated by the brakes together with rolling resistance effects, bearing friction and drive train drag.

If an additional variable for linear deceleration, d, is defined such that

$$d = -\ddot{x} \qquad (14.12)$$

then equation 14.11 becomes

$$Md = T_f + T_r + D + P \sin \theta = T \qquad (14.13)$$

in which T is the sum of all those forces that contribute to the overall braking effort.

By considering the case of constant deceleration, straightforward and fundamental relationships can be derived that yield an appreciation of the physics which governs all braking events.

From equation 14.13, the linear deceleration of a vehicle can be expressed as

$$d = \frac{T}{M} = -\frac{dv}{dt} \qquad (14.14)$$

in which v is the forward velocity of the vehicle. Since the deceleration is assumed constant, then the total brake force is also constant and so equation 14.14 can be integrated with respect to time between the limits of the initial velocity, v_0, to the final velocity, v_f, to determine the duration of the braking event, t_b.

On rearranging, equation 14.14 becomes

$$\int_{v_0}^{v_f} dv = -\frac{T}{M} \int_{0}^{t_b} dt \qquad (14.15)$$

which leads to

$$v_0 - v_f = \frac{T}{M} t_b \qquad (14.16)$$

The fact that velocity and displacement are related by $v = dx/dt$ permits an expression for stopping distance to be derived from equation 14.14 through substitution for dt and integration between v_0 and v_f as before. On rearranging, equation 14.14 becomes

$$\frac{T}{M} \int_{x_0}^{x_f} dx = -\int_{v_0}^{v_f} v\, dv \qquad (14.17)$$

which leads to

$$\frac{T}{M}(x_f - x_0) = \frac{Tx}{M} = \frac{v_0^2 - v_f^2}{2} \qquad (14.18)$$

where x is the distance travelled during the brake application.

When considering a stop, the final velocity v_f is zero and so the stopping distance x is, from equation 14.18, given by

$$x = \frac{Mv_0^2}{2T} \qquad (14.19)$$

and the time, t_b, taken to stop the vehicle is, from equation 14.16,

$$t_b = \frac{Mv_0}{T} = \frac{v_0}{d} \qquad (14.20)$$

Thus from equation 14.19 the distance required to stop the vehicle is proportional to the square of the initial velocity and, from equation 14.20, the time taken to stop the vehicle is proportional to the initial velocity.

To achieve maximum deceleration and hence minimum stopping distance on a given road surface, each axle must simultaneously be on the verge of lock. If this is so then, if z is the deceleration as a proportion of g, $z = \dfrac{d}{g}$, and the brake force T is equal to the product of the vehicle weight and the coefficient of tyre ground adhesion, P_μ, then from equation 14.13,

$$P_z = T = P_\mu \qquad (14.21)$$

from which it can be deduced that

$$z = \mu \qquad (14.22)$$

This represents a limiting case in which it is clear that the maximum deceleration cannot exceed the value of tyre-ground adhesion. A deceleration in excess of 1 g therefore implies that the tyre-ground adhesion has a value greater than unity and this is quite realizable with certain types of tyre compound.

As already indicated, the primary source of retardation force arises from the foundation brake. Secondary forces which contribute to the overall braking performance include:

- Rolling resistance, expressed by a coefficient of rolling resistance. The total rolling resistance is independent of the load distribution between axles and the force is typically equivalent to a nominal 0.01g deceleration;
- Aerodynamic drag which depends on dynamic pressure and is proportional to the square of the vehicle speed. It is negligible at low speeds, however aerodynamic drag may account for a force equivalent to 0.03g when travelling at high speed;
- Gradient makes either a positive (uphill) or negative (downhill) contribution to the total braking force experienced by a vehicle. This force is simply the component of the total vehicle weight acting in the plane of the road;
- Drivetrain drag may either help or hinder the braking performance of a vehicle. If the vehicle is decelerating faster than the components of the drivetrain would slow down under their own friction then a proportion of the brake torque generated by the wheel brakes must be used to decelerate the rotating elements within the drivetrain. Thus the inertia of the elements of the drivetrain effectively add to the mass of the vehicle and so should be considered in any rigorous brake design programme. Conversely, the drivetrain drag may be sufficient to decelerate the rotating elements and so contribute to the overall vehicle braking effort and this is often the case during braking manoeuvres involving a low rate of deceleration.

14.3.3 Tyre–road friction

The brake force, F_b, which acts at the interface between a single wheel and the road is related to the brake torque, T_b, by the relationship

$$F_b = \frac{T_b}{r} \qquad (14.23)$$

where r is the radius of the wheel. The brake force on a vehicle can be predicted using equation 14.23 as long as all the wheels are rolling. The brake force F_b cannot increase without bound as it is limited by the extent of the friction coupling between the tyre and the road.

The friction coupling that gives rise to the brake force characteristic reflects the combination of tyre and road surface materials together with the condition of the surface. The best conditions occur on dry, clean road surfaces on which the brake force coefficient, defined as the ratio of brake force to vertical load, μ_b, can reach values between 0.8 and unity. Conversely, icy surfaces reflect the poorest conditions and on ice the brake force coefficient can lie between 0.05 and 0.1. On wet surfaces or on roads contaminated by dirt, the brake force coefficient typically spans the range 0.2 to 0.65.

Hysteresis and adhesion are the two mechanisms responsible for friction coupling. Surface adhesion comes about from the intermolecular bonds which exist between the rubber and the aggregate in the road surface. Hysteresis, on the other hand, represents an energy loss in the rubber as it deforms when sliding over the aggregate. Each of these mechanisms rely on a small amount of slip taking place at the tyre-road interface and so brake force and slip coexist. The longitudinal slip of the tyre is defined as a ratio:

$$\text{slip} = \frac{\text{slip velocity in contact patch}}{\text{forward velocity}}$$

$$= \frac{v - \omega r}{v} \tag{14.24}$$

where v is the forward velocity of the vehicle, ω is the angular velocity of the wheel and r the wheel radius.

Useful information can be obtained by plotting brake force coefficient against slip, Figure 14.7. During straight line braking, no lateral forces are generated which means that all of the force that is potentially available within the tyre-ground contact patch can be used to decelerate the vehicle. The uppermost characteristic illustrates the brake coefficient derived from both the

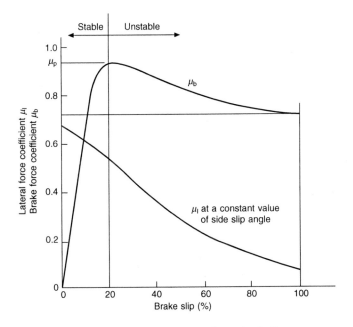

Figure 14.7 Brake force against wheel slip

adhesive and hysteretic mechanisms and it increases linearly with increase in slip up to around 20% slip. On dry roads, the adhesion component dominates the production of friction coupling. The peak coefficient, denoted by μ_p, defines the maximum braking force that can be obtained for a given tyre-road friction pair. At higher values of slip this coefficient decreases to its lowest value of μ_s at 100% slip, which represents the full lock condition. The maximum brake force, corresponding to μ_p, is a theoretical maximum as the system becomes unstable at this point. Once a wheel is decelerated to the point at which μ_p is achieved, any disturbance about this point results in an excess of brake torque that causes the wheel to decelerate further. This leads to an increase in slip and this in turn reduces the brake force leading to a rapid deceleration to the full lock condition. It is worthwhile to note that anti-lock brake systems make use of this phenomenon.

The negotiation of a bend requires a vehicle to develop a lateral force in the tyre-ground plane through the deformation of the tyre carcass brought about by a slip angle. Lateral forces, characterized by the coefficient μ_l, being a function of slip angle, and longitudinal forces, characterized by μ_b thus coexist and compete for the finite amount of force that is available within the contact patch. A typical lateral force coefficient is shown on Figure 14.7 for a given value of slip angle. It has a maximum value when the brake slip is zero and falls with increase in brake slip. The minimum value occurs when the wheel has locked and is unable to generate further lateral force.

14.4 Brake proportioning and adhesion utilization

The vertical loads carried by the front and rear wheels of a rigid, two axle vehicle are not, in general, equal. In order to efficiently utilize the available tyre–road adhesion the braking effort must be apportioned between the front and rear of the vehicle in an intelligent and controlled fashion. Failure to do so could result in any of the following:

- A vehicle being unable to generate the necessary deceleration for a given pedal pressure;
- Front axle lock, in which the vehicle remains stable yet suffers from a loss of steering control;
- Rear axle lock that causes the vehicle to become unstable.

The terms front wheel lock and rear wheel lock can be alternatively labelled front axle over brakes and rear axle over brakes respectively. In this section factors influencing the fore–aft axle loads are identified and their effect on braking examined, concentrating on vehicles that have a fixed brake ratio. Devices used to modify the braking ratio are discussed in the latter part of this section and their effect on braking performance evaluated.

The theory developed in Sections 14.4.2, 14.4.3, 14.4.4 and 14.4.8 is applied to the analysis of the braking requirements of a fictitious two axle road vehicle. This vehicle is assumed to be unladen and is described by the parameters given in Table 14.1.

14.4.1 Static analysis

A simple representation of a two axle road vehicle is shown in Figure 14.8, which has a mass,

Table 14.1 Prototype vehicle parameters

Parameter		Symbol	Units	Value
Mass		M	kg	980
Wheel base		l	m	2.45
Height of centre of gravity above ground		h	m	0.47
Static axle loads (% of total)	Front	F_f	–	65
	Rear	F_r	–	35
Fixed brake ratio	Front	x_f	–	75
$R = \dfrac{x_f}{x_r} \ (\%)$	Rear	x_r	–	25

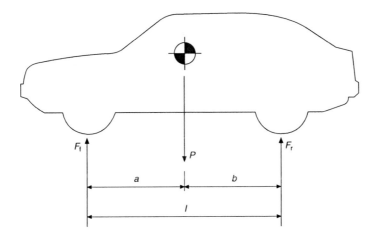

Figure 14.8 Static axle loads

M, concentrated at the centre of gravity. The centre of gravity is assumed to lie on the longitudinal centreline of the vehicle and the road surface is flat with no camber. As a consequence of this, the loads on the two wheels mounted on each axle are equal and so the following analyses treat axle loading rather than the individual wheels.

The vertical load due to the vehicle mass is simply

$$P = Mg \tag{14.25}$$

where g is the acceleration due to gravity.

Taking moments about the rear tyre–road contact gives

$$F_f = \frac{Pb}{l} \tag{14.26}$$

and moments about the front tyre–road contact provides a value of the vertical load acting at the rear axle given by

$$F_r = \frac{Pa}{l} \qquad (14.27)$$

The influence of the fore-aft location of the centre of gravity on the vertical wheel loads, due to change in loading conditions is readily apparent from equations 14.26 and 14.27.

The maximum payload of a passenger car forms only a fraction of the unladen vehicle mass and the inherent space restrictions limit the extent to which the centre of gravity can move. For a light rigid truck such as a transit type van or pickup, there exists considerable scope for varying the payload and so the change in mass from unladen to laden can be severe. The distribution and size of the payload, which also governs the location of the centre of gravity, may change on a daily basis or throughout a delivery cycle in which the payload is discharged in stages.

14.4.2 Braking with a constant brake ratio

The object of this analysis is to show how braking affects the vertical loads carried by the front and rear axles. This in turn leads to a way of determining the maximum deceleration attainable by a vehicle under specified conditions that does not result in axle lock.

If the front and rear axles are to be on the point of locking, then the braking forces T_f and T_r acting at each axle must be in proportion to the vertical loads being carried, R_f and R_r. The magnitude of the braking force generated by each axle, up to the point at which it locks, is a function of the design of the braking system.

Changes in load transfer between the front the rear axles occur during braking and so a variable brake effort ratio is required to provide ideal braking. The situation is in reality complicated by the following:

- Change in vehicle weight;
- Change in weight distribution;
- The effect of gradients (positive and negative);
- Cornering, in which some proportion of the total force present at the tyre-ground interface is used to generate lateral forces;
- Varying road surfaces and weather conditions;
- Split friction surfaces where the coefficient of adhesion changes from port to starboard.

Consider the case for a vehicle with a fixed brake ratio, $R = \dfrac{x_f}{x_r}$, on a flat road that has a uniform coefficient of tyre–ground adhesion μ. In the analysis that follows, the governing equations of motion for the decelerating vehicle are derived by the direct application of Newton's Second law to the free body diagram of the vehicle rather than through D'Alembert's method which is adopted in the EEC Council Directive 71/320/EEC. This permits easy extension of the model to embrace additional degrees of freedom or to take account of the presence of a trailer. The free body diagram of the decelerating vehicle is shown in Figure 14.9.

In the x direction

$$M\ddot{x} = \Sigma F_x = -D - T_f - T_r - T_{fr} - T_{rr} \qquad (14.28)$$

which through combination with equation 14.12 becomes

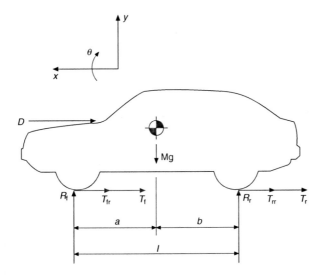

Figure 14.9 Free body diagram of the decelerating vehicle

$$Md = D + T_f + T_r + T_{fr} + T_{rr} \tag{14.29}$$

If the aerodynamic drag and rolling resistance forces are assumed to be negligible, then equation 14.29 reduces to

$$Md = T_f + T_r \tag{14.30}$$

By defining z to be the vehicle deceleration as a proportion of g

$$z = \frac{d}{g} \tag{14.31}$$

then equation 14.30 takes the form

$$Mgz = T_f + T_r = Pz \tag{14.32}$$

In the vertical direction

$$M\ddot{y} = \Sigma Fy = R_r + R_f - Mg = 0 \tag{14.33}$$

as $y = 0$ ms^{-2}, whilst in the θ direction, taking moments about the centre of gravity of the vehicle leads to

$$I\ddot{\theta} = \Sigma M_{cg} = R_f a - R_r b - T_f h - T_r h = 0 \tag{14.34}$$

as $\ddot{\theta} = 0$ *rad s^{-2}*.

Manipulation of equations 14.33 and 14.34 results in the following expressions for the front and rear dynamic axle loads

$$R_f = \frac{Mgb}{l} + \frac{h}{l}(T_f + T_r)$$

$$(14.35), (14.36)$$

$$R_r = \frac{Mga}{l} - \frac{h}{l}(T_f + T_r)$$

which can be combined with equation 14.32 and the static axle loads, equations 14.26 and 14.27, to give

$$R_f = F_f + \frac{Pzh}{l}$$

$$(14.37), (14.38)$$

$$R_r = F_r - \frac{Pzh}{l}$$

The above are in accord with those given in the EEC Directive and they show that a change in axle load in favour of the front axle occurs during a braking manoeuvre. In order for each axle to be simultaneously on the verge of locking, the brake force generated at each axle must be in direct proportion to the vertical axle load. This means that to fully utilize the available tyre–ground adhesion, the braking system must support an infinitely variable brake ratio.

Consider first the case of a vehicle in which the brake ratio is fixed. If the ratio has been set so that the front axle locks in preference to the rear, then the brake force generated at the front axle when about to lock is given by

$$T_f = \mu R_f$$

$$= \mu \left(F_f + \frac{Pzh}{l} \right) \qquad (14.39)$$

During the same braking event, the rear axle is also generating a brake force that has not exceeded its limiting value and this is found by considering the vehicle brake ratio

$$R = \frac{x_f}{x_r} = \frac{T_f}{T_r} \qquad (14.40)$$

from which

$$T_r = T_f \frac{x_r}{x_f} = \mu \left(F_f + \frac{Pzh}{l} \right) \frac{x_r}{x_f} \qquad (14.41)$$

leading to a total brake force of

$$T = Pz = T_f + T_r$$

$$= \mu \left(F_f + \frac{Pzh}{l} \right) + \mu \left(F_f + \frac{Pzh}{l} \right) \frac{x_r}{x_f} \qquad (14.42)$$

which reduces to

$$T = Pz = \mu \left(F_f + \frac{Pzh}{l} \right) \frac{1}{x_f} \qquad (14.43)$$

This equation can be rearranged to yield the maximum value of deceleration as a proportion of g as

$$z = \frac{\mu F_f}{P(lx_f - \mu h)} \tag{14.44}$$

If, however, the ratio has been set so that the rear axle locks in preference to the front, then the brake force generated at the rear axle when about to lock is given by

$$T_r = \mu R_r$$

$$= \mu\left(F_r - \frac{Pzh}{l}\right) \tag{14.45}$$

In this case, the brake force that is generated at the front axle is not necessarily the limiting value and its magnitude is found from the brake ratio as

$$T_f = T_r \frac{x_f}{x_r}$$

$$= \mu\left(F_r - \frac{Pzh}{l}\right)\frac{x_f}{x_r} \tag{14.46}$$

which leads to a total brake force of

$$T = Pz = T_f + T_r$$

$$= \mu\left(F_r - \frac{Pzh}{l}\right)\frac{x_f}{x_r} + \mu\left(F_r - \frac{Pzh}{l}\right) = \mu\left(F_r - \frac{Pzh}{l}\right)\frac{1}{x_r} \tag{14.47}$$

and this can be solved for the deceleration as a proportion of g to be

$$z = \frac{l\mu Fr}{P(lx_r + \mu h)} \tag{14.48}$$

Direct solution of equations 14.44 and 14.48 for z is straightforward, however, greater insight to the mechanics of the braking process can be gained through the following graphical solution that deals with each axle in turn.

The adhesion force acting between the front tyres and ground depends upon the ratio of the tangential forces at the front and rear wheels due to the brake torques. It is therefore linked to the fixed brake ratio and so the front adhesion force, T_f, as a proportion of the total is given by

$$T_f = x_f T$$

$$= x_f Pz \tag{14.49}$$

which when normalized to the vehicle weight, P, becomes

$$\frac{T_f}{P} = x_f z \tag{14.50}$$

Equation 14.50 is shown on Figure 14.10 labelled as T_f/P. The total available braking force at the front of the vehicle is, from equation 14.43,

$$Tx_f = \mu\left(F_f + \frac{Pzh}{l}\right)$$

(14.51)

which when normalized to the vehicle weight becomes

$$\frac{Tx_f}{P} = \frac{\mu}{P}\left(F_f + \frac{Pzh}{l}\right)$$

(14.52)

This represents the maximum braking force, expressed as a proportion of the total vehicle weight, that could be sustained between the front tyres and road surface for a given set of vehicle parameters. It is shown on Figure 14.10 as the line *Txf/P*.

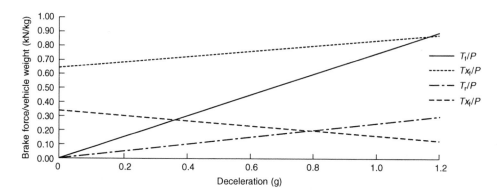

Figure 14.10 Normalized brake force against deceleration

Application of the same procedure to the rear axle of the vehicle results in the following normalized expression for the rear adhesion force T_r

$$\frac{T_r}{P} = X_r z$$

(14.53)

and this is labelled *Tr/P* on Figure 14.10. Similarly, the total available braking force at the rear axle, normalized to the vehicle weight, is, from equation 14.47

$$\frac{Tx_r}{P} = \frac{\mu}{P}\left(F_r \frac{Pzh}{l}\right)$$

(14.54)

and this is the line *Txr/P* on Figure 14.10.

The point of intersection of lines *Tf/P* and *Txf/P*, labelled *a*, represents the solution to equation 14.44 and the point of intersection of lines *Tr/P* and *Txr/P*, labelled *b*, is the solution to equation 14.48. The data used to generate Figure 14.10 is that of the prototype vehicle defined in Section 14.4.4. The vehicle is assumed to be braking on a road that has a tyre–ground adhesion coefficient of unity. The lower value of deceleration, *b*, is that value of deceleration that would first give rise to wheel lock for the given value of tyre–ground adhesion. In this instance, the rear axle will lock first and the tyres will be unable to generate the braking force required by the rear brakes at higher levels of deceleration. Direct solution of equation 14.48 leads to a limiting value of deceleration of $z = 0.79g$.

14.4.3 Braking efficiency

The efficiency with which a brake system uses the available tyre–ground adhesion, η, can be conveniently defined as the ratio of the deceleration, z, to the tyre–ground adhesion coefficient, μ.

$$\eta = \frac{z}{\mu} \tag{14.55}$$

There are two expressions for η; one for the case in which the front axle is about to lock and the other for the case in which the rear axle is about to lock. To determine which is applicable recall that

$$\eta \leq 1.0 \tag{14.56}$$

For the case of front axle lock, η can be written using equation 14.44 as

$$\eta = \frac{z}{\mu}$$

$$= \frac{\dfrac{l\mu F_f}{P(lx_f - \mu h)}}{\mu}$$

$$= \frac{F_f}{P\left(x_f - \dfrac{\mu h}{l}\right)} \tag{14.57}$$

For the rear axle lock case, application of equation 14.48 results in the second expression for efficiency

$$\eta = \frac{z}{\mu}$$

$$= \frac{\dfrac{l\mu F_r}{P(lx_r + \mu h)}}{\mu}$$

$$= \frac{F_r}{P\left(x_r + \dfrac{\mu h}{l}\right)} \tag{14.58}$$

A measure of the efficiency with which a vehicle, having a particular brake design with fixed brake ratio, performs over a variety of road surfaces can be shown by a graph of efficiency, η, against tyre–ground adhesion coefficient, μ, Figure 14.12. This has been generated using the prototype vehicle data. Both axles are on the verge of lock and the system is 100% efficient when the vehicle is braked on a road surface that has a tyre–ground adhesion coefficient of 0.52

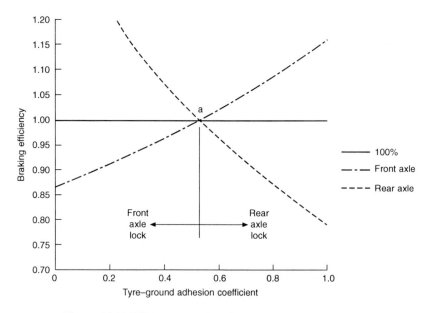

Figure 14.11 Efficiency as a function of tyre–ground adhesion

indicated by the point **a**. On road surfaces below this value of adhesion, the vehicle is limited by front axle lock, equation 14.57, and the efficiency falls to a minimum of 87%, whilst on road surfaces with an adhesion coefficient greater than 0.52 the vehicle is limited by rear axle lock, equation 14.58, and this falls to a minimum of 79%. Data presented above the line defining 100% efficiency has no physical meaning and can be ignored.

An alternative means of presenting efficiency data can be achieved by plotting deceleration, z, against tyre–ground adhesion, μ, for the cases of front and rear axle lock defined by equations 14.44 and 14.48 respectively. This method of presentation has the advantage that the brake engineer can obtain a comparison of possible deceleration levels attainable on different road surfaces along with a measure of the system efficiency. In this case, the efficiency is the gradient of the curve drawn on the deceleration–adhesion space. A line with unit gradient represents optimum performance.

Recall that for front wheel lock,

$$z = \frac{l\mu F_f}{P(lx_f - \mu h)} \qquad (14.59)$$

and for rear wheel lock,

$$z = \frac{l\mu F_r}{P(lx_r + \mu h)} \qquad (14.60)$$

The two curves that define the limiting deceleration of the front and rear axles, derived from equations 14.59 and 14.60 respectively, are shown in Figure 14.11 for the prototype vehicle. They intersect the optimum line at the point *a* which indicates 100% efficiency. Meaningful

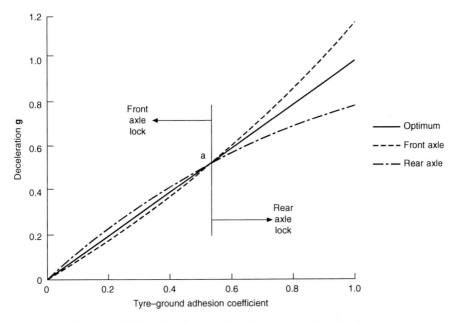

Figure 14.12 Deceleration as a function of type–ground adhesion

information is taken from those portions of the curves that lie below the optimum line and it is clear that the vehicle is governed by rear axle lock on road surfaces with a high tyre–ground adhesion coefficient and that the system is least efficient in this area.

14.4.4 Adhesion utilization

Adhesion utilization, f, is the theoretical coefficient of adhesion that would be required to act at the tyre–road interface of a given axle for a particular value of deceleration. It is therefore the minimum value of tyre–ground adhesion required to sustain a given deceleration and is defined as the ratio of the braking force to the vertical axle load during braking.

For the front of the vehicle the adhesion utilization is defined by

$$f_f = \frac{T_f}{R_f} \tag{14.61}$$

The vertical axle load is defined by equation 14.37 and the front axle brake force, expressed as a proportion of the total is $x_f Pz$, leads to

$$f_f = \frac{x_f Pz}{F_f + \dfrac{Pzh}{l}} \tag{14.62}$$

Similarly, for the rear of the vehicle

$$f_r = \frac{T_r}{R_r} \tag{14.63}$$

The vertical axle load is defined by equation 14.38 and the rear axle brake force, expressed as a proportion of the total is $x_r Pz$, leads to

$$f_r = \frac{x_r Pz}{F_r - \frac{Pzh}{l}} \tag{14.64}$$

Using the data that describes the prototype vehicle leads to Figure 14.13a. The optimum line has unit gradient and defines the ideal adhesion utilization characteristic in which the brake system remains 100% efficient over all possible values of deceleration. The upper limit on allowable adhesion utilization, defined in the *EEC Braking Directive*, Section 14.2, is shown for reference purposes. The remaining two lines define the axle adhesion characteristics for the vehicle. The point labelled *a*, at which the curves cross, intersect the optimum line of adhesion utilization indicating that at this value of deceleration both axles are on the verge of lock. The axle having the highest adhesion utilization coefficient for a given value of deceleration is that which limits the braking performance of the vehicle and in this case, braking is limited by front axle lock up to a deceleration of 0.52 g. Thereafter braking is limited by rear axle lock. It is also possible to find from this diagram the maximum deceleration for a given coefficient of adhesion utilization.

Comparison of the adhesion utilization diagram derived for the prototype vehicle with the legislative requirements outlined in Section 14.2 shows that the vehicle brake system does not meet the minimum standard, as the front axle adhesion curve does not lie above that of the rear axle for all values of deceleration between 0.15 g and 0.8 g. This can be remedied by changing the fixed brake ratio in favour of the rear axle and this causes the point *a* to move up the optimum adhesion line. The limiting deceleration is set at 0.8 g which leads to a new fixed brake ratio of $\dfrac{x_f}{x_r} = \dfrac{0.803}{0.197}$. This in turn results in the modified adhesion diagram shown in

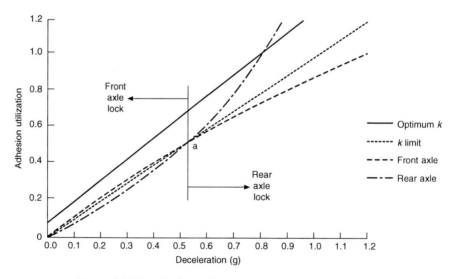

Figure 14.13a Adhesion utilization, datum prototype vehicle

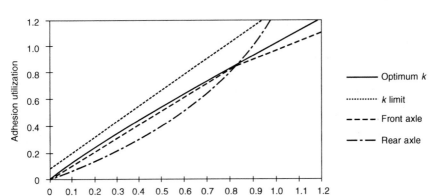

Figure 14.13b Adhesion utilization, modified prototype vehicle

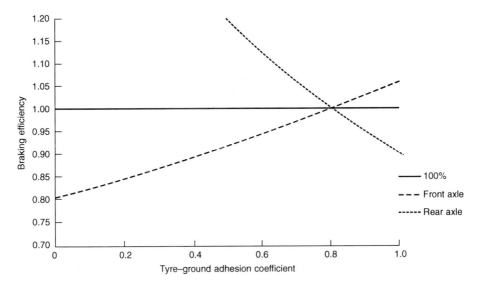

Figure 14.13c Brake system efficiency of modified prototype vehicle

Figure 14.13b and this satisfies the legislative requirements. The modified vehicle is governed by front axle lock up to a deceleration level of $0.8g$, achieved at the expense of the overall brake system efficiency (Figure 14.13c).

14.4.5 Wheel locking

The role of tyre–ground friction and the dependency of the brake force coefficient on the degree of longitudinal slip has been outlined in Section 14.3.3. From 0% to approximately 20% longitudinal slip, the magnitude of the brake force coefficient increases in a roughly linear fashion to its maximum value, μ_p, at 20% longitudinal slip. Further increase, due to increase in applied brake torque, causes the wheel to decelerate rapidly to a condition of full lock and the brake force coefficient takes a value of μ_s at 100% longitudinal slip. The ratio of μ_p/μ_s depends upon the nature of the road surface in question and it takes its highest value under wet or icy conditions.

This leads to a possible scenario in which a vehicle is capable of generating its maximum braking potential when one axle is locked and the second is on the verge of lock. This contrasts with the generally accepted idea that maximum deceleration occurs when the first axle is about to lock and is dependent upon the vehicle weight distribution and the fixed braking ratio.

If the front axle is locked and the rear axle is about to lock then the total brake force is given by

$$Pz = \mu_s \left(F_f + \frac{Pzh}{l} \right) + \mu_r \left(F_r - \frac{Pzh}{l} \right)$$
(14.65)

Similarly, if the rear axle is locked and the front axle is on the verge of lock, then the total brake force is

$$Pz = \mu_p \left(F_f \frac{Pzh}{l} \right) + \mu_s \left(F_r - \frac{Pzh}{l} \right)$$
(14.66)

14.4.6 Effect of axle lock on vehicle stability

When an axle locks, there is reduced friction in both the longitudinal and lateral directions and so the ability of the vehicle to generate the lateral forces required to maintain directional control and stability is severely impaired.

Irregularities in the road surface or lateral forces can cause the vehicle to deviate from its direction of travel. The nature of the ensuing motion, which is rotational about the vehicle vertical axis, depends on which axle has locked together with the vehicle speed, tyre–ground friction coefficient, yaw moment of inertia of the vehicle body and the vehicle dimensions. By considering the two cases of front and rear axle lock it is possible to derive useful insight into the stability problem:

Front axle lock
Any disturbance in the lateral direction due to gradient, sidewind or left to right brake imbalance produces a side force F_y that acts through the centre of gravity of the vehicle, as shown in Figure 14.14.

The resultant force F_R that is due to the inertia force F_x caused by the braking event and the lateral force F_y gives rise to a slip angle α. This slip angle represents the difference between the longitudinal axis of the vehicle and the direction in which the vehicle centre of gravity is moving. The lateral force F_y must be balanced by the side forces generated in the tyre–ground contact patches. As the front axle is locked, no side force is generated by the front wheels and the resulting side force is developed solely by the still rolling rear wheels. This gives rise to a total moment of $S_r b$. This yaw moment has a stabilizing effect since it causes the longitudinal axis of the vehicle to align with the direction of travel thereby reducing the initial slip angle α. Thus when the front axle is locked, the vehicle is unable to respond to any steering inputs and so its forward motion continues in a straight line.

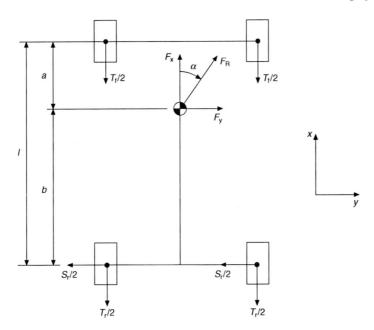

Figure 14.14 Front axle lock

Rear axle lock
Assume now that the fixed brake ratio associated with the same vehicle has been changed such that the rear axle locks in preference to the front as depicted in Figure 14.15. If the vehicle is subject to the same lateral disturbance, then this can only be reacted by a side force generated between the front wheels and ground and the resulting moment about the vehicle centre of gravity has a magnitude of $S_f a$. In contrast, this yaw moment now has a destabilizing effect as it causes the longitudinal axis of the vehicle to move away from the direction of travel, thereby increasing the vehicle slip angle α. This in turn leads to a rise in lateral force at the front of the vehicle causing an increase in yaw acceleration.

It is thus preferable, from a safety point of view, for the front axle to lock in preference to the rear as this is a stable condition and the driver is able to regain directional control of the vehicle simply by releasing the brakes. If the rear axle has locked and the vehicle has begun to spin, driver reaction must be rapid if control of the situation is to be regained.

In a collision situation, a frontal impact, linked to front axle lock, will usually result in less serious occupant injury than the possible side impact that could well be associated with the uncontrolled yawing of the vehicle that results from rear axle lock.

It is therefore feasible to apply the preceding ideas to the formulation of a fixed brake ratio that will invariably lead to front axle lock and this is commonly applied to the design of brake systems found on passenger vehicles. The fixed brake ratio is chosen such that for the unladen case both front and rear axles are on the verge of lock when the vehicle undertakes a $1g$ stop on a road surface that has a tyre–ground adhesion coefficient of unity. Under such conditions, the brake ratio is equal to

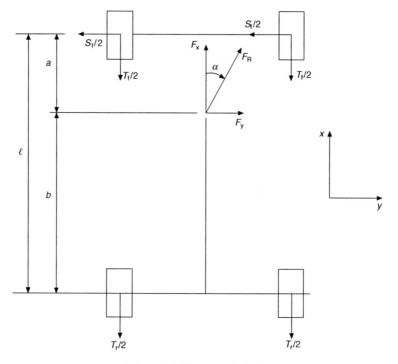

Figure 14.15 Rear axle lock

$$\frac{x_f}{x_r} = \frac{F_f + \dfrac{Ph}{l}}{F_r - \dfrac{Ph}{l}} \qquad (14.67)$$

and on all surfaces where the tyre–ground adhesion is less than unity, the braking will be limited by front axle lock.

The effect of axle lock on vehicle stability may also be assessed through the formal derivation of the equation of motion associated with the yawing of the vehicle. Analysis of the same cases of axle lock lead to identical conclusions regarding the behaviour of the vehicle with the added benefit that measures of yaw acceleration, velocity and displacement can be deduced.

14.4.7 Pitch motion of the vehicle body under braking

The transfer of load from the rear to the front axle that takes place during a braking event will cause the vehicle body to rotate about its lateral axis. This pitching motion also results in a change in the height of the vehicle centre of gravity. Both of these quantities can be determined as a function of vehicle deceleration using the notation in Figure 14.16. The following analysis assumes the vehicle body to be rigid and that the front and rear suspension spring rates, k_f and k_r, are linear. The spring rates used are the axle rates.

The opposed spring forces generated during a braking event are equal to the load transfer that takes place and so are equal to

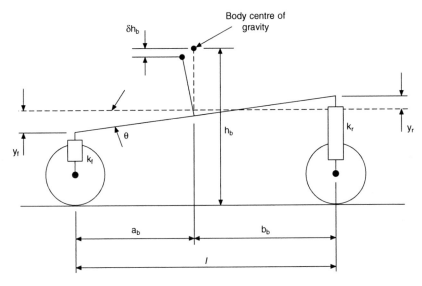

Figure 14.16 Determination of vehicle body pitch angle

$$\pm \frac{Pzh}{l}$$

and this causes the vehicle to go down at the front and move upwards at the rear as shown in Figure 14.16. Thus, on the assumption of linear springing, the compression travel at the front is

$$y_f = \frac{\dfrac{Pzh}{l}}{k_f} \qquad (14.68)$$

and the corresponding travel at the rear is

$$y_r = \frac{\dfrac{Pzh}{l}}{k_r} \qquad (14.69)$$

The pitch angle, θ, in degrees, adopted by the vehicle body is therefore given by

$$\theta = \left(\frac{y_f + y_r}{l} \right) \times \frac{360}{2\pi} \qquad (14.70)$$

Vertical and longitudinal movement of the vehicle body centre of gravity occurs as a result of the body pitch motion and this in turn causes a small change in the overall centre of gravity of the vehicle. The extent of movement of the vehicle body centre of gravity, initially located a distance a_b from the front axle at a height h_b above ground, depends upon its location within the structure, the suspension rates and the rate of deceleration. An indication of the extent of this movement can be seen in Figure 14.16. Under severe braking conditions, the vertical displacement, δh_b, of the vehicle body centre of gravity equates to approximately 5% of its original height. A detailed account of the relevant theory can be found in (Reimpell, 1996) and from this the change in height is given by

$$\delta h_b = -y_f \frac{F_{bf}}{F_b} + y_r \frac{F_{br}}{F_b} \tag{14.71}$$

where

$$F_{bf} = F_{sf} + F_{af} \tag{14.72}$$

$$F_{br} = F_{sr} + F_{ar} \tag{14.73}$$

$$F_b = F_{bf} + F_{br} \tag{14.74}$$

in which F_b is the vehicle body weight, $F_{bf,r}$ are the brake reaction loads applied to the front and rear of the vehicle body, $F_{af,r}$ are the unsprung weights of the front and rear axles and $F_{sf,r}$ are the front and rear axle loads. If the loads due to the unsprung axle masses are ignored then a corresponding expression for the change in the height of the overall centre of gravity of the vehicle, δ_h, can be found using

$$\delta_h = -y_f \frac{F_{sf}}{P} + y_r \frac{F_{sr}}{P} \tag{14.75}$$

in which P is the total vehicle weight.

14.4.8 Braking with a variable braking ratio

If a vehicle is to achieve maximum retardation, equal to the value of the tyre–ground adhesion coefficient, equation 14.22, then the brake system must be designed with a continuously variable brake ratio. This must be equal to the ratio of the dynamic load distribution between the front and rear for all values of deceleration. Thus the variable brake ratio, R_v, is defined as

$$\begin{aligned} R_v &= \frac{x_{fv}}{x_{rv}} \\[2mm] &= \frac{R_f}{R_r} \\[2mm] &= \frac{F_f + \dfrac{Pzh}{l}}{F_r - \dfrac{Pzh}{l}} \end{aligned} \tag{14.76}$$

from which it can be shown that

$$x_{fv} = \frac{F_f}{P} + \frac{zh}{l} \tag{14.77}$$

and

$$x_{rv} = \frac{F_r}{P} - \frac{zh}{l} \tag{14.78}$$

A situation giving rise to the need for a variable braking ratio might result from a given vehicle design in which the maximum deceleration using a fixed braking ratio is too low. In practice the introduction of a regulating valve into the braking system helps to optimize the braking efficiency over a wide range of operating conditions. Although such devices do not permit a continuously variable braking ratio, they do offer a means of improving the overall braking performance. Mathematical models of deceleration sensitive pressure regulating valves are now derived.

Deceleration sensitive pressure limiting valve

A typical valve design is shown in Figure 14.17. At a predetermined deceleration, determined by the mass of the ball and the angle of installation, the inertial force acting on the ball causes it to roll up the valve body and close the valve thereby isolating the rear brakes. These valves are gradient sensitive but do act in a favourable manner. On a rising slope the valve closes at higher levels of deceleration allowing the rear brakes to contribute more to the total braking effort, whilst on a falling slope the rear brakes are isolated sooner reflecting the load transfer to the front of the vehicle caused by the gradient.

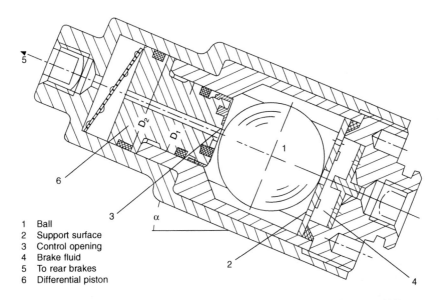

1 Ball
2 Support surface
3 Control opening
4 Brake fluid
5 To rear brakes
6 Differential piston

Figure 14.17 Deceleration sensitive pressure limiting valve (Limpert, 1992)

The effect on performance brought about by the inclusion of a regulating valve in the rear brake line can be assessed by deriving equations which define the brake ratio for all possible values of deceleration. These may then be used in the equations for efficiency and adhesion utilization, derived earlier, which quantify the brake system performance. In the following analysis it is assumed the valve isolates the line to the rear brakes when the vehicle deceleration has reached a certain value of deceleration, z_v. Note that the mechanism through which cut-off is achieved depends upon the chosen valve type and this determines the actual value of z_v.

Figure 14.18. shows a typical front to rear brake force characteristic. For all values of deceleration less than z_v, the brake force is apportioned between the front and rear axles in the fixed ratio R. Once the deceleration has exceeded z_v, the line pressure to the rear brakes is held constant and so they can no longer generate additional braking force. Consequently the brake ratio changes from its original value.

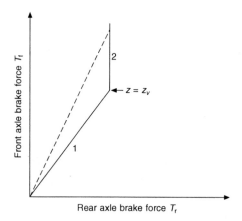

Figure 14.18 Typical limiting valve brake force distribution

In region 1, for $z \leq z_v$, the proportion of braking effort at the rear of the vehicle is

$$x_{rv} = x_r \qquad (14.79)$$

The proportion of braking effort at the front of the vehicle is therefore

$$x_{fv} = 1 - x_{rv}$$

$$= 1 - x_r = x_f \qquad (14.80)$$

and so the brake ratio R_v is

$$R_v = \frac{x_{fv}}{x_{rv}} = \frac{x_f}{x_r} \qquad (14.81)$$

In region 2, $z > z_v$, the valve actuates and isolates the rear brakes from any further increase in line pressure and the rear brakes can no longer generate additional braking force. As the front brakes are able to respond to further increase in line pressure then the rate of deceleration can increase above z_v and the brake ratio changes, being equal to the slope of the dashed line. When in region 2, the brake force at the rear, T_r, is constant and is

$$T_r = Pz_v x_r \qquad (14.82)$$

Simultaneously, the brake force acting at the front of the vehicle, T_f, increases and is equal to the difference between the total brake force, Pz, and that sustained at the rear axle

$$T_f = Pz - Pz_v x_r \qquad (14.83)$$

Thus in region 2, the brake ratio is defined by

$$R_v = \frac{T_f}{T_r} = \frac{Pzx_{fv}}{Pzx_{rv}} = \frac{Pz - Pz_v x_r}{Pz_v x_r} \qquad (14.84)$$

from which

$$x_{fv} = \frac{z - z_v x_r}{z} \qquad (14.85)$$

and

$$x_{rv} = \frac{z_v x_r}{z} \tag{14.86}$$

The incorporation of such a valve into the brake system of the prototype vehicle results in improved adhesion utilization and efficiency. Biasing the fixed brake ratio in favour of rear axle lock improves front axle adhesion up to the point of lock. This can be set, through the fixed

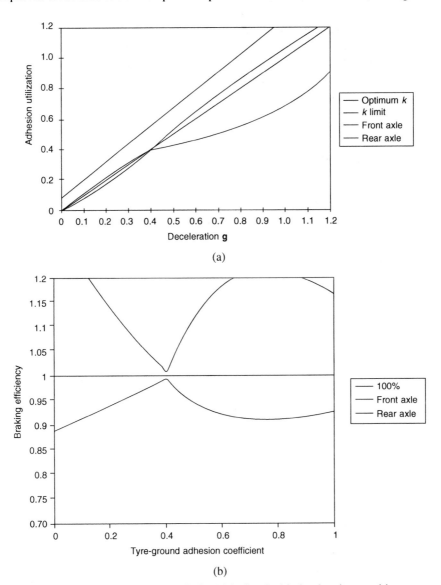

(a)

(b)

Figure 14.19 (a) Adhesion utilization, modified vehicle fitted with deceleration sensitive pressure, (b) Brake system efficiency, modified vehicle fitted with deceleration sensitive

brake ratio, to lie within the deceleration range of 0.35 g to 0.45 g. The deceleration sensitive pressure limiting valve is chosen to actuate at the point of lock and this results in an adhesion utilization diagram that has the form shown in Figure 14.19a. The fixed brake ratio has been changed to $R = \dfrac{x_f}{x_r} = \dfrac{0.73}{0.27}$ which results in a critical deceleration of 0.4 g and the valve is assumed to actuate at this level of deceleration. With reference to Figure 14.19a, the vehicle is now governed by front axle lock over all values of deceleration as the front axle adhesion lies above that of the rear. The brake system now makes much better use of the available adhesion when executing low/moderate g stops as, in comparison to Figure 14.13b, the front axle adhesion now lies closer to the optimum line. However at higher rates of deceleration, the front axle adhesion utilization has reduced. The beneficial effect of the valve on efficiency can be seen through comparison of Figure 14.19b (valve fitted) to Figure 14.13c (no valve).

Deceleration sensitive pressure modulating valve

A pressure modulating valve, or reducer valve, differs from a pressure limiting valve as once the activation point has been exceeded, they do not isolate the rear brakes but for higher pressures the rear brake pressure increases at a lower rate than that of the front brakes. The main advantages of this type of valve are that the rear pressure can be increased even after the front brakes have locked and the front and rear line pressures lie close to the optimum values. A typical pressure modulating valve characteristic is shown in Figure 14.20.

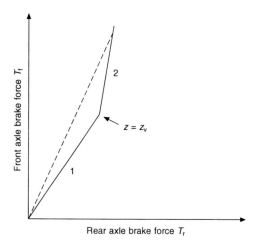

Figure 14.20 Typical modulating valve brake force distribution

    The effect on system performance of a pressure modulating valve may be assessed in a similar fashion to that of a limiting value. As before, it is assumed that the valve is actuated once the vehicle deceleration has exceeded a certain value of deceleration, z_v, and the exact value of z_v is determined by the mechanics of the valve.

In region 1, the vehicle deceleration is less than z_v and so the vehicle brakes in accordance with the fixed brake ratio assigned to the system,

$$x_{rv} = x_r \tag{14.87}$$

$$x_{fv'} = x_f \tag{14.88}$$

and

$$R_v = \frac{x_{fv}}{x_{rv}} = \frac{x_f}{x_r} \tag{14.89}$$

In region 2, the vehicle deceleration is greater than z_v and the valve has actuated. The front and rear brake forces increase in accordance with the slope of the valve characteristic and this causes the overall vehicle brake ratio to vary, being equal to the slope of the dashed line. If the slope of the brake force characteristic in region 2 is defined as $\dfrac{x_{f2}}{x_{r2}}$ then the brake force at the rear axle, T_r, is

$$T_r = Pz_v x_r + (Pz - Pz_v)\, x_{r2} \tag{14.90}$$

and the brake force at the front axle is

$$T_f = T - T_r$$
$$= Pz - (Pz_v x_r + (Pz - Pz_v)x_{r2}) \tag{14.91}$$

Thus, the overall brake ratio, defined by the slope of the dashed line, is

$$R_v = \frac{T_f}{T_r} = \frac{Pzx_{fv}}{Pzx_{rv}} = \frac{Pz - (Pz_v x_r + (Pz - Pz_v)x_{r2})}{Pz_v x_r + (Pz - Pz_v)x^2} \tag{14.92}$$

from which

$$x_{fv} = \frac{z - z_v x_r - (z - z_v)x_{r2}}{z} \tag{14.93}$$

and

$$x_{rv} = \frac{z_v x_r + (z - z_v)x_{r2}}{z} \tag{14.94}$$

The substitution of the modulation valve for the limiting valve into the brake system of the prototype vehicle enables improvements to be made to the adhesion utilization at high rates of deceleration. This is due to the ability of the valve to increase the line pressure to the rear brakes at a reduced rate. By appropriate choice of the slope of the brake force characteristic in region 2, the front axle adhesion curve can be forced to move closer to the optimum. In this example, setting the ratio for region 2 to be $\dfrac{x_{f2}}{x_{r2}} = \dfrac{0.88}{0.12}$ causes the front and rear adhesion curves to cross at a deceleration of 0.8 g and gives rise to the adhesion utilization diagram of Figure 14.21a.

The adhesion behaviour is identical to that shown in Figure 14.19a up until valve actuation. Thereafter the front axle adhesion converges on that of the optimum at 0.8 g. Decelerations greater than 0.8 g lead to rear axle lock but this is strictly admissible according to the adhesion

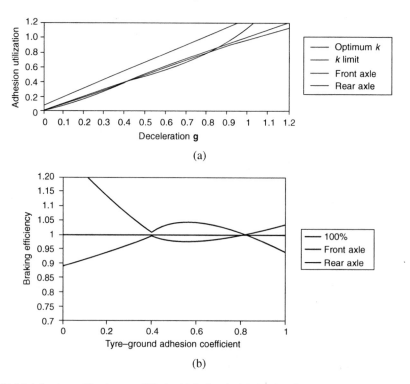

Figure 14.21 (a) Adhesion utilization, modified vehicle fitted with deceleration sensitive pressure modulating valve, (b) brake system efficiency, modified vehicle fitted with deceleration sensitive

utilization requirement specified in the governing EEC braking directive. The fact that the front axle adhesion now deviates little from the optimum illustrates the positive advantage that can be gained through the introduction of a bias valve into the brake system. Comparison of Figures 14.11, 14.13c, 14.19b and 14.21b illustrate this point by showing the progressive refinement of the brake system efficiency during the design process.

14.5 Materials design

14.5.1 Materials requirements for braking systems

In any conventional foundation brake, the relative rotation of the so-called 'friction pair' under the action of the brake system activating force is responsible for generating the frictional retarding torque required to slow the vehicle. Most friction pairs consist of a hard, usually metallic, rotating component and a relatively compliant 'friction' material in the form of a brake pad or shoe. The materials requirements for the rotating and stationary components of the friction pair are therefore quite different as discussed below.

Any rotor material must be sufficiently stiff and strong to be able to transmit the frictional torque to the hub without excessive deformation or risk of failure. However, the stresses arising from thermal effects are much higher than purely mechanical stresses and are more likely to give concerns over disc integrity. Thus the rotor material should have high volumetric heat capacity ($\rho \cdot c_P$) and good thermal conductivity (k) in order to absorb and transmit the heat generated at the friction interface without excessive temperature rise. Furthermore the maximum operating temperature (MOT) of the material should be sufficiently greater than the maximum expected temperature rise to ensure integrity of the rotor even under the most severe braking conditions. Ideally the rotor material should have a low coefficient of thermal expansion (α) to minimize thermal distortions such as 'coning' of a disc. It should also have low density (ρ) to minimize the unsprung mass of the vehicle. It should be resistant to wear since generally it is far easier and cheaper to replace the friction pads or shoes than the rotor itself. Finally, and most importantly, the rotor should be cheap and easy to manufacture.

The brake pad or shoe represents the stationary part of the foundation brake assembly. Normally a proprietary composite friction material is bonded to a steel backing plate or shoe platform. The primary function of the friction material is generally considered to be the production of a stable and predictable coefficient of friction to enable reliable and efficient braking of the vehicle over a wide range of conditions. In fact, it is the combined tribological characteristics of both rotor and stator materials (i.e. the 'friction pair') which are responsible for the generation of the frictional torque. As for the rotor, the friction material must have sufficient structural integrity to resist the mechanical and thermal stresses. This is particularly important for the bond between the friction material itself and the steel structure which supports it, as a complete failure here could have disastrous consequences. The friction material should have a relatively high MOT to prevent thermal degradation of the surface although, due to the nature of its composition, the MOT of the pad material will always be lower than that of the disc. A low conductivity for the pad or shoe material is desirable to minimize conduction of heat to other components of the system, in particular to the hydraulic fluid. The material should be reasonably wear resistant but not excessively so since wear can be beneficial in promoting a uniform contact pressure distribution and preventing 'hot spotting'. Likewise the elastic modulus of the material should be relatively low to give good conformity with a roughened or thermally distorted rotor surface. Finally, as for the rotor, the friction material should be cheap and easy to manufacture.

The friction material selected to meet the above requirements is invariably a complex composite consisting of a variety of fibres, particles and fillers bonded together in a polymeric matrix such as phenolic resin. For many years, asbestos fibres were an important element of friction materials due to their excellent thermal and friction properties. For health and safety reasons, asbestos has now largely been replaced by other less harmful fibres, e.g. Kevlar. The exact composition of any friction material must be tailored to the application and knowledge of the formulation is proprietary to the supplier.

14.5.2 Cast iron rotor metallurgy

The overwhelming majority of rotors for conventional automotive brakes are manufactured from grey cast iron (GI). This material, also known as flake graphite iron, is cheap and easy to cast and machine in high volumes. It has good volumetric heat capacity due mainly to its relatively high density, and reasonable conductivity due largely to the presence of the graphite (or carbon) flakes. The coefficient of thermal expansion is relatively low and the material has a MOT well in excess of 700 °C (but note that martensitic transformations at high temperatures can lead to hot judder problems). Although the compressive strength is good, the tensile strength is relatively low and the material is brittle and prone to microcracking in tension. As the proportion of flake graphite in GI is increased, the tensile strength reduces but thermal conductivity increases as shown in Table 14.2. Note that spheroidal graphite iron (SG) has a higher tensile strength than GI but a much reduced conductivity which explains why it is rarely used for brake rotors.

Table 14.2 Tensile strength and conductivity of some common cast irons

Grade	Min. tensile strength (MPa)	Thermal conductivity at 300 °C (W/m K)
400/18 SG*	400	36.2
250 GI	250	45.4
200 GI	200	48.1
150 GI	150	50.5

*Spherical graphite iron

Currently GI grades used for disc brakes fall into two categories reflecting two different design philosophies (MacNaughton, 1998):

1. Medium carbon GI (e.g. Grade 220)
These irons are used for small diameter discs such as on small- and medium-sized passenger cars. Such discs will run hot under extreme conditions, and good strength and thermal crack resistance at high temperatures are therefore required.

2. High carbon GI (e.g. Grade 150)
These grades tend to be used for larger vehicles where space constraints are not as content limited. Discs are larger and, with the improved conductivity due to the high carbon, will run cooler. Strength retention at high temperature is therefore not as critical and manufacturability improves with the higher carbon content.

Alloying elements can be applied to all grades of cast iron with the general effect of improving strength but at expense of thermal properties and manufacturability. The most commonly used elements and their effects are as follows:

- chromium – increases strength by stabilizing pearlitic matrix at high temperatures (preventing martensitic transformations) but tends to promote formation of bainitic structures which cause casting/machining difficulties and can reduce pad life;
- molybdenum – similar to chromium;
- copper – increases strength without causing manufacturing difficulties;
- nickel – as for copper but more expensive;
- titanium – reported to influence friction performance but rarely used at significant levels.

14.5.3 Alternative rotor materials

Although GI is a cheap material with good thermal properties and strength retention at high temperature, its density is high and, because section thickness must be maintained for both manufacturability and performance, cast iron rotors are heavy. Currently there are significant incentives to reduce rotor weights in order (a) to reduce emissions by improving the overall fuel consumption of the vehicle, and (b) to aid refinement and limit damage to roads by reducing the unsprung mass. Thus much effort has been directed at investigating light-weight alternatives to cast iron. Two such alternatives which have received serious attention are aluminium metal matrix composites (MMCs) and carbon–carbon composites, typical properties for each of which are displayed in Table 14.3. together with corresponding properties for a high carbon cast iron [Grieve, 1995].

Table 14.3 Physical properties of three candidate disc materials

Disc material	ρ kg m^{-3}	C_p J kg^{-1} K^{-1}	$\rho \cdot C_p$ kJ m^{-3} K^{-1}	k W m^{-1} K^{-1}	α $\times 10^{-6}$ K^{-1}
High carbon cast iron	7150	438	3132	50	10
Generic 20% SiC-reinforced Al MMC	2800	800	2240	180	17.5
Carbon–carbon composite	1750	1000	1750	40–150	0.7

Aluminium MMCs normally incorporate 10–30% by volume silicon carbide particle reinforcement within a silicon-containing alloy matrix. The resulting composite has much lower density than cast iron and much improved conductivity. Thus the thermal diffusivity $(k/\rho \cdot c_p)$ is much higher which opens the possibility of lighter discs running cooler by being able to rapidly conduct heat away from the friction interface. However, aluminium MMCs have a low MOT (c. 500 °C) and there are serious consequences if this MOT is exceeded since complete surface disruption may then occur leading to extremely rapid pad wear. Ideally higher reinforcement contents or alternative reinforcing materials (e.g. alumina) should be used to increase the MOT but the former causes severe casting difficulties whilst alumina reinforcement results in poorer thermal properties.

It can be seen from Table 14.3 that carbon–carbon composites have an even lower density than aluminium MMCs and can have a conductivity almost as high. Their MOT is also very high, raising the possibility of using thin rotors which run much hotter and lose heat by radiation as well as by conduction/convection. Also the very low coefficient of thermal expansion of carbon minimizes thermal distortions. Thus there is the potential for very significant weight savings with carbon–carbon composite discs. However, the material has a poor low temperature friction performance and moreover is currently much more expensive than metallic alternatives. Hence it is likely to remain confined to high performance race car applications for the foreseeable future.

When considering alternative materials or designs for disc brakes, reference can be made to the so-called 'bucket-and-hole' analogy in which the rate of water flow into the bucket is taken to represent the heat flow into the disc and the height of the water level in the bucket represents the maximum temperature of the disc surface. A hole in the bucket represents the ability of the disc to lose heat to the surroundings. The volume of the bucket is therefore the heat capacity of the disc whilst the height is the MOT of the disc material. The question then is how close does the level of water in the bucket get to overflow!

Consideration of the 'bucket-and-hole' analogy and with reference to the typical material properties of Table 14.3, three distinct strategies for brake rotor materials can be identified (Grieve, 1996):

Strategy I
Large diameter and relatively deep bucket with small hole (see Figure 14.22a). This implies a high volumetric heat capacity to store heat during braking and a relatively high MOT but only moderate conductivity of heat away from the rubbing surfaces. Current grey cast iron discs represent such a system but some steels may also meet these criteria.

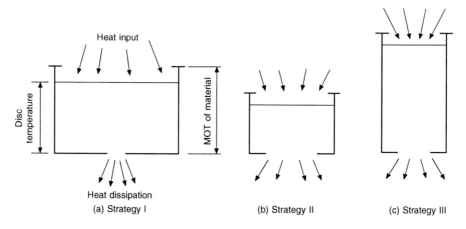

Figure 14.22 The 'bucket and hole' analogy

Strategy II
Smaller diameter and relatively shallow bucket but large hole (see Figure 14.22b). This implies smaller volumetric heat capacity and a relatively low MOT. Hence it is important to have high conductivity to transfer heat to other parts of the rotor and thence the surroundings in order to prevent temperature build-up at the rubbing surfaces. Aluminium MMC may meet these criteria but recent research (Grieve, 1998) suggests that this can only be successfully achieved for currently available MMCs if the brake rotor is redesigned to increase its thermal mass and cooling capability. Other materials that may be successful with appropriate development include high reinforcement content MMCs and coated alloy discs but again there are manufacturing, integrity and cost issues to be resolved.

Strategy III
Even smaller diameter bucket but much deeper with moderately-sized hole (see Figure 14.22c). This implies a material with high MOT which can be allowed to run much hotter than current designs and so lose significant amounts of heat by radiation as well as more moderate amounts by conduction/convection. Carbon–carbon composites are a possibility here but, as mentioned above, these are currently too expensive for mass produced vehicles. High temperature steels with good strength retention at temperatures well in excess of 1000 °C may also be candidate materials under this heading. Such discs could perhaps be made much thinner and without vents, and therefore also save significant weight. However, there would be concerns over compatible friction materials and heat transfer to other components in the underbody wheel arch area if discs were allowed to run much hotter than is currently the practice with cast iron.

14.5.4 Disc materials/design evaluation

Ultimately, any new brake material or design must be validated by experimental trials on actual vehicles to allow accurately for model-specific parameters such as the effect of body trim on rotor cooling. However much can be learnt about potential new rotor materials or designs by numerical simulations of critical brake tests using finite element (FE) analysis. Such techniques require the rotor and/or stator geometry to be broken down into a number of small non-overlapping regions known as elements which are assumed to be connected to one another at certain points known as nodes. A 2D axisymmetric finite element idealization can be used as a first approximation but, for more accurate simulation of the heat flow and stresses, a 3D model is desirable such as the 10° segment model of a brake disc and hub shown in Figure 14.23. Note that in order to accurately simulate the heat loss from the rotor, it is sometimes necessary to include the wheel and other components in the model.

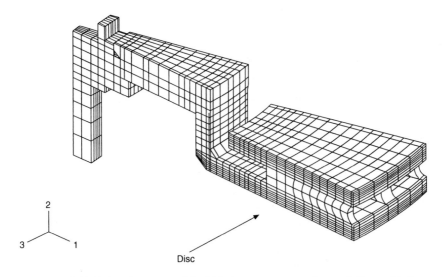

Figure 14.23 Finite element model of 10 degree segment of vented disc and hub

The heat input to the system is estimated from theoretical consideration and applied over the rubbing surface. The heat loss to the surrounding is specified by convective and sometimes radiative heat transfer conditions along relevant boundaries of the model. The temperatures predicted by a thermal analysis can be used as input conditions to a structural analysis in order to predict thermal deformations and stresses. If the pad is included in the model, the contact pressure distribution (and hence the distribution of heat input) can be estimated leading to the possibility of a fully coupled thermal-structural analysis (Brooks, 1994).

In addition to details of geometry and material properties, accurate date on heat loss to other components and to the atmosphere are vital to allow accurate predictions of rotor temperatures using FE methods. Such data can be generated by conducting so-called 'cooling tests' on actual vehicles fitted with representative brake rotors carrying rubbing or embedded thermocouples. The rotor surface is first heated to a predetermined temperature by dragging the brakes and then allowed to cool whilst the vehicle is driven at constant velocity. By comparing the experimental rate of cooling with that predicted by the finite element simulation for different boundary conditions, optimized heat transfer coefficients can be derived which are then assumed to apply for different rotor materials and factored for the varying air stream velocity under different test conditions.

Two very different vehicle brake tests are often simulated to critically examine the maximum temperatures and integrity of new rotor materials or designs: (i) a long slow Alpine descent during which the brakes are dragged and the vehicle is subsequently left to stand at the end of the descent; (ii) a repeated high speed autobahn stop with the rotor allowed to cool only moderately between stops. The former test determines the ability of the design to limit temperature build-up in the rotor by heat transfer to the atmosphere whilst the high speed repeated stop examines the ability of the rotor material to withstand repeated thermal cycling and the ability of the friction pair to resist 'fade' under these severe conditions.

Friction performance cannot easily be predicted by the FE approach and there remains a requirement for dynamometer testing to determine the fade and wear characteristics of every new friction pair. The dynamometer can either be a full-scale device or a small sample rig in which the geometry and loading conditions are scaled to give an accurate representation of the actual brake. These tests will not only give data on friction performance over a wide range of conditions but can also be used to determine the MOT of the pad and rotor materials by progressively increasing the temperature at the rubbing interface until some form of failure occurs (Grieve, 1996).

14.6 Advanced topics

14.6.1 Driver behaviour

The driver of a vehicle plays a key role during any braking event since his/her reactions to external stimuli have a direct bearing on his/her ability to maintain complete control over the vehicle trajectory and deceleration rate. A knowledge of how the driver interacts with these external stimuli and the way in which the vehicle responds to the control signals generated by the driver is vital to the future development of safe road transport systems.

Many experimental studies, including Newcomb (1974, 1981), Mortimer (1976) and Spurr

(1972), have been undertaken that have led to improved understanding of driver behaviour during braking. These have focused on the study of limb dynamics, pedal effort, braking kinematics and response to external stimuli such as obstacles and road signs. This has given rise to the development of mathematical models that embody a representation of the driver into a model of the vehicle dynamics. Any such model, typified by McLean (1976), contains elements that describe the dynamics of the vehicle, the braking system, the neuro-muscular system and force characteristics of the driver and finally the motion detection system/sensory characteristics of the driver together with feedback loops as appropriate to the model in question. The adaptive nature of the driver that is captured in such models requires enhancement but simulation of vehicle braking performance with the driver can yield deceleration characteristics that match closely those from experiment.

14.6.2 Brake by wire

The driver behind brake-by-wire systems has arisen from the ongoing development of modern braking systems such as anti-lock and traction control systems along with the need to effect their seamless integration within the overall chassis control strategy. There are two strategies currently receiving attention.

The first utilizes a conventional hydraulically actuated braking system, that includes the brake fluid, brake lines and conventional actuators, together with a significant number of electro-hydraulic components (Jonner, 1996).

The second relies upon a full electro-mechanical system (Bill, 1991), (Maron, 1997), (Schenk, 1995) in which the brake force is generated directly by electro-mechanical foundation brake actuators. The electro-mechanical system potentially requires little maintenance due to the removal of the hydraulic fluid as the means of energy transmission and this conveniently combines with a reduction in the amount of hardware demanded by the brake system which in turn leads to an overall weight reduction. Such systems may also contribute towards the enhancement of passenger safety as the location of the pedal assembly within the vehicle can be optimized so that the likelihood of lower leg injury is minimized during impact events. As with all advanced control systems, it is the control unit, its associated software and the array of sensors that combine to define the overall effectiveness of the system. The controller must operate in closed loop fashion, be able to take into account the in-use variation of the system parameters and fail safe.

14.6.3 Anti-lock braking systems (ABS)

Under normal braking conditions, the driver of a vehicle makes use of the linear portion of the brake slip vs brake force characteristic, Figure 14.7. The brake force coefficient, μ, builds from zero in the free rolling state to a maximum, μ_p, at around 20% slip and within this region the wheel is both stable and controllable. When braking under extreme conditions the driver may demand a brake torque that is greater than that which is capable of being reacted by the wheel. This results in a torque imbalance that causes the wheel slip to increase and the wheel rapidly decelerates to the full lock condition and in this state, the brake force coefficient is approximately $0.7\mu_p$. If the front wheels have locked, then steering control is lost and if rear wheel lock takes place then the vehicle becomes unstable. Simultaneously, the ability of the vehicle to generate side force markedly reduces, Figure 14.7, and this explains why limiting wheel slip, thereby avoiding wheel lock, is more critical for steering and directional stability of the car than for stopping distance alone.

The purpose of ABS is to control the rate at which individual wheels accelerate and decelerate through the regulation of the line pressure applied to each foundation brake. The control signals, generated by the controller and applied to the brake pressure modulating unit, are derived from the analysis of the outputs taken from wheel speed sensors. Thus, when active, the ABS makes optimum use of the available friction between the tyres and the road surface. The role of ABS as an automotive control system is reviewed in Section 11.6 of this text.

14.6.4 Traction control systems (TCS)

Traction control systems aim to control and maintain vehicle stability during acceleration manoeuvres, by, for example, preventing wheel spin when accelerating on a low friction surface or on a steep up-grade. This is achieved by the optimization of individual wheel torques through the control of some combination of fuel mixture, ignition and driven wheel brake torque. TCS are able to utilize components used in ABS and integration of the two systems is becoming commonplace. TCS are reviewed in Chapter 11 of this text.

14.7 References and further reading

Automotive brake systems. Pub. Robert Bosch GmbH, (1995). Distributed by SAE, ISBN 1-56091-708-3.

Bill, K. (1991) Investigations on the behaviour of electrically actuated friction brakes for passenger cars. EAEC 91021.

Brake handbook. Alfred Teves GmbH, 2nd edition 1981.

Brake technology and ABS/TCS systems. SP-1413, Pub SAE, 1999. ISBN 0-7680-0345-8.

Brooks, P.C., Barton, D.C., Crolla, D.A., Lang, A.M. and Schafer, D.R. (1994). A study of disc brake judder using a fully coupled thermo-mechanical finite element model. Proceedings FISITA 94 Conference, Paper No 945042, Beijing, October.

Grieve, D., Barton, D.C., Crolla, D.A., Buckingham, J.T. and Chapman, J. (1995). Investigation of light weight materials for brake rotor applications. IoM Conference on Materials for Lean Weight Vehicles, University of Warwick, November.

Grieve, D., Barton, D.C., Crolla, D.A., Buckingham, J.T. and Chapman, J. (1996). Investigation of light weight materials for brake rotor applications, in Advances in Automotive Braking Technology, Barton D.C. (ed.), PEP.

Grieve, D., Barton, D.C., Crolla, D.A., Chapman, J. and Buckingham, J.T. (1998). Design of a light weight automotive disc brake using finite element and Taguchi techniques. Proc IMechE Part D, pp 245–254.

Jonner, W.-D., Winner, H., Dreilich, L. and Schunck, E. (1996). Electrohydraulic brake system – the first approach to brake-by-wire technology. SAE 960991, Detroit.

Limpert R. (1992). Brake design and safety. SAE, ISBN 1-56091-261-8.

MacNaughton, M.P. and Krosnar, J.G. (1998). Cast iron – a disc brake material of the future? In *Automotive Braking : Recent Developments and Future Trends*, Barton, D.C. and Haigh, M.J. (eds), PEP.

Maron, C., Dieckmann, T., Hauck, S. and Prinzler, H. (1997). Electromechanical brake system: Actuator control development system. SAE 970814, Detroit.

McLean, D., Newcomb, T.P., Spurr, R.T. (1976). Simulation of driver behaviour during braking. Proc IMechE Conference on Braking of Road vehicles, paper C41/76.

Mortimer, R.G. (1976). Implications of some characteristics of drivers for brake system performance. Proc ImechE Conf Braking of Road Vehicles, Loughborough p.187–195.

Newcomb, T.P. (1981). Driver behaviour during braking. SAE/IMechE Exchange Lecture 1981, SAE 810832.

Newcomb, T.P and Spurr, R.T. (1967). *Braking of Road Vehicles*. Chapman and Hall Ltd,. Out of print.

Newcomb, T.P and Spurr, R.T. (1974). Aspects of driver behaviour during braking. XV Congress FISITA, paper A.1.4.

Puhn, F. (1995). *Brake handbook*. HP. Books ISBN 0-89526-232-8.

Reimpell, J. and Stoll H. (1996). *The Automotive Chassis: Engineering Principles*. Arnold.

Schenk, O.E., Wells, R.L. and Miller, I.E. (1995). Intelligent braking for current and future vehicles. SAE 950762, Detroit.

Spurr, R.T. (1972). Driver behaviour during braking. Proc. Symp. on Psychological Aspects of Driver Behaviour, paper 1-2, Holland.

15. Failure prevention – The role of endurance and durability studies in the design and manufacture of reliable vehicles

F.L. Jones, BSc, PhD, MIM, CEng,
R. Scott, BSc, MIM, CEng
D.E. Taylor, GRIC, PhD, MIM, CEng

The aim of this chapter is to:

- Understand the importance of mechanical failure in the design process;
- Aid the designer to appreciate methods by which mechanical failure can be controlled by design;
- Illuminate practical methods by which mechanical failure can be analysed;
- Aid the design process by giving examples of vehicle design failures and their remedies.

15.1 Introduction

If hindsight could be developed as a fully credited scientific discipline it would be a major asset in the field of failure prevention. However, this is an unrealistic proposition and in the real world plausible alternative techniques are devised to avoid bad design features, poor material selection and unsustainable loads. These alternative techniques have been developed and refined over the past hundreds of years and require that the performance of a component should be predictable. Consequently, as more and more factors affect the performance of a component it becomes more difficult to predict the life of the component. In fact, in real working environments there is a need to adopt a probabilistic approach to allow for potential variations of service stress, service environment, material quality and the quality of various manufacturing processes such as welding, casting, forming and machining. It is often the unpredictable or incorrectly predicted behaviour of the materials or environmental conditions that leads to failure.

The creation of these 'plausible alternative techniques' has tested the skills and ability of several generations of engineers and it is a credit to these engineers that in general the current reliability levels in automotive products are relatively high. For example, for a successful car engine there could be over 250 000 operational units each with 8 exhaust valves. It would be unacceptable for more than 1 or 2 engines to fail per year due to exhaust valve problems. This requires a reliability of less than 10^{-7} failures per component, per year (i.e. 1 failure in 10 million, per year!).

In very general terms conservative targets of stress and the use of high quality materials and methods of manufacture usually guarantee high reliability and low risk of failure. In the converse circumstances (usually found during design optimization) where stress levels are high and 'value analysis' techniques dictate low material and manufacturing costs, then the risk of failure could be significant. The application of this logic provides the blindingly obvious solution to preventing failures, which would be a 'safe design' with low working stresses and using techniques of manufacture and inspection that would ensure high integrity regardless of cost. Unfortunately this would be unsustainable in the automotive industry.

In this short chapter which reviews the principles applied in the automotive industry to avoid failures it is difficult to do justice to such an important subject or to outline the in-depth knowledge needed to carry out effective failure analysis. Therefore wherever possible the reader is directed to the specialist literature on the subject.

15.2 Important aspects of failures in the real engineering world

15.2.1 Defining failure and reliability

The term 'failure' is used to describe a wide variety of phenomena. In general terms we all understand the word failure both in a personal sense and when applied to objects and events in every-day life. It may be a system failure (e.g. mechanical system, or an electrical system), or we may be considering the failure of an engineering component, possibly involving fracture. (This classification is somewhat arbitrary, since component failure is often the root cause of system failure, i.e. this classification simply reflects how closely we are prepared to look at a given failure.)

In an engineering context BS 4778 provides the universally accepted engineering definitions for 'failure' and 'failure classification' used in the quality assurance field. For example:

'Failure' is defined as the termination of the ability of an item to perform a required function.
'Failure cause' is described as the circumstances during design, manufacture, assembly, installation, or use that have led to failure.
'Failure mode' is the effect by which failure is observed.
'Reliability' is the ability of an item to perform a required function under stated conditions for a stated period of time.

An equally important list of definitions is associated with 'failure classifications' where an appropriate adjective is added to the word failure. The most important terms used are (Powell, 1979):

'critical failure', where failure may jeopardize life or result in significant financial loss.
'major failure', where failure would immobilize a vehicle.
'minor failure', where failure would not require immediate replacement.

These classifications allow the identification of key components and thus appropriate allocation of resources to develop the test procedures, inspection practices and application of the required level of quality assurance practices.

The consequential financial loss caused by a failure can be significantly greater than the cost of the actual failed component. For example, in the late seventies and early eighties there were a number of truck wheel stud failures that resulted in wheels leaving vehicles which caused several fatalities (Dickson-Simpson, 1983; Wright, 1989). The difference between the cost of the wheel studs and the costs associated with the accidents was several orders of magnitude.

When defining failure it is also necessary to consider the actual point at which failure is judged to have occurred. If we consider the failure of a brake pipe, it is apparent that failure will be considered to occur when sufficient of the fluid is lost that the braking system no longer operates. Similar examples, often involving component fracture, may be identified, e.g. failure of light bulb filaments. However, in many cases the point at which failure occurs is not so clearly defined. For example, with brake pads (or for car tyres) the performance deteriorates with time through 'wear and tear'. In order to define failure it is, therefore, necessary to define a point at which the performance has deteriorated to a sufficient level that the system may no longer be regarded as performing satisfactorily. At this point, failure may be considered to have occurred. In the UK it is difficult to obtain data on vehicle break down rates. One published source of the causes of vehicle immobilization has been provided by the RAC (Bidgood, 1989) based on 160 000 call outs. Their analysis of this data can be seen in Figure 15.1.

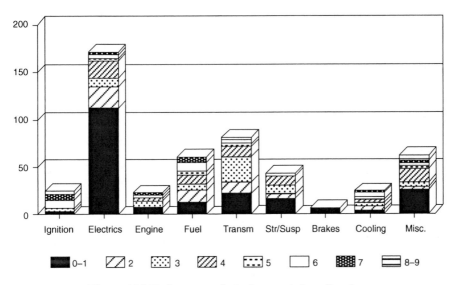

Figure 15.1 Fault report – fault characteristics, all makes

The demarcation between satisfactory and unsatisfactory performance can be a difficult judgement unless there is a clearly defined product design specification (PDS) or component design specification (CDS), agreed by all involved parties at the initial design stage or component procurement stage. This will be a statement of the main functions, performance characteristics and endurance targets and many books on product design give a full description of the procedures for preparation of the PDS and CDS (Pugh, 1991). Figure 15.2 illustrates the way in which a PDS would be a key document in a component test procedure.

The actual point of failure, and whether a component achieved its design life, is not the only

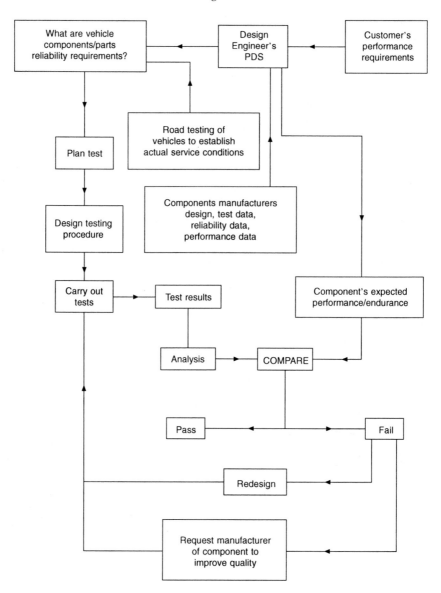

Figure 15.2 Component test philosophy diagram

aspect that needs to be considered in many manufacturing industries such as the automotive industry. Consideration must also be given to the effect of the failure/deterioration processes on the 'value' and 'aesthetics' of the product. Hence, a car that shows corrosion will quickly depreciate in value, even though the corrosion may occur in an area that does not seriously influence the performance of the car. Noise vibration and harshness (NVH) may also be considered in this category, where the subjective view of the user could lead to unacceptable performance and the response being judged as a failure.

15.2.2 Engineering responsibility and product liability

Poorly engineered products may lead to expensive legal settlements. It is the responsibility of the engineer to ensure that his products do not fail in a catastrophic manner, and this may necessitate 'knowledge' of a wide range of disciplines. In fact, this was pointed out by Wolff (Wolff, 1967) over forty years ago in a paper on design:

> 'The designer only operates by the application of known data, known methods, known materials and known art to his design problems. If this is not so, the designer cannot say definitely that an article manufactured to his instructions will satisfy the special requirements.'

Engineering judgement is, therefore, an important factor in the design and manufacture of reliable components. A typical design project would require detailed information, a good level of understanding and clear decision making on aspects of the following:

- The fabrication behaviour of a material;
- The performance and endurance in service of a material;
- Details of the service loads, working environment and potential failure modes;
- The cost basis of raw materials, manufacturing processes and assembly;
- The design life, maintenance opportunities, and the consequences of failure.

This simplified list confirms the observation often made that 'applied knowledge' is the essence of the design and manufacture of reliable products. The modern appraisal of this concept is now referred to as 'know-how' and represents the total capability of a company to produce profitable reliable products. When there is a flaw in this knowledge, or an error in the application of the knowledge we have a potential failure. For example, when Al-Li alloys were first introduced into the manufacture of airframes (low ductility produced problems in manufacture), and the introduction of new manufacturing processes (welding) in the production of the Liberty ships during the 1940s (Open University Video, 1984) (resulting in failures), both illustrate incomplete knowledge, i.e. not enough 'know-how'. In addition, the sheer complexity of design necessary for some products (e.g. space shuttle) can often make failures inevitable.

'Know-how' is often closely associated with commercial confidentiality and consequently secrecy that occasionally prevents open communication between engineers. In the field of failures, especially where potential compensation payments are involved, engineers should be aware that all the human emotions are at work to avoid accepting responsibility and any associated financial penalty. This typically involves being 'economical with the truth' and taking what would normally appear to be an intransigent position regarding what would appear to be clear scientific facts. Under these conditions all engineers should be aware of their professional status and try to remain constructive whilst also being realistically aware of their own limitations, as problems change from metallurgy to rotodynamics to tribology. Under these circumstances the engineer must involve additional expertise or if the engineer has sufficient confidence begin the background work to acquire the required specialist knowledge.

Patent law also limits the design engineer and the avoidance of a competitor's patent, which is often the simple and easy solution, can result in complex designs with a potential risk of failure.

Appropriate knowledge must be made easily accessible to the design engineer and in a format that can be easily applied within the time frame available. This well identified need has resulted in the development of several computer-based techniques, information databases and expert systems. In addition validated data and established calculation methods provided by the Engineering Sciences Data Unit (ESDU) (Barret, 1981) have been built up over a period of nearly 60 years.

When failures do occur it is important to carry out an appropriate analysis of the failure to establish the cause and to then translate this into improved design knowledge to ensure that the failure does not re-occur. There are several well documented procedures for undertaking failure investigations such as those found in *ASM Metals Handbook*, Volume 10 (*Metals Handbook*, 1988).

As well as the obvious economic loss to the company, especially with failures within the warranty period, losses can be significantly amplified by potential product liability. The EC Directive on product liability (85/374/EEC) was adopted on 25 July 1985 and made the producer responsible for damage caused by defective products (EEC Directive, 1985). The potential threat of claims for engineering negligence has had a major impact on improving the records kept during the design and testing of products. In addition it has highlighted the need for detailed Quality Assurance procedures, Failure Modes and Effects Analysis (FMEA) and the application of 'state of the art' inspection and test procedures. Documented evidence of having used these 'best practices' can have a positive effect during any subsequent legal proceedings.

In a publication in the late 1970s Breen (Breen and Wene, 1974), then working at International Harvester, introduced the term 'Parts Performance Analysis'. This was the preferred description for 'failure analysis' because of the legal/litigation implications of the term 'failure'. The concept of PPA was based on the well known need to have 'closed loop' control of the design cycle, the detailed analysis of failures providing the data to build better quality machines. This publication presents PPA studies on several automotive components (e.g. engine crank shafts, rear axle shafts, gear teeth, engine con-rods, differential crosses).

The achievement of reliability in service means ensuring that reliability has been thoroughly proven at the design stage and well established at the mass production stage. Late design changes or vehicle recalls usually means that potential business profit is lost (Pye, 1992). BS5760 (1993) provides a comprehensive introduction to reliability testing with details for reliability programme management, reliability assessment and examples of reliability practices, including those used in the automotive industry.

One of the early accounts of reliability applied to automobile manufacture was published in 1973 and described how it was carried out at Rover-Leyland (Allen, 1973). The successful application of reliability concepts requires an engineering knowledge of the mechanical aspects of the components, the expected failure modes and an appreciation of statistical techniques.

15.2.3 Major engineering failures

There are several major texts covering the failure of engineering components which provide useful insight into the methods and procedures involved in failure investigations and prevention of failures (Reynolds, 1977; Hutchings and Unterweiser, 1981; Colangelo and Heiser, 1974; Borer and Peters, 1970; Engel and Klingele, 1981; McCall and French. 1978; German, 1956, Wulpie, 1985; and ASM Sourcebook, 1974). In addition, the recent CD ROM compiled by the

American Society of Metals (ASM) provides the first comprehensive technical database on failures (ASM CD ROM, 1997). However, perhaps one of the most memorable illustrations of the importance of component failure is given by Harrison (Harrison, 1980), when relating a story from the Welding Institute.

> 'A short story concerning Dr Geoff Egan, who is well known in the fracture field illustrates the type of problem that may confront those at the forefront of fracture research. Dr Egan was flying back to Europe from a conference in the USA. He awoke as the plane crossed the coast of France and saw a group of people looking out of one of the windows. The Captain approached him and said "Dr Egan, I understand that you are an expert on fatigue and fracture, I would like to ask your advice about a problem we have, could you come and look out of this window." From the window a crack about 1 m long could be seen in the top of the wing. Dr Egan said "I can give you two pieces of advice: firstly get this plane on the ground as fast as you can; and secondly slow down!" It is to be hoped that advice from experts in fracture is generally less conflicting. The crack in question grew another 75 mm during the flight.'

Fortunately, Dr Egan did not witness the 'in-service' failure of the aircraft wing, and hopefully none of us are put in the position of testing our ability to give 'less conflicting' advice. However, there is no doubt that we would certainly come away from such an experience with a much clearer understanding of the importance of avoiding failure in engineering components.

In most failure investigations one of the initial steps is to identify similar events in the published literature and to identify key publications on the technology of the particular component. This would include engineering design calculations, previously used materials and their limitations and the methods of manufacture, assembly, maintenance and operating conditions.

Often the literature refers to the different modes of failure. These are considered in more detail later in this chapter (Section 15.2.5), although they will now be briefly described.

(i)   **Ductile fracture** – fracture accompanied by large amounts of plastic deformation.
(ii) **Brittle fracture** – fracture accompanied by limited, or no, plastic deformation.
(ii) **Fatigue** – fracture during service in a dynamically stressed component.
(iv) **Creep** – fracture during service in a stressed component operating at high temperature.
(v)   **Corrosion** – environmental degradation of the integrity of a material during service.
(vi) **Wear** – localized removal of material from contacting surfaces during service.

These terms have been widely used to describe many of the well known failures, responsible for creating the current scientific foundation associated with failure analysis. For example the failure of the Liberty ships referred to earlier which were built to carry strategic supplies from the USA to the UK during the Second World War, led to a better understanding of brittle fracture in welded structures (Open University, 1984). Cracks were found to initiate at the sharp corners on the hold entries which highlighted the engineering concept of stress concentration, a factor which all engineers should now understand. Within a decade we had failures in the first passenger aircraft fitted with jet engines – the Comet (Riederer 1979). These failures caused severe commercial damage to the UK aerospace industry although subsequent research and development laid the foundation of our understanding of fatigue crack growth. With the Comet failure, fatigue cracks had initiated at the rectangular designed windows due to stress concentration, 'a factor which all engineers should now understand'.

Further examples of failures in structures and other mechanical systems has received much attention over the years, with many of the more notorious examples being discussed in detail in the literature, e.g.

(i)   Flixborough (Swann, 1976) – the failure of a temporary bypass pipe in a chemical plant near Humberside. Twenty eight people died.
(ii)   3-mile Island (Kaku, 1983) – a USA PWR nuclear power reactor released radioactive water and came close to melt down due to a faulty indication of a valve position indicator.
(iii) Alexander Kielland (Easterling, 1983) – an oil platform overturned with the loss of life. This was caused by structural failure due to fatigue crack propagation near a bracket added to the structure by a sub-contractor.

Most of these failures are well-known events and considerable time, money and effort (usually by government authorities), has been spent on studying such events so that future possible failures are prevented. A common objective is that failures need to be understood. However, it is often a specialist task to establish the true causes of failure in complex systems. This frequently requires a multidisciplinary approach and expertise in mechanical engineering, material science, metallurgy, mathematics and non-destructive testing. The above well-known examples of failures demonstrate the need for a systematic, comprehensive approach to failure investigations.

In the automotive industry there has not been such public analysis of failures, for obvious commercial reasons. However, some examples may be found in the literature. One particular example can be found (Devlukia and Davies, 1985), where, although it was not disclosed in the technical paper, the problem considered had actually involved over £1 million in warranty costs. It is also prudent to point out that there are several published papers on automotive failures that hide the fact that they involved undisclosed large financial loss. If motor companies spend significant funds on projects the justification is financial reward, i.e. the solution to problems that are currently causing financial loss, either in terms of warranty claims or loss of customer confidence.

15.2.4 Statistical nature of reliability and failures

Preventing failures and guaranteeing reliability has been a major human endeavour since prehistoric times. The pyramids at Gisa are often quoted as an example of stone-age durability. The Chariots of Tutenkamun in the Cairo museum confirm the long history of axle and hub design.

The scientific foundation of reliability came much later and is attributed to Von Braun during the Second World War development of the V2 rocket. A further military association with the history of reliability was provided in the Korean War where USA electronic equipment was found to lack adequate reliability. After the war this led to significant expenditure to improve the reliability of electronic systems and components (it may be noted from Figure 15.1, that the reliability of electrical systems is still an important consideration).

It is important to appreciate that a large volume of the background to reliability was established in the electronics field. Several mechanical engineers subsequently tried to adapt the techniques to mechanical survivability although it was established that with mechanical components there were some important differences (Carter, 1972; Bompas-Smith 1973). Random failures could be eliminated in mechanical components by correct design. However, there was much greater uncertainty regarding the duty cycle imposed on mechanical components, which would influence the statistical distribution of failure times.

The awareness that we are dealing with failure distributions (known as scatter) emphasizes that the whole area of life predictions is not deterministic but must be based on probabilistic models. Whilst some of the scatter may be attributed to poor experimental equipment and methodology, even when these areas are improved it is often still found that scatter persists, although it may be less pronounced. This remaining scatter may be attributed to a number of facts, e.g. lack of control of surface finish, of fabrication, or poor control of material purity.

It is also apparent that in many components the service loading is not constant, particularly in dynamic systems (e.g. under fatigue conditions). Thus the capability of the material and the service duty are both variable. In order to deal with this complex situation it is necessary to examine the various distributions contributing to the overall behaviour using statistical methods, such that reliability, e.g. for safety warranty periods, may be assessed quantitatively. An example of the need to use statistical methods is the volume of data collected by the use of wheel sensors to monitor the road load input into vehicles (Pountney and Dakin, 1998) and the monitoring of road surface longitudinal profile measurements (Kondo *et al.*, 1993; Belfiore *et al.*, 1996; Mimuro *et al.*, 1996). The service load/potential load input distributions obtained from these measurements are important because of the serious stresses that can be generated in the suspension and body structure due to road roughness. A thorough treatment of the statistical procedures, especially those associated with reliability, is beyond the scope of this chapter, but many excellent treatments can be found in the literature (Lipson and Sheth, 1972; O'Connor, 1981; Condra, 1993).

An additional important development to assist with guaranteeing mechanical reliability has been the developments in condition monitoring (Davies, 1998; Neale, 1975). For many years there has been an indication of water temperature and oil pressure but now there is a wide range of sensors monitoring the vehicle, for example, brake lining sensors, tyre pressure monitors and anti-knock accelerometers mounted on the engine.

The importance of a statistical approach may be outlined by considering a set of fatigue data generated under laboratory conditions and using a constant stress amplitude (Figure 15.3a). For nominally constant test conditions the number of cycles to failure is found to vary over a considerable range. Thus, for a typical set of strain input, producing an average fatigue life of 570 hrs, in the order of 25% of the specimens may be expected to have failed after 440 hrs. The cumulative probability plot for this data is shown in Figure 15.3b. When a more realistic service regime is considered, where the strain regime is varying on the 'standard', represented by the input values then the situation becomes far more complex and more advanced mathematics and statistics need to be adopted.

A simple picture of this situation may be obtained by considering the material to be characterized by the fatigue limit, or endurance limit (the stress level below which the component will not fail which is, of course, a distribution of values) and assuming the imposed stress amplitude to be characterized by a second distribution. If these distributions are completely separated, as shown in Figure 15.4, then no failures will occur. However, this is generally the exception rather than the rule. In most cases some overlap of the distributions will occur, which is also shown in Figure 15.4. In this case some failures must be expected to occur. Thus the reliability is not 100%, and this situation must be analyzed quantitatively before the safety, or cost effective

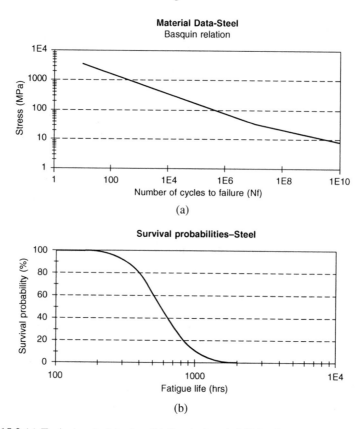

Figure 15.3 (a) Typical materials data (b) Survival probabilities for a typical strain history

warranty period, can be assessed. Based on the predicted separation or overlap (interference) the safety margin can be calculated.

This simple model allows insight into the need to control the strength of a component to a narrow distribution to enable good predictions of service performance. (cf Taguchi in the 1990s (Roy, 1990; Dehnad, 1986) or Shewhart in the 1930s (Cutler, 1998) with the concept of stable and predictable manufacture). More complex analyses can be used to consider more realistic situations and excellent treatments are given in the literature.

Finally, it should be mentioned that the shape of the failure distribution curve can help identify the mode of failure. In practice failures with normal distributions are usually related to wear out failures. Fatigue failures are usually log normal on test pieces but can be extreme value distributions on real components.

15.2.5 The origins of failure and failure modes

The origins of failure
Throughout the history of failure investigations, numerous attempts have been made to provide classifications of the exact origin of failures. However, it should be appreciated that a failure

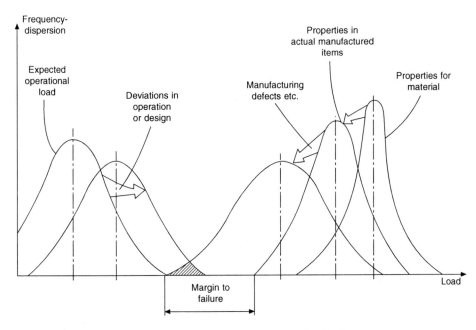

Figure 15.4 Chart of failure/strength distributions

usually has several contributing factors and it is often difficult to place it in one specific category. The main use of categorizing failures is to pin-point areas which require due diligence during the design and test periods of product development. They are also used to make a comparison of the failure rates in various industries (Drummer *et al.*, 1997) or for the reasons for the failure of a particular product such as a gear box (Penter and Lewis, 1990). In addition we can use these classifications in a positive way, since they clearly identify potential problems that exist at the design stage and by using better methods of design or manufacture, we can achieve improved reliability and less failures. (e.g. changing from gravity die casting to squeeze casting for the manufacture of an aluminium wheel rim.)

The majority of failures are the result of some type of fracture process that has initiated from an initial weakness or flaw in the component. Frequently failures contain sufficient evidence to indicate the cause, however, it often requires a detailed metallurgical analysis to identify the exact cause or causes. Well-known origins of failure may be listed as follows:

- Errors in design;
- Errors in materials selection;
- Materials defects;
- Fabrication/manufacturing defects;
- Deterioration of properties in service;
- Operator error/service abuse.

Each of these aspects will now be considered in more detail.

Errors in design

The influence of cost on design is universal. Cost not only controls materials and manufacture, but may also dictate and restrict the resources available at the design stage. Severely limited product development budgets often lead to designs that are extrapolations of past practice and service experience (often gained by trial and error at the expense of the user!). Thus in many instances failures are frequently the result of a lack of resources at the design stage.

Perhaps the most significant demonstration of this was presented in 'Prescription for disaster' by Trento (Trento, 1987). Trento points out that during the 'Cold war' period, President Kennedy set the challenge to be first on the moon to try to recover from being second into space. NASA was asked to prepare the budget. They established that $10 billion was required, which was increased to $20 billion by the Appropriations Committee to guarantee success. Ten years later a similar budget was requested for the Space Shuttle. The super power confrontation was on the decline. NASA's initial estimate for the space shuttle was $10 billion although only $5 billion was eventually approved. The subsequent failure of 'Challenger' has been blamed on this squeeze of project funds.

It would be impossible to imagine to a lesser extent, that similar situations do not occur in the automotive sector with financial control in the hands of financial executives and engineers involved in project planning and design.

One published account in the automotive sector concerns the initial versions of the Range Rover. Financial constraints led the original models to use a gear box developed for military vehicle applications despite the objections of the designer who disliked the decision since it reflected a lack of refinement for such a prestige vehicle.

Designs that are simply developments of existing forms are the most common. However, in some cases the design engineer may be required to work in areas where there is little past experience to draw upon and an incomplete knowledge of service requirements. This situation, typified by the Comet failures mentioned previously, can give wide scope for dramatic design error failures.

Designs often evolve with each repetition, correcting previous inadequacies. Automotive gears and gear boxes (Willin and Love, 1969; Boxter, 1990), drive shafts (Russel 1969; Walton and Prayoonrat, 1986) and CV joints (Macieinski, 1969; Dickenson, 1985; Sharpe, 1985) are good examples of design evolution, where failure analysis has identified the inherent weaknesses of particular components.

At a basic design level well known deficiencies which have constantly frustrated the design engineer include the following:

Designs include stress concentrations

Sharp notches should be avoided and generous radii should be used at a change of section. Similar guidelines are applicable for components that require heat treatment, or are manufactured as castings. Heat treatment stresses generated during the hardening process can result in cracking or unfavourable residual stress distributions that can be damaging in service. Some of the heat treatments such as carburizing, nitriding and induction hardening are actually selected because they can, in addition to local hardening, generate compressive residual stresses at the component surface which substantially improves the fatigue strength. The 'world best practice' for heat treatment quoted by Schronerberg (Schronerberg, 1986) is to avoid it, if at all possible, because it is an extra operation. However, in many automotive components it cannot be avoided.

Guidelines for avoidance of distortion and cracking during surface hardening have been outlined by Child (Child, 1982). Castings require similar considerations to ensure freedom from unacceptable internal porosity, segregation and surface cracking. These improved quality levels can be achieved by using processes such as the Cosworth Process (Lavington, 1986) for up-hill pouring of aluminium cylinder blocks, or shell moulding for brake discs and cam shafts (Curtis, 1992), which has recently increased in popularity.

Inadequate knowledge of service loads and working environments
It was not until the early 1950s that appropriate consideration was given to the actual measurements of service loads (Dally, 1991). As outlined in Section 15.4.2, there are now several methods to measure service loads.

Stress analysis of complex parts and complex service stress systems
Section 15.4.2 outlines some of the recent developments in computer based analysis techniques. Some of these programmes such as finite element based fatigue damage maps for full vehicle bodies require the computing power of large mainframe Cray computers (Kim *et al.*, 1992).

Errors in materials selection
In many classification systems this aspect would simply be considered under the classification of 'design error failures'. However, materials selection has traditionally been neglected or left to a late stage in the design process and is often an optimization exercise with one of the major constraints being cost. This has, of course, led to a wide range of failures due to the selection of inappropriate materials, or material combinations, some examples relevant to the automotive industry being outlined below.

- The SAE and ASM journals contain a number of technical papers on automotive gear failures which demonstrate the problems with incorrect material choice and provide guidelines to appropriate material selection (Alban, 1985).
- The material for the friction face of synchromesh cones has been well established as being a surface coating of molybdenum (Rosen *et al.*, 1970), however, failures still occur when this method is not adopted.
- Materials used inside engines for pistons, piston rings, valves, cylinder liners, etc which require tribological wear resistance have all required material optimization to ensure reliable engines (Monaghan, 1989; Banks and Lacy, 1989; Bovington and Hubbard, 1989; Bell and Delargy, 1989; Hill et al., 1996; Erickson, 1980).
- Steels for automotive body structures have undergone considerable development to ensure freedom from corrosion (Uchida, 1991; Matthews, 1991; Teulon, 1991), fatigue failure (Lewis, 1996), and to ensure good spot welding capability (Rivett, 1979; Johnson, 1975; Mitchell, 1975; Powell, 1995), especially as the competition from aluminium becomes more intense (Davies and Goodyer, 1991).
- Steel selection for drive line half shafts to ensure optimum resistance from brittle failure and fatigue failure often demonstrate the 'trial and error' approach. Breen (Breen and Wene, 1975) concluded that induction hardened medium carbon steel was superior to through hardened alloy steel, whilst Ross concluded that a 0.5% carbon 1% chromium, 0.2% vanadium spring steel, through hardened to 1500 Nmm^{-2} and locally nitrided in the splined area, was the optimum choice for a racing car drive shaft (Ross, 1995).

In addition it should be remembered that material selection is often strongly linked with other aspects of manufacturing such as surface engineering and heat treatment, which are continually being improved. For example, in the past 10 years the available techniques of surface engineering, to provide improved wear and corrosion resistance, have undergone significant development which has in many ways made the selection task more difficult.

Material selection has also often been optimized with 'trial and error' procedures, although in the view of many engineers, there is too much choice. Even though materials rationalization can provide substantial cost savings, it has been suggested that there is little economic justification for changing a specification which does not show a 20% margin of improvement in some vital property over an alternative (Lofthouse, 1969). Recent developments of quantitative material selection procedures, backed up with comprehensive software packages (Cebon and Ashby, 1992; Waterman *et al.*, 1992) should allow integration of materials selection into the iterative design process. It is hoped that this will reduce the number of failures due to poor materials selection.

Materials defects

All engineering materials are produced in a price competitive environment and have to be sold at a profit. This often means that they are not 'pure' or premium grade but are commercial grade and often contain various impurities or inclusions. These inclusions can have a major influence on the behaviour of the material (for example manganese sulphide inclusions in steels may act as sites for the initiation of cracks that may eventually lead to fracture). Consequently, processes have been developed to control the quantity, size and shape of inclusions, and steel making methods have been developed to reduce the number of inclusions, although this generally results in a more expensive material. Thus, suitable materials must be specified and used if failure is to be avoided, whilst still maintaining a competitive price.

Inclusions were found to play a part in the failure of the Alexander Kielland offshore oil platform mentioned earlier. However, a more relevant example for the automotive industry would be the development of high fatigue endurance bearing steels, steels for cold forming and improved sheet steel to enable weight saving and improve fuel consumption (Lagneborg, 1991; Matthews, 1994).

Examples of defective automotive wheel bearings have been highlighted by several manufacturers recalls (*Daily Telegraph Book of Motoring*, 1997) such as the 1994 Skoda Favorit recalls (i.e. two) for potential wheel bearing failures. Also in 1994 Suzuki Vitara (Sept 1993–July1994) recalled vehicles due to potential wheel bearing failures.

It has been well established that the endurance of ball and roller bearings is dependent on the volume fraction and type of non-metallic inclusions. In service the rolling action of the balls and rollers against the raceway generates Hertizian stresses which result in tensile shear stresses just below the surface. Thus, inclusions which generate local stress fields due to major differences in thermal expansion coefficients, such as alumina, are particularly damaging and are avoided during the making of bearing steels.

The automotive industry requires high quality high endurance bearings which means that the bearing steel must be of high quality. Frequently the total vacuum fusion oxygen content is used to indicate the level of oxide inclusions that remain in the steel. Twenty years ago 20 ppm would be acceptable but now levels of less than 10 ppm are specified with some recent specifications as low as 6 ppm. In 1988 American Society of Testing and Materials (ASTM) published a major review of manufacturing and endurance and causes of failure of bearing steels (Hoo, 1988).

One of the difficulties in bearing selection is that the B10 life is often shown on bearing endurance data. The B10 value indicates when 10% of the bearings have failed, although for warranty claims the designer will need to know the operating conditions for 0.5% and 1% bearing failures. A Weibull probability plot can provide a better display of fatigue data and the intercept on the x-axis can be used to predict the first failure with the slope providing an indication of the distribution of failure times (see Figure 15.5).

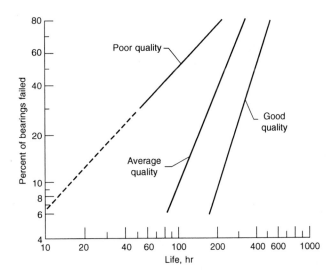

Figure 15.5 Fatigue life distribution of roller bearings made from carburized steel grouped into three quality ratings based on ultrasonic response

Fabrication/manufacturing defects
Fabrication/manufacturing operations involve a wide range of processes, and given this it is hardly surprising that numerous possibilities exist for error. For example the casting temperature may be incorrect, the heat treatment temperature or time may be incorrect, bad welds may be produced. The majority of these problems should be identified prior to service if good quality control systems are used. In addition it should be the responsibility of the design engineer to select methods of manufacture that can provide the correct quality level. However, in the automotive industry even when acknowledged best methods of manufacture are chosen and combined with good inspection methods, some defective items find their way to the customer. This is demonstrated by about 20 official recalls of vehicles per year (Daily Telegraph Book of Motoring, 1997).

To reduce the number of recalls it is necessary to have better quality products through effective product and process improvement (Bralla, 1986). This would involve improved process capability (Owen, 1989), and possibly better Non Destructive Examination (NDE) methods.

Deterioration of properties in service
In the majority of cases it is not sufficient for a component to operate satisfactorily when initially put into service, it must also operate satisfactorily over a specified period of time,

known as the service life. A number of factors may be identified which can produce a degradation in properties during the service life. These include:

(i)   environmental factors (e.g. corrosion, elevated temperature);
(ii)   wear;
(iii)   fatigue (i.e. dynamic, or variable, loading conditions).

These factors will be considered in more detail in the next section. However, to illustrate the effect of 'in service' deterioration of properties, consider the corrosive conditions to which a car, running in typical British conditions, is subjected, conditions which are very complex and difficult to predict. Hence, it is hardly surprising that many automotive failures have resulted from corrosion. For example, during the early 1960s Fiat cars acquired a notoriety for excessive rusting. Fiat now guarantees no rust through for eight years. The Tipo, Punto, Brava and Bravo are all manufactured from part electro-galvanized and part hot dip galvanized steels.

Operator error/service abuse
Operator error, or service abuse, contributes to a surprisingly large number of industrial failures. Typical examples range from the dumper trucks which have had additional supports welded on to allow extra loads to be carried (but resulting in premature fatigue failures of axles) to poorly maintained lubrication systems producing bearing failures. Once again, the lack of expenditure particularly on aspects of essential maintenance is often the root cause of these failures!

Summary
The various examples outlined above demonstrate the need to use appropriate design skills, accurate materials information, detailed manufacturing specifications and maintenance procedures. It must be remembered that even small component failures, not resulting in major injury or loss of life, can have major influences on the reputation of a company, from which it is often difficult to recover.

The modes of failure
The different modes of failure have been introduced previously, and are extensively described in the literature (*Metals Handbook*, 1988), such that in the present text only a brief explanation of each type of failure will be given.

Plastic instability
This type of failure results when deformation (plastic and/or elastic) produced by the service loading is sufficient that the system can no longer perform satisfactorily, although fracture does not occur. A typical example could be the plastic deformation of gear teeth that have inadequate strength or the buckling of car door impact bars.

Ductile fracture
Ductile fracture is considered to occur when extensive plastic deformation accompanies the fracture. In many cases this plastic deformation occurs on a macroscale, that is in a large volume of material, although in some cases only a localized region is affected. Obviously the first case is more easily identified, such that remedial action may be taken before actual fracture occurs.

Since fracture involves a large amount of plastic deformation, it is generally associated with service overload.

Brittle fracture

Brittle fracture is the opposite end of the spectrum to ductile fracture. In the worst case fracture occurs well below the yield stress, while the material is still exhibiting elastic deformation. This is the case for certain types of material, e.g. glasses and ceramics, but for metallic materials small amounts of plastic deformation are generally involved, although this deformation may occur on a very localized scale. In industrial failures brittle fractures are often found to be associated with flaws, e.g. cracks produced by welding, or high levels of residual stress.

It is often necessary to differentiate between the engineer's use of the term brittle fracture and the definition based on the metallographic appearance of the fracture surface at high magnification. When the engineer views the macroscopic appearance of the fracture and no plastic deformation is present the engineer will refer to the fracture as brittle fracture. However, at a microscopic level the metallographer, using the scanning electron microscope (SEM) at high magnification, can identify microstructural features which confirm either ductile fracture mode (Dimples) or a brittle fracture mode (Cleavage). A metallographer may, therefore, refer to a fracture with little macroscopic deformation as ductile if the microscopical features indicate this mechanism, even though the engineer may refer to the fracture as being brittle.

The absence of large scale plastic deformation means that impending fracture is rarely expected and many descriptions of catastrophic fractures can be found in the literature (Vander Voort, 1976).

Fatigue

When a material is subjected to a varying load it may eventually fracture, even when the maximum stress is considerably less than the macroscopic yield stress measured in a tensile test. This behaviour is known as fatigue and it is apparent in many areas of modern life, ranging from the failure of paper clips being absent mindedly bent, to the high technology components in modern aircraft and automobiles. In fact there are few areas of engineering where fatigue is not relevant and it has been estimated that some 80–90% of failures in metallic components arise from fatigue. In view of the importance of this mode of failure in the automotive sector the Society of Automotive Engineers has published a *Fatigue Design Handbook* (Rice *et al.*, 1988). This book provides comprehensive details on all aspects of the fatigue damage of automotive components.

Fatigue fracture may be considered to take place in three stages, namely

Crack nucleation
Crack growth
Fracture

Initially, very localized plastic deformation nucleates cracks, often at external or internal stress concentrations (e.g. surface machine grooves or inclusions inside the material). These cracks grow, in service, until a critical size is reached, leading to final fracture in a ductile or brittle manner. In some cases it is possible that cracks already exist in the material, e.g. cracks resulting from poor welding operations, such that the nucleation stage may be by-passed, with a consequent reduction in the service life.

Examination of fatigue fractures reveals many characteristic features, which help to identify the type of fracture in a given failure analysis. The first notable feature (shown in Figure 15.6) is the surface often appearing to the naked eye, to be flat, with little plastic deformation being apparent. A second feature, is the appearance of lines on the fracture surface. These lines are often referred to as 'beach marks' or 'clam shell marks', and are thought to be due to corrosion occurring during times when the crack is not propagating. For example, periods between shifts in a machine used on a production line. A typical example of these beach marks is shown in Figure 15.6.

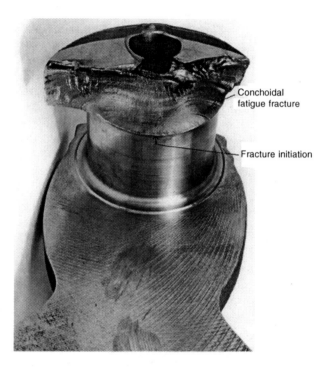

Figure 15.6 Engine crankshaft failure showing the position and type of fracture

Service conditions may be such that the fatigue is combined with additional factors, such as corrosion, and this may have a considerable influence on the service life.

Creep
When a stress is applied to a material a virtually instantaneous strain occurs, followed by a time dependent strain, which may eventually lead to fracture. This time dependent deformation is known as creep. However, whilst creep can occur at all temperatures it is only found to be significant at temperatures greater than approximately $0.4\ T_m$ (where T_m is the melting temperature in Kelvins). It may, therefore, be noted that for most metallic and ceramic materials creep is only significant at temperatures well in excess of room temperature. Hence, for the automotive industry creep in metals and ceramics is relatively unimportant. However, many plastic materials

have much lower melting temperatures, such that creep may need to be considered when dealing with this group of materials, e.g. in underbonnet conditions.

One example of the creep in metals causing problems in the Automotive industry has been reported by Kaye and Street (Kaye and Street, 1982). They report that Volkswagen, who pioneered the use of magnesium die-casting alloys in the automotive industry, found problems with crank cases manufactured from magnesium alloys. As engine power outputs were increased, the crank cases of air cooled engines were raised to temperatures of the order 150 °C, which is in the region where creep can occur in these relatively low melting point alloys. When this elevated temperature is combined with high stresses around steel stud bolts, due to the difference in thermal expansion between the magnesium alloy and the steel, creep deformation occurs, resulting in loosening of the bolts.

These problems led to the development of magnesium alloys with greater creep resistance, such as AS21 (containing 1.9–2.5%Al, 0.15–0.20%Zn, 0.7–1.2%Si and maximum levels of 0.35% and 0.4% for Mn and Cu respectively).

Wear
When materials are in contact and subjected to relative motion, deterioration and material removal can take place in the contact zone. The amount of wear that takes place will depend upon the materials involved, the pressure/load exerted on the materials, the surface roughness, the presence of oxide or lubricant, the geometry, the relative speed of the components and the environmental conditions such as temperature and atmosphere.

Wear is recognized as an important mode of failure in metallic systems and the prevention of wear, or the repair of worn components can provide large economic benefits (Hutchings, 1992 and *ASM Source Book*, 1975). An obvious solution to prevent wear is the use of an appropriate lubricant to ensure that the surfaces do not make contact. In many situations this is not possible to achieve and an understanding of the range of wear mechanisms that can take place is an essential step in minimizing the wear that takes place. Many of the methods to do this have been developed by T. S. Eyre and associates during the past 40 years. These include metallographic examination of the wear zone (Wilson and Eyre, 1969), wear testing (Eyre and Davis, 1993; Plint and Partners, 1997) and wear mapping (Kato *et al.*, 1994). Surface engineering techniques play a major role by allowing the development of wear resistant and low coefficient of friction surfaces.

In the automotive context a very interesting quote from 1967 by W. J. Arrol (Arrol, 1967) is appropriate.

> 'Probably the unkindest statement of the effect of material science on design is the well-known remark that "the internal combustion engine represents the triumph of metallurgy over mechanical design". This is the sort of statement which is partially true; sufficiently so to be uncomfortable. In the last 30 years the ordinary petrol engine of about 1–1.5 litres has gone from the state where it would give 20 bhp/litre to the present day of 45 or upwards. In 1935 it was common to rebore the engine and to put in new pistons and new valves and by then the engine had been decarbonized several times. Modern engines frequently do 50 000 miles without being opened up at all. This clearly represents a big improvement.'

The improved level of performance, endurance and reduced wear of current engines confirms the success achieved by the engineers and companies involved with engine development.

Several of the components inside the engine and transmission have undergone detailed development to minimize wear.

- Cylinder liners and piston rings (Banks and Lacy, 1989; Bell and Delargy, 1989; Hill *et al.* 1996; Erickson, 1980; Hartfield Wunsch *et al.*, 1996; Ting, 1996; Alcraft, 1984);
- Valve train including tappet and camshaft (Bovington and Hubbard, 1989; Dowson *et al.*, 1990; Dowson *et al.*, 1992; Dowson, 1996);
- Engine valves (Matlock, 1980; Beddoes, 1983);
- Crankshafts and associated components (Monaghan, 1989);
- Clutch materials (Burton and Senior, 1969);
- Synchromesh cones (Rosen *et al.*, 1970; Haigh *et al.*, 1989; Laird and Gregory, 1990);
- Gears, bearings and shafts (Willin and Love, 1969; Russel, 1969; Walton and Prayoonrat, 1986; British Gear Association, 1992; Miwo *et al.*, 1987).

The basic wear mechanisms can be described as follows:

Adhesion – This process occurs during sliding where local plastic deformation and the heat due to high frictional forces can result in tiny areas of 'welding' followed by subsequent breaking which can result in metal transfer.

Abrasion – This is often sub-divided between two particle abrasion and three particle abrasion. In two particle abrasion the asperities of the harder material abrade the softer material. The harder surface shows little signs of damage whereas the softer surface shows grooves, scores or scratches. In three particle abrasion 'foreign' particles are trapped between two moving surfaces which are in contact and abrade both surfaces. Soft surfaces can allow the 'foreign' particles to become embedded, this occurs for example in tin and lead based hydrodynamic bearings. A form of low stress abrasion can also be referred to as erosion where the stresses are mainly due to velocity. Examples include applications where components such as pipes are handling sand slurries or parts are exposed to airborne abrasives or flowing liquids containing abrasive particles. The term gouging wear implies the removal of relatively large particles by a method similar to metal machining or grinding.

Delamination – This process occurs to the softer material which, after repeated asperity contact becomes smoother and then cracks nucleate in the sub-surface zones causing thin flat delaminations often detached from the surface. The harder surface undergoes surface plastic deformation.

Pitting fatigue – This occurs during rolling action between two components such as in cams, gears and bearings which generates sub-surface Hertizian stresses. This occurs during gear tooth contact where the once per revolution repeated cyclic loading/unloading causes sub-surface shear stresses which can initiate small fatigue cracks which propagate to the surface and give rise to the removal of small particles of metal. This pitting causes a deterioration in the gear profile and requires either a limit to be placed on the level of contact stress or the use of surface hardening techniques such as carburizing, induction hardening or nitriding.

Fretting – Fretting occurs when two surfaces in contact under load and nominally at rest with respect to one another are caused to undergo slight periodic motion of a very small amplitude. The amplitude of the movement would be typically 20 micron and would be caused by the transmission of vibration. The major importance of this type of failure is that it often leads to the development of fatigue cracks (Taylor, 1996).

One of the major problems of wear is that, being a surface reaction, only a small amount of wear is required to completely disable a component. Wear frequently leads to additional displacements in gears, bearings and shafts that can introduce further dynamic stresses and other failure modes. It has been calculated that an automobile weighing about 1000 kg can be classed as 'worn out' when only 1 kg has been worn from the sliding surfaces (Grawne, 1996).

Corrosion

Corrosion is the result of a chemical reaction at the interface between the material and the operating environment. Most metals are found in nature combined with either oxygen or sulphur and undergo extraction and refining processes to separate the metal. To achieve this separation, energy in the form of heat and chemical potential, are supplied. In service metals frequently revert back to the combined state by the processes known as corrosion. If we can effectively separate the metal from the environment by the use of paints and coatings we can prevent the corrosion. The use of paint systems to protect automotive products is a well-established technique. However, R&D work is still undertaken by major automotive companies to validate their paint systems. Nissan recently announced that they had dedicated six man-years to the effect of different bird droppings throughout the world on the paint systems used on their cars. Because birds feed on different fruits and deposit the digested waste on cars in different temperature and humidity regimes it was necessary to validate the paint systems in all conditions where their vehicles may be sold.

In corrosion reactions we are dealing with a chemical reaction, the application of chemical thermodynamics and reaction kinetics which can predict the rate, likelihood and extent of a corrosion reaction. Consequently, corrosion engineering has a strong scientific basis that is combined with an empirical approach to document corrosion rates for a wide range of metals in an extensive range of environments. Most environments are corrosive but the actual corrosion rate varies for a given metal and for a given location. In general terms rural inland environments cause the lowest corrosion rates whereas coastal, industrial environments would provide the worst corrosion rates.

An Institute of Mechanical Engineers conference held in 1974 (ImechE Conference, 1974) contained several papers that highlighted the corrosion problems experienced at that time, many of which are now solved, due to the significant improvements in the 80s and 90s (Trethewey and Chamberlaine, 1995).

The American Society of Metals (ASM) classifies the various forms of corrosion as:

Wet corrosion – Taking place in the presence of moisture.
Dry corrosion – Occurring in the absence of liquids. Dry corrosion would be the result of vapours and gases and is generally associated with high temperatures.
Direct corrosion – The direct combination, usually at high temperatures, between a metal and non metal such as steam, oxygen, sulphur and chlorine.
Electrochemical corrosion – Taking place in the presence of an electrolyte such as water, sea water, acids, etc. In electrochemical corrosion part of the metal behaves as an anode and introduces ions into the electrolyte (corrodes) and part of the metal or a metal in electrical contact acts as a cathode and accepts and disposes of electrons by a cathodic reaction such as hydrogen evolution. The major cause of corrosion damage is achieved by this mechanism.

In addition the common types of corrosion found in service could be listed as:

Uniform or general corrosion – This is the form of corrosion that most metals experience in the atmosphere. Corrosion rates and expected service life are well documented together with protective measures by paints and coatings.

Galvanic corrosion – When dissimilar metals are in electrical contact in an electrolyte the less noble metal acts as an anode (corrodes) and the noble metal has a degree of protection. These conditions arise as a result of poor design and should be avoided. However, this technique is used to advantage in automotive engineering by coating steel with zinc to create cathodic protection of the steel. This is extensively used to protect car bodies from corrosion. Dissimilar metals in the cooling system can also suffer galvanic corrosion (Baboian *et al.*, 1988).

Concentration cell corrosion – This occurs in an electrolyte due to different concentrations of ions at different locations. The most important corrosion mechanism of this type is known as differential aeration due to different oxygen levels in the electrolyte.

Pitting corrosion – This is a very localized corrosion and can cause failure by perforation in a relatively short period of time. There are a number of causes of pitting such as local inhomogeneities, mechanical or chemical rupture of a passive film or protective oxide, or the creation of a concentration cell beneath a deposit.

Dezincification – This occurs in brasses which contain less than 85% copper. The zinc is selectively corroded to leave a porous residue of copper with little structural strength.

Intergranular corrosion – This occurs when grain boundaries become more chemically active than the body of the grain and corrode preferentially. The classic example is the sensitizing of austenitic stainless steel due to carbide precipitation and thus chromium depletion in the Heat Affected Zone (HAZ) of welds.

Stress corrosion cracking – This is the combination of tensile stress and a specific corrosion environment.

Erosion-corrosion – The combination of mechanical wear and corrosion.

Crevice corrosion – The formation of a crevice at a mechanical joint or beneath a deposit provides the conditions for a concentration cell. It is usually associated with differential aeration. This results in very rapid corrosion and should be avoided.

It is clear from the above that even if it is possible to produce defect free components and structures, in many cases defects can be produced in service by processes such as fatigue or corrosion. It would, therefore, be useful if a methodology existed by which the engineer could assess the possible damaging effect of a given defect. Realistic estimates could then be made for the service life and the maintenance schedule.

Over the last 20–30 years attempts have been made to develop this methodology, using what has become known as 'fracture mechanics' (Broek, 1994). This topic, which has grown out of the pioneering work on brittle fracture by Griffith, allows the engineer to make a realistic estimate of the effect of a defect of a given size on the performance of a given component or structure. However, extensive calculations are generally necessary and it is not surprising to find that the development of this topic has closely followed developments in computer hardware.

15.3 Testing and failure prediction

15.3.1 Test philosophy

Baker (1986) has provided an excellent general review of the background to testing for quality and reliability. One of the important aspects of any test is the balance between 'relevance' and 'cost of the test'. It is important to strive for tests which are relevant to a component's future performance and can be carried out at a relatively low cost. Figure 15.7 summarizes this concept.

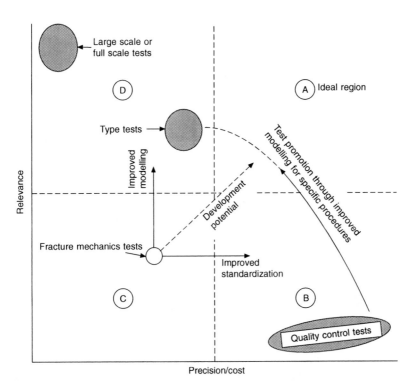

Figure 15.7 Chart showing the balance between 'relevance' and 'cost' of the test

In addition it is important to understand the broad range of tests that are available to the engineer and can be used to ensure reliability and to prevent failures. The main methods can be summarized as follows:

- *Materials tests* These are carried out to characterize the properties of a material and to measure the variability with various processing conditions. This would include normal mechanical properties such as strength, toughness, hardness, fatigue strength. The tests are carried out on specific specimen geometry and controlled test conditions. As a result the relevance of these measured values on the performance of a component can be vague.

- *Manufacturing and fabrication tests* These are necessary to validate the methods of manufacture such as welding, metal forming and machining. In some cases they can be mandatory tests such as in welding, where it is essential to obtain both welder and weld procedure approval.
- *Design validation test.* This typically involves the testing of prototype components. It has the objective of providing proof that the design will provide reliable service performance and endurance. This is a key area in the product evaluation stage. The British Standard 5760 provides guidance on procedures. When it is necessary to accelerate the tests special precautions must be taken (Condra, 1993).
- *Quality assurance tests.* These tests establish that the materials used are to the appropriate specification requirements and that all the specified design and manufacturing requirements have been met. It also involves testing the significance of defects (usually using Fracture Mechanics methods) to determine whether it is appropriate to remove the defects and to carry out a repair, to reject and replace the component or indeed to accept the component (BSI PD6493).

The above comments provide a useful means of classification to discriminate between various tests. However, a number of problems may immediately be identified:

- How can the actual service conditions be simplified for testing?
- How are the 'simplified' service conditions related to actual service?
- For service over a number of years, can the test be speeded up?
- If accelerated testing is carried out is the test still relevant to service?
- How reproducible are the results?

It is intended that these and other questions to do with testing and service behaviour be considered in the present section.

15.3.2 Materials testing

In an ideal world it would be possible to have a number of simple tests to measure material properties, from which the behaviour of the material could be predicted under any given service conditions. Unfortunately, this situation can only be realised in a few simple cases, normally additional factors operating in service tend to make the behaviour more complex and difficult to predict. However, there are a number of laboratory tests that aim to test whether materials conform to their purchase specification. In addition they are used to characterize materials so that quantitative information can be used in material selection.

A wide variety of standard mechanical and corrosion tests are available to measure material properties. These range from the well established test procedures such as tensile testing and hardness testing to the more modern methods such as fracture toughness testing, crack growth and CTOD (crack tip opening displacement).

Extensive descriptions of the test equipment, methods and parameters obtained from the test are given in the appropriate British or ASTM standard.

15.3.3 Automotive component/vehicle testing

Present-day market conditions are forcing manufacturers to reduce the time to develop and market new products, to reduce costs with value analysis and to produce higher standards of reliability to satisfy the expectations of consumers. Modern analytical techniques may go some way to easing the problems faced by manufacturers, but complex products are often difficult or impossible to model accurately. Thus, products still need to be tested in order to back up predictions of the analytical models and to answer questions that they simply cannot answer satisfactorily. However, testing has had to undergo a major rethink to accommodate the modern approach to manufacturing. The traditional method of testing a number of prototypes, often under somewhat idealized conditions, has been (or, in some cases, is still being) replaced by testing under more realistic 'service conditions' determined from refined data acquisition techniques.

Two forms of product testing may be identified, namely, performance testing and endurance testing. Performance testing evaluates the functional and performance characteristics of the product against the product design specification (PDS) and unless these characteristics are fully met no further testing is carried out. Endurance testing evaluates the durability characteristics of the product. This may involve:

- Laboratory testing;
- Proving ground testing;
- Field testing.

A recent review of the current methods of accelerated vehicle durability tests has been carried out by L. J. Niemand *et al.* (Niemand and Wannenburg, 1997) which compares:

(i)   Pascar used in North America.
(ii) Carlos a European standardized laboratory load sequence.
(iii) Telco used in India and based on CG accelerations.
(iv) A Japanese programme using strain gauge data and rainflow counting.
(v)   The Two Parameter approach based on bivariate statistics (the km/day and the data/km).

The main conclusion was a preference for the Pascar and the Two Parameter approach.

Laboratory testing
Testing is generally performed on individual components and sub-assemblies, using specially designed test fixtures. These fixtures are generally designed to accelerate the time to failure, so that deficiencies may be identified and corrected quickly. However, the test conditions must ensure that the failure mode in the laboratory test is the same as in service, otherwise the laboratory test is not relevant to service behaviour.

Much of the testing involves consideration of fatigue and modern-day test programmes attempt to simulate the service conditions as accurately as possible, although some data manipulation may still be required in order to accelerate failure.

The generally accepted rules for accelerated fatigue testing can be summarized as follows:

- The low range of load cycles of less than 50% of the endurance limit could be eliminated from the test signal. Some sources such as Gleeson in gear applications, suggest that load cycles up to 75% of the endurance limit can be removed from the test signal without affecting the result. A recent SAE publication has confirmed the 50% value for steel components (DuQuesray *et al.*, 1996);
- The frequency of the stress/load cycles can be increased for metal components. For plastic components internal heating would alter the damage mechanism;
- A comparison of the relative damage of the two load levels, the service load and the reduced test load, can be made using Miner's summation of cumulative damage;
- The limit to the degree of acceleration is that a similar failure must occur with each level of stress. In addition there must be an awareness that 'coaxing' caused by the low range of stress cycles may increase the level of fatigue strength in some steels by up to 50%.

Proving ground testing (PGT)

Proving grounds are used to provide a simulation of 'worst case' field events. The major problems with proving ground testing are that not all field events can be simulated, individual components and subassemblies cannot be tested and generally the cost of testing is very high.

It is, therefore, apparent that this type of testing is somewhere between full field testing and laboratory testing. This has advantages in that the testing approaches aspects of the service behaviour, however the relationship with service behaviour may be difficult to assess if the frequency of the 'worst case' events in service is not precisely known.

A number of published methods allow a correlation to be established between PGT and field service data, i.e.

(i)   Linear programming techniques.
(ii)   Bivariate statistics or the two parameter method (Wannenburg, 1993).
(iii) Accelerometer plus the use of the least squares technique (Stockley and Devlukia, 1984).
(iv)   Historical company experience.

One of the most important principles in the use of these techniques is that they take time to develop. However, in general the actual production schedule cannot wait until everything is perfect. It is important to be able to assist with accelerated techniques even if they are based on incomplete information. It is under these circumstances that the principles of failure analysis become essential analysis techniques.

The bivariate technique looked initially attractive, however in some initial studies to apply this technique it was found to be difficult unless converging distributions were found.

A popular technique is to base accelerated tests on calculated fatigue damage using specialized software and Miner's rule.

Field testing

This is obviously the ideal way of testing products. However, this type of testing is very expensive and can take long periods of time (particularly for products designed to have long lives). For example, the design life of a heavy duty truck may be of the order of one million miles, which is obviously impractical for field testing.

It is a prudent practice to place prototype components into vehicles owned by haulage companies or the vehicles of a large customer and to monitor the load histories carefully to identify any deficiencies. This method requires good co-operation and confidentiality and an agreement to pay all costs associated with a failed component. This is a method often used in Europe to minimize the cost of component testing.

15.3.4 Failure prediction

To validate new component designs an ideal method is the use of computer aided techniques, which require appropriate models, algorithms and software.

Fatigue design philosophy has changed from one based on an 'endurance limit' and an 'infinite life' concept to methods based on establishing an assessment of the actual fatigue life. These methods have been developed to overcome the two classic problems in fatigue:

1. It is now too expensive to eliminate failures by 'over design' using large safety margins.
2. It is commercially damaging to have unexpected service failures.

These new analysis methods are computer based and therefore allow large volumes of data to be analyzed quickly and at low cost. The methods include:

(i)   Conventional S/N methods using a mean stress correction factor and component S/N fatigue curves.
(ii)   Local strain life algorithms based on the SAE methods developed in the 1970s. These methods can also be based on the resolved principle stress from tri-axial strain gauges.
(iii) Bi-axial fatigue algorithms which were developed to overcome the problems associated with the SAE local strain approach.
(iv)   Fatigue crack growth algorithms developed by Paris and subsequently modified to allow for crack closure and variable service loads.
(v)   Algorithms based on power spectral density and the proven relationship with rainflow counts.
(vi)   The combination of FEM and the various fatigue algorithms.

Each of these techniques has been developed to meet a particular problem in the broad field of potential fatigue failures. Aerospace, nuclear and power generation have preferred the crack growth methods based on damage tolerance concepts. Non -Destructive Testing (NDT) techniques establish the maximum defect size present and crack growth methods predict the service life where the crack would represent an imminent potential failure via brittle fracture. Power generation engineers have developed the R-code procedures for pressure vessels which predict whether a vessel will fail safely by developing a leak rather than an explosive fracture.

In general the automotive sector has preferred the other methods and has had a reluctance to accept the concept of cracked components and cracks growing. The importance of the crack nucleation stage for typical automotive applications (and hence the development of the local strain approach to fatigue) has been confirmed by recent studies that have found for high cycle fatigue about 85% of the cycles to failure are required for crack initiation whereas for low cycle fatigue 45% of the load cycles are required.

The combination of data acquisition and data analysis together with the use of engineering software, such as gear design software and computer based fatigue software, can provide accurate durability predictions. However, under complex load situations such as the rear axle of a vehicle, more complex analyses are involved. For example, in this case vertical forces are generated due to road load and horizontal forces arise from braking which can be uncorrelated and out of phase, such that bi-axial fatigue algorithms should be used. The use of computer software to predict fatigue life has undergone considerable developments in recent years, and some important aspects of these developments can be found in Section 15.4.2.

15.4 Automotive technology and the importance of avoiding failures

15.4.1 The automotive business climate

To understand the role of failure investigations in the automotive industry it is important to appreciate some of the relevant factors that are unique to this industry. Vehicles are designed and built for a wide range of applications and working environments. Vehicles range from high performance luxury cars to small compact cars and even Formula 1 high performance racing cars. In the commercial world there are a wide variety of vehicles, four wheel off-road, military, earth moving, trucks of up to 50 tonne as well as buses and coaches. The worldwide operating conditions are equally diverse with temperatures ranging from –50 deg to +50 deg C together with a wide variation in other climatic conditions, operating terrain and service duty. In fact the corrosive nature of the automotive working environment has been judged as one of the most severe of all engineering environments (Trethewey and Chamberlain, 1995).

The history of British car manufacture is well displayed in the British Heritage Museum. There are also several books which describe the eventful history of the British car industry and one of the most factual regarding technical and industrial developments is 'Cars in the UK' by Graham Robson (Robson, 1997).

This book follows the decline of the automotive industry in Britain from being the second largest manufacturer of cars in the 1950s, to the low levels of the 1970s and 1980s. The market share of UK manufacturers is shown in Table 15.1.

A similar decline occurred within the bus, coach and truck industry creating a constantly changing commercial environment.

The actual events behind these statistics are beyond the scope of this chapter, however, there are examples where the reliability of components has been poor, excessive failures have occurred and the reputations of manufacturers have been adversely affected.

Engine failures were a major factor in the demise of Jensen in the 1970s. A Lotus 2 litre engine was used on the Jensen-Healey for the American market. The subsequent unreliability of the Lotus engine damaged the image of the company and sales declined (Georgano, 1995).

In addition, similar problems were experienced with the Triumph Stag. Sales of the car dwindled after the 3 litre single overhead cam V8 engine displayed the classic symptoms of unreliability. The engine blew cylinder head gaskets and the single timing chain stretched in service eventually leading to major engine damage (Georgano, 1995).

In the 1950s the British bulldozer the Vickers Vigor VR180 was in direct competition with the Caterpillar D7 for several major roadway projects. The Vickers Vigor had transmission problems (Johnson, 1997) was quickly identified as unreliable and faded into obscurity. Meanwhile

Table 15.1 The market share of UK manufacturers from 1965 to 1979 (Robson, 1997)

Calendar Year	UK Market Share (%)
1965	44.5
1966	45.2
1967	40.7
1968	40.6
1969	40.2
1970	38.1
1971	40.2
1972	33.1
1973	31.9
1974	32.7
1975	30.9
1976	28.2
1977	24.3
1978	23.5
1979	19.6

the D7 was a commercial success and led to future generations of D8, D9 and company prosperity.

The complex nature of the automotive market, where subjective judgements of aesthetics, ride quality, handling behaviour, durability, fuel economy, safety and cost often determines which vehicles are purchased, has been the downfall of many automotive companies. Market research by truck and coach companies show that the individual driver can often have a major input into the final vehicle choice of even commercial vehicles. Consequently press reviews by magazines such as *What Car* and the motoring journalists can have a major impact on our purchasing habits. The launch of a new vehicle can involve expenditure of up to £500 million and therefore to obtain a return on investment, sales of over 400 000 units are needed. Failure to achieve the required sales, for whatever reason, can have a terminal effect on business prospects.

The automotive industry is now an industrial sector where failures are not tolerated and a quality level at least matching the competition is a basic requirement. A number of additional factors must also be considered.

(i) The very competitive market
The impact of the notion of 'the survival of the fittest' in the commercial environment was uniquely described by Nichols and Pye (Pye, 1992).

> 'Two industry executives were in the African bush, discussing the relative performance of their companies, when a ravenous lion appeared on the horizon. 'Run for it', said one executive. The other took a pair of running shoes from his rucksack and bent down to put them on. 'What are you doing', said the first executive. 'You can't out-run the lion'. 'I don't need to out-run the lion,' said the kneeling businessman 'I need to out-run you'."

By the year 2002 it is predicted that there will be 30 percent overcapacity in automotive manufacturing. Only the best companies will prosper and survive. Therefore the user perceived cost/performance/reliability are key factors. An example of the competitive nature of the market can be gained by considering coach prices. Currently coaches can be built and sold in the Far East for about £100K compared to £170K in Europe.

(ii) Significant environmental pressure
This environmental pressure necessitates:

- Good fuel economy with lightweight structures employing lightweight materials and efficient structural design;
- Reduced engine emissions;
- Planned material recycling.

These factors ensure that change has to occur to achieve improved business performance. Senior executives are searching for innovative parts, components and methods of manufacture but cannot accept any risk of reduced reliability. Therefore considerable attention has been given to understanding the service requirements, material capabilities and the development of computer software capable of predicting endurance.

The impressive performance of the Japanese motor industry during the late seventies and early eighties completely re-shaped European and American views on design, testing, reliability targets and manufacturing techniques. Prior to this new era the production of a new model required a 5–6 year period with several prototypes. The Japanese applied modern efficient methods and reduced this period to 3 years and thus acquired a substantial cost advantage.

The co-operation between Honda and Rover (by then virtually the only surviving UK major car producer) during the mid-eighties played a significant role in enabling Rover to achieve World Class performance. A published engineering review confirms the significant role Honda played in changing the philosophy of design and testing to help prevent component failures (Bertodo, 1989).

Rover was acquired by British Aerospace but continued to face hard times and in 1991 admitted defeat in the USA because of quality problems with their new 800 series and completely withdrew from the USA market (Stockley and Devlukia, 1984).

With the ownership of Rover being passed to BMW in 1996, and Leyland trucks going out of business with DAF, the UK owned companies are virtually restricted to component suppliers. However, Leyland Van did survive as a management takeover from the receiver and has steadily grown, announcing (in 1997) a joint design and development agreement with Daewoo. The demise of the British owned car industry could be judged a major industrial failure in itself, especially in view of the long history of British vehicle manufacture and the large increase in car building in the UK by foreign owned companies.

A further example of the dynamic nature of the Far East is the entry of Samsung into car manufacture where they have no previous experience. With incredible boldness they progressed from earthmoving off-road vehicles to truck and then, with bought in help from Nissan, to car manufacture (7.66 billion Yen in 1994). They have complete confidence in the capability, quality and performance of their engineers partly stimulated by good training and motivation. Unfortunately the collapse of the Asian currencies during the early part of 1998 has terminated many of these ambitious plans.

15.4.2 Failure investigations and failure prevention

The engineer involved in automotive technology requires an understanding of failure investigations for application in the following areas:

(i)   Forensic investigations in order to understand and establish liability and blame, and to ensure it does not recur.
(ii) During component/vehicle testing for performance and endurance.
(iii) Quality control testing during manufacture.
(iv) R&D for improved component manufacture, new materials, value engineering and the introduction of new products and manufacturing processes.

A current trend has been to focus attention on the 'Capital Asset' value associated with 'know-how'. A new concept in displaying the knowledge in an easily assimilated format is referred to as 'mind maps'. Figure 15.8 shows a mind map developed to outline the relationship between failure investigations and other aspects of the business.

Intelligent test equipment
The design, selection and use of test equipment has always been an important aspect of component and vehicle validation. Laboratory simulation testing of components has developed into the preferred test method of many vehicle and component manufacturers. Modern generation test equipment now involves the investment of large financial resources. Test equipment manufacturers now provide a vast range of equipment ranging from engine dynamometers, vehicle climatically controlled wind tunnels, transmission test rigs, road simulators, CV joint simulation rigs and a full range of component test rigs. These units aim to accurately reproduce the operating conditions (based on measured road load data) and are microprocessor controlled with automatic data acquisition, analysis and report generation.

Major software developments
The integration of Finite Element Methods (FEM) and Experimental Modal Analysis (EMA) (Brughmans *et al.*, 1994) and the linking of FEM to modern fatigue algorithms has allowed accurate modelling of in-service behaviour of components. For example, the use of 'rainflow counts', and the ability to convert these to Power Spectral Density (PSD) curves has allowed realistic service behaviour to be input to FEM models. (Fatigue damage of components depends upon both the range of stress variation and the actual mean value. The rainflow count was developed in the early seventies and is an algorithm used to generate a three-dimensional histogram plot of frequency of occurrence of the various combinations of range and mean strain generated from strain gauges. The various fatigue algorithms can usually use the rainflow count as input data and can predict the fatigue damage and component life resulting from the measured input of stress/strain. A typical analysis procedure would generate the data shown in Figure 15.11.)

Proving grounds
The use of proving grounds and the accurate correlation between real vehicle use and proving grounds and simulation tests. An example of the financial investment/commitment needed to

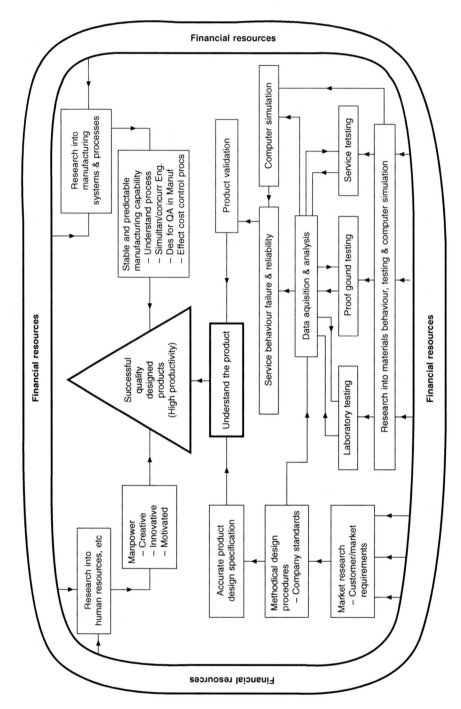

Figure 15.8 A mind map that has been developed to outline the relationship between failure investigations and other aspects of business

provide modern generation world class R&D and test facilities in the automotive industry is the new technical centre build by Hyundai at Namyang. Figure 15.9 shows the layout of the facility which represents an investment of over $400 million.

Figure 15.9 View of Hyundai Research Centre at Namyang

Fatigue software
The use of fatigue life prediction software to allow strain gauge data from in service components to be edited, evaluated and by the use of various fatigue algorithms to provide an estimate of the components expected fatigue life provided a much improved method to study mechanical reliability (Musiol *et al.* 1981, Morrow and Socie, 1981). The current methods include the classical S/N method, the local strain method based on strain vs reversals, Biaxial methods, power spectral density methods using the Dirlec equation to link the PSD to the rainflow count and crack growth techniques based on either linear elastic methods or elastic/plastic fracture mechanics (Broek, 1994).

15.5 Case studies – typical examples of automotive failures

The best way to achieve growth and prosperity in the automotive industry is to make products that do not fail in service. We have outlined in previous parts of this chapter that failure prevention requires testing to evaluate the performance and endurance of components. This type of testing inevitably results in failure. A crucial step in product improvement is the 'failure analysis' which aims to establish the cause and to make cost effective recommendations to prevent similar failures. This is referred to as a deterministic approach to reliability and requires a detailed knowledge of the product and the specialist skills of failure examination.

The following case studies demonstrate many of the aspects already covered and are intended to give insight into the engineering difficulties faced in the automotive industry.

These examples are:

- Vehicle structural durability. This demonstrates the importance of control of weld quality. In addition, it outlines the difficulty of establishing test procedures which can realistically simulate service behaviour.
- Wishbone link failure. The provision of replacement and spare parts is a business sector of substantial value. Unfortunately the OEM (original equipment manufacturer) parts are often in competition with lower quality, lower cost replacement parts. This example outlines a failure caused by very poor quality none OEM spare parts.
- Drive line durability and engine crankshaft failure. This example outlines the complex combination of material, methods of manufacture, design and dynamic service stresses that can make failure investigations difficult. The failure of an engine crankshaft occurred after 120 000 km compared to the target durability of 500 000 km.
- Recovery vehicle boom failure. This example outlines a poor quality repair carried out on the boom of a recovery vehicle.

15.5.1 Vehicle structural durability

Introduction
During proving ground prototype testing of a mini-bus on Belgian Pave various cracks developed adjacent to weld deposits on the upper part of the window pillars and around the door pillar areas. The test had been carried out for over 1000 miles and the unexpected appearance of the cracks caused serious concern.

The immediate task was to establish whether it was:

(i)   A design problem
(ii)   A manufacturing problem
(iii) A test severity problem

The investigation involved a detailed information review, visual and metallographic inspection and trial work with a strain gauged bus, followed by fatigue modelling, using computer software.

Vehicle body structures are required to provide guaranteed structural durability for substantial periods of time. Although warranties are typically 3–6 years, service lives can be over 12 years for cars and up to 15 years for buses, coaches and trucks. The vehicle structures are exposed to severe corrosive conditions combined with complex dynamic stresses.

The design of vehicle body structures has gradually evolved over the past 20 years partly due to the developments in computer hardware and software that has allowed structural optimization and refinement and partly due to improved instrumentation for vibration and fatigue evaluation.

The major factors in vehicle structures are:

- To provide sufficient stiffness to keep distortion to acceptable limits which allows control over the alignment of mechanical and suspension components;

- To keep stress levels low to avoid fatigue;
- To minimize structural weight to allow improved fuel consumption;
- To provide good NVH (noise vibration and harshness) performance. This requires high torsional and bending stiffness to prevent excitation from the load input by the road and the power train;
- To satisfy impact, crashworthiness and corrosion requirements.

Technical background

There is a substantial volume of relevant literature both historical and at the leading edge of technology. In the mid-70s Daimler Benz (Mischke, 1978) published a structural comparison and analysis of different bus structures. This work was one of the early applications of FEM and was used to evaluate the static and dynamic stresses in the structure. The vertical load inputs from the road irregularities occurred at frequencies of 0 to 20 Hz.

As computing power developed these concepts were further refined and more detailed predictions became possible. In 1986 DAF reported using the same software, Nastran, and similar load input (i.e. 0–20 Hz) to calculate the torsional and bending mode shapes that were likely (Van Asperen and Voets, 1992). This work demonstrated the effect of sidewall stiffness on the overall expected level of bus structural stiffness, and the level of vertical acceleration and allowed the selection of optimum torsion and bending stiffness. An important aspect of this work was the calibration of the FEM model with actual accelerometer measurements with the accuracy of the model being for static loads within 10% and the natural frequencies within 12%. Further papers by Mercedes-Benz (Schneider, 1992) demonstrated that the technology was capable of fully linking CAD to FEM, and full vibration, stress, and acoustic simulation and analysis. More recently, in 1994, LMS of Belgium claimed that they have effectively linked FEM and EMA (experimental modal analysis) creating an even more powerful simulation environment (Brughmans *et al.*, 1994).

Although this technology exists, it is very expensive to adopt. However, often a detailed study of this background work provides a much greater understanding of the potential problems, although major automotive companies with this type of problem will often be able to call on these resources.

Hungarian work by Matolcsy provided an excellent collection of papers over a period of 10 years from 1977 (Matolcsy, 1977; Matolcsy, 1984; Matolcsy, 1986). These papers provide details of several relevant aspects of the causes of cracking, the expected endurance performance of various joint design and welding techniques and, the important stress systems. These include:

(i)   The fatigue of window columns and door frames are mostly due to torsion.
(ii) The endurance limit for spot welded joints can be only 10–15% of the endurance limit of the basic material. With CO_2 welding this ratio can be 70–80%.
(iii) Typical bus lives can be over 10 years or 500 000 km. Accelerated testing on Pave introduces much greater than normal vertical bumps. Horizontal forces due to braking and acceleration, side loads due to manoeuvring as well as the change in passenger load are neglected. If performed incorrectly then an inadequate simulation would result.

An additional factor in the testing has been outlined by work in 1984. This work demonstrated that passengers were an active pay load and reduced the level of dynamic peak stress levels by 20–30% compared to a similar artificial pay load (e.g. iron weights) (Michelberger, 1984).

A similar review of factors affecting the production of consistent, high quality joints, examines welding procedures (including spot welding and CO_2), weld test procedures (including the strength of spot welds) and weld monitoring techniques. This work also included weld geometry, the problems with welding coated steels and the fatigue and corrosion fatigue properties of welds.

Data acquisition from the strain gauged mini-bus

Figure 15.10 shows the bus structure and the position of the strain gauges. A quarter bridge configuration with 360 ohm gauges was used. The strain gauge positions were based on previous experience of the likely positions for cracks to develop. Other methods could have been used such as FEA or brittle lacquers to establish the high stress areas. Timoney (Timoney, 1986) described a laboratory method where a suspension component from a heavy vehicle was lacquered then hydraulically loaded in the laboratory, and strain gauged to correctly record the maximum service stresses.

Figure 15.10 Strain gauge positions on bus

Dynamic strain data was measured at a sampling rate of 300 Hz and for a measuring period of about 100 secs during test drives on the Belgian Pave. Several runs were made to ensure representative samples were taken.

An analysis of the data was carried out using proprietary fatigue analysis software and the results are presented in Figure 15.11.

Metallurgical examination

Figure 15.12 shows a schematic drawing of the sample taken for metallurgical examination from cracks formed during the testing of the mini-bus. The structure was fabricated from roll

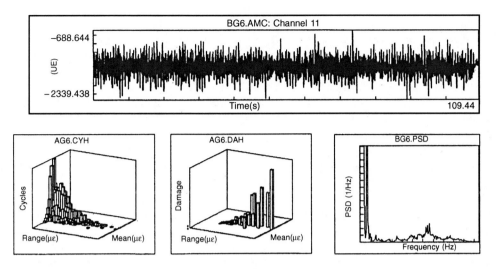

Figure 15.11 Results of a fatigue analysis using proprietary software

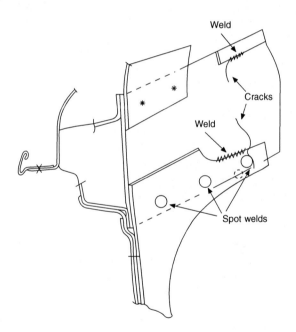

Figure 15.12 Position of cracks near window corner of bus structure

formed and pressed sections. The welds were either spot welds, or gas shielded puddle welds and fillet welds when access from both sides of the structure could not be obtained. Cracks were seen to have initiated from spot, puddle and fillet welds.

Metallographic work showed a very coarse grain size in the HAZ of the weld. The work indicated that the cracking problems were probably linked with poor quality welds.

Conclusions

The investigation established that the weld quality and the corrosion protection of the welds was not of a good standard. In addition the combination of vehicle speed and severity of the Pave in introducing resonance, high levels of torsion and bending into the structure. This represented a severe test compared with actual duty cycles for this mode of failure.

The problem was solved by the development of better weld procedures and ensuring better weld quality and establishing a better correlation between the test procedure and the service duty cycle.

15.5.2 Wishbone suspension failure

Introduction

The automotive spares market represents a substantial profitable market. The original equipment manufacturers (OEM) usually have a policy of cost plus a good commercial profit margin.

These attractive profit margins attract a number of lower cost inferior products which have been 'reverse engineered' from the original OEM part. Usually these companies do not have the resources, the metallurgical expertise or the test equipment to validate a component. Under these conditions 'rogue' replacement components can enter the supply chain. For example, during 1997 Vauxhall motors publicized that a number of replacement steering support components which had failed were not their official validated replacement part. The part contained an inadequate number of spot welds.

The failure examination described here was performed on a suspension wishbone for a BMW Series 3 car. To qualify for a Ministry of Transport (MoT) certificate the car required new suspension wishbones. Based on a strong endorsement by the supplier the garage had purchased non-OEM parts at about 80% of the price of OEM parts.

Subsequent to fitting the parts a short test drive was attempted. However, after a short drive of 20 metres to a speed of 15 mph followed by the application of the brakes one of the suspension wishbones fractured causing the wheel to engage the wheel arch and causing damage. Presented with damage to a client's car, repair costs for the damage and new wishbones, plus late delivery of the vehicle, the garage decided to send the parts for a failure examination.

Metallurgical examination

Figure 15.13 shows the suspension link. The fracture had occurred at the mid section between two of the weight reduction cut-outs. The crystalline appearance of the fracture indicated a brittle fracture. For comparison purposes the OEM part that had been removed was recovered. Samples for metallographic examination and for chemical analysis were taken from both links.

Basic metallographic analysis showed that the OEM part was a forged medium carbon steel and that the failed part was a shell cast ductile cast iron, as shown in Figure 15.14(a) and (b), respectively.

In addition, the failed ductile iron was of poor quality and showed a brittle subsurface carbide structure shown by the 'needle-like' structure in Fig 15.14(b).

Conclusions

This failure study did not require a detailed knowledge of the component's service requirements.

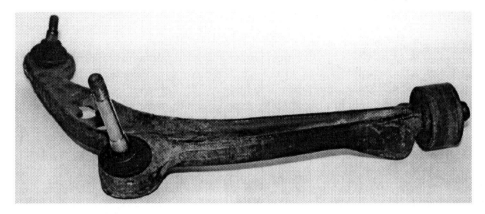

Figure 15.13 Lower wishbone from BMW Series 3 car

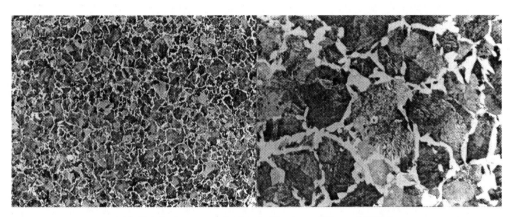

(a) Original component (magnifications 100× and 500×)

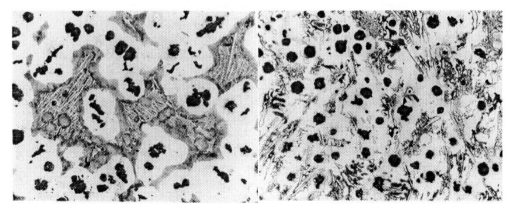

(b) Failed ductile iron (magnifications 400× and 100×)

Figure 15.14 Photo micrographs of the OEM and replacement components

The inferior toughness of the poor quality ductile iron shell casting compared to the forged medium carbon steel was obviously a major factor in the failure.

In view of the potential serious consequences of this type of failure, for instance if it had been delayed until the vehicle was at 70 mph and manoeuvring in busy traffic, a letter was sent to the supplier advising him of procedures that should be followed. The garage received full consequential damages and free issue OEM parts, although, there was no reply to the letter describing the serious nature of the problem.

15.5.3 Drive line durability and the study of a failed crankshaft

Introduction
Newly formed commercial vehicle manufacturers frequently begin manufacture under license from an established maker followed by the development of their own range of vehicles. The durability testing of the 'first off' vehicles of their own design can be an onerous task due to the significant number of proprietary components (engine, gearbox, rear axle, drive shaft). Although each of these items in turn should have an established test history to validate the endurance, the testing of the assembled parts would also be vital. This can be further complicated by the fact that some of the component suppliers are developing their own components and the correct validation testing of these items is unclear.

The development of realistic test procedures for checking the production of commercial vehicles presents several major problems.

- The durability is of such a high level, long test periods are required to establish reliability.
- All parts of the drive line accumulate damage at different rates.
- The service duty of the customer-owned vehicles can vary considerably.
- The attempt to accelerate a test procedure can begin to enter a regime of abusive testing and be deemed invalid.

Drive lines are complex composite systems which should be designed from well-tested proprietary components. Ever since the 1930s drive axle, clutch, bearing and transmission manufacturers have carried out laboratory testing of their products and several reliable designs have been established. There are many published sources of information to consult on the important design features of individual components. This example considers the evaluation of a premature failure in a crankshaft.

During a vehicle endurance test an engine crankshaft failed. The truck had a 305 BHP engine that had achieved about 100 000 km, i.e. substantially below the target of 500 000 km.

Metallurgical examination
(a) the crank pin from the gearbox end and the fracture had progressed across the complete section of the web. The general form and appearance of the fracture is illustrated in Figure 15.6, which clearly shows 'conchoidal' or 'beach' markings on the fracture surface. The remainder of the shaft appeared to be in a good condition and there did not seem to be any excessive wear either on the bearing surfaces or on the bearings shells. However, some circumferential machining marks were in evidence and on the failed pin surface a machine mark was observed running in line with the fracture origin in the fillet radius.

(b) Crack detection. Die penetrant crack detection was carried out on the remaining journal and pin surfaces but there were no indications of any further cracking.

(c) Bearing surface measurements. A series of measurements were carried out on the bearing surface dimensions and from the results given in Table 15.2 it can be seen that there is virtually no variation in the size of each journal and pin, and ovality was negligible.

Table 15.2 Engine bearing surface measurements

	Journal No.	Diameter (mm)	Pin No.	Diameter (mm)
Drive end	1st	101.08	1	84.25
	2nd	101.05	2	84.20
	3rd	101.08	3	84.20
	4th	101.08	4	84.20
	5th	101.05	5	84.20
Gear box end	6th	101.08	6	84.25

Therefore wear has been even, consistent and minimal.

(d) Chemical examination. Samples were taken for chemical analysis and gave the following results:

C%	Si%	Mn%	S%	P%	Ni%	Cr%	Mo%	Al%	Cu%	Nb%	V%
0.38	0.36	0.74	0.034	0.011	0.155	0.94	0.23	0.007	0.16	0.001	0.002

This analysis compares closely to material conforming to BS970 709M40, or S.A.E. 4140 etc. This is a chromium–molybdenum steel often used for crankshaft manufacture.

(e) Metallographic examination. A section was cut through the failed area in a longitudinal direction and as close to the origin of the fracture as possible. Preparing the surface through the pin and macroetching, clearly revealed that the pins had been induction hardened. The depth of hardening was of the order of 3 mm and the extent of the hardened surface included the whole bearing surface but did not extend into and around the fillet radius. The general flowline pattern which had been produced during manufacture was favourable for this type of component.

A section was taken from the fillet area for microscopic examination and this revealed a structure consisting of a mixture of ferrite and fine carbides. There were numerous sulphide and silicate non-metallic inclusion stringers in the microstructure.

(f) Hardness survey. A hardness survey carried out at various points of the shaft gave the following results:

Web material	–	22 Rc
Fillet radius	–	23 Rc
Induction hardened surface	–	50–55 Rc

These results are considered normal for a forged crankshaft of the given composition.

Discussion
The form of 'conchoidal' fracture shown in Figure 15.6 is very typical of fatigue cracking occurring under cyclic loading conditions. The fact that the fatigue crack had progressed almost through the full section of the shaft before final failure occurred, demonstrates that the fracture had been in progress over a period of time without any significant overloading.

It is well established that the fatigue or endurance limit of a steel material under alternating stress occurs at levels well below the yield stress of the material. It is also generally accepted that the presence of any stress concentration factors will considerably reduce the fatigue limit of the component. Stress concentration may take the form of sharp changes in section, or sharp bottomed machining or scoring marks. Therefore the fatigue crack advances under a low level of stress and if the material is not constantly overloaded or shock loaded, this fatigue crack will extend through almost the whole section of the shaft until sudden failure occurs on a much reduced cross-sectional area.

The fatigue limit of a component can be improved by introducing compressive stresses into the surface layers at the vulnerable points. This can be done by either

(a) mechanical deformation, i.e. cold working, or
(b) thermal treatment, i.e. heat treatment.

In the present case, improvement in fatigue life of the pin could have been enhanced by extending the area treated by the induction hardening process to include the fillet radius instead of only hardening the bearing surfaces. This would be a standard procedure for some crankshaft manufacturers.

The initiation and propagation of the fatigue crack would require cyclic rotating bending or torsional stresses of sufficient magnitude. These stresses could be caused by either the dynamic loading characteristics associated with the shaft design or any imbalance that may have been present.

Conclusion
The shaft has failed prematurely by fatigue fracture. This was initiated at the root of a fillet radius and had been aggravated by the presence of a small machined groove which had caused stress concentration. There were no indications of abuse or overloading of the shaft in service. Attention should be given to the method of manufacture.

15.5.4 Recovery vehicle boom buckling

Introduction
A heavy vehicle recovery company was unable to operate their heavy recovery facilities due to damage and buckling of the main arm of the under lift, fitted to a IVECO 190.38 truck.

The failure had resulted in significant consequential financial losses and an expensive repair bill, together with further delays while the repairs were performed. Samples were cut from the boom for a metallurgical examination.

Background
During 1992 the elevation hydraulic actuator on the under lift fitted to the IVECO 190.38 truck

required repair. Quotations were received for this work and a suitable company was chosen to perform the work. The vehicle was repaired and then put back into service.

During subsequent months in service a 'crack' developed in the vertical hollow section near to the area where the repair had been carried out. This defect significantly reduced the load carrying capability of the structure, 'buckling' occurred and the vehicle had to be withdrawn from service.

During subsequent discussions the company that performed the repair work denied any responsibility for the subsequent failure.

Metallurgical examination

Visual
To repair the hydraulic actuator the company that had carried out the repair had burnt an access hatch in order to remove the top of the actuator. The overall inspection hatch was 36 mm × 19 mm which was also cut into two pieces the upper 19 mm × 22 mm and the lower 14 mm × 19 mm.

Confirmation that both pieces were cut at this time can be obtained by two observations.

(i)   During replacement the outside weld metal was ground away/dressed prior to painting. The vertical seams show a continuous path of 'grind marks' around the full extent of the inspection hatch.
(ii) The same inadequate welding style with lack of side-wall fusion and intermittent welding was evident around the full 36 mm × 19 mm inspection hatch.

The observed 'crack' was actually attributed to the separation of the inadequately welded bottom edge of the 36 mm × 19 mm access hatch. Examination of the surface after removal showed a total lack of side wall fusion with less than 10% of the thickness of the plate actually being welded.

Hardness
The hardness was 213 on the Brinell Hardness Scale.

Chemical analysis
The chemical analysis was typical of a BS4360 Grade 43 strength rolled section. The weldability of this steel would be classed as very good.

C	Si	Mn	S	P	N	Cr	Mo	Cu	Sn	Ti	Al
0.15	0.162	1.4	0.012	0.017	0.005	0.084	0.010	0.070	0.005	0.001	0.024

Microstructural examination
A specimen was prepared for microstructural examination and etched in a nital etchant. The microstructure was ferrite/pearlite typical of a low carbon structural steel.

Discussion
As previously mentioned the company that performed the repair took the decision to cut out an

access hatch. The owner of the vehicle was not requested to give permission for the removal of the hatch. The removal and subsequent replacement was therefore the total responsibility of the company that performed the repair.

There was inadequate consideration of the engineering load carried by the vertical hollow section. The subsequent replacement had been performed for cosmetic purposes (for the outside view) rather than structural integrity. The weld should have included weld preparation and a proven practice capable of giving good side wall fusion.

The manufacturer's drawing shows that all welds on the original structure were 'of sound welded construction with min. 6 mm fillet welds or full penetration'. If this type of welding procedure had been followed during the replacement of the cut-out this failure could have been avoided.

Conclusions
(i)   The steel used to make the vertical section was to a BS4360 Grade 43 specification and was of good quality and easily weldable.
(ii)   The Company responsible for the actuator repair gave insufficient consideration to the weakening effect of the boom structure prior to cutting the access port. In addition, permission to make the access port had not been requested.
(iii) The welding that had been performed to replace the access port had been done with an inadequate weld procedure and a lack of weld preparation.

References and further reading

ASM (1974). Source book in Failure Analysis. ASM.
ASM CD ROM (1997). CD ROM on Failures. ASM.
Alban, L.E. (1985). Systematic analysis of gear failures. ASM, ISBN 0-87170-200-2. American Society for Metals.
Alcraft, D.A. (1984). Ensuring the reliability of diesel engine components, Chapter 10, p. 317. *Design and Application in Diesel Engineering*, ed Sam Haddad. Ellis Horwood Ltd. ISBN 0-85312-733-6.
Allen, E.T. (1973). Reliability engineering as applied to the production of a quality motor vehicle. IMechE C90/73.
Arrol, W.J. (1967). Future trends in the selection of materials and design (the engineering view), p. 69. The Institution of Metallurgists, Iliffe Books.
BS4778 (1987). Glossary of terms used in quality assurance.
BS5760 (1993). Guide to Reliability Programme Management.
Baboian, R. *et al.* (1998). Galvanic corrosion on Automobiles. Galvanic Corrosion ASTM STP 978. Hp. P. hack ed., p. 249.
Baker, R.G. (1986). Testing for Quality. Metals & Materials, May, p. 268.
Banks, T.J. and Lacy, D.J. (1989). The application of analysis to piston ring performance. IMechE C375/003.
Barret, A.J. (1981). The refinement of data resources to assist engineering design. IMechE C231/81.
Beddoes, G.N. (1993). Valve Materials and Design. *Surface Engineering*, Vol. 9, no. 1, p. 44.
Belfiore, D. *et al.*, (1996). Vehicle axle accelerations due to road roughness for accelerated life testing. SAE 930256.
Bell, J.C. and Delargy, K.M. (1989). Lubrication influence on the wear of piston-ring coatings. Proceedings of the 16th Leeds–Lyon Symposium on Tribology. 5–8th Sept, Paper 21.

Bertodo, R. (1989). Human resource deployment for design excellence. IMechE C377/009 ICED 89, Harrogate.

Bidgood, J.F.S. (1989). The interactive implications for manufacturers, owners and rescuers of new concepts in the design of passenger cars. IMechE C392/022, p. 163.

Bompas-Smith, J.H. (1973). *Mechanical Survival: the Use of Reliability Data.* McGraw-Hill. ISBN 0-708-4411-9.

Borer, R.B. and Peters, B.F. (1970). *Why Metals Fail.* Gordon and Breach, New York.

Bovington, C.H. and Hubbard, A. (1989). Lubricant additive effects on valve train friction and wear. IMechE C375/021.

Boxter, D.F. (1990). Users like steels new look. Advance Materials & Processes 8/90 p.17 (see page 24 on Diado Super Gear steel development. Also SAE 890531).

Bralla, J. (ed.) (1986). *Handbook of Product Design for Manufacture – a practical guide to low cost production.* McGraw-Hill Book Company. ISBN 0-07-007130-6.

Breen, D.H. and Wene, E.M. (1975). Fatigue in Machines and Structures – Ground Vehicles, p. 57.

British Gear Association (1992). Teaching pack on Gear Technology. British Gear Association, Tel: (+44) 0121 456 3445, Fax: (+44) 0121 456 3161.

Broek, D. (1994). *The Practical Use of Fracture Mechanics.* Kluwer Academic Publishers. ISBN 90-247-0223-0. Also Software for damage tolerance analysis from Fracturesearch. Tel/Fax: (1) 740 965 2999.

Brughmans, M. *et al.* (1994). The application of FEM-EMA correlation and validation techniques on body-in-white. ImechE, C487/003/94.

Burton, R. and Senior, L.P. (1969). Clutch Proof Test Rigs. Proc. Ins. Mech. Eng. Vol. 184, pt 31, p. 859.

Carter, A.D.S. (1972). Mechanical Reliability. McMilan Press. SBN 333 138317.

Cebon, D. and Ashby, M.F. (1992). Computer aided materials selection for mechanical design. *Metals & Materials*, January, p. 25.

Child, H.C. (1982). *Surface Hardening of Steel.* Engineering Design Guide,. Oxford University Press.

Colangelo,V.S. and Heiser, F.A. (1974). *Analysis and Metallurgical Failures.* John Wiley, New York.

Condra, L.W. (1993). *Reliability Improvements with Design of Experiments* (DOE). Marcel Decker. ISBN 0-8247-8888-5.

Curtis, M.W. (1992). The Shell Process in the nineties and beyond. Hepworths Minerals and Chemicals Ltd Publication.

Cutler, A.N. (1998). Modern Statistical Ideas in Fatigue Testing. *Journal of the Engineering Integrity Society*, January, No. 3, p. 14.

Daily Telegraph (1997). *The Daily Telegraph Book of Motoring Answers.* Robinson. ISBN 1-85487-912XX.

Dally, J.W. (1991). *Experimental Stress Analysis*, Third Edition. ISBN 0 07100825 X, Chapter 6.

Davies, A. (ed.) (1998). *Handbook of Condition Monitoring Techniques and Methodology.* Chapman & Hall. ISBN 0-412-61320-4.

Davies, G.M. and Goodyer, B.G. (1991). Aluminium in automotive applications. *Metals & Materials*, February, p. 86.

Dehnad, K. (1986). Quality Control, Robust Design and the Taguchi Method. Wadsworth & Brooks/Cole, ASQC (American Society for Quality Control).

Devlukia, J. and Davies, J. (1985). Failure analysis of a vehicle structural component under biaxial loading. IMechE C104/85.

Dickenson, J.I. (1985). Vehicle data acquisition and its application in drive shaft endurance testing. IMechE C4/84.

Dickson-Simpson, J. (1983) Road Wheel Security. IMechE C294/83.

Dowson, D. *et al.* (1990). Experimental observations of lubrication film state between a cam and bucket follower using electrical resistivity techniques. Proceedings of the Japan Int. Tribology Conference, p. 119.

Dowson, D. *et al.* (1992). a transient elasto-hydro-dynamic lubrication analysis of a cam and followers. *Journal of Applied Physics* 25, pp. 313–320.

Dowson, D. (Ed) (1996). *The Third Body Concept.* Elsevier Science.

Dummer, G.W.A. *et al.* (1997). *An Elementary Guide to Reliability.* 5th edition p. 38, Butterworth-Heinemann. ISBN 0-7506-3553-3

Du Quesray, D.L. *et al.* (1996). Fatigue life prediction for variable amplitude strain histories. SAE 930400.

EEC (1985). Directives on Product Liability.

Easterling, K. (1983). *Introduction to the Physical Metallurgy of Welding.* Butterworths, p. 203. ISBN 0-408-01352-4.

Engel, L. and Klingele, H. (1981). *An Atlas of Metal Damage.* Wolfe Publishing, London.

Erickson, R. (1980). A study of the wear behaviour of cast iron cylinder liners. Wear & Fracture Prevention. ASM Proc of conf 21–22 May, Peoria, Illinois, p. 261.

Eyre, T.S. and Davis, F.A. (1993). The testing and evaluation of materials in tribology.

Georgano, N. (ed.) (1995). Britain's Motor Industry. The first hundred years. G.T. Foulis & Company 1995. ISBN 0-85429-923-8.

Grawne, D.T. (1996) The mechanisms and control of wear. *Surface Engineering Case Book*, p. 203. Woodhead Publishing. ISBN 1 85573 2602.

Haigh, M.J. *et al.* (1989). Interfacial contact behaviour of transmission synchronisers. IMechE. C382/021.

Harrison, J.D. (1980). The brittle fracture story PXIX. Engineering applications of fracture analysis. Proceedings of the first National Conference on Fracture held in Johannesburg, South Africa, 7th–9th November, 1979 Pergamon Press. ISBN 0-08-025437-3.

Hartfield Wunsch, S.E. *et al.* (1996). Development of a bench wear test for the evaluation of engine cylinder components, SAE 932693.

Hill, S.N. *et al.* (1996). Bench wear testing of common gasoline engine cylinder bore surface/piston ring combinations. Presented at the 51st Annual Meeting, Ohio, May 19–23, Society of Tribologists and Lubrication Engineers, Paper 63.

Hoo, J.J.C (ed.) (1988). Effect of steel manufacture on the quality of bearing steels. ASTM STP 987, American Society for Testing and Materials, Philadelphia.

Hutchings, F.R. and Unterweiser, P.M. (1981). Failure Analysis: The British Engine. Technical Reports, Am. Soc. Metals.

Hutchings, I.M. (1992). *Tribology: Friction and Wear of Engineering Materials.* Edward Arnold. ISBN 0-340-56184-X.

IMechE Conference (1974). Corrosion of motor vehicles (CP18). 13–14th November. ISBN 0-85298-329-8.

Johnson, B. (1997). *Classic Plant Machinery.* McMillan, p. 53. ISBN 07522 13407.

Johnson, K.I. (1975). Realistic mechanical tests for resistance welded structures. The Welding Inst. Res. Bulletin, June, p. 157.

Kaku, M. (1983). *Nuclear Power Both Sides.* Pub. W.W. Norton & Co, p. 25. ISBN 0-393-01631-5.

Kato, H., Eyre, T.S. and Ralph, B. (1994). Mapping wear behaviour of nitrided steel. *Materials World*, July.

Kaye, A. and Street, A. (1982). *Die Casting Metallurgy.* Butterworths. Monographs in Materials.

Kim, J.J. *et al.* (1992). Integrated fatigue analysis at Kia Motors. Cray Channels, Winter, p. 7

Kondo, Y. *et al.* (1993). Prediction methods of rough-road-load applied to vehicle body. SAE 900665.

Lagneborg, R. (1991) New steels and steel applications for vehicles. Materials and Design, Vol. 12, No.1, February, p. 3.

Laird, M.P. and Gregory, R.P. (1990) Dog clutches for rapid gear changes in automotive gear boxes. IMechE C404/016.

Lavington, M.H. (1986) The Cosworth Process – a new concept in aluminium alloy casting production. *Metals & Materials.* November, p. 713.

Lewis, C. (1996). Fatigue performance of fusion welded automotive high strength steels. *Welding & Metal Fabrication*, July, p. 275.

Lipson, C and Sheth, N.J. (1972). *Statistical Design and Analysis of Engineering Experiments*. McGraw Hill Book Company.

Lofthouse, J.A. (1969). Value for money. *Metals and Materials*, March, p. 73.

Macieinski, J.W. (1969). Propellor shaft and universal joint characteristics and methods of selection. Proc. Inst. Mech. Eng., Vol. 184, Pt 31, p. 516.

Matlock, W.H. (1980). Development of an iron base hardfacing alloy for internal combustion engine valves. Wear and Fracture Prevention, p. 277, ASM Proc of conf., 22–23 May, Preoria, Illinois.

Matolcsy, M. (1977). Bus research and development at Autokut Institute, Hungary. ImechE, C145/77.

Matolcsy, M. (1984). Fatigue life distribution of vehicle frame structures. ImechE, C169/84.

Matolcsy, M (1986). The use of fracture mechanics in the Hungarian Bus Industry. ImechE, C222/86.

Matthews, A. (1991). The future of coated sheet. Materials & Design, Vol. 12, No. 6, Dec, p. 340.

Matthews, A. (1994). Achieving Fuel Efficiency with sheet steel automobiles. *Materials World*, March, p. 133.

McCall, J.L. and French, P.M. (eds) (1978). *Metallography in Failure Analysis*. Plenum Press, New York.

Metals Handbook (1988). Vol 10, Failure analysis and presentation. 9th Edition, ASM.

Michelberger, P. *et al.* (1984). Modelling problems in the dynamics design of autobuses. ImechE, C174/84.

Mimuro, T. *et al.* (1996). Development and application of the road profile measuring system. SAE 930257.

Mischke, A. (1978). Common and distinguishing factors in the development of light and heavy trucks. ImechE, C208.

Mitchell, J.W. (1975). Resistance – spot welding of microalloyed steels for automotive applications. Micro-Alloy 75, p. 599.

Miwo, Y. *et al.* (1987). Carbo-nitriding and hard shot peening for high strength gears. SAE 880666.

Monaghan, M.L. (1989). Putting friction in its place. IMechE C375/KN1.

Morrow, J.D. and Socie, D.F. (1981). The evolution of fatigue crack initiation life prediction methods. Proceedings of Fatigue 81, Soc. Environmental Engineers, p. 3, Westbury House. ISBN 086103 0427.

Musiol, C. *et al.* (1981). Advances in computer aided design against fatigue. C234/81.

Neale, M.J. (1975). Component failures, maintenance and repair. *A Tribology Handbook*, Newnes-Butterworth. ISBN 0408000821.

Niemand, L.J. and Wannenburg, J. (1997). Vehicle durability programmes. EIS, Issue 2, July, p. 22.

O'Connor. P. (1981). *Practical Reliability Engineering*. John Wiley. ISBN 0471926965.

Open University (1984). Open University Visual Aids, T353, Last of the Liberties.

Owen, M. (1989). SPC and continuous improvement. IFS Publications UK. ISBN 0-94850-795-8.

Penter, A.J. and Lewis, D.C. (1990). Build quality inspection of repaired and new gearboxes. ImechE, C404/042.

Plint and Partners Ltd. (1997). Tribology Update Issue 6, p. 11.

Polushkin, E.P. (1956). *Defects and Failures of Metals*. Elsevier, Cat No. 5511998.

Pountney, R. and Dakin, J. (1998). Road loads for durability using wheel force transducers and component characterisation. EIS Conference on 'Smarter Testing & Simulation', *Journal of the Engineering Integrity Society*, No. 4, July, p. 20.

Powell, D.E. (1979). Vehicle Component fatigue testing. SEECO, Volume 5, May 9–11th, p. 87.

Powell, H.J. (1995). Resistance spot welding the guest for best practice. *Welding & Metal Fabrication*, March, p. 9.

Pugh, S. (1991). *Total Design, Integrated Methods for Successful Product Engineering*. Addison Wesley.

Pye, A. (1992). Getting your process right. Engineering, September 1992, p. 14.

Reynolds, K.A. (1997). The Role of the Metallurgist in the Investigation of Component Failures and Vehicle Accidents. *Journal of Inst. of Automative Eng. Assoc.*

Rice, R.C. *et al.* (ed.) (1988). Fatigue design handbook, second edition. SAE. ISBN 0-89883-011-7.

Rivett, R.V. (1979). The failure of spot welds in low carbon mild steel. The Welding Inst. Res. Bulletin, Aug, p. 235.

Riederer, D. (1979). Case Histories in Failure Analysis. ASM. ISBN 0-87170-078-6.

Robson, G. (1997). Cars in the UK, Volume 2 1971–1995. ISBN 1899870164.

Rosen, I. *et al.* (1970). Synchromesh Mechanisms: Experience of heavy truck gearboxes. Driveline 1970. *Proc. Inst Mech Eng.* Vol. 184, p. 31 and p. 438.

Ross, B. (1995). *Investigating Mechanical Failures. The Metallurgists Approach.* Chapman & Hall, ISBN 0-412-54920-4.

Roy, R. (1990), *A Primer on the Taguchi Method.* Van Nostrand Reinhold, ISBN 0-442-23729-4.

Russel, J.E. (1969). Strength of Transmission Shafting. *Proc. Inst, Mech. Eng.*, Vol. 184, Pt 31, p. 477.

Schneider, B.R. (1992). Structural analysis: an integrated tool of commercial vehicle developments. ImechE, C389/341.

Shronerberg, R. (1986). *World Class Manufacturing.* Butterworths.

Sharpe, F.T. (1985). Techniques for endurance testing of automotive driveshaft components. IMechE C102/85.

Stockley, B. and Devlukia, J. (1984). Durability Route Correlation. ImechE, C177/84.

Swann, P.R. (1976). Flixborough – The Metallurgical Implications. The Metallurgist and Materials Technologist, October.

Taylor, D.E. (1996). Fretting and fretting fatigue, incidence and alternation. Surface Engineering Casebook, p. 191. Woodhead Publishing. ISBN 1 85573 2602.

Teulon, H. (1991). Auto manufacturers choice of coated steel – What & Why!. *Materials & Design*, Vol. 12, No. 6, Dec, p. 338.

Timoney, S.S. (1986). Fatigue life predictions from short duration tests. SAE 850368.

Ting, L.L. (1996). Development of a reciprocating test rig for tribological studies of piston engine moving components. Parts I and II, SAE 930685 and SAE 930686.

Trento, J.J. (1987). *Prescription for Disaster.* Harrop Ltd. ISBN 0245-546154.

Trethewey, K.R. and Chamberlain, J. (1995). *Corrosion for Science and Engineering*, 2nd Edition, pp. 12 and 13. Longman.

Uchida,Y. (1991). Zinc coating developments in Japan. *Materials & Design*, Vol. 12, No. 6, Dec, pp. 331.

Van Asperen, F.G.J. and Voets, H.J.M. (1992) Optimisation of the dynamic behaviour of a city bus structure. ImechE, C389/341.

Vander Voort, G.F. (1976). Analysing Ductile and Brittle Failures. Metals Eng. Quart., August.

Walton, D. and Prayoonrat, S. (1986). A knowledge based system for shaft design. IMechE C276/86.

Wannenburg, J. (1993). Probabilistic establishment of vehicle durability test requirements based on field failure data. *Environmental Engineering*, September, p. 18.

Waterman, N.A. *et al.* (1992). Computer based materials selection systems. *Metals and Materials*, January, p. 9.

Willin, J.E. and Love, R.J. (1969). Preparations of automobile speed gear teeth. *Proc. Inst. Mech. Eng.*, Vol. 184, Pt 31, p. 292.

Wilson, F. and Eyre, T.S. (1969). Metallographic aspects of wear. Metals and Materials. March, p. 86.

Wolff, P.H.W. (1967). The design function, Page 3. Selection of materials and design. The IOM London, Iliffe Books Ltd.

Wright, D.H. (1989). Road Wheel Security on heavy goods vehicles. IMechE C392/025.

Wulpie, D.J. (1985). Understanding How Components Fail. ASM. ISBN 0-87170-189-8.

Further reading

Deformation and Fracture Mechanics of Engineering Materials – R.W. Hertzberg. Published by John Wiley & Sons, New York, 1989. ISBN 0-471-61722-9.

This final year undergraduate/postgraduate text gives a wide ranging review of most aspects of material deformation and fracture. It considers both fracture mechanisms and fracture mechanics, presenting numerous worked examples and problems. Particularly useful chapters are to be found on fatigue. Several case studies are presented, although these are not specifically related to the automotive industry.

Corrosion Engineering – M.G. Fontana and N.D. Green. Published by McGraw-Hill International Book Company, 1982. ISBN 0-07-066288-6.

This standard undergraduate text on corrosion engineering has long been the yardstick by which all new corrosion texts are judged. It covers the principles and types of corrosion, illustrating many of the ideas with practical examples. Two aspects of particular interest to Automotive Engineers are well covered in the text, namely corrosion testing and corrosion prevention.

An Atlas of Metal Damage – L. Engel and H. Klingele. Published by Wolfe Science Books, 1981. ISBN 0-7234-0750-9.

This book presents details of all the major failure mechanisms, combined with detailed scanning electron microscope (SEM) photographs of metal fractures and damaged surfaces. The book also provides a brief introduction to the background metallurgy and an essentially practical guide to how metal is damaged.

Automotive Components and Systems: The Role of Full-Scale Fatigue Testing – S.J. Hill. Published by Butterworths, 1989.

The most relevant part of this book is Chapter 3, by E.K.J. March. A very interesting historical review of automotive testing and test procedures is provided. The current trends in automotive testing are also outlined together with some interesting case studies.

Fatigue Design Handbook AE10, Second Edition. Published by Society for Automotive Engineers, 1988. ISBN 0-8988-011-7.

This is a very authoritative handbook, covering all aspects of fatigue design for automotive components. It was prepared under the auspices of the Design Handbook Division of the SAE Fatigue Design Committee.

An Elementary Guide to Reliability (5th edn) – G.W. A. Dummer, M.H. Tooley and R.C. Winton. Published by Butterworth-Heinemann, 1997. ISBN 0-7506-3553-3.

This elementary text explains what is meant by reliability and the factors affecting reliability. It is not a mathematical text, but concentrates on developing an understanding of important concepts in reliability. The text is recommended for a variety of diploma level courses and contains many self assessment questions, in order that the reader may more easily gauge their grasp of the topic. A number of more advanced texts are referenced at the end of the book, for those wishing to take the subject further.

16. Future trends in automobile design

J. Happian-Smith, PhD, MSc BTech, MSAE
Eric Chowanietz, PhD, BSc, MInstP, MIED, CEng

The aim of this chapter is to:

- Demonstrate the mechanical and electrical possibilities for future vehicle design;
- Indicate how current advances will create fundamental design changes for future vehicles.

16.1 Introduction

The design of modern cars has already reached the stage where the Ford Fiesta has more computing power than Space Shuttle. There is definitely going to be a great expansion in electrical control and its attendant systems. To make this distinction more obvious this chapter has been split into electrical and mechanical future possibilities, but this does in no way mean that these two futures are separate, they are inevitably intertwined. There are many future possibilities but now more than ever before they are dependent on the development of future technologies, such as electrical systems using light as a carrier medium.

16.2 Mechanical possibilities

This section is split into six areas where significant changes will occur in the near future, however this does not preclude changes in other areas of vehicle design that may have a profound effect on these designs. There is inevitably an inter-linking between mechanical and electrical possibilities, so where this occurs the emphasis has been placed on the electrical side as this is the area that will show most change over the coming years.

16.2.1 Design possibilities

There are approximately 15 000 components that make up an automobile. Each of these components needs to be designed for efficient use of materials and costed down to a price. Each one must have a suitable reliability since any failure will inevitably result in customer dissatisfaction. This results in the total component number continuously dropping to increase reliability and drive costs down.

Systems such as FMEA (Failure Modes and Effects Analysis), ENTRA and weighted objectives methods (Cross, 1989) can be used effectively to optimize design and reduce component numbers. A recent example is a development of an engine design where the basic layout remained but a redesign reduced the number of components by seventy. This was in spite of the fact that the redesign contained new features such as liquid cooling and four valves per cylinder.

It should be noted that the quality of the final assembly is always a function of the number of component parts.

The use of these processes will increase, especially with the aid of computerized integrated design packages. There are many possibilities here but several may well be:

(a) design packages where the designer is continually updated on all the legal implications of each design change, thereby avoiding international vehicle acceptance problems.
(b) at the concept design stage previous models and current competitor vehicles could be analysed to create design envelopes for a new vehicle. These envelopes, which could be external shape, drive system design or interior layout, could then be explored to create new generations of vehicles.
(c) in a similar vein, new component parts could be developed with a computer system such that an existing part can be manipulated (by stretching, compressing, etc.) to quickly create a suitable new component. Obviously a life and reliability analysis would have to be performed but it avoids the blank sheet of paper situation.

It could be argued that these systems could remove innovative ideas; however, it usually takes less time to develop an existing idea than create the idea from scratch. However, there will always be a need to practically test designs to confirm the computed outcomes.

This is of increasing importance as the times for vehicle programmes reduce. It currently takes approximately three years for a new vehicle to be produced from concept, but two years could well be on the horizon. This means the design and development programme has to be thoroughly integrated so that resources and manpower are available exactly as required. An increasing use of project planning and constant appraisal is also indicated, as any delay could be catastrophic for the project. Again, computers will be essential here not only to run the design and development programme but also to predict likely problem areas.

An increasing role is being played by parts suppliers as they take on the design and development of their products. This leaves the vehicle manufacturer to collect and collate the design data to ensure the product as a whole will have the required performance. Thus, a complex computer network is being developed to enhance the design lead-time. This brings about the essential requirement that the design is constantly updated as changes are made, so that every group involved with the project is aware of the updates as they happen and can comment as necessary. It also provides a record so that back-tracking can occur when necessary.

The recycling of vehicle parts will always be an issue, which again needs a continual interlinking of the groups concerned with design issues. The cost of strip-down and ease of recycling is reduced when there are compatible materials used in component assemblies; for example, the complete dash-panel should be made of compatible plastics so that no sorting or breaking down of the panel is required for recycling.

Prototype models of design variants can now and increasingly will be made from computer models by the use of Rapid Prototyping processes (Jacobs, 1996; Venus, 1997; Kalpakjian, 1997). The time for these processes will reduce and will mean that tests such as coolant flow visualization in engines can be performed before any metal has been cut. This will also aid the design of electrical systems against electromagnetic interference, which will become an increasingly important issue as the whole vehicle becomes computer controlled. These on-board computers could also be used to relay vehicle performance and reliability information back to the manufacturers so that the vehicle could be updated before any potential problems (and expensive recalls) occur. This could be a two-way process as vehicle transmitter/receiver (not necessarily of a radio type) could also receive updates as well.

Further methods for decreasing design and development time revolve around modular design systems. This process has been used for a long time, but the advent of cheap high-powered computers brings far more scope to this reality. Various manufacturers currently use this system for engine design, but this system could easily be brought into body design. This could be achieved, for example, by the use of lightweight space frame chassis structures (platforms) to which plastic clip-on body panels are attached. These panels could readily be changed in shape using injection moulding techniques, which could provide relatively inexpensive model re-vamps.

New technology will inevitably bring with it new design possibilities, which could well be almost endless. These developments will mainly come from manufacturing processes becoming available with faster and improved final product properties. There is a definite trend towards lightweight materials, such as aluminium, where improved processing techniques will eliminate most of the machining processes. Styling will inevitably change to the limits of these processes, so for example the replacement of carbon in rubber with silicon or the use of prisms for mirrors could bring about a plethora of new design concepts. It is up to the design teams to realize the potential of these new technologies when they occur and exploit them fully.

16.2.2 Advances in manufacturing methods

Most automotive manufacturers currently have links of various forms with other automotive manufacturers. These ties will strengthen in the future mainly due to the economy of scale, especially with power systems and transmissions. There has also been a tendency for body styling to be updated regularly but the basic platform to remain similar to previous models, e.g. Ford Ka and Toyota Classic. This intermixing will create many new versions of basic vehicles without the costs of a completely new vehicle, so manufacturers will have a tendency for linking to create the broadest coverage of specifications for their vehicle range. This has further repercussions where, at least, first tier suppliers would be involved at all stages of the vehicle design. They could then supply and be responsible for complete systems. The main problem with this scenario is confidentiality with suppliers to more than one manufacturer. This can be alleviated by having a sole supplier relationship, patenting or allowing other manufacturers the use of designs for a fee.

These arrangements could result in the vehicle manufacturer having overall control of the design and development of its vehicles, but just being an assembler of component parts. Some manufacturers are already well down this path, and have a good showing within the market place. This is probably partly due to the extra control the vehicle assembler has on the quality of its bought-in parts. Some of these relationships are being formalized now in 'Target Agreements' which create a legal as well as a working relationship. There is now necessarily a logical end to this process as it depends on the strength of the linking, but would unquestionably mean that the limit would be the stage where the vehicle manufacturer feels that their individuality is brought into doubt, as this differentiates one company's products from those of another. This will create, however far down this route this process goes, a mutual responsibility which will have knock-on effects for all the workforces involved and could lead to 'service support systems' relating the vehicle manufacturer to its supply companies. The inter-linking of automotive manufacturers may well lead to there being only three or four major players in the world market.

An extension to this, that some companies are already arranging, is to create a 'supplier park' where first tier companies are geographically located close to the assembly plant. This may be an obvious move for new companies, but for established ones it may not be so simple, especially when 'Just in Time' (JiT) systems are becoming the industry norm. This will result in clusters of automotive industry within countries and indeed, the world.

Labour costs will vary from country to country over a period of time which means that labour intensive activities will be based on a short term basis geographically. A knock-on effect of modularization is that robotization is more effective which may bring an end to these short term policies for most production plants. This would seem to be the case especially when the spiralling of future transport costs of components are considered.

Part of the emphasis in computer controlled manufacturing is that feedback loops could exist to continually reduce production times. This could be built into the manufacturing system so that the processes are continually monitored and optimized. Mathematical modelling of production lines could also be used as part of this feedback loop.

These processes could also be used to reduce the time taken to obtain a required production rate from a new line, or even for a new vehicle. Currently a new vehicle takes about thirty days from the start-up of the line to the required rate for most manufacturers. However, a few companies manage this in one day with efficient planning and feedback systems incorporated into their computer management systems.

Further use of computerized control systems could be used to improve various stages of production within the complete cycle. An example of this could be the elimination of hot engine testing from the production process. Once the characteristics of an engine have been evaluated with bandwidths set for the control variables, engines could be cold tested at low cranking speeds to check factors such as crankshaft torque to turn, compression analysis, cylinder block airflow and liquid cavity leaks. Even NVH tests could be performed at, say, 1500 r.p.m. Again, the use of modularization would make such systems more economically feasible.

Another process that demonstrates the use of overall computer control by using experimental data as an initialization is painting by electrostatic deposition. These systems totally avoid the use of solvents by using water as a carrier. Flake orientation (for metallic paints) can be determined by the charge and the position of the spray gun, which could be computer variables. The carrier water could then either be re-circulated after cleaning (which involves large cost and potential contamination problems), or passed to filter beds and then into the local river system. This whole process could be automated once the system characteristics have been determined, so that these systems require systematic development but once set up could be self correcting for all eventualities. Implementation of such processes would greatly reduce costs and also minimize plant equipment costs. Such simplified control systems could be implemented throughout automotive production plants to improve final quality and reduce costs and provide innovative finishes.

It is certain that there will be an ever increasing technological advance that cannot be predicted, but these advances usually come through the application of older processes with new materials and control systems. An example of this is the use of hydra-forming. This process has been used for many years in the brass musical instrument manufacturing industry for forming complex curved tubes at fairly low fluid pressures. Advances in sealing technology mean that this process is currently used for the manufacture of camshafts and exhaust manifolds, but there is no reason why sills, rails and posts could not be manufactured by this process. This would avoid flanges and could thus make optimum use of the space taken by the current designs. This process of up-dating old processes will continue on the back of advances in appropriate technology.

16.2.3 Materials advances

The use of lightweight materials within road vehicles has been considered for at least twenty years (*Automotive Engineering*, 1991) but it is only recently that a few manufacturers have produced low volume mass production vehicles that use a substantial volume of these materials. As yet there are no truly mass produced lightweight (approx. 500 kg) road vehicles, this is bound to change. Current projected vehicles propose 40% by weight of aluminium for a vehicle weighing 1 Mg. that would return 100 m.p.g. using a hybrid power system. Part of the reason for the concentration on a range of medium/large sized vehicles is that the return on development costs would be greater. Aspects of true mass production can also be explored with this type of vehicle before any major small car mass production takes place.

However, aluminium is not the only contender for lightweight structures. The magnesium industry predicts (*Automotive Engineering*, 1993) a 15–20% annual growth within the automotive industry over the next ten years. Castings would be the most likely form of usage, and doors and dash-panels using this method have already been developed. Further uses could be for the nodes (lugs) of a space frame chassis and engine mountings. Other possible materials are highly ductile stainless steels with a yield stress over 800 MPa and high tensile steels. However, body shells need to be designed specifically for these materials and a direct replacement of current steel systems would not be appropriate. Part of this re-design would have to be a re-consideration of NVH properties. This could be aided by suitable positioning of sandwich construction panels, which is currently used as 'sound deadened' steels, but the principle could be applied to other metals than steel, once the bonding technology has been developed.

A further development would be to use plastics for the complete body shell including the windows, however at the moment their inherent brittle transition phase is hindering this advance. A metal–plastic composite could be the answer here. It is reckoned that an aluminium body shell could weigh 60% of a steel equivalent and a plastic one could be even less than that. The assumption here is that there is sufficient package space to absorb this intrusion.

Plastics currently have a very efficient application in the absorption of impact energy in race cars in the form of honeycomb structures. This technology is extremely expensive, but there are continuing pressures to develop this technology to reduce costs. This would be a great advance in road vehicle design and should, if implemented, radically alter body shapes and lengths.

Development prototype engines have been built where plastics have been the major material, up to about 80%, with the use of mainly ceramic coatings by various companies for racing and motorcycle use. Both these uses require lightweight and fast throttle response whereas the latter was also considered as a 'throw-away' engine. These engines have never got beyond the development stage to date, like many other prototype engines before them. However, plastics are finding their way below the bonnet as heat exchangers, cam covers, inlet manifolds and electric motor housings, so there could well be in the near future, a leap to plastic engines.

Part of these advances is the development of processing technologies. Currently work is being centred on metal matrices and long fibre injection processes. Metal matrices have the advantage that their properties can be tuned to suit the particular application, where not only fibre or whisker density is altered but also their orientation. This tuning of properties would be of great use for engine components such as pistons and connecting rods.

Combinations of metals, plastics and ceramics will be combined to create specific use composites for vehicle building of the future and just needs the technology to advance so that efficient mass-manufacturing processes can be developed.

16.2.4 Energy conservation

Currently global ecological issues are to the fore where motor vehicles are concerned. As part of this, most manufacturers are looking at 'life cycle assessment' to create an ecologically sound awareness within the manufacturing organization and the general public. This considers issues such as global warming, eco-toxicity, end of life disposal, manufacturing waste, resource depletion, climate changes and health risks as potential issues that must be considered when design decisions are made. These issues have to be ranked by each manufacturer so that designs can be compared using systems such as a weighted objectives system, and this ranking of relative merits would be different from company to company. This will become more of an issue as more of the world's countries become motorized.

In spite of this there is no real conscious effort by manufacturers and the general public to look at truly ecologically sound fuel efficient vehicles. There are a few exceptions to this, such as France and Japan, where vehicle tax regimes exist to encourage small engined vehicles. There has been a general acceptance that vehicles with engines, for example, of around 250 kW for personal transport are a thing of the past, but they are still being manufactured in large quantities and there is still a demand for them. This will only change by public demand, so it has to be the task of the manufacturers to create this demand imaginatively, and this will be the challenge of the 21st century.

Hybrid engine technology and re-generative braking systems are being used by some bus and truck manufacturers. These vehicles will be the test beds for this technology so that these systems are practically optimized and the mass manufacturing technology can be developed. These systems are currently of the pneumatic–hydraulic type which are inherently expensive, so alternative operating systems must be investigated. Currently hybrid systems are considered to be expensive, have energy storage deficiencies, have unknown reliability and have not, as yet, a mass production system in place where component parts can be easily organized. However, systems using bi-fuels, and efficient diesel engines acting as generators powering electric motors on the wheels look promising. Such systems could be developed where costs would then be driven downwards and then make these systems appropriate for smaller vehicles. These should give lighter loading conditions on the power train and thereby create lighter vehicles. This spiral of optimization could then create a mass production vehicle less than 500 kg, with a fuel efficiency greater than 100 m.p.g.

Alternative power sources have been considered for many years, including electric powered vehicles. There have been limited advances over this period of time in battery design to reduce weight and increase output. Examples of this include nickel–metal hydride and zinc–air batteries, both these systems being lighter than the conventional lead-acid type. Again, this could be the ideal choice for lightweight vehicles, but the emphasis to date has been to convert current fossil-fuelled vehicles on a mass production basis.

Technology will continually develop new energy saving systems. Current and near future systems include:

(a) fibre optics in place of copper wiring looms, which could reduce the cost, weight and space taken by current systems;
(b) low rolling resistance tyres are continually being developed; the addition of silicon may well contribute towards this aim;
(c) controllable aerodynamic systems for use as the vehicle is on the move, which could greatly aid fuel efficiency.

16.2.5 Power systems

Some engine manufacturers use a modular approach to the design of engines, which means that changes in size, and number of cylinders can be easily arranged. It also simplifies the amount of design analysis required, and will become more widespread. The emissions from engines is widely controlled by regulation and the environmental side is the main initiator for change. Should this be developed much further the main problem will be sensing the exhaust component gases on board. This will require complex feedback loops and system mapping, and the simplest method for control would be direct injection of the fuel into the cylinder.

The use of direct injection would open up many new opportunities for engine design. Two-stroke engines with direct injection, stratified charge and lean burn could become a reality once lean burn NO_x catalysts have been developed to a commercial level. Part of the control of particulates and NO_x could be achieved by advances in fuel spray technology, where the length and shape of the injection pulse could be determined by mapping techniques. Further controls could be achieved by valving systems within the exhaust.

Advances in electronic functionality will be needed to enable vehicle manufacturers to meet increasingly stringent environmental legislation and ever higher consumer expectation. Stricter exhaust gas emission limits in Europe and the US will require more sophisticated engine management systems. In certain regions of the US manufacturers will be expected to introduce Low Emission Vehicles (LEVs) from 2000 and, later, Ultra Low Emission Vehicles (ULEVs). The demand will therefore increasingly be for compact engines (0.5–1.5 litre) delivering a relatively high power output and producing very little pollution. These conflicting requirements will call for the adoption of radically new combustion technology with associated sensing and control electronics. Ultra lean-burn engines, using direct injection and variable valve timing together with direct sensing of combustion gas temperature and pressure are perhaps the most likely outcome. Interestingly several Japanese manufacturers (Honda and Mitsubishi, for example) are already working on the development of such engines.

Detail developments of engine design will follow technology, but some foreseeable advances are: electro-magnetically operated poppet valves and diesel injectors, engine oil thermal conductivity used to predict engine oil changes, and efficient computer controlled continuously variable transmissions. These advances will all require computer operation and it is evident that the progress of the design of vehicles and computers are intertwined.

16.2.6 Vehicle sales

At the moment there is the potential for manufacturing globally to have an over-capacity with the possibility of producing twenty million more vehicles than the world wants by the year 2002. The Asian and Pacific manufacturers will carry about 40% of this burden. Part of the solution to this problem is to ensure that vehicles are designed to fulfil customers' real and perceived desires. Added to this is the increasing chance for manufacturers to create a demand through the almost global television and communication networks and to some extent manipulate customers' desires to make them suit the company ethic.

The use of computerization will lead to a faster turn-around of car designs which will lead to a probable three-fold increase in car launches over the next ten-year period. These systems applied to the sale of vehicles means that logistics can be improved to give the customer the satisfaction that once an order has been placed the car will be in the showroom within a five-day period, and that includes the manufacture of the car!

The globalization of the car market is inevitable and major sales will occur in the Russian and Chinese Republics; however, the sales will only match the investment in road systems and will not occur until an effective road system exists. Major manufacturers are currently designing and building prototype utility vehicles for these markets and expect a major return on these investments.

16.3 Electrical and electronic possibilities

The ever-increasing use of electronic systems on motor cars seems to be a trend set to continue for at least the next two decades; the electronics industry continues to reduce costs and increase sophistication, and car manufacturers are becoming more skilful in using the technology to gain competitive advantage. Indeed, it is likely that sometime in the near future the automobile will be considered as an electronic consumer product with some mechanical components; viewed in perhaps the same way as we now regard a video cassette recorder.

Although the number of electronic features is set to grow rapidly, the actual value of electronic components installed on a vehicle may not change significantly, the reason being the steady fall in the relative cost of electronic hardware. Steadily increasing circuit integration is leading to smaller, lighter and more energy-efficient electronic assemblies. Falling semiconductor memory prices and more powerful microcontrollers will allow more complex processing algorithms, and the development of semiconductor switching devices with high power capacity will eliminate the need for electromechanical relays. As the cost of electronic systems falls, sales volumes will increase which will, in turn, allow suppliers to reduce prices still further.

Perhaps one of the most exciting future developments, the potential of which we are only just beginning to glimpse, is what is known as MicroSystems Technology (MST). MST, which is based on existing electronic integrated circuit (IC) technology, potentially allows the fabrication of microscopic electromechanical systems combining complex mechanical sensing or actuating elements with intelligent electronics. The application of MST to airbag systems could, for example, result in a complete crash sensor and airbag control system smaller than a postage stamp. The future exploitation of MST could allow the mass production of very small, lightweight and highly reliable automotive control systems which will provide a level of functionality hitherto undreamed of.

Already in Europe safety is taking a higher profile with customers. Heightened purchaser awareness of vehicle safety issues, backed by EU publicity initiatives, has led to a rapid increase in the fitment rate for airbags. In the next few years it is very likely that, perhaps coupled with legislation, there will be universal fitment of frontal impact airbags and far more widespread use of side impact airbags.

Some of the greatest gains from electronics have yet to be seen, and they present vehicle manufacturers with marvellous opportunities to secure competitive advantage and make their products safer and more environmentally acceptable. There are great dangers too. For the majority of vehicle manufacturers electronics has not, so far, been regarded as a 'core' activity and the majority do not manufacture their own components, rather they source assemblies from specialist automotive electronics companies. One of the great challenges to vehicle manufacturers is therefore to continue to maintain full control of the vehicle design and this will doubtless require the establishment of specialist in-house electronic engineering departments. In general, since all electronic control systems have a broadly similar architecture it should be possible for one group of hardware and software engineers to oversee developments in diverse areas such as braking systems, engine management, body electronics and so forth. Over and above this, it is likely that the automotive industry will need to establish universally recognized standards and protocols for the design and testing of electronic systems, especially for safety-critical applications.

16.3.1 Electronic advances in powertrain design

Engine management
Progressively more stringent world-wide legislation governing automotive emissions and fuel consumption will lead the development and installation of increasingly powerful electronic engine management systems over the next few decades. Already the European Union (EU) is proposing that emission limits for 2005 should be around 50% lower than those in force in 1996 (Table 16.1).

Table 16.1 Intended EU regulations for petrol engine emissions 2000–2005 (g km^{-1})

	2000	2005
CO	2.3	1.00
HC	0.20	0.10
NO$_X$	0.15	0.08

Along with the requirement to reduce emissions comes the need to provide On-Board Diagnostics (OBD) to continuously monitor and log the operation of all emissions-related equipment. In 1990, for example, the engine management system of a typical US-market vehicle had to monitor and control about 20 separate functions, by 2000 this has approximately doubled to 40 functions and by 2010 it is likely to double again to 80 functions. In order to keep pace with these demands, vehicle manufacturers are using increasingly powerful microprocessors with large amounts of memory; for the 1998 model year some manufacturers are already using very fast Reduced Instruction Set Computer (RISC) microcontrollers. The size of the computer program contained within the engine management system (i.e., the control software) is also growing quickly; in 1990 a typical engine management program contained about five million instructions, but in 2000 that grew to about 15 million instructions, with similar growth predicted to continue through to at least 2010.

Providing OBD will demand more sensors than have hitherto been required – for example, the number of sensors installed per vehicle is expected to approximately double between 2000 and 2010. In particular it is likely that sensors will be installed in each cylinder to enable combustion quality to be continuously monitored, thereby permitting more accurate fuelling and also enabling higher compression ratios to be used to increase thermal efficiency whilst safeguarding against knock and detonation. One of the most promising techniques here is the use of the spark ionization decay monitoring system, developed by NKT of Japan, which enables combustion monitoring by measuring the decay of a small current which flows between the spark plug electrodes just after the spark ends.

Electronically controlled valve actuation
The most radical improvement in engine design and performance over the next two decades is likely to arise through the introduction of electronically-controlled valve actuation (EVA). Through the ability to open and close inlet and exhaust valves under microcomputer control, quickly, at any time and by any lift up to about 12 mm, it is possible to transform an engine's performance and efficiency. For example, at low engine speeds short opening times with low lift are required to maintain inlet gas velocity and improve efficiency, whilst at high speeds long opening times and high lift are required to maximize performance. Similarly, by having total control of the phasing of the inlet and exhaust valves it is possible to continuously modify the fuel charge and the amount of exhaust gas recirculation (EGR) so as to optimize engine performance and economy to the prevailing driving conditions. Experimental work on EVA engines has already demonstrated increases in peak power and torque of up to 50%.

A further advantage of EVA, which may not be so readily apparent, is the improvement of engine efficiency through the elimination of throttle pumping losses. Using the facility of EVA to continuously control the inlet valve lift, it becomes possible to 'throttle' the engine by varying the valve opening, thereby eliminating the need for the butterfly flap and placing the engine management system in control of the engine speed ('electronic throttle'). This facility alone has been shown to increase low-speed efficiency by up to 10% whilst at the same time giving a much sharper throttle response since the inlet valve opening can be varied instantaneously according to the accelerator pedal position.

An extra feature which EVA could also offer, at virtually no additional cost, is the ability to selectively deactivate cylinders so that a four-cylinder engine could, say, function on just two-cylinders when at idle. Such strategies have been tried before using conventional engines (by GM in the US, for example) but failed because the deactivated cylinders soon cooled down, increasing friction and emissions. However, by using EVA it becomes possible to sequentially deactivate different cylinders on each engine cycle; a process termed 'skip firing'. Idle speed emission and fuel consumption improvements of up to 50% have already been achieved on research engines.

Although a number of vehicle manufacturers have sought to develop EVA engines many have been defeated by the additional cost and complexity of the prototype systems. Developments are continuing, however, and now show promise for the next decade. Several manufacturers appear to be adopting a system in which a solenoid valve is used to direct high-pressure oil either side of a small piston which is connected to the valve. The valve timing and lift is therefore controlled by the electrical impulses sent from the engine management system to the solenoid valve, whilst the energy to move the valve is derived from the oil pump.

Other companies, for example Siemens, are known to be investigating direct electrical actuation of valves using a pair of electromagnets and balance springs, Figure 16.1. With no power applied, the balance springs hold the valve half open. When power is applied to the 'closing' coil, the core is drawn upwards, progressively closing the valve, similarly, the 'opening' coil controls that aspect of the operation. Such a system is able to provide adequate valve lift and response times for most engines up to about 7000 rpm. A further advantage of the system is that the elimination of camshaft, rockers, timing gears and belt makes for a shorter and more compact, lightweight engine — important for small front-wheel-drive cars.

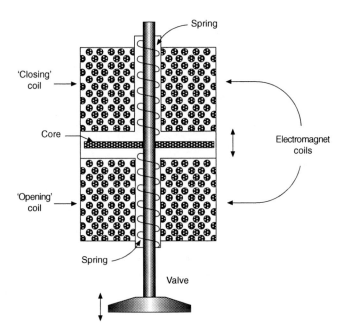

Figure 16.1 Section through an electromagnetic valve actuator (no power applied)

Electronically-controlled diesel engines

Although diesel-engined cars are rarely sold on the US and Japanese home markets they are quite common in Europe. In the EU as a whole, around 30% of all new cars are fitted with diesel engines; in some countries, such as France, the figure is almost 50%. Presently, around two-thirds of these engines are indirect-injection (IDI) units with minimal electronic control. Indeed, for some customers the lack of electronic control may be seen as an advantage! The greater efficiency and performance of direct injection (DI) engines is, however, leading to a growth in their popularity and it is anticipated that by 2010 virtually all diesel engines will be of this type. One of the drawbacks of DI engines is their potentially high levels of exhaust emissions, leading to the requirement for very high fuel pressures (to ensure thorough atomization of the fuel spray) and electronic control for precise injection timing and fuel volume delivery.

Several manufacturers, such as Lucas Varity, have already developed suitable DI systems using electronically-controlled solenoid valves which orchestrate the delivery of pressured fuel supplied from a cam-operated pump. More advanced 'continuous rail' (CR) systems, allowing greater flexibility, are under development by Fiat, Mercedes-Benz and Siemens and are likely to be standard fitment by 2005. CR systems are analogous to electronic petrol injection systems, however they use conventional diesel injectors fitted with piezo-activated delivery valves and have a much higher operating pressure (1500–2000 bar). Ingenious use of electronic control allows better engine operation, for example by injecting a 'priming' pulse of fuel just before the main delivery – this gives a smoother burn and so reduces noise and harshness.

Electronic transmission control
Although automatic transmissions are popular in North America and Japan, where they are fitted to over 80% of vehicles, they tend to be much less popular in Europe and are typically fitted to less than one tenth of new cars. European manufacturers have therefore tended to place less emphasis on automatic transmission development and hence many of the more sophisticated stepped automatic transmissions originate from the US and Japan. Japanese manufacturers, in particular, have led in the development of electronically controlled automatic transmissions (EAT). Mitsubishi are particularly noteworthy for their early adoption of fuzzy logic control for EATs (Chowanietz, 1995) in order to provide a transmission which adapts to the driver's style. In 1997 Mitsubishi took a further step by introducing an EAT controlled by neural network software. The neural network attempts to 'learn' a driver's response to given driving circumstances by continuously monitoring inputs such as speed, acceleration, turning angle, engine torque and so forth. It then attempts to predict the driver's next action. Any error between the prediction and what the driver actually does is then fed back and used to modify and refine the neural network. Over a period of time the neural network can therefore 'learn' the driver's habits. This approach is effective in providing greater driving pleasure and efficiency and is likely to become a standard feature of automatic transmissions in the next decade. In the longer term, the introduction of very powerful microcontrollers will probably lead to the development of integrated powertrain management systems employing artificial intelligence (AI) with the ability to instantly optimize all powertrain parameters for the prevailing driving conditions. Transmission developments in Europe are likely to centre on continuously variable transmissions (CVT), where manufacturers are globally competitive, and also on the automation of standard manual transmissions.

Electronically controlled CVTs are already widely available and the German companies LuK and ZF are in the process of refining their electronic control systems to improve drivability. With the introduction of fuzzy logic control and, later, neural networks, the CVT appears the most promising transmission solution for small cars and is likely to quickly gain in popularity.

Automated manual transmissions, obtained by installing electrical or electrohydraulic actuators on a standard manual transmission, have now been developed to quite a high level of sophistication (Chowanietz, 1995) and are popular with those who have tried them. Once again, the falling cost of advanced electronic control hardware is likely to see the functionality of these transmissions increase. For example, it is likely that in the next decade transmissions with driver-selectable 'manual', 'semi-automatic' and 'fully-automatic' options will be available; BMW already offer the 'Sequential M' transmission as a precursor to this interesting development.

16.3.2 Electronic advances in safety systems

Electronics is likely to be the main enabling technology for the improvement of car safety, both active and passive, in the next century. Airbags and anti-lock braking systems (ABS) are already well established in many countries; market penetration for both systems has reached almost 100% in the US by 2000, and about 75% in Europe and Japan. By 2010 both features are likely to be standard world-wide.

Other active safety features such as collision avoidance and road-following radar are now being developed and are likely to come onto the market early in the next century.

Airbag developments
Electronic developments for frontal impact airbags are centring on reducing the size of the electronic control units (ECUs) and integration of the crash sensors into the ECU itself to eliminate the use of cables and connectors to improve reliability. Increasingly, microengineered accelerometers are being used to sense deceleration and they are now being designed with a variety of self-test features to ensure fail-safe operation (Bryzek *et al.*, 1994; Olbrich *et al.*, 1994).

Although airbags have traditionally used a solid propellant to generate nitrogen gas (Chowanietz, 1995) there is now a trend towards using a cylinder of compressed inert gas (for example argon) which is connected to the airbag via an electronically-controlled solenoid valve. Such a regime allows the speed of inflation to be carefully controlled in order to minimize possible facial damage caused by the bag deploying into the face of a child or small adult. Coupled with occupant weight or height sensors built into the front seats, such a system could radically improve the performance of airbags in many circumstances.

Approximately one-third of all road accidents involve a side impact and therefore side impact airbag systems are likely to rapidly gain in popularity as they fall in price. One of the difficulties facing designers of side airbag systems is the rapid speed of deployment which is required. A frontal airbag typically needs to fully deploy within about 100 ms of impact, whereas a side airbag must inflate within 20 ms to be effective. The key to rapid deployment is fast and accurate side impact sensors. Various manufacturers have developed microengineered accelerometers specifically for this purpose; often several sensors are used both on the door case and on the vehicle chassis in order to guard against false deployment. One promising development is the use of a fast-responding microengineered pressure sensor for mounting inside the door cavity; it responds to changes in air pressure within the door caused by the start of an impact.

Collision avoidance systems
The future control of road vehicles could well follow the lines of modern aircraft in that the whole vehicle could be controlled automatically with the driver (pilot) taking control only in emergency conditions or in heavy traffic conditions, which would involve the driver being made aware of road conditions by alarms or similar devices. This would virtually be a 'hands-off' driving system. There will inevitably be much emphasis on computer systems and this could not be countenanced unless the costs could be considerably reduced from their present levels.

Airbag systems can be very effective in preventing injury but are very much a technology of last resort – it is much better to prevent a crash occurring. The 1990s have already seen the remarkable fusion of radar, laser and microcomputer technology to produce sophisticated collision avoidance systems. Several of these systems have been integrated with cruise control systems to provide what is termed 'intelligent cruise control', enabling the vehicle to stay in a given lane and at a safe following distance, entirely without driver intervention. As the price of miniature laser, microwave and computer components continues to fall it is likely that such systems will progressively gain more customers. More sophisticated signal processing will also enable such systems to follow curves (as already demonstrated by Mercedes-Benz and Honda), detect pedestrians about to step into the road and avoid obstacles on the road. Much of the required development work has already been undertaken via the EU's Prometheus programme, see Figure 16.2. Coupled with navigation and traffic information systems it seems likely that at some time within the next two decades it will be possible to complete a car journey without any driver intervention.

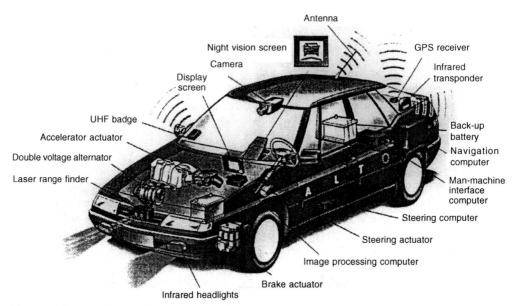

Figure 16.2 'Prometheus' project car featuring a camera and laser-radar for collision avoidance, GPS navigation system, automatic lane guidance system and traffic information system (PSA Peugeot Citroen)

An extension of this is being currently developed where erratic driving, or doze monitoring, is continually assessed. The problem here is how to feed the information back to the driver without any startling effects.

Materials have a great input into energy absorption in vehicle safety. Foams are being developed that can maximize the displacement of the impactor and provide a safe level of restraint. This is incorporated into the design of the vehicle interior so that items such as loudspeakers and armrests are designed to absorb energy safely. This could avoid the use of extra strengthening in side collisions, which are undesirable because they make the deceleration levels to be absorbed higher.

16.3.3 Electronic developments in chassis systems

Chassis systems are those components of the vehicle structure which control the speed and motion of the car; essentially the steering, braking and suspension systems. Electronic developments in steering and suspension systems are at a very early stage, with much promise for the future. In braking systems, however, the 1990s have already seen strong growth in the market for antilock braking systems (ABS) and the technology is now comparatively mature; further developments are likely to focus on integrating chassis systems to provide a higher level of functionality with minimal cost penalty.

Once the vehicle is entirely under electronic control, there is no reason for the conventional mechanical linkages between the driver's controls and the engine compartment. These linkages could be either electrical or optical connections, although there would need to be allowances made for the public perception of these types of systems.

Developments in anti-lock braking (ABS)
ABS is now well developed and most braking system suppliers are therefore seeking to use additional software and extra sensors to allow ABS to provide extra facilities. Once an ABS system is installed it becomes a relatively simple matter to add a traction control facility (TCS) by detecting wheelspin during acceleration and momentarily applying a braking force to limit it. Since the marginal cost of adding this facility is small, manufacturers are likely to go for the widespread adoption of such a strategy to maintain a competitive edge.

Looking beyond TCS, the trend will be to integrate active stability control (ASC) into ABS. ASC works by having an electronic control unit determine the degree of understeer or oversteer during cornering and then use the ABS to apply a carefully determined braking force to one or more wheels in order to restore neutral handling and maximize vehicle stability. To implement ASC it is necessary to install some additional sensors; a lateral acceleration sensor to measure cornering force, and a gyroscope to detect vehicle yaw. Using the rapidly developing technology of microengineering it will soon be possible to fabricate such sensors at minimal cost. Once this is achieved ASC will progressively spread down market, from executive cars to superminis.

Electric braking
A number of braking companies are already developing electric braking systems for introduction in the next decade. Full electric braking does away with the hydraulic system and uses an electronic control unit to directly actuate small electric motors, mounted in the wheel assemblies, which apply the required braking force. Software in the controller can monitor driving conditions and individually control the braking force on each wheel to compensate for variations in load, road grip, steering angle, and so on.

The chief obstacle to the introduction of electric braking is the difficulty of manufacturing small, powerful and fast responding electric motors which can equal the performance of existing hydraulics. Another problem is the potential for catastrophic failure of the system because of a simple electrical fault; for these reasons electric braking systems are unlikely to be introduced before 2010. In the meantime hybrid electric/hydraulic braking systems, with the electric unit operating the less highly stressed rear brakes, will be introduced. Such a system has already been trialed on the GM EV1 electric vehicle.

Electric steering systems

Electric power-assisted steering (EPAS) has been used for over ten years on Japanese microcars and is now common on mid-engined sports cars where it replaces conventional hydraulic PAS (for example on Honda's NSX and Rover's MGF). Its advantages are such that it is likely to become standard on most vehicles by 2010.

Full electric steering (or 'drive-by-wire') has been demonstrated by a number of manufacturers and potentially offers many advantages – not least the elimination of the injury-causing steering column. Unfortunately, the safety critical aspects of drive-by-wire mean that it is unlikely to be introduced before years, perhaps decades, of satisfactory tests have been completed.

16.3.4 Electronic developments in body systems

Body systems are all of those systems which are installed within the passenger compartment of the vehicle to improve security, comfort and convenience. The last few years have already seen rapid technological expansion in this area and this process is likely to continue as consumer expectations rise and the cost of hardware falls.

Security

Remote transponder-based immobilizer and alarm systems are now in volume production and have proved effective in reducing car crime. The next step is for the car key to be replaced by a so-called 'smart card'. Numerous companies are currently working on this technology, which involves the use of an integrated circuit (IC) and transponder circuitry which is buried within a credit-card size plastic holder. The IC's memory contains data used to unlock the doors, activate the engine management system and starting circuits, as well as a set of details relating to the driver, including preferences for cabin temperature and seat, mirror and steering column positions. Once smart cards are established there is no reason why manufacturers should not eliminate the traditional mechanical door locks and steering column lock. These are relatively bulky and unreliable and can readily be replaced by electrically-energized alternatives.

Comfort features

There is an increasing amount of effort being expended in the area of occupant comfort. Currently seat position and angles can be controlled upon entry of a vehicle, but technology could soon enable the shape of the seat to be continuously changed by using a pressure mapping system within the seat. Indeed, a modular seating approach is not too far away, where everything the occupant could require is ergonomically arranged upon entry to the vehicle.

Air conditioning is installed on around 90% of US vehicles and is becoming more popular in Europe. Air conditioning technology is now very well developed and the trend is to refine the control of the system, generally by the addition of more sensors. For example, having an array of solar radiation and temperature sensors distributed around the cabin allows cold air to be directed exactly where it is required and so avoids 'hot-spots'.

A completely new comfort feature, which may reach production in the next 10–15 years, is that of active noise cancellation (ANC). ANC seeks to reduce cabin noise by using microphones in the passenger compartment to sample the background sound level, and, using powerful loudspeakers, 'feed back' an inverted copy of the sampled sound. The effect is to cancel out background noise. Such a technique works best with low-frequency noise (generally in the range 50–200 Hz) and is therefore most effective in reducing body 'boom' and low-frequency vibration emanating from the engine.

Multiplexing systems
In the last fifteen years the length of wire used in an automobile has more than doubled, prompting manufacturers to adopt multiplexing. Multiplexing is currently quite expensive, mainly due to the current high cost of the IC's required; however, as the cost of the hardware falls it will become relatively cheaper than conventional wiring. This factor, along with the reduction in weight and simplicity of installation will ensure the universal adoption of multiplex wiring systems by 2010.

Power supply developments
The vast amount of additional equipment now installed on cars has placed an increased burden on the electrical power supply system. Fortunately, it has so far been possible to develop more efficient alternators, however, as the power requirement increases, other strategies may be adopted. One possibility is the use of variable-ratio drive systems for alternators, so that they can be turned at high speed even when the engine is running slowly. Another option is the use of higher voltages (24 or 48 volts) which would give a proportionate reduction in current and therefore permit the use of thinner wires and smaller motors for an equivalent power delivery. In the longer term, as full electric braking and steering are introduced, there will be a requirement for a totally secure power supply; this will probably involve the installation of two separate electrical supply systems on the vehicle.

Dashboard display technology
For many years the motor industry has been predicting the introduction of head-up displays (HUDs) in which the instrument panel is projected at the windscreen so that it appears 'floating' in the driver's field of view. Unfortunately, the introduction of HUDs has only occurred on one or two vehicle models, mainly because of the difficulty in viewing the image in strong sunlight. Research is continuing, however, and it is likely that HUDs will eventually reach the mass market.

Of more immediate interest is the trend towards large-area programmable liquid-crystal display (LCD) instrument panels. These essentially comprise a large area flat panel display which is covered with a fine matrix of coloured LCD dots that can be turned on or off. A microcontroller drives the display, and is able to generate images according to software instructions. For example, the display could show speedometer and tachometer images during normal driving, but perhaps replace the tachometer with a temperature gauge image if the engine temperature were to rise above normal. Using such a display panel has advantages for the manufacturer, including the ability to reconfigure the display for different markets and vehicle models simply by altering software.

Unfortunately, although suitable LCD display panels have recently been manufactured in small volumes, they remain prohibitively expensive for automotive applications. Manufacturing technology is improving all the time, though, and it is likely that low-cost panels will be available by 2005.

16.3.5 Vehicle information and navigation systems

An area likely to show strong development, both in the short and long term, is that of automobile information and navigation systems. This is a broad area but, briefly, it includes traffic information, navigation systems and mobile telephony (including fax and data transmission).

Navigation systems
Currently, the Japanese vehicle manufacturers lead the world in the introduction of vehicle navigation systems, which are available on several high-line vehicles including the 1997 Nissan Cima, Honda Legend and Mitsubishi Galant. Technical specifications vary, but in general the systems are based around a Global Positioning System (GPS) satellite receiver and a highly sensitive vehicle-mounted gyroscope. GPS signals can often be obscured by high buildings and so additional inputs, for example from an electronic compass and wheel-speed sensors, may be used to improve accuracy in cities.

A complex computer system then uses the input data to compute the location of the vehicle and display it on a full-colour LCD panel. Surrounding streets and landmarks are shown on an overlay map which is downloaded from a CD-ROM; navigational instructions can then be conveyed to the driver via sequences of symbols and arrows. These systems have been shown to work well and sales currently exceed half a million per annum, but unfortunately they are very expensive at the moment due to the high cost of the LCD panels. As sales volumes rise and manufacturing technologies improve, however, it is likely that display prices will fall sharply and such systems will come into common usage.

The situation in Europe and the US is slightly different from that in Japan. Customers in these countries prefer to receive spoken directional instructions delivered by a voice synthesis unit. GPS and CD-ROM input is still used, however. A number of these systems have recently been introduced, for example the TravelPilot (Bosch), Carin (Philips), Telepath 100 and OnStar (Delco). The OnStar is a particularly interesting system and perhaps a portent to future trends; in addition to navigation it offers a breakdown help button (communicating the vehicle's position to a local repair workshop), an emergency help button to summon the emergency services (activated by the driver or by deployment of the airbag) and a radio tracking facility should the car be stolen. The cost of this system is only US$1000 and is likely to fall further as GM offers it across their US model range.

Route guidance
Perhaps the area destined for greatest growth is that of driver information systems which allow vehicles to be accurately navigated to a desired destination whilst avoiding areas of congestion. A major factor holding back progress in many countries is the high cost of establishing the required information infrastructure.

For optimum effectiveness a vehicle navigation system needs to present the car driver with the quickest route to his destination, taking account of traffic conditions. To do this a national road information infrastructure is required, something that most governments have so far been unwilling to fund. Up until now only the Japanese have established a working system, the Visual Information and Communications System (VICS). VICS takes a vast assortment of traffic and weather data from a central control room and broadcasts it to passing vehicles by means of roadside radio and infra-red links. Data is also 'piggybacked' onto commercial radio broadcasts. A vehicle equipped with a VICS receiver can then pick-up and decode the data and provide the driver with information on local roadworks, the availability of parking spaces, accidents or obstacles and bad weather. The information is also fed into the vehicles navigation system to enable it to display the most suitable route to the desired destination.

Studies by the Japanese government predict that VICS has the potential to reduce the country's petrol consumption by about 5%, with an accompanying reduction in pollution, it is for these reasons that the VICS service is offered free to the consumer. Because of the environmental benefits of route guidance many other countries have taken an interest in VICS and it is likely that similar systems will proliferate in congested urban areas around the developed world.

References and further reading

Automotive Engineering. (1991). All Aluminium Car, **99**.
Automotive Engineering. (1993). Design of a magnesium/Aluminium Door frame, May.
Bryzek, J. and Smith, G., (1994). On-sensor-chip electronics integration technical and economic considerations, *Proc. of Nexus Workshop, Sept. 1994, Toulouse.*
Chowanietz, E. (1995). *Automobile Electronics.* Butterworth-Heinemann. ISBN 0-7506-1878-7.
ENTRA, Failure Modes and Effects Analysis, EITB Training Module, with video.
Cross, N. (1989). Engineering Design Methods, Wiley, ISBN 0471942286.
Jacobs, P. (1996). Stereolithography and other Rapid Prototyping and Manufacturing Technologies. SME.
Kalpakjian, S. (1997) *Manufacturing Processes for Engineering Materials*, Addison Wesley. ISBN 0 201 823705.
Olbrich, T. *et al.*, Built-in self-test and diagnosis in microsystems, MST News 11, p. 8, 1994.
SAE Paper 960581 'Camless Engines'.
Venus, A.D. An overview of the major rapid prototyping systems and their capabilities, 1997, Vol. 1, No. 2.

Recommended texts

1. Womack, Jones and Roos, (1990). *The Machine That Changed the World*, Rawson Associates, ISBN 0 892563508.
   This is based on the Massachusetts Institute of Technology five million dollar, five year study of the future of the automobile. The book which introduced the term 'lean manufacture' to the world!
2. Womack, Jones, (1996). *Lean Thinking*, Schuster, ISBN 0-684-81035-2.
   Written by two of the authors of '1', this provides clear definitions, route maps and case histories of the implementation of Lean Thinking.
3. The magazines of the Society of Automotive Engineers of Detroit and The Institution of Mechanical Engineers, *Automotive Engineering* and the *Automotive Engineer* respectively are good sources for current thinking and trends within the automotive industry.

Index

Please note that:
Figure numbers are given the prefix f (e.g. f2.2)
Table numbers are in bold and are given the prefix t (e.g. **t2.1**)
Equation numbers are in italics and are given the prefix e (e.g. *e6.1*)

Printed in the United Kingdom
by Lightning Source UK Ltd.
128668UK00001B/1-4/A

9 780750 650441

MANCHESTER
CITY COUNCIL

D1313108

Please return/renew this item
by the last date shown.
Books may also be renewed by
phone or the internet.

Tel: 0161 254 7777

www.manchester.gov.uk/libraries